NATEF Standards Job Sheets for Performance-Based Learning

by
Chris Johanson
ASE Certified Master Technician

Publisher
The Goodheart-Willcox Company, Inc.
Tinley Park, Illinois
www.g-w.com

The Goodheart-Willcox Company, Inc. Brand Disclaimer: Brand names, company names, and illustrations for products and services included in this text are provided for educational purposes only and do not represent or imply endorsement or recommendation by the author or the publisher.

The Goodheart-Willcox Company, Inc. Safety Notice: The reader is expressly advised to carefully read, understand, and apply all safety precautions and warnings described in this book or that might also be indicated in undertaking the activities and exercises described herein to minimize risk of personal injury or injury to others. Common sense and good judgment should also be exercised and applied to help avoid all potential hazards. The reader should always refer to the appropriate manufacturer's technical information, directions, and recommendations; then proceed with care to follow specific equipment operating instructions. The reader should understand these notices and cautions are not exhaustive.

The publisher makes no warranty or representation whatsoever, either expressed or implied, including but not limited to equipment, procedures, and applications described or referred to herein, their quality, performance, merchantability, or fitness for a particular purpose. The publisher assumes no responsibility for any changes, errors, or omissions in this book. The publisher specifically disclaims any liability whatsoever, including any direct, indirect, incidental, consequential, special, or exemplary damages resulting, in whole or in part, from the reader's use or reliance upon the information, instructions, procedures, warnings, cautions, applications, or other matter contained in this book. The publisher assumes no responsibility for the activities of the reader.

Introduction

Within the National Institute for Automotive Service Excellence (ASE) is an independent organization called The National Automotive Technicians Education Foundation, or NATEF. One of NATEF's duties is to evaluate technician training programs according to an established set of standards. As part of its program standards, NATEF has developed a task list that includes tasks from each of the eight ASE automotive certification areas. All tasks on the list are assigned a priority number: P-1, P-2, or P-3.

Hands-on performance of the NATEF tasks, as well as knowledge of the theory behind each task, provides crucial training for employment in the automotive service field and tells potential employers that the new technician is both knowledgeable and skilled.

Using the Manual

This manual contains thorough coverage of the tasks listed in the NATEF Task List. Each job in this manual is a hands-on activity that corresponds to one or more of the NATEF tasks. The jobs have been carefully organized and developed to increase the student's chances of passing the related ASE tests by having the student apply what has been learned in the classroom.

The jobs are not correlated to specific textbook chapters, but will be assigned when the instructor determines that the student has sufficient knowledge to complete them. This manual steps the student through all essential NATEF tasks including inspecting, testing, and diagnosing vehicle systems; removing and replacing self-contained components; and removing, overhauling, and reinstalling major components.

When performing the jobs, the student should:
- Follow the instructions carefully and proceed through the steps in the order that they are presented.
- Carefully read all notes, cautions, and warnings before proceeding.
- Record specifications and the results of diagnostic procedures in the spaces provided.
- Consult the instructor when directed by the text, and when unsure of how to proceed.
- Never attempt a step if unsure of how to perform it correctly.

Features of the Manual

This manual is divided into eight areas corresponding to the eight ASE certification areas and the eight areas of the NATEF Task List. Each area of the manual is further subdivided into a number of projects, or collections of closely related jobs. The eight areas of the manual, the projects, and jobs they contain are color coded to make it easy to locate specific content in the manual.

The projects in this manual include a brief introduction about the type of service being performed, a list of the jobs included in the project, and a tools and materials list for the jobs.

The jobs in this manual are designed to be accomplished in one or two lab sessions. Check boxes are provided in the left-hand column of the job so the student can mark off tasks as they are performed. Blanks are provided for recording service-related information.

In addition, three types of special notices appear throughout the jobs in this manual. These notices point out special information or safety considerations for the task being performed. They are color coded according to the type of information being provided:

Note

 Note features provide additional information, special considerations, or professional advice about the task being performed. Note features are blue.

Caution

 Caution features appear near critical steps in a service procedure. These features warn the reader that failure to properly perform the task can lead to equipment or vehicle damage. Caution features are yellow.

Warning

 Warning features also appear near certain critical steps in the service procedure. These features warn the reader that failure to properly perform the task could result in personal injury. Warning features are red.

If properly implemented, this manual will help the student to do well in his or her courses, pass the ASE certification tests, and find a job in the automotive industry.

Chris Johanson

NATEF Standards Job Sheets for Performance-Based Learning

Table of Contents

Introduction   3

Engine Repair

Project 1—Preparing to Service a Vehicle   11
Job 1—Perform Safety and Environmental Inspections   13
Job 2—Interpret Vehicle Numbers, Use Service Information, and Prepare a Work Order   23

Project 2—Removing and Replacing Gaskets, Seals, and Bearings   31
Job 3—Remove and Install a Gasket   33
Job 4—Remove and Install a Seal   37
Job 5—Remove and Replace a Bearing   41

Project 3—Performing Oil Changes and Checking an Engine for Leaks   45
Job 6—Change Oil and Filter   47
Job 7—Inspect an Engine for Leaks   51

Project 4—Diagnosing Engine Problems   55
Job 8—Perform a Vacuum Test and a Power Balance Test   57
Job 9—Perform a Cranking Compression Test, Running Compression Test, and a Cylinder Leakage Test   61
Job 10—Perform an Oil Pressure Test   67
Job 11—Verify Warning Indicator Operation and Retrieve Trouble Codes   69

Project 5—Removing and Disassembling an Engine   73
Job 12—Remove an Engine   75
Job 13—Disassemble an Engine   81

Project 6—Performing Bottom End Service   85
Job 14—Inspect the Block, Crankshaft, and Bearings   87

Job 15—Repair Damaged Threads   91
Job 16—Measure Cylinder Wear and Hone Cylinders   95
Job 17—Inspect Connecting Rods   99
Job 18—Inspect Pistons and Replace Piston Rings   101
Job 19—Inspect Balance Shafts, Vibration Damper, and Oil Pump   107

Project 7—Servicing the Cylinder Head(s) and Valve Train   109
Job 20—Disassemble and Inspect a Cylinder Head   111
Job 21—Inspect and Service Valve Train Parts   117
Job 22—Resurface and Inspect Valves and Valve Seats   125
Job 23—Assemble a Cylinder Head   129
Job 24—Inspect Valve Timing Components   131
Job 25—Adjust Valves   133
Job 26—Replace Valve Seals with the Cylinder Head Installed   137

Project 8—Reassembling and Reinstalling an Engine   141
Job 27—Reassemble the Engine's Bottom End   143
Job 28—Install the Cylinder Head(s) and Valve Train   149
Job 29—Complete the Engine Reassembly   153
Job 30—Reinstall the Engine   157

Project 9—Servicing a Cooling System   159
Job 31—Inspect and Test a Cooling System; Identify Causes of Engine Overheating   161
Job 32—Flush and Bleed a Cooling System   169
Job 33—Replace Belts and Hoses   171
Job 34—Replace a Coolant Pump   175
Job 35—Replace a Radiator   179
Job 36—Remove and Replace a Thermostat   181

Automatic Transmission/ Transaxle Service

Project 10—Inspecting, Replacing, and Aligning Powertrain Mounts   185
Job 37—Inspect Powertrain Mounts   187
Job 38—Replace and Align Powertrain Mounts   191

Project 11—Diagnosing Transmission and Transaxle Problems 195
Job 39—Use the Seven-Step Diagnostic Process to Identify Transmission/Transaxle Problems 197
Job 40—Perform Transmission/Transaxle Pressure Tests 203

Project 12—Servicing a Transaxle Final Drive 207
Job 41—Service a Planetary-Type Final Drive 209
Job 42—Service a Helical Gear Final Drive 213
Job 43—Service a Hypoid Gear Final Drive 217

Project 13—Servicing an Automatic Transmission or Transaxle 221
Job 44—Change Transmission/Transaxle Oil and Filter 223
Job 45—Adjust Transmission Linkage and Bands and Neutral Safety Switch/Range Switch 227
Job 46—Service an Oil Cooler and Lines 231
Job 47—Service Speedometer Drive and Driven Gears 235
Job 48—Service Electronic and Electrical Components 237
Job 49—Service Vacuum Modulators and Governors 241

Project 14—Removing an Automatic Transmission or Transaxle 245
Job 50—Remove an Automatic Transmission 247
Job 51—Remove an Automatic Transaxle 251

Project 15—Rebuilding an Automatic Transmission 255
Job 52—Disassemble and Inspect an Automatic Transmission 257
Job 53—Service Automatic Transmission Components and Rebuild Subassemblies 265
Job 54—Reassemble an Automatic Transmission 269

Project 16—Rebuilding an Automatic Transaxle 271
Job 55—Disassemble and Inspect an Automatic Transaxle 273
Job 56—Service Automatic Transaxle Components and Rebuild Subassemblies 281
Job 57—Reassemble an Automatic Transaxle 285

Project 17—Installing an Automatic Transmission or Transaxle 287
Job 58—Install an Automatic Transmission 289
Job 59—Install an Automatic Transaxle 293

Manual Drive Train and Axles

Project 18—Correcting Noise, Vibration, and Harshness Problems 297
Job 60—Diagnose a Drive Train Problem 299
Job 61—Check and Correct Drive Shaft Runout 303
Job 62—Check and Adjust Drive Shaft Angles 305
Job 63—Check and Adjust Drive Shaft Balance 309

Project 19—Performing Manual Drive Train Service 311
Job 64—Diagnose and Service Linkage and Switches/Sensors 313
Job 65—Diagnose Drive Train Component Leakage 319

Project 20—Servicing Drive Shafts and CV Axles 323
Job 66—Remove, Inspect, and Reinstall U-Joints and a Drive Shaft 325
Job 67—Remove, Inspect, and Reinstall a CV Axle 331
Job 68—Diagnose and Service CV Axle Shaft Bearings and Seals 337

Project 21—Removing a Manual Transmission or Transaxle 341
Job 69—Remove a Manual Transmission 343
Job 70—Remove a Manual Transaxle 347

Project 22—Servicing a Clutch 351
Job 71—Diagnose Clutch Problems 353
Job 72—Adjust Clutch Pedal Free Play 355
Job 73—Bleed a Hydraulic Clutch System 357
Job 74—Remove a Clutch 359
Job 75—Inspect and Repair a Clutch 363
Job 76—Install a Clutch 367

Project 23—Rebuilding and Installing a Manual Transmission or Transaxle 371
Job 77—Disassemble and Inspect a Manual Transmission 373
Job 78—Reassemble and Install a Manual Transmission 379
Job 79—Disassemble and Inspect a Manual Transaxle 385
Job 80—Reassemble and Install a Manual Transaxle 389

Project 24—Servicing a Transfer Case   395
Job 81—Inspect, Remove, and Replace a
   Transfer Case   397
Job 82—Overhaul a Transfer Case   403
Job 83—Service Transfer Case Shift Controls,
   Locking Hubs, and Wheel Bearings   409

**Project 25—Servicing Rear-Wheel Drive
Axles and Differentials   419**
Job 84—Remove, Service, and Install Retainer-Type
   and C-Lock Axles   421
Job 85—Remove and Replace an Independent Rear
   Axle Shaft   429
Job 86—Remove a Rear Axle Assembly   433
Job 87—Disassemble and Inspect
   a Differential   435
Job 88—Reassemble and Adjust
   a Differential   439
Job 89—Install a Rear Axle Assembly   443

Suspension and Steering

**Project 26—Diagnosing and Servicing
General Steering and Suspension
Concerns   445**
Job 90—Diagnose Steering and
   Suspension Problems   447
Job 91—Lubricate Steering and Suspension   453
Job 92—Service Tapered Roller
   Wheel Bearings   455

**Project 27—Servicing the
Steering System   459**
Job 93—Inspect a Steering Column   461
Job 94—Service a Steering Column   465
Job 95—Remove and Replace a Rack-and-Pinion
   Steering Gear and Inner Tie Rods   471
Job 96—Remove and Replace Steering Linkage
   Components   475
Job 97—Test a Power Steering System   479

**Project 28—Servicing a Suspension
System   487**
Job 98—Replace Strut Rods, Stabilizer Bars,
   and Bushings   489
Job 99—Replace Ball Joints, Control Arms,
   and Bushings   493
Job 100—Replace Coil Springs   499
Job 101—Inspect and Replace MacPherson
   Strut Components   503
Job 102—Inspect and Replace
   Shock Absorbers   509

**Project 29—Performing Alignment and
Tire Service   513**
Job 103—Perform an Alignment   515
Job 104—Inspect Tires and Rims   523
Job 105—Remove and Install a Tire on a Rim;
   Repair a Punctured Tire   529
Job 106—Rotate and Balance Tires   535

Brakes

**Project 30—Performing General Brake
System Diagnosis and Service   537**
Job 107—Diagnose a Brake System   539
Job 108—Inspect and Replace Wheel Studs   545
Job 109—Service Wheel Bearings   547
Job 110—Service a Parking Brake   551

**Project 31—Servicing a Brake
Hydraulic System   557**
Job 111—Diagnose Brake Hydraulic
   System Problems   559
Job 112—Replace Brake Lines, Hoses,
   and Valves   565
Job 113—Service a Master Cylinder   571
Job 114—Service a Power Assist System   575
Job 115—Bleed a Brake System   579

Project 32—Servicing Drum Brakes   585
Job 116—Service Drum Brakes   587
Job 117—Machine a Brake Drum   593

Project 33—Servicing Disc Brakes   597
Job 118—Service Disc Brakes   599
Job 119—Overhaul a Disc Brake Caliper   603
Job 120—Machine a Brake Rotor   605

**Project 34—Servicing Anti-lock Braking and
Traction Control Systems   609**
Job 121—Identify and Inspect ABS and
   TCS Components   611
Job 122—Bleed an ABS System   615
Job 123—Adjust Speed Sensor Clearance   617
Job 124—Diagnose ABS/TCS Problems   619

Electrical/Electronic Systems

**Project 35—Performing General Electrical
System Diagnosis and Service   623**
Job 125—Check for Parasitic Battery Load   625
Job 126—Service Wiring and Wiring
   Connectors   627
Job 127—Inspect and Test a Battery   631
Job 128—Service a Battery   635
Job 129—Jump Start a Vehicle   639

Project 36—Diagnosing and Servicing a Starting System   641

Job 130—Inspect, Test, and Diagnose a Starting System   643

Job 131—Replace a Starter Relay or Solenoid   649

Job 132—Replace an Ignition Switch or Neutral Safety Switch   651

Job 133—Replace a Starter   653

Project 37—Diagnosing and Servicing a Charging System   657

Job 134—Inspect, Test, and Diagnose a Charging System   659

Job 135—Replace an Alternator   663

Project 38—Diagnosing and Servicing a Lighting System   665

Job 136—Diagnose Lighting System Problems   667

Job 137—Replace and Aim Headlights   675

Project 39—Diagnosing and Servicing Gauges and Warning Lights   679

Job 138—Diagnose Instrument Panel Warning Light Problems   681

Job 139—Check Electromechanical Gauge Operation   685

Job 140—Check Electronic Gauge Operation   689

Job 141—Replace a Warning Light Sensor or Gauge Sensor   691

Project 40—Diagnosing and Servicing Horns and Windshield Washers/Wipers   693

Job 142—Check and Service a Horn   695

Job 143—Check and Correct the Operation of Windshield Washers and Wipers   699

Project 41—Diagnosing and Servicing Accessories   703

Job 144—Diagnose and Service Power Windows and Power Door Locks   705

Job 145—Diagnose a Cruise Control Problem   711

Job 146—Diagnose Problems with a Radio/Sound System   715

Job 147—Disable and Enable an Air Bag System   719

Job 148—Diagnose an Air Bag System   721

Job 149—Use a Scan Tool to Determine a Body Electrical Problem   723

Job 150—Diagnose a Body Control Module and Module Communication Errors   727

Job 151—Check the Operation of Anti-Theft and Keyless Entry Systems   731

Heating and Air Conditioning

Project 42—Diagnosing and Servicing an Air Conditioning System   735

Job 152—Diagnose an Air Conditioning System   737

Job 153—Diagnose Heating and Air Conditioning Control Systems   745

Job 154—Recover Refrigerant, and Evacuate and Recharge an Air Conditioning System   751

Job 155—Remove and Replace a Compressor and Compressor Clutch   759

Job 156—Remove and Replace Air Conditioning System Components   765

Job 157—Manage Refrigerant and Maintain Refrigerant Handling Equipment   771

Project 43—Diagnosing and Servicing a Heating and Ventilation System   775

Job 158—Inspect and Service Heater Components   777

Job 159—Inspect and Service Air Handling Components   783

Engine Performance

Project 44—Diagnosing General Engine Concerns   787

Job 160—Diagnose Engine Performance Problems   789

Job 161—Use a Scan Tool to Access and Reprogram a Vehicle Computer   793

Job 162—Use a 4- or 5-Gas Analyzer to Check Emission System Operation   799

Job 163—Diagnose an NVH Problem   803

Project 45—Diagnosing and Repairing an Ignition System   809

Job 164—Inspect and Test an Ignition System's Secondary Circuit Components and Wiring   811

Job 165—Inspect and Test an Ignition System's Primary Circuit Components and Wiring   815

Project 46—Diagnosing and Servicing Fuel and Air Induction Systems   821

Job 166—Test Fuel Quality   823

Job 167—Test and Replace Fuel Supply System Components   827

Job 168—Inspect and Service a Throttle Body   835

Job 169—Check Idle Speed Motor or Idle Air Control   839

Job 170—Check Fuel Injector Waveforms and Perform a Fuel Injector Balance Test   841

Job 171—Remove and Replace a Fuel Injector   845

Job 172—Check Supercharger and Turbocharger Operation   847

Project 47—Diagnosing and Servicing an Exhaust System   853

Job 173—Inspect and Test an Exhaust System   855

Job 174—Replace Exhaust System Components   859

Project 48—Diagnosing and Servicing Emissions Control Systems   863

Job 175—Test and Service an EGR Valve   865

Job 176—Test and Service an Air Pump and Air Injection System   869

Job 177—Test and Service a Catalytic Converter   875

Job 178—Test and Service Positive Crankcase Ventilation (PCV) and Evaporative Emissions Control Systems   879

NATEF Correlation Charts

Maintenance and Light Repair (MLR)   885

Automobile Service Technology (AST)   898

Master Automobile Service Technology (MAST)   919

Project

Preparing to Service a Vehicle

Introduction

Thousands of technicians are injured every year while on the job, and some are even killed. Most of these technicians were breaking basic safety rules before their accidents. The technicians that survived learned to respect safety precautions the hard way by experiencing a painful, but instructive, injury. Know how to avoid accidents—by studying and following shop safety rules.

The service technician is responsible for maintaining a safe workspace and performing service safely, but is also responsible for protecting the environment. You must follow all applicable environmental laws or risk heavy fines. In addition to the legal obligations, proper waste disposal and recycling will save money. In Job 1, you will locate and identify safety equipment and hazards in the shop. You will also identify the types of wastes generated by the shop and describe the proper procedures for waste disposal.

Certain tasks must be completed before a vehicle can be properly serviced. For example, it is no longer sufficient to open the hood and visually identify an engine. You must use identifying numbers to determine the type of engine and drive train used, specifications, part numbers, and what service operations can be performed. Some vehicle systems cannot be successfully serviced without finding and interpreting vehicle numbers. In Job 2, you will locate and interpret vital vehicle numbers.

Once the vehicle and its systems have been properly identified, you must locate the appropriate service information. No twenty-first century technician can do his or her job without knowing how to find, read, and understand service information. Modern service information comes in many forms: printed service materials, compact discs, and Internet websites. In Job 2, you will also locate and interpret service information and prepare a work order.

Project 1 Jobs

- Job 1—Perform Safety and Environmental Inspections
- Job 2—Interpret Vehicle Numbers, Use Service Information, and Prepare a Work Order

Tools and Materials

The following list contains the tools and materials that may be needed to complete the jobs in this project. The items used will depend on the make and model of vehicle being serviced.

- One or more vehicles.
- Applicable printed service manual.
- A computer and the appropriate service-related compact discs (CDs).
- Internet access.
- Technical service bulletin(s).
- Vehicle service history when available.
- One or more material safety data sheets (MSDS).
- Other service information as needed.
- Safety glasses and other protective equipment as needed.

Safety Notice

Before performing this job, review all pertinent safety information in the text and review safety information with your instructor.

Job 1—Perform Safety and Environmental Inspections

After completing this job, you will be able to locate the shop's fire extinguishers, fire exits, and eye wash stations. You will be able to locate and properly use safety glasses and other shop safety equipment. You will also learn the general safety rules of an auto shop. You will learn the methods of preventing environmental damage through environmentally friendly work procedures.

Procedures

> **Warning**
>
> ⚠ Before performing this job, review all pertinent safety information in the text and discuss safety procedures with your instructor.

Personal Protective Equipment

☐ 1. Eye protection (safety glasses or goggles) should be worn during any operation that could injure your eyes. See **Figure 1-1**. This includes, for example, hammering, drilling, grinding, sandblasting, using compressed air, carrying a battery, or working around a spinning engine fan.

List five common tasks that require the use of safety goggles: _____

_____

_____

Where are the safety glasses and goggles kept in your shop? _____

_____

_____

Figure 1-1. Eye protection should be worn while working in the shop. A—Safety glasses. B—Safety goggles. C—Face shield.

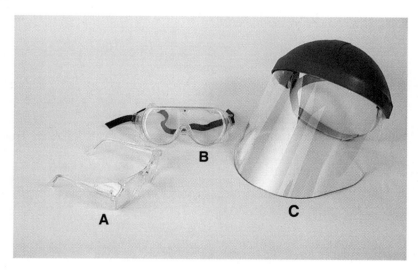

☐ 2. Hearing protection (earplugs or protective earmuffs) should be worn during any loud activities. These include tasks like hammering, operating pneumatic tools, and grinding.

List five common tasks that require the use of hearing protection: _____

_____

_____

Where are the earplugs and protective earmuffs kept in the shop? _____

_____

_____

☐ 3. Dust masks and respirators may be required performing certain tasks or working with certain chemicals.

Where are respirators and dust masks kept in the shop? _____

_____

Material Safety Data Sheets

☐ 1. Walk through the shop and familiarize yourself with the chemicals stored there.

☐ 2. Find the material safety data sheets for the chemicals in the shop. Look up information about the chemicals stored in the shop.

Were any chemicals improperly stored according to the material safety data sheets?
Yes ___ No ___
If Yes, describe the chemicals and inform your instructor:_____

_____

Were material safety data sheets missing for any of the chemicals in the shop?
Yes ___ No ___
If Yes, describe the chemicals and inform your instructor:_____

_____

☐ 3. Obtain an MSDS from your instructor.

☐ 4. List the following information from the MSDS.

General class of material covered by the MSDS (if applicable): _____
Trade (manufacturer's) name for the material: _____
Chemical or generic name of the material:_____
Known breathing hazards of the material: _____

_____

Known fire hazards of the material: _____

_____

Known skin damage hazards of the material:_____

_____

Proper storage and disposal methods:_____

_____

Emergency response to spills of the material: _____

_____

Job 1—Perform Safety and Environmental Inspections (continued)

Fire and Shop Safety

☐ 1. Walk around the shop and locate all of the fire extinguishers, the fire exit, and fire alarms. Such information is crucial in the event of an emergency.

How many fire extinguishers and alarms are there in your shop? _____

_____

Where are the fire extinguishers located? _____

_____

_____

What types of fire extinguishers are available in the shop? _____

_____

_____

Where are the fire alarms? _____

_____

_____

How do you leave the shop in case of a fire? _____

_____

_____

☐ 2. To help prevent an emergency, memorize these important fire prevention tips:
   • Always take actions to prevent a fire.
   • Store gasoline-soaked and oily rags in safety cans, **Figure 1-2**.
   • Wipe up spilled gasoline and oil immediately.
   • Hold a rag around the fitting when removing a car's fuel line, **Figure 1-3**.

Figure 1-2. Special safety cans should be used to store oily rags.

Figure 1-3. This technician is using a shop rag to prevent fuel from leaking as he disconnects a fuel line.

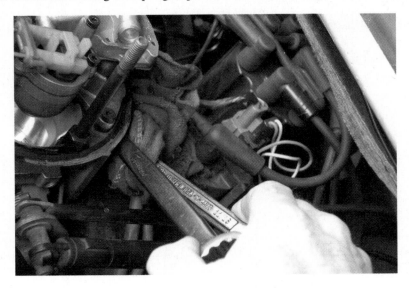

3. Identify the location of safety equipment and special hazard areas within the shop.

Where are the eye wash stations located?_____

_____

_____

Where are first aid stations located? _____

_____

_____

What other special hazard areas exist within the shop? _____

_____

_____

How are the special hazard areas identified? _____

_____

_____

Electrical Safety

1. Check the shop for unsafe electrical conditions, such as damaged electrical cords and overloaded outlets.

Were any unsafe electrical conditions found? Yes ___ No ___

If Yes, describe them in as much detail as possible:_____

_____

_____

2. Make sure that all electrically operated tools and equipment with three-prong electrical plugs have their grounding prongs intact, **Figure 1-4**.

Do any tools or equipment have the grounding prong removed? Yes ___ No ___

If Yes, list the items: _____

_____

Warning

⚠ Do *not* use tools or equipment on which the grounding prong has been removed.

Job 1—Perform Safety and Environmental Inspections (continued)

Figure 1-4. The grounding prong has been broken off the electrical plug on the right. Do *not* use a piece of equipment if the grounding prong has been removed.

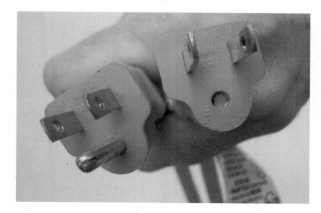

 3.   Obtain the use of a vehicle.

☐ 4.   Check the following:
- Battery and battery terminal condition.
- Battery bracket and tray corrosion.
- Ignition secondary wiring condition.
- Condition of all visible wiring.
- Condition of hybrid vehicle high-voltage system wiring.

> **Warning**
>
> ⚠ Hybrid high-voltage wires are orange, **Figure 1-5**. Follow all safety precautions when inspecting the insulation and connectors. Remember that these wires carry several hundred volts, and can deliver a lethal shock!

Were any defects found? Yes ____ No ____

If Yes, describe: _____

_____

Figure 1-5. Be extremely careful when working around the orange high-voltage wires found on hybrid vehicles.

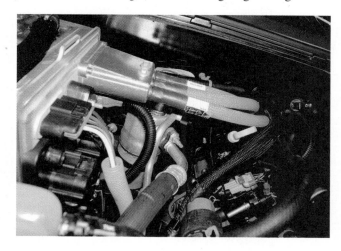

☐ 5. Describe the dangers posed by the following:

Damaged high-voltage wiring insulation: _____

_____

_____

Loose or corroded connections: _____

_____

_____

Battery gases: _____

_____

_____

Short-circuited wiring: _____

_____

_____

Careless handling of hybrid high-voltage system wiring: _____

_____

_____

Compressed Air Safety

☐ 1. Locate the shop's compressed air supply.

What is the air pressure setting on the shop compressor? _____

Describe the dangers posed by compressed air: _____

_____

_____

_____

Shop Cleanliness

☐ 1. Check the shop floor for unsafe conditions, including spills and trip hazards.

List any unsafe conditions: _____

_____

_____

☐ 2. Check the shop tools for cleanliness and organization.

List any ways that tool storage can be improved:_____

_____

_____

Job 1—Perform Safety and Environmental Inspections (continued)

Clothing Safety

☐ 1. Different types of gloves are required for different service procedures.

List tasks that require leather gloves: _____

_____

List tasks that require nitrile gloves: _____

_____

List tasks that require electrically insulated rubber (electric lineman's) gloves: _____

_____

List tasks for which mechanic's gloves would be appropriate: _____

_____

_____

☐ 2. Examine your clothing from a safety standpoint.

List and explain any changes that would make your clothing safer or better-suited in the shop:

_____

_____

Carbon Monoxide

☐ 1. Locate the exhaust hoses in the shop.

Where are they located? _____

_____

☐ 2. Locate the controls to operate the shop's exhaust fans.

Where are they located? _____

_____

_____

Grinder and Drill Press Safety

☐ 1. Go to the electric grinder and inspect it closely. Locate the power switch. Observe the position of the tool rest and face shield. Also, check the condition of the grinding wheel.

Is the electric grinder in the shop safe? Yes ___ No ___

If No, explain: _____

_____

☐ 2. Locate and inspect the operation of the shop's drill press. Find the on/off button, feed lever, chuck, and other components.

List the safety precautions associated with operating a drill press: _____

_____

_____

Floor Jack and Jack Stand Safety

☐ 1. Check out a floor jack and a set of jack stands.

☐ 2. Without lifting a vehicle, practice operating a floor jack. Close the valve on the jack handle. Pump the handle up and down to raise the jack. Then, lower the jack slowly. It is important that you know how to control the lowering action of the jack.

In what direction must you turn the jack handle valve to raise the jack? _____
To lower the jack? _____

> **Warning**
> Ask your instructor for permission before beginning the next step. Your instructor may need to demonstrate the procedures to the class.

☐ 3. After getting your instructor's approval, place the jack under a proper lift point on the vehicle (frame, rear axle housing, suspension arm, or reinforced section of the unibody), **Figure 1-6**. If in doubt about where to position the jack, refer to a service manual for the particular vehicle. Instructions will usually be given in one of the front sections of the service information.

Where did you position the floor jack?_____
_____

☐ 4. To raise the vehicle, place the transmission in Neutral and release the emergency brake. This will allow the vehicle to roll as the jack goes up. If the vehicle cannot roll and the small wheels on the jack catch in the shop floor, the vehicle could slide off the jack.

☐ 5. As soon as the vehicle is high enough, place jack stands under the suggested lift points. Lower the vehicle onto the stands slowly. Check that they are safe. Then, remove the floor jack and block the wheels. It should now be safe to work under the car.

Where did you position the jack stands? _____

☐ 6. Raise the vehicle. Remove the jack stands. Lower the vehicle and return the equipment to the proper storage area.

Figure 1-6. One manufacturer's recommended lift points are shown here. Consult the proper service literature to determine the lift points for the specific vehicle you are working on.

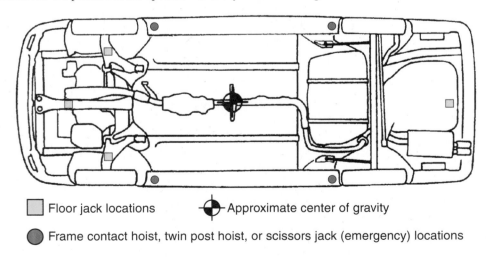

☐ Floor jack locations   ✛ Approximate center of gravity

⬤ Frame contact hoist, twin post hoist, or scissors jack (emergency) locations

Job 1—Perform Safety and Environmental Inspections (continued)

Hydraulic Lift Safety

☐ 1. Obtain the use of one of the shop's hydraulic lifts.

☐ 2. Without lifting a vehicle, practice operating the lift controls.

☐ 3. Raise the lift and ensure that the lift's safety lock operates properly.

  Does the lock operate properly? Yes ___ No ___

  If Yes, go to step 4.

  If No, consult your instructor before proceeding.

> **Note**
>
> The lift may be equipped with more than one safety lock.

☐ 4. After getting your instructor's approval, drive a vehicle onto the lift.

> **Note**
>
> If the rack is a drive-on type, skip steps 5 through 9.

☐ 5. Consult service information to determine the proper lift points for the vehicle.

☐ 6. Position the pads under the vehicle's lift points.

☐ 7. Raise the vehicle so that the pads lightly contact the vehicle's lift points.

☐ 8. Recheck the lift points to ensure that the pads are contacting them properly.

  Are the pads contacting the frame at the proper points? Yes ___ No ___

  If Yes, go to step 9.

  If No, lower the rack and repeat steps 6 and 7.

☐ 9. Raise the vehicle until the safety lock is engaged. It should now be safe to work under the vehicle.

☐ 10. Make sure all personnel are out from under the lift. Release the safety lock.

☐ 11. Lower the lift.

Environmental Protection

☐ 1. Carefully observe all areas of the shop to see how wastes are produced and stored.

☐ 2. Locate and list the types of solid waste produced.

  How are solid wastes disposed of? _____

  _____

☐ 3. Locate and list the types of liquid waste produced.

  How are liquid wastes disposed of? _____

  _____

☐ 4. Locate and list types of gases or airborne particles produced.

How are these contaminates prevented from entering the atmosphere?_____

_____

_____

☐ 5. From the lists above, identify the types of solid and liquid waste that could be recycled:

_____

_____

Identify the materials that could be returned for a core deposit: _____

_____

Do any Environmental Protection Agency (EPA) regulations apply to the wastes generated in the shop? Yes ___ No ___
If Yes, briefly summarize them: _____

_____

Do any local and state regulations apply to the wastes generated by the shop? Yes ___ No ___
If Yes, briefly summarize them: _____

_____

List any of the shop's waste disposal practices that require improvement: _____

_____

Explain what improvements could be made: _____

_____

_____

☐ 6. Clean the work area and return any equipment to storage.

☐ 7. Did you encounter any problems during this procedure? Yes ___ No ___
If Yes, describe the problems: _____

_____

What did you do to correct the problems? _____

_____

☐ 8. Have your instructor check your work and sign this job sheet.

Performance Evaluation—Instructor Use Only

Did the student complete the job in the time allotted? Yes ___ No ___
If No, which steps were not completed?_____
How would you rate this student's overall performance on this job?_____
5–Excellent, 4–Good, 3–Satisfactory, 2–Unsatisfactory, 1–Poor
Comments: _____

_____

_____

INSTRUCTOR'S SIGNATURE _____

Job 2—Interpret Vehicle Numbers, Use Service Information, and Prepare a Work Order

After completing this job, you will be able to locate and interpret vehicle and vehicle subassembly numbers. You will also be able to locate, identify, and use service information for a specific vehicle from a variety of sources. Using this information, you will prepare a work order that includes customer information, vehicle identifying information, customer concern, related service history, cause, and correction.

Procedures

☐ 1. Obtain a vehicle to be used in this job. Your instructor may specify one or more vehicles to be used.

Locate the Vehicle Identification Number (VIN), Emissions Certification Label, and Refrigerant Identification Label

☐ 1. Locate the vehicle identification number (VIN). On all vehicles built after 1968, the VIN will be visible in the lower driver's side corner of the windshield, **Figure 2-1**. On most vehicles built before 1968, the VIN will be located in the driver's side front door jamb, **Figure 2-2**.

Write the VIN here:_____

☐ 2. Open the vehicle hood or engine compartment cover and locate the vehicle's emission certification label.

☐ 3. Answer the following questions:

What is the recommended spark plug gap?_____

Can the ignition timing be adjusted? Yes ___ No ___

Is other service information listed on the label? Yes ___ No ___

If Yes, list: _____

Figure 2-1. On all vehicles made after 1968, the VIN is located on the driver's side of the dashboard, where it can be clearly seen through the windshield.

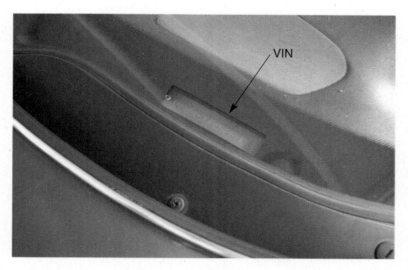

Figure 2-2. The VIN on this older vehicle is installed inside the driver's side door jamb.

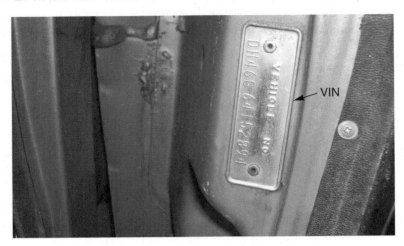

> ### Note
> If the vehicle is not equipped with air conditioning, skip the following step. Also, some older air-conditioned vehicles may not have a refrigerant identification label.

☐ 4. Return to the engine compartment and locate the refrigerant identification label, **Figure 2-3**. The refrigerant identification label may be on an air conditioner component or on the inner fender or strut housing.

What type of refrigerant is used in this vehicle? _____

> ### Note
> Identification labels vary by manufacturer. Not all vehicles will have the same labels. Typical labels may be called calibration labels, safety certification labels, or vehicle service parts labels. Needed information may be included in the certification label rather than on separate labels.

Figure 2-3. The refrigerant label will be installed under the hood, but not necessarily near any air conditioner components.

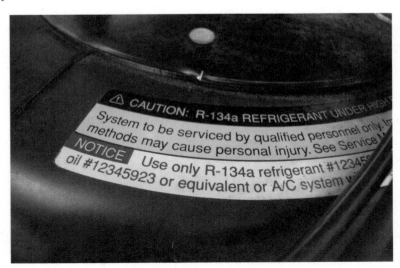

Job 2—Interpret Vehicle Numbers, Use Service Information, and Prepare a Work Order (continued)

☐ 5. Locate all of the vehicle identification labels on the vehicle. Labels may be found on the passenger side door or door pillar, under the vehicle hood, on the underside of the vehicle trunk lid, the tailgate of a station wagon, and on the spare tire cover. In your own words, describe the types of labels and the information that they contain.

_____

_____

_____

Locate Major Component Identification Numbers

☐ 1. Ask your instructor for information concerning which vehicle component to use for locating a component number.

☐ 2. Locate the component and record the component identification number(s).

_____

☐ 3. In your own words, describe the purpose of these numbers.

_____

_____

_____

☐ 4. Close the vehicle hood, trunk, and door and park the vehicle where indicated by your instructor.

Use Printed Service Literature

☐ 1. Ask your instructor to assign a vehicle problem to diagnose or a service procedure to perform.

Describe the problem or procedure assigned:_____

_____

☐ 2. Consult your instructor and obtain service literature relating to the assigned task. The service literature can be a service manual, diagnostic manual, troubleshooting chart, or any other printed service information.

Title of printed literature: _____

☐ 3. Review the formats of the service literature you have and complete the following sections by placing a check mark next to each type of literature you have:

Type of literature:

Manufacturer's service manual:___ Manufacturer's diagnostic manual:___

Manufacturer's troubleshooting chart:___ Aftermarket supplier's manual:___

Technical service bulletin:___ Other type of manual (list):_____

_____

Format:

Chapters:___ Sections:___ Subsections:___ Other (list):_____

_____

Page numbering:

By chapter and section:___ From first page:___ Other (list): _____

Index:

By job topic:___ By vehicle subassembly:___ Other (list): _____

_____

> **Warning**
>
> ⚠ When reviewing service information, always pay close attention to whether the specifications are given in metric or US Customary units. Mixing up the units can result in part failure and unsafe conditions.

☐ 4. Using the table of contents or the index, locate the chapter or section relating to the task assigned by your instructor.

Describe where this information is located: _____

On what page does the information begin?_____

☐ 5. Read the service information and then describe it in your own words in the spaces provided.

_____

_____

_____

_____

_____

_____

Diagram the Service Procedure—Optional Task

☐ 1. Using the gathered service information, draw a diagram showing the steps in the service procedure.

Use a Computer, CDs, and the Internet to Obtain Service Information

☐ 1. Ask your instructor to assign a vehicle problem. Describe the type of problem assigned.

_____

Job 2—Interpret Vehicle Numbers, Use Service Information, and Prepare a Work Order (continued)

☐ 2. Make arrangements to use the shop computer or another school computer as determined by your instructor. Then, answer the following questions:

Who manufactures the computer?_____

What is the computer speed (if known)? _____

What drives does the computer have? Check all that apply:

- Hard drive(s). ___
- CD drive(s). ___
- 3.5″ floppy drive(s). ___
- Other drives. ___

> **Note**
> If the computer is running, skip the following step.

☐ 3. Start the computer and ensure that it boots up properly. In your own words, describe how you can tell that the computer has started correctly.

_____

_____

_____

☐ 4. Use the computer to access the required program.

Software program to be used: _____

Software program maker: _____

> **Note**
> If the program is installed on the computer's hard drive, ignore the references to CDs in steps 5 through 8.

☐ 5. Obtain the needed CDs, as determined by your instructor.

☐ 6. Insert the first CD into the computer. In many cases, this will be the menu or master disc.

☐ 7. Using the computer keyboard and/or mouse, perform the following tasks:

- Locate the needed files on the CD(s).
- Locate service information related to the assigned task.

If asked to do so by your instructor, print out the selected information and attach it to this activity sheet or summarize the information in the space provided. _____

_____

_____

_____

_____

_____

_____

☐ 8. Exit the program and remove the CD from the drive.

☐ 9. Follow your instructor's directions to access the Internet.

☐ 10. Obtain the needed website address from your instructor.

☐ 11. Call up the website. If you do not know how to access the World Wide Web, ask your instructor for directions.

☐ 12. Locate the required information on the website. If asked to do so by your instructor, print out the selected information and attach it to this activity sheet or summarize the information in the space provided.

_____

_____

_____

_____

_____

_____

_____

☐ 13. Exit the website, close your web browser, and follow your instructor's directions to terminate the Internet connection, if necessary.

☐ 14. If you are instructed to shut down the computer, follow your instructor's directions closely to ensure that you shut it down properly.

Use an MSDS to Determine the Safety and Environmental Hazards of a Substance

☐ 1. Locate an MSDS or obtain an MSDS from your instructor.

☐ 2. List the following information from the MSDS:

General class of material covered by the MSDS (if applicable): _____

Trade (manufacturer's) name for the material: _____

Chemical or generic name of the material: _____

Known breathing hazards of the material: _____

Known fire hazards of the material: _____

Known skin damage hazards of the material: _____

Proper storage and disposal methods: _____

_____

Emergency response to spills of the material:

Derive Information from Vehicle Service History—Optional Task

☐ 1. Obtain a vehicle service history from your instructor. This may consist of receipts for previous service, a journal of maintenance and repairs performed, or another type of service record.

☐ 2. Read the vehicle service history and answer the following questions.

Was the vehicle properly maintained? Yes ___ No ___

How can you tell? _____

_____

_____

Does anything in the vehicle service history indicate a developing problem? Yes___ No___

If Yes, what is the problem? _____

_____

☐ 3. Return any printed literature, material safety data sheets, and CDs to storage.

Job 2—Interpret Vehicle Numbers, Use Service Information, and Prepare a Work Order (continued)

Complete a Work Order

> **Note**
>
> You will be asked to fill out additional work orders for other jobs in this manual.

- [ ] 1. Obtain a shop work order from your instructor.
- [ ] 2. Enter today's date in the proper space on the work order.
- [ ] 3. Enter the customer's name and address in the proper work order spaces. If the vehicle is a shop unit, enter the name and address of your school.
- [ ] 4. Record phone number(s) where the customer may be reached. If the vehicle is a shop unit, use the school phone number or create a number.
- [ ] 5. In the proper spaces on the work order, enter information about the vehicle, including the following:
    - Make.
    - Model.
    - Year.
    - Mileage.
- [ ] 6. Enter the vehicle's VIN in the proper space on the work order.
- [ ] 7. In the proper work order spaces, record the customer complaint(s) or other relevant customer remarks. Be brief but record all necessary information.
- [ ] 8. Record any vehicle service history that may help with diagnosis.
- [ ] 9. When repairs are complete, list the following in the appropriate work order spaces:
    - Cause of the vehicle problem.
    - Corrections made (repairs or adjustments performed).
    - Parts installed and their price.
- [ ] 10. Add the parts prices and write the total in the parts total space on the work order.
- [ ] 11. Determine the total labor time of the above operations and write it in the appropriate space on the work order.
- [ ] 12. Multiply the total labor time by the labor rate for your area and write the total in the labor price space on the work order. Your instructor may furnish a labor rate to use.
- [ ] 13. If necessary, add the cost of shop supplies such as lubricants and shop towels in the appropriate space on the work order.
- [ ] 14. Add the parts, labor, and shop supply totals and write this amount in the proper space.
- [ ] 15. Calculate the sales tax:
    - Determine the sales tax rate for your area.
    - Multiply the parts, labor, and shop supply total by the tax rate to obtain the tax.
- [ ] 16. Add the tax to the parts and labor total and enter this total in the proper space on the work order.
- [ ] 17. Make other remarks on the work order as necessary. Examples of other remarks are the need for additional work or fasteners that must be retorqued after several use cycles.
- [ ] 18. Recheck your work and submit the completed work order to your instructor.

☐ 19. Did you encounter any problems during this procedure? Yes ___ No ___

       If Yes, describe the problems: _____

       _____

       What did you do to correct the problems? _____

       _____

☐ 20. Have your instructor check your work and sign this job sheet.

Performance Evaluation—Instructor Use Only

Did the student complete the job in the time allotted? Yes ___ No ___

If No, which steps were not completed?_____

How would you rate this student's overall performance on this job?_____

5–Excellent, 4–Good, 3–Satisfactory, 2–Unsatisfactory, 1–Poor

Comments: _____

_____

_____

INSTRUCTOR'S SIGNATURE _____

Project

Removing and Replacing Gaskets, Seals, and Bearings

2

Introduction

Many service jobs involve replacing gaskets, seals, and/or bearings. Gaskets are used throughout the vehicle to hold in engine oil, transmission fluid, gear oil, or grease. Gaskets are generally durable, but may eventually lose their flexibility and begin to leak. Also, over tightening fasteners can damage a gasket. In Job 3, you will remove and install a gasket.

Lip seals are used throughout the vehicle to seal rotating shafts. A lip seal will be found in each location where a rotating shaft exits a stationary vehicle component. Eventually lip seals wear out and begin leaking. While lip seal replacement is relatively simple, the job must be done properly or the new seal will leak. In Job 4, you will replace a lip seal.

Transmissions, transaxles, transfer cases, and other drive train parts contain antifriction bearings. Antifriction bearings can be grouped into three main classes: ball, roller, and tapered roller. Bearing replacement is part of many vehicle service and overhaul procedures. Antifriction bearing replacement procedures must be followed exactly. Improper procedures will ruin the new bearing and may damage the shaft or bearing housing. In Job 5, you will remove and replace an antifriction bearing.

Project 2 Jobs

- Job 3—Remove and Install a Gasket
- Job 4—Remove and Install a Seal
- Job 5—Remove and Replace a Bearing

Tools and Materials

The following list contains the tools and materials that may be needed to complete the jobs in this project. The items used will depend on the make and model of vehicle being serviced.

- One or more vehicles.
- Applicable service information.
- Drain pan.
- Gasket scraper.
- Replacement gasket.
- Gasket sealer or adhesive.
- Housing in need of pressed-in lip seal replacement.
- Lip seal removal tool.
- Seal driver.
- Replacement seal.
- Nonhardening sealer, if needed.
- Shaft or housing containing an antifriction bearing.
- Hydraulic press.
- Bearing driver.
- Bearing adapters.
- Replacement bearing.
- Hand tools as needed.
- Air-powered tools as needed.
- Safety glasses and other protective equipment as needed.

Safety Notice

Before performing this job, review all pertinent safety information in the text and review safety information with your instructor.

Job 3—Remove and Install a Gasket

After completing this job, you will be able to remove and install a gasket on an automotive part.

Procedures

☐   1.   Obtain a vehicle to be used in this job. Your instructor may direct you to perform this job on a shop vehicle or engine.

☐   2.   Gather the tools needed to perform the following job. Refer to the project's tools and materials list.

☐   3.   Make sure all lubricant has been drained from the unit containing the part and gasket to be removed.

> **Note**
>
> Skip the preceding step if the lubricant does not cover the part to be removed when the vehicle is not running. Care should be taken that the lubricant is not contaminated with gasket material or other foreign material during this task.

☐   4.   Remove the part covering the gasket.

☐   5.   Scrape all old gasket material from both sealing surfaces.

☐   6.   Thoroughly clean the removed part and remove any gasket material or other debris on the related components in the vehicle.

☐   7.   Inspect the removed part and the sealing surface in the vehicle for gouges, cracks, dents and warped areas.

Describe the condition of the removed part and the sealing surface in the vehicle. _____

_____

_____

☐   8.   Repair or replace the parts if defects are found.

☐   9.   Obtain a replacement gasket and compare it with the part's sealing surface. See **Figure 3-1**. Make sure that the following conditions are met:
- The replacement gasket is the correct size and shape. ____
- The gasket material is correct for this application. ____

☐   10.   Place a light coat of sealer or adhesive, as applicable, on the removed part to hold the gasket in place during reinstallation.

> **Note**
>
> Skip the preceding step if the gasket is being installed inside an automatic transmission or transaxle. Sealer also may not be needed on other applications.

☐   11.   Place the gasket in place on the part.

☐   12.   Reinstall the part, being careful not to damage or misalign the gasket.

☐   13.   Install the fasteners. Do not tighten any fastener until all fasteners are started.

Figure 3-1. Compare the new gasket to the part-sealing surface to ensure that it is correct.

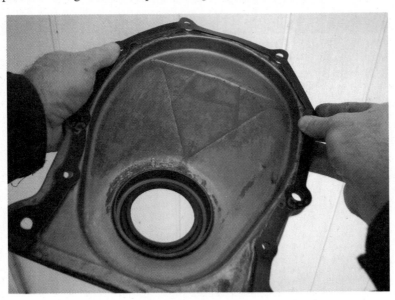

14. Tighten all fasteners to the correct torque. Follow these rules to avoid gasket damage and leaks:
    - Tighten the fasteners in the manufacturer's sequence. If a sequence is not available, follow the procedures shown in **Figure 3-2**. If the part is irregularly shaped, start from the inner fasteners and work outward.
    - Tighten each fastener slightly, then move to the next fastener. Repeat the sequence until final torque is reached.

Figure 3-2. A—When tightening a square or rectangular pan or cover, always start from the center and work outward. This allows the gasket to spread out, improving the seal. B—When installing the fasteners on a round cover, tighten in a star or crisscross pattern.

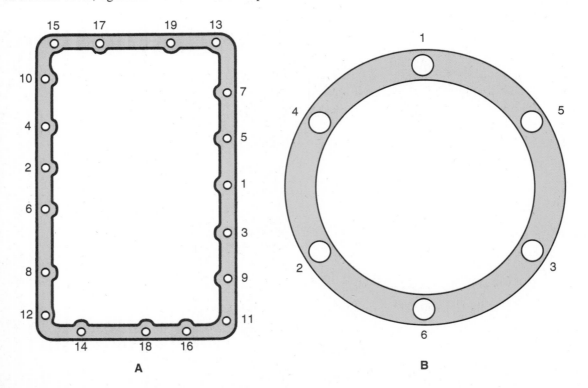

A

B

Job 3—Remove and Install a Gasket (continued)

- Be especially careful not to overtighten fasteners holding sheet metal components.
- Rubber gaskets, sometimes called "spaghetti gaskets," require very low torque values. **Figure 3-3** is a typical spaghetti gasket. Be extremely careful not to overtighten these types of gaskets.

☐ 15. Reinstall the correct type and amount of fresh lubricant into the unit as needed.

> **Note**
>
> If there is any possibility that the lubricant was contaminated during the gasket replacement process, it should be replaced.

☐ 16. Operate the vehicle and check for leaks.

☐ 17. Clean the work area and return any equipment to storage.

☐ 18. Did you encounter any problems during this procedure? Yes ___ No ___

   If Yes, describe the problems: _____

   _____

   What did you do to correct the problems? _____

   _____

☐ 19. Have your instructor check your work and sign this job sheet.

Figure 3-3. A spaghetti gasket used on an intake manifold is shown here. Some spaghetti gaskets are simply sections of flexible rubber that strongly resemble spaghetti. They are extremely easy to over tighten.

Performance Evaluation—Instructor Use Only

Did the student complete the job in the time allotted? Yes ____ No ____

If No, which steps were not completed?_____

How would you rate this student's overall performance on this job?_____

5–Excellent, 4–Good, 3–Satisfactory, 2–Unsatisfactory, 1–Poor

Comments: _____

_____

_____

INSTRUCTOR'S SIGNATURE _____

Job 4—Remove and Install a Seal

After completing this job, you will be able to replace a pressed-in lip seal.

Procedures

☐ 1. Obtain a vehicle to be used in this job. Your instructor may direct you to perform this job on a shop vehicle or engine.

☐ 2. Gather the tools needed to perform the following job. Refer to the project's tools and materials list.

☐ 3. Remove shafts and other parts that restrict access to the lip seal to be replaced.

> **Caution**
> ⬦ Drain oil to a level below the seal if necessary.

☐ 4. Remove the lip seal by one of the following methods:
   - Prying the seal from the housing. See **Figure 4-1**.
   - Driving the seal from the back side of the housing.
   - Using a special removal tool to remove the seal.

   Describe the method used to remove the lip seal._____

   _____

☐ 5. Obtain a replacement seal and compare it with the old seal.

   Is the replacement seal correct? Yes ___ No ___

   If No, what should you do next?_____

   _____

   _____

Figure 4-1. If the seal will not be reused, pry it from the housing with a large screwdriver or pry bar. Be careful not to damage the seal housing during removal.

6. Thoroughly clean the seal housing to remove oil, sludge and carbon deposits, and old sealer.

7. Inspect the seal housing for cracks, gouges, and dents at the sealing area. **Figure 4-2** shows a cracked seal housing.

Describe the condition of the housing and the old seal: _____

_____

_____

_____

8. Inspect the shaft at the sealing area. Look for nicks, burrs, and groove wear.

Describe the condition of the shaft at the sealing area: _____

_____

_____

_____

Note

Defective housings should be repaired or replaced. Defective shafts should also be replaced. Consult your instructor to determine what additional steps, if any, need to be taken.

---

9. Lightly lubricate the lip of the seal as shown in **Figure 4-3**. Use the same type of lubricant as is used in the device being serviced.

10. If specified by the seal manufacturer, lightly coat the outside diameter of the seal with nonhardening sealer.

11. Install the new seal using one of the following methods:
    - Drive the seal into place with a seal driver, **Figure 4-4**.
    - Carefully tap the seal into place with a hammer.

Figure 4-2. This seal housing is visibly cracked. A damaged seal housing must be properly repaired or replaced.

Crack in housing

Job 4—Remove and Install a Seal (continued)

Figure 4-3. Lubricate the sealing lip before installation.

Figure 4-4. A typical seal driver.

 Caution

When installing a seal by tapping it into place with a hammer, lightly tap on alternating sides of the seal.

Describe the method used to install the seal: _____

_____

_____

> **Caution**
>
> Some technicians prefer to drive the seal into place using a hammer and a block of wood. This should only be done when no other method is available. Wood fragments can enter the bore and cause damage.

12. Reinstall the shaft and other parts as necessary.

13. Check that the shaft turns freely after all parts are installed.

14. Add fluid as needed. Then, operate the unit and check for leaks.

15. Clean the work area and return any equipment to storage.

16. Did you encounter any problems during this procedure? Yes ___ No ___

    If Yes, describe the problems: _____

    _____

    What did you do to correct the problems? _____

    _____

17. Have your instructor check your work and sign this job sheet.

Performance Evaluation—Instructor Use Only

Did the student complete the job in the time allotted? Yes ___ No ___

If No, which steps were not completed?_____

How would you rate this student's overall performance on this job?_____

5–Excellent, 4–Good, 3–Satisfactory, 2–Unsatisfactory, 1–Poor

Comments: _____

_____

_____

INSTRUCTOR'S SIGNATURE _____

Job 5—Remove and Replace a Bearing

After completing this job, you will be able to remove and replace an antifriction bearing.

Procedures

☐ 1. Obtain a vehicle to be used in this job. Your instructor may direct you to perform this job on a shop vehicle or engine.

☐ 2. Gather the tools needed to perform the following job. Refer to the project's tools and materials list.

Replace a Shaft-Mounted Bearing

☐ 1. Remove the bearing and shaft from the vehicle.

☐ 2. Place the bearing and shaft or housing in a suitable press for removal.

☐ 3. Support the bearing and the shaft or housing, as applicable. Place the correct adapters under the press-fit race. See **Figure 5-1**.

☐ 4. Note whether the bearing design indicates that the bearing should be installed facing a particular direction. Some bearings have external marks indicating the direction of installation.

☐ 5. Using the proper bearing driver, operate the press to remove the bearing from the shaft or housing.

Describe the condition of the bearing and the shaft or housing: _____

_____

☐ 6. Obtain a replacement bearing. Compare the old bearing to its replacement to ensure that the replacement bearing is correct.

☐ 7. Pack the bearing with grease, if necessary, using manufacturer's recommended type and grade of bearing lubricant.

Type of lubricant used: _____

☐ 8. If applicable, make sure the bearing is installed facing the proper direction.

☐ 9. Place the replacement bearing over the shaft.

Figure 5-1. The correct and incorrect procedures for removing a bearing from a shaft. Pressure should be applied to the press-fit race only. Applying force through the rolling members can damage the balls, rollers, or races. (FAG Bearings)

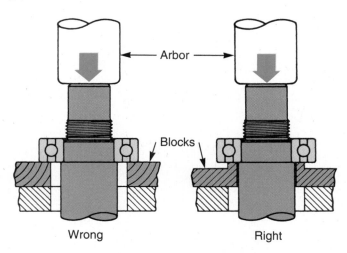

☐ 10. Support the bearing and the shaft with correct adapters under the press-fit race. See **Figure 5-2**.

Figure 5-2. The correct and incorrect ways to install bearings. Do not apply force through the rolling elements when installing a bearing. (FAG Bearings)

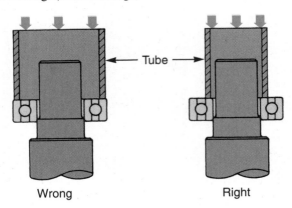

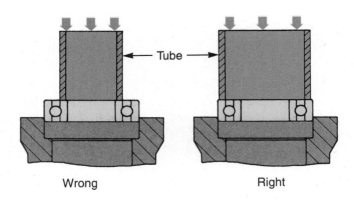

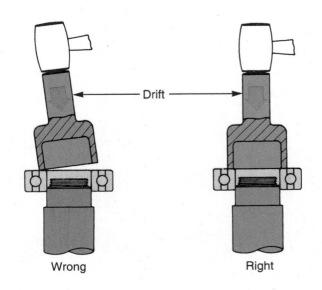

Job 5—Remove and Replace a Bearing (continued)

☐ 11. Using the proper bearing driver, operate the press to install the bearing onto the shaft or into the housing.

☐ 12. Check the bearing for free rotation and correct any problems.
- If the newly installed bearing is binding, press the bearing off the shaft or out of the housing, as described in steps 2 through 4.
- Inspect the bearing for visible signs of damage and to make sure it matches the bearing being replaced. Replace the bearing if there are any signs of damage or if it does not precisely match the bearing it is replacing.
- Rotate the bearing in your hands and feel for any resistance or unusual vibration. If you detect excessive resistance or unusual vibrations, discard the bearing and replace it.
- If the bearing passes inspection, reinstall the bearing as described in steps 6 through 10, being especially careful that the bearing is aligned squarely on the shaft or in the bore.

☐ 13. Reinstall the bearing and shaft on the vehicle.

☐ 14. Check bearing operation.

Replace Tapered Roller Bearings and Races

☐ 1. Remove the bearing assembly dust cover, cotter keys, and/or locking nuts as necessary.

☐ 2. Remove the outer bearing from the hub or housing as necessary.

☐ 3. Remove the hub or housing from the vehicle.

☐ 4. Remove the inner bearing and seal as necessary.

☐ 5. Inspect the bearings for wear or damage.

List any signs of bearing damage:_____

_____

Can this bearing be reused? Yes ____ No ____

> **Note**
>
> If any part of the bearing assembly requires replacement, replace all parts of the bearing assembly. Never reuse old races with new bearings, or vice versa.

☐ 6. Clean the bearing races and check the races for wear and damage.

List any signs of race damage: _____

_____

Can this race be reused? Yes ____ No ____

> **Note**
>
> If the old race turns inside the hub, replace the rotor or hub.

☐ 7. If a bearing race must be replaced, turn the drum or housing over and locate the back of the race. There will usually be one or more notches to allow the race to be driven out.

☐ 8. Using a punch, drive out the race by alternately tapping it at several places around its diameter.

> **Caution**
> If the race will be reused, use a brass drift instead of a punch.

☐ 9. Inspect the hub or housing for dents, nicks, or other defects that would keep the new race from seating properly. Remove small defects with a file. Major defects require that the hub or housing be replaced.

☐ 10. Using a special installer tool or a brass drift, install the new race. If a special tool is not available, install the race by alternately tapping it at several places around its diameter, not on the area where the bearing contacts the race.

> **Caution**
> Do not allow the race to become cocked in the bore. The race and the hub or housing will be ruined.

☐ 11. Clean and regrease the bearings as necessary.

☐ 12. Reinstall the inner bearing and seal if necessary.

☐ 13. Reinstall the hub or housing on the vehicle.

☐ 14. Reinstall the outer bearing in the hub or housing.

☐ 15. Adjust the bearing preload.
- Tighten the nut to approximately 100 foot pounds to seat all of the components.
- Back off the nut until it is only finger tight.
- Tighten the nut to the specified torque.
- Measure the preload by determining the hub endplay or by measuring the force needed to turn the hub.
- Tighten or loosen the nut as necessary, then recheck the preload.
- Ensure that the hub assembly turns freely.

☐ 16. Install cotter keys and/or locking nuts and the bearing assembly dust cover.

☐ 17. Clean the work area and return any equipment to storage.

☐ 18. Did you encounter any problems during this procedure? Yes ___ No ___

If Yes, describe the problems: _____

_____

What did you do to correct the problems? _____

_____

☐ 19. Have your instructor check your work and sign this job sheet.

Performance Evaluation—Instructor Use Only

Did the student complete the job in the time allotted? Yes ___ No ___
If No, which steps were not completed?_____
How would you rate this student's overall performance on this job?_____
5–Excellent, 4–Good, 3–Satisfactory, 2–Unsatisfactory, 1–Poor
Comments: _____

_____

INSTRUCTOR'S SIGNATURE _____

Project
Performing Oil Changes and Checking an Engine for Leaks

3

Introduction

Periodic oil and filter changes can greatly prolong the life of an engine. Oil and filter changes are relatively simple, but must be done correctly. It is also vital that any oil or coolant leaks be found and promptly fixed. In Job 6, you will drain and replace the engine oil and change the oil filter. In Job 7, you will observe the engine for oil, coolant, and fuel leaks.

Project 3 Jobs

- Job 6—Change Oil and Filter
- Job 7—Inspect an Engine for Leaks

Tools and Materials

The following list contains the tools and materials that may be needed to complete the jobs in this project. The items used will depend on the make and model of vehicle being serviced.

- Vehicle in need of service.
- Applicable service information.
- Oil filter wrench.
- Oil drain pan.
- Oil filter.
- Correct type and grade of oil.
- Leak detection equipment.
- Shop towels.
- Hand tools as needed.
- Safety glasses and other protective equipment as needed.

Safety Notice

Before performing this job, review all pertinent safety information in the text and review safety information with your instructor.

Notes

Job 6—Change Oil and Filter

After completing this job, you will be able to perform an oil and filter change.

Procedures

☐   1.   Obtain a vehicle to be used in this job. Your instructor may direct you to perform this job on a shop vehicle or engine.

☐   2.   Gather the tools needed to perform the following job. Refer to the project's tools and materials list.

Change Oil and Filter

> **Caution**
>
> Before performing any service or test driving any vehicle, always place disposable covers over the upholstery and steering wheel and place mats on the carpet to prevent stains.

☐   1.   Obtain the correct oil filter and type and grade of motor oil.

       Brand, grade, and weight of oil: _____

       Brand and stock number of oil filter: _____

☐   2.   Run the engine until it reaches operating temperature.

☐   3.   Raise the vehicle on a lift or raise it with a jack and secure it on jack stands. The vehicle should be level when raised to allow all of the oil to drain from the pan.

> **Warning**
>
> The vehicle must be raised and supported in a safe manner. Always use approved lifts or jacks and jack stands. See **Figure 6-1**.

☐   4.   Place the oil drain pan under the engine oil pan. **Figure 6-2** shows a typical oil drain pan.

☐   5.   Remove the oil drain plug. A few engines have two drain plugs.

Figure 6-1. When raising a vehicle, make sure the lift pads are positioned at recommended lift points.

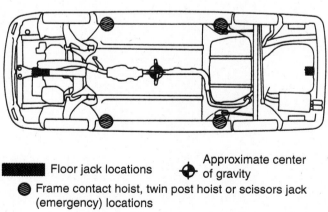

■ Floor jack locations

✛ Approximate center of gravity

● Frame contact hoist, twin post hoist or scissors jack (emergency) locations

Figure 6-2. The oil pan should be large enough to fit under both the oil drain and the oil filter.

> ## Warning
> ⚠ Avoid contact with hot oil. You could be severely burned.

- [ ] 6. Inspect the drain plug. The seal should be undamaged to prevent leaks. Check the drain plug's threads for damage. If the drain plug is damaged, replace it.

    Are the drain plug threads and seal OK? Yes ___ No ___

- [ ] 7. Loosen the oil filter using a filter wrench and remove the filter from the engine. If necessary reposition the oil drain pan under the filter or use a second drain pan.

- [ ] 8. Wipe off the oil filter mounting base and check that the old filter seal is not stuck on the base.

- [ ] 9. Check the new filter to ensure that it is the correct replacement.

- [ ] 10. Fill the new filter with the correct grade of motor oil. If the oil filter is mounted with the opening down, skip this step.

- [ ] 11. Smear a thin film of clean engine oil on to the new oil filter rubber seal as shown in **Figure 6-3**.

- [ ] 12. Screw on the new oil filter and hand tighten it.

- [ ] 13. Turn the filter an additional 1/2 to 3/4 turn.

- [ ] 14. Install and tighten the oil drain plug.

- [ ] 15. Remove the oil filler cap from the engine.

> ## Caution
> ◇ Before performing any work under the hood of a vehicle, always place covers over the fenders to prevent scratches and dings.

- [ ] 16. Open the first oil container and pour the oil into the filler opening in the engine. Carefully monitor the filler opening to ensure that oil does not spill out.

Job 6—Change Oil and Filter (continued)

Figure 6-3. Use clean oil to lightly lubricate the filter seal.

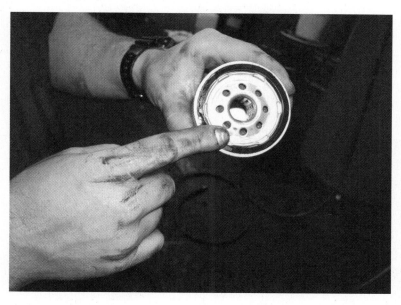

☐ 17. Repeat step 16 until the engine is filled to the proper level.

☐ 18. Replace the filler cap and wipe any spilled oil from the engine.

☐ 19. Start the engine and watch the oil pressure light or gauge. The light should go out or the gauge should begin to register within 10 to 20 seconds. If it does not, stop the engine immediately and locate the problem.

☐ 20. Raise the vehicle and check for leaks from the oil filter and drain plug.

Are there any leaks? Yes ___ No ___

If you found leaks, what did you do to correct them? _____

_____

_____

_____

☐ 21. Lower the vehicle and stop the engine.

☐ 22. Allow the engine to sit for several minutes.

☐ 23. Recheck the oil level.

Is the level correct? Yes ___ No ___

If the level is low, add oil until the dipstick reads full.

 Note

Before returning a vehicle to a customer, always inspect the work area and passenger compartment to ensure that all tools, parts, and debris related to the service have been picked up. Some companies may want the disposable floor mats and protective covers left in place until the customer takes possession of the vehicle. Always prepare the vehicle in accordance with your company or school's policy.

24. Clean the work area and return any equipment to storage. Dispose of the old oil filter properly, **Figure 6-4**.

25. Did you encounter any problems during this procedure? Yes ___ No ___

    If Yes, describe the problems: _____

    _____

    What did you do to correct the problems? _____

    _____

26. Have your instructor check your work and sign this job sheet.

Figure 6-4. Always dispose of the old oil filter in special oil filter containers. The oil filters are disposed of without damaging the environment.

Performance Evaluation—Instructor Use Only

Did the student complete the job in the time allotted? Yes ___ No ___
If No, which steps were not completed?_____
How would you rate this student's overall performance on this job?_____
5–Excellent, 4–Good, 3–Satisfactory, 2–Unsatisfactory, 1–Poor
Comments: _____

_____

_____

INSTRUCTOR'S SIGNATURE _____

Job 7—Inspect an Engine for Leaks

After completing this job, you will be able to check an engine for leaks.

Procedures

☐ 1. Obtain a vehicle to be used in this job. Your instructor may direct you to perform this job on a shop vehicle or engine.

☐ 2. Gather the tools needed to perform the following job. Refer to the project's tools and materials list.

Visual Inspection

☐ 1. Inspect the top of the engine for leaks. Such leaks will usually be visible at the point of leakage.

- Common upper-engine oil leak points are valve covers, timing covers, and the oil filler cap.

Were any oil leaks found? Yes ___ No ___

If Yes, list the locations of the oil leaks: _____

_____

- Common upper-engine fuel leak points are fuel fittings, hose clamps, and pressure regulators.

Were any fuel leaks found? Yes ___ No ___

If Yes, list the locations of the fuel leaks: _____

_____

- Common upper-engine coolant leak points are hoses, hose fittings, radiator seams, and the radiator cap.

Were any coolant leaks found? Yes ___ No ___

If Yes, list the locations of the coolant leaks: _____

> **Note**
> Pressure testing the cooling system is covered in Job 31.

☐ 2. Obtain a drop light or other source of illumination.

☐ 3. Examine the underside of the vehicle for evidence of oil or grease. Slight seepage is normal.

Was excessive oil or grease observed? Yes ___ No ___

If Yes, where does the oil/grease appear to be coming from? _____

_____

> **Note**
> Airflow under the vehicle will blow leaking oil backwards. The leak may be some distance forward from where the oil appears.

Powder Method

☐ 1. Thoroughly clean the area around the suspected leak.

☐ 2. Apply talcum powder to the clean area.

☐ 3. Lower the vehicle from the lift and drive it several miles or carefully run it on the lift for 10–15 minutes.

☐ 4. Raise the vehicle (if necessary) and check the area around the suspected leak.

> **Warning**
>
> ⚠ The vehicle must be raised and supported in a safe manner. Always use approved lifts or jacks and jack stands.

Does the powder show streaks of oil, fuel, or coolant? Yes ___ No ___

If Yes, what type of fluid is leaking and where does it appear to be coming from? _____

_____

Black Light Method

☐ 1. Ensure that the engine has enough oil or coolant. Add oil or coolant as needed.

☐ 2. Add fluorescent dye to the unit through the filler plug, being careful not to spill dye on the outside of the engine.

☐ 3. Lower the vehicle from the lift and drive it several miles, or carefully run it on the lift for 10–15 minutes.

☐ 4. Raise the vehicle, if necessary.

> **Warning**
>
> ⚠ The vehicle must be raised and supported in a safe manner. Always use approved lifts or jacks and jack stands.

☐ 5. Turn on the black light and direct it toward the area around the suspected leak. See **Figure 7-1**.

Does the black light show the presence of dye? Yes ___ No ___

If Yes, where does the dye appear to be coming from? _____

_____

☐ 6. Consult your instructor about the steps to take to correct the leak. Steps may include some of the following:
   - Tightening fasteners.
   - Replacing gaskets or seals. See Jobs 3 and 4.
   - Replacing a cracked, broken, or punctured part.

   Steps to be taken: _____

   _____

   _____

☐ 7. Make the necessary repairs.

☐ 8. Clean the work area and return any equipment to storage.

☐ 9. Did you encounter any problems during this procedure? Yes ___ No ___

   If Yes, describe the problems: _____

   _____

Job 7—Inspect an Engine for Leaks (continued)

Figure 7-1. A black light can be used to locate leaks when necessary. (Tracer Products Division of Spectronics Corporation)

What did you do to correct the problems? _____
_____

☐ 10. Have your instructor check your work and sign this job sheet.

Performance Evaluation—Instructor Use Only

Did the student complete the job in the time allotted? Yes ___ No ___
If No, which steps were not completed?_____
How would you rate this student's overall performance on this job?_____
5–Excellent, 4–Good, 3–Satisfactory, 2–Unsatisfactory, 1–Poor
Comments: _____
_____
_____

INSTRUCTOR'S SIGNATURE _____

Notes

_____
_____
_____
_____
_____
_____
_____
_____
_____
_____
_____
_____
_____
_____
_____
_____
_____
_____
_____
_____
_____
_____
_____
_____
_____
_____
_____
_____
_____
_____
_____
_____
_____
_____
_____
_____
_____
_____
_____
_____
_____

Project

Diagnosing Engine Problems

4

Introduction

Underneath all of the electronic devices on a modern engine are mechanical parts that would be recognized by a technician from 100 years ago. Before assuming that a problem is in one of the high-tech areas of the vehicle, it is often necessary to use one of the tests covered in this project to eliminate the engine's internal parts as the source of problems.

It is almost impossible to service modern vehicles without knowing how to use a scan tool. Every vehicle system is controlled by, or sends input to, the onboard computer. This information can only be accessed with a scan tool. Even if a problem is in a system not monitored by the onboard diagnostics, the scan tool can still be useful in helping the technician narrow the field of possible causes.

In the following jobs, you will test engine components and interpret test results. You will perform a vacuum test and power balance test in Job 8. In Job 9, you will test check engine compression and cylinder leakage. In Job 10, you will check engine oil pressure. You will use a scan tool to retrieve diagnostic trouble codes in Job 11.

Project 4 Jobs

- Job 8—Perform a Vacuum Test and a Power Balance Test
- Job 9—Perform a Cranking Compression Test, Running Compression Test, and a Cylinder Leakage Test
- Job 10—Perform an Oil Pressure Test
- Job 11—Verify Warning Indicator Operation and Retrieve Trouble Codes

Tools and Materials

The following list contains the tools and materials that may be needed to complete the jobs in this project. The items used will depend on the make and model of vehicle being serviced.

- Vehicle in need of service.
- Applicable service information.
- Oil filter wrench.
- Oil drain pan.
- Oil filter.
- Correct type and grade of oil.
- Leak detection equipment.
- Compression gauge.
- Vacuum gauge.
- Scan tool.
- Scan tool operating instructions.
- Shop towels.
- Hand tools as needed.
- Safety glasses and other protective equipment as needed.

Safety Notice

Before performing this job, review all pertinent safety information in the text and review safety information with your instructor.

Job 8—Perform a Vacuum Test and a Power Balance Test

After completing this job, you will be able to perform a vacuum test and a power balance test as part of the diagnostic process.

Procedures

☐ 1. Obtain a vehicle to be used in this job. Your instructor may direct you to perform this job on a shop vehicle or engine.

☐ 2. Gather the tools needed to perform the following job. Refer to the project's tools and materials list.

Engine Vacuum Test

☐ 1. Attach a vacuum gauge to the engine. The easiest way to do this is to remove an engine vacuum line and attach the vacuum gauge hose. If a vacuum line cannot be removed or if removing any hose would affect engine operation, the gauge should be connected to the vacuum circuit without interrupting it. This can be accomplished by removing one of the engine's vacuum lines, inserting a length of vacuum hose with a T fitting, and then attaching the vacuum gauge hose to the open nipple on the fitting. It can also be accomplished by removing a fitting from the intake manifold and temporarily installing a hose fitting that allows the gauge to be connected between the vacuum line and the manifold.

☐ 2. Start the engine and allow it to idle.

☐ 3. Refer to the chart in **Figure 8-1** to diagnose engine problems. If necessary, increase engine speed to make further checks.

   What conclusions can you make from reading the vacuum gauge?_____

   _____

   _____

   _____

   _____

   _____

☐ 4. Turn off the engine.

☐ 5. Remove the vacuum gauge and reattach vacuum hoses or fittings as necessary. If you used a T fitting to attach the vacuum gauge hose, remove the T fitting and reconnect the vacuum line.

Power Balance Test

☐ 1. Attach a power balance testing device. This can be a scan tool or engine analyzer with the capability of shutting down the ignition or fuel injector on individual cylinders.

☐ 2. Start the engine and set the engine speed to fast idle before beginning the test.

☐ 3. Kill (disable) each cylinder in turn and record the engine speed as each cylinder is disabled.

   Cylinder # 1:_____rpm.   Cylinder # 2:_____rpm.

   Cylinder # 3:_____rpm.   Cylinder # 4:_____rpm.

   Cylinder # 5:_____rpm.   Cylinder # 6:_____rpm.

   Cylinder # 7:_____rpm.   Cylinder # 8:_____rpm.

   Cylinder # 9:_____rpm.   Cylinder # 10:_____rpm.

   Cylinder # 11:_____rpm.   Cylinder # 12:_____rpm.

Figure 8-1. Compare the vacuum gauge readings for the engine you are testing with the readings shown here.

Vacuum Gauge Readings

Note: White needle indicates steady vacuum. Red needle indicates fluctuating vacuum.

Needle steady and within specifications.

Cause: Normal vacuum at idle.

Needle very low and steady.

Cause: Vacuum or intake manifold leak.

Needle normal at idle, but fluctuates as engine speed is increased.

Cause: Weak valve spring.

Needle jumps to almost zero when throttle is opened and comes back to just over normal vacuum when closed.

Cause: Normal acceleration and deceleration reading.

Needle slowly drops from normal as engine speed is increased.

Cause: Restricted exhaust (compare at idle and 2500 rpm).

Needle steady, but low at idle.

Cause: Improper valve or ignition timing.

Needle has small pulsation at idle.

Cause: Insufficient spark plug gap.

Needle occasionally makes a sharp fast drop.

Cause: Sticking valve.

Needle regularly drops 4 to 8 inches.

Cause: Blown head gasket or excessive block-to-head clearance.

Needle slowly drifts back and forth.

Cause: Improper air-fuel mixture.

Needle drops regularly; may become steady as engine speed is increased.

Cause: Burned valve, worn valve guide, insufficient tappet clearance.

Needle drops to zero when engine is accelerated and snaps back to higher than normal on deceleration.

Cause: Worn piston rings or diluted oil.

Job 8—Perform a Vacuum Test and a Power Balance Test (continued)

☐ 4. Return the engine to its normal idle speed and turn it off.

☐ 5. Compare the rpm drops for all cylinders.

Did the rpm fail to drop when any cylinder was disabled? Yes ___ No ___

If Yes, what was the number of the cylinder(s) that did not drop?_____

What do you think could cause the readings that you observed?_____

_____

_____

_____

_____

☐ 6. Remove the test equipment leads from the engine.

☐ 7. Clean the work area and return any equipment to storage.

☐ 8. Did you encounter any problems during this procedure? Yes ___ No ___

If Yes, describe the problems: _____

_____

What did you do to correct the problems? _____

_____

☐ 9. Have your instructor check your work and sign this job sheet.

Performance Evaluation—Instructor Use Only

Did the student complete the job in the time allotted? Yes ___ No ___

If No, which steps were not completed?_____

How would you rate this student's overall performance on this job?_____

5–Excellent, 4–Good, 3–Satisfactory, 2–Unsatisfactory, 1–Poor

Comments: _____

_____

_____

INSTRUCTOR'S SIGNATURE _____

Notes

Job 9—Perform a Cranking Compression Test, Running Compression Test, and a Cylinder Leakage Test

After completing this job, you will be able to perform a compression test and a cylinder leakage test as part of the diagnostic process.

Procedures

☐ 1. Obtain a vehicle to be used in this job. Your instructor may direct you to perform this job on a shop vehicle or engine.

☐ 2. Gather the tools needed to perform the following job. Refer to the tools and materials list at the beginning of the project.

Cranking Compression Test

☐ 1. Remove all spark plugs.

☐ 2. Disable the ignition and fuel systems by any of the following methods:
- **Ignition system:** ground the coil wire or remove the ignition fuse.
- **Fuel system:** remove the fuel pump relay or relay control fuse.

> **Note**
>
> It is not necessary to disable the fuel system on engines with carburetors.

☐ 3. Block the throttle valve in the open position.

☐ 4. Install the compression tester in first spark plug opening.

☐ 5. Crank the engine through four compression strokes.

Record the reading:_____

☐ 6. Repeat steps 4 and 5 for all other engine cylinders.

Cylinder # 1:_____   Cylinder # 2:_____

Cylinder # 3:_____   Cylinder # 4:_____

Cylinder # 5:_____   Cylinder # 6:_____

Cylinder # 7:_____   Cylinder # 8:_____

Cylinder # 9:_____   Cylinder # 10:_____

Cylinder # 11:_____   Cylinder # 12:_____

☐ 7. Compare the compression readings for all cylinders.

Is the compression excessively low or high on any cylinder(s)? Yes ___ No ___

If Yes, which cylinders have abnormal readings? _____

What do you think could cause the readings that you observed? Refer to **Figure 9-1** for information about diagnosing compression problems. _____

_____

_____

_____

_____

_____

Figure 9-1. The causes of the these compression readings apply to any engine. Consult the manufacturer's service literature for any special problem.

Compression Gauge Readings		
Compression Test Results	**Cause**	**Wet Test Results**
All cylinders at normal pressure (no more than 10–15% difference between cylinders)	Engine is in good shape, no problems.	No wet test needed.
All cylinders low (more than 20%)	Burned valve or valve seat, blown head gasket, worn rings or cylinder, valves misadjusted, jumped timing chain or belt, physical damage to engine.	If compression increases, cylinder or rings are worn. No increase, problem caused by valve train. Near zero compression caused by engine damage.
One or more cylinders low (more than 20% difference)	Burned valve, valve seat, damage or wear on affected cylinder(s).	If compression increases, cylinder or rings are worn. No increase, problem caused by valve train. Near zero compression caused by engine damage.
Two adjacent cylinders low (more than 20% difference)	Blown head gasket, cracked block or head.	Little or no increase in pressure.
Compression high (more than 20% difference)	Carbon build-up in cylinder.	Do not wet test for high compression.

8. If any cylinder had a low compression reading, pour approximately one teaspoon (1 ml) of engine oil in the cylinder and retest the compression.

Did the compression improve? Yes ___ No ___

What is a possible cause? _____

_____

_____

_____

9. Return the throttle valve to the closed position and replace the fuel pump relay or relay control fuse if it was removed.

10. Reinstall the spark plugs and reconnect the ignition system.

Running Compression Test

1. Remove the spark plug of the cylinder to be tested. If the spark plugs were removed for the cranking compression test, reinstall all of them except the plug of the cylinder to be tested.

2. Install the compression tester in the spark plug hole.

3. Start the engine.

4. Compare the compression reading at idle with the static compression reading. Is the compression approximately 50% of the static compression reading? Yes _____ No _____

5. Quickly snap the throttle open and closed. Does compression rise to about 80% of the static compression reading? Yes _____ No _____

Job 9—Perform a Cranking Compression Test, Running Compression Test, and a Cylinder Leakage Test (continued)

☐ 6. Repeat steps 4 and 5 for all other engine cylinders.

> **Note**
> Reinstall each plug before moving to the next cylinder. Leaving out any plug other than the plug to be tested will affect readings.

Cylinder # 1: _____ Cylinder # 7: _____

Cylinder # 2: _____ Cylinder # 8: _____

Cylinder # 3: _____ Cylinder # 9: _____

Cylinder # 4: _____ Cylinder # 10: _____

Cylinder # 5: _____ Cylinder # 11: _____

Cylinder # 6: _____ Cylinder # 12: _____

☐ 7. Compare the compression readings for all cylinders.

Is the running compression excessively low or high on any cylinder(s)? Yes _____ No _____

If Yes, which cylinders have abnormal readings? _____

What do you think could cause the readings that you observed? Possible causes include worn camshaft lobes, excessive carbon deposits on the valves, and worn rings. _____

_____

_____

_____

_____

_____

☐ 8. Ensure that all plugs have been properly reinstalled and torqued.

☐ 9. Remove all test equipment.

Cylinder Leakage Test

> **Note**
> A cylinder leakage test is usually performed when a compression test indicates low compression on a cylinder. Your instructor may direct you to perform this test on a randomly selected cylinder in good condition.

☐ 1. Disable the ignition system to prevent accidental starting.

☐ 2. Attach a shop air hose to the leak tester.

☐ 3. Adjust the tester pressure to the recommended setting. This is usually about 5–10 psi (35.5–68.9 kpa).

Manufacturer's pressure setting:_____ psi or kpa (circle one).

☐ 4. Remove the spark plug from the suspect cylinder.

☐ 5. Bring the suspect cylinder to the top of its compression stroke by feeling for compression as the piston comes up. Some technicians use a whistle to indicate when the compression stroke is occurring.

☐ 6. Install the leak tester hose in the spark plug opening.

☐ 7. Open the valve allowing regulated pressure into the cylinder.

☐ 8. Observe the pressure. It should be at or near the regulated pressure.

Pressure reading:_____ psi or kpa (circle one).

Does this pressure match the regulated pressure? Yes ___ No ___

If No, what is a possible cause? _____

_____

_____

_____

_____

☐ 9. If the pressure is low, refer to the chart in **Figure 9-2**.

Figure 9-2. The engine cylinder is supposed to be almost airtight on the top of its compression stroke. Air escaping anywhere is caused by a defect.

Cylinder Leakage Test Results	
Condition	**Possible Causes**
No air escapes from any of the cylinders.	Normal condition, no leakage.
Air escapes from carburetor or throttle body.	Intake valve not seated or damaged.
	Valve train mistimed, possible jumped timing chain or belt.
	Broken or damaged valve train part.
Air escapes from tailpipe.	Exhaust valve not seated or damaged.
	Valve train mistimed.
	Broken or damaged valve train part.
Air escapes from dipstick tube or oil fill opening.	Worn piston rings. Worn cylinder walls.
	Damaged piston.
	Blown head gasket.
Air escapes from adjacent cylinder.	Blown head gasket.
	Cracked head or block.
Air bubbles in radiator coolant.	Blown head gasket.
	Cracked head or block.
Air heard around outside of cylinder.	Cracked or warped head or block.
	Blown head gasket.

After consulting the chart, list the possible engine problem: _____

_____

_____

_____

_____

Job 9—Perform a Cranking Compression Test, Running Compression Test, and a Cylinder Leakage Test (continued)

- [ ] 10. Remove the tester hose and reinstall the spark plug.
- [ ] 11. Repeat steps 4 through 10 on all suspect cylinders.
- [ ] 12. Analyze the readings of all tested cylinders.
  What conclusions can you make? _____
  _____
  _____
  _____

- [ ] 13. Remove the shop air hose from the tester.
- [ ] 14. Clean the work area and return any equipment to storage.
- [ ] 15. Did you encounter any problems during this procedure? Yes ___ No ___
  If Yes, describe the problems: _____
  _____

  What did you do to correct the problems? _____
  _____

- [ ] 16. Have your instructor check your work and sign this job sheet.

Performance Evaluation—Instructor Use Only

Did the student complete the job in the time allotted? Yes ___ No ___
If No, which steps were not completed? _____
How would you rate this student's overall performance on this job? _____
5–Excellent, 4–Good, 3–Satisfactory, 2–Unsatisfactory, 1–Poor
Comments: _____
_____
_____

INSTRUCTOR'S SIGNATURE _____

Notes

Job 10—Perform an Oil Pressure Test

After completing this job, you will be able to perform an oil pressure test as part of the diagnostic process.

Procedures

☐ 1. Obtain a vehicle to be used in this job. Your instructor may direct you to perform this job on a shop vehicle or engine.

☐ 2. Gather the tools needed to perform the following job. Refer to the tools and materials list at the beginning of the project.

☐ 3. Locate the oil pressure sender and remove it.

☐ 4. Install an oil pressure gauge in the sender opening.

> **Caution**
> Position the oil pressure hose away from moving parts and the exhaust system.

☐ 5. Start the engine.

☐ 6. Measure the oil pressure at idle.

Pressure:_____ psi or kpa (circle one).

> **Warning**
> If the oil pressure is zero, stop the engine immediately and determine the cause.

☐ 7. Raise the engine speed to 2500 rpm and measure the oil pressure.

Pressure:_____ psi or kpa (circle one).

☐ 8. Compare the pressure readings to the manufacturer's specifications.

What conclusions can you make?_____

_____

_____

_____

☐ 9. Shut off the vehicle. Remove the gauge from the pressure sender opening.

☐ 10. Reinstall the pressure sender.

☐ 11. Clean the work area and return any equipment to storage.

☐ 12. Did you encounter any problems during this procedure? Yes ___ No ___

If Yes, describe the problems: _____

_____

What did you do to correct the problems? _____

_____

☐ 13. Have your instructor check your work and sign this job sheet.

Performance Evaluation—Instructor Use Only

Did the student complete the job in the time allotted? Yes ___ No ___

If No, which steps were not completed?_____

How would you rate this student's overall performance on this job?_____

5–Excellent, 4–Good, 3–Satisfactory, 2–Unsatisfactory, 1–Poor

Comments: _____

_____

_____

INSTRUCTOR'S SIGNATURE _____

Job 11—Verify Warning Indicator Operation and Retrieve Trouble Codes

After completing this job, you will be able to use a scan tool to retrieve diagnostic trouble codes.

Procedures

☐ 1. Obtain a vehicle to be used in this job. Your instructor may direct you to perform this job on a shop vehicle.

☐ 2. Gather the tools needed to perform the following job. Refer to the tools and materials list at the beginning of the project.

> **Note**
> Although the charging system light and ammeter do not monitor engine condition, they are included in the checking procedures below because they are commonly associated with the oil and temperature lights, and because a defective charging system will eventually keep the engine from cranking.

Check the Operation of Engine Warning Lights

> **Note**
> Some newer vehicles may monitor an engine condition with both a light and a gauge in the instrument panel. If both warning systems are used, each should be checked.

Oil Pressure Lights

☐ 1. Turn the ignition switch to the *on* position but do not start the engine. Is the oil pressure light on? Yes _____ No _____

☐ 2. Start the engine. With the engine running, does the oil pressure light go out? Yes _____ No _____

Engine Temperature Light

☐ 1. Turn the ignition switch to the *start* position. Does the temperature light come on? Yes _____ No _____

☐ 2. Release the ignition switch. Does the temperature light go off? Yes _____ No _____

Coolant Level Light

☐ 1. Turn the ignition switch to the *on* position but do not start the engine. Is the coolant level light on? Yes _____ No _____

> **Note**
> The coolant level light may be on or off with the ignition switch in the *on* position. The light is working properly if it moves from one state to the other (on to off or off to on) when the engine is cranked.

☐ 2. Turn the ignition switch to the *start* position. Is the coolant level light on? Yes _____ No _____

Charging System Light

☐ 1. Turn the ignition switch to the *on* position but do not start the engine. Is the charging system light on? Yes ____ No ____

☐ 2. Start the engine. With the engine running, does the charging system light go out?
Yes ____ No ____

☐ 3. If any of the lights do not operate as indicated, check with your instructor to determine what further diagnosis and service steps to perform.

Check the Operation of Engine Gauges

Oil Pressure Gauge

☐ 1. Turn the ignition switch to the *on* position but do not start the engine. Does the oil pressure gauge read zero oil pressure? Yes ____ No ____

☐ 2. Start the engine. With the engine running, does the oil pressure gauge begin to rise?
Yes ____ No ____

Engine Temperature Gauge

☐ 1. Ensure that the vehicle engine has not been operated in the last 4 to 6 hours.

☐ 2. Start the engine. Does the temperature gauge read approximately ambient (surrounding air) temperature? Yes ____ No ____

☐ 3. Allow the engine to warm up. Does the temperature gauge begin to rise after a few minutes of operation? Yes ____ No ____

Charging System Ammeter

☐ 1. Turn the ignition switch to the *on* position but do not start the engine. Does the ammeter show a value at or below battery voltage? Yes ____ No ____

☐ 2. Start the engine. With the engine running does the ammeter show a value above battery voltage? Yes ____ No ____

☐ 3. If any of the gauges do not operate as indicated, check with your instructor to determine what further diagnosis and service steps to perform.

Retrieving Diagnostic Trouble Codes with a Scan Tool

> **Note**
>
> There are many kinds of scan tools. The following procedure is a general guide to scan tool use. Always obtain the service instructions for the scan tool that you are using. If the scan tool literature calls for a different procedure or series of steps from those in this procedure, always perform procedures according to the scan tool literature.

☐ 1. Obtain a scan tool and related service literature for the vehicle selected.
Type of scan tool: _____

☐ 2. If necessary, attach the proper test connector cable and power lead to the scan tool.

☐ 3. Ensure that the ignition switch is in the *off* position.

☐ 4. Locate the correct diagnostic connector.
Describe the location of the diagnostic connector: _____
_____

☐ 5. Attach the scan tool test connector cable to the diagnostic connector. If necessary, use the proper adapter to connect the scan tool.

Job 11—Verify Warning Indicator Operation and Retrieve Trouble Codes (continued)

> **Note**
>
> The connector should be accessible from the driver's seat. If you cannot locate the diagnostic connector, refer to the vehicle's service literature.

☐ 6. Attach the scan tool's power lead to the cigarette lighter or battery terminals as necessary.

> **Note**
>
> OBD II scan tools are powered from terminal 16 of the diagnostic connector, and no other power connections are needed.

☐ 7. Observe the scan tool's screen to ensure that the scan tool is working properly. Most scan tools will complete an internal self-check and notify the technician if there is a software or communication problem.

☐ 8. Enter vehicle information as needed to program the scan tool:
   • Most OBD II scan tools automatically read the vehicle identification number (VIN) when the ignition switch is turned to the *on* position. This gives the scan tool the information needed to check for codes and perform other operations.
   • Older scan tools are programmed with the proper vehicle information by entering the vehicle year, engine type, and other information. This information is usually contained in certain numbers and letters in the VIN.

☐ 9. Turn the ignition key to the *on* position.

☐ 10. Observe the scan tool to determine whether any trouble codes are present.

   Are any trouble codes present? Yes ___ No ___

   If Yes, go to step 13.

   If No, consult your instructor. Your instructor may wish to produce a code for this job by temporarily disconnecting a sensor or output device.

☐ 11. List all trouble codes in the space provided:

   Trouble codes: _____

   _____

☐ 12. Use the scan tool literature or factory service manual to determine the meaning of the codes.

   Code   **Defect**

   _____   _____

   _____   _____

   _____   _____

   _____   _____

   _____   _____

☐ 13. Using the trouble code information in step 14, write a brief description of what you think might be wrong with the vehicle.

   _____

   _____

   _____

☐ 14. If your instructor directs you to do so, perform the following actions:
- Make further checks to isolate the problem(s) revealed by the trouble codes.
- Correct the problem(s) as necessary. Refer to other jobs as directed.
- Use the scan tool to confirm that the codes do not reset.

☐ 15. When the test is completed, turn the ignition switch to the *off* position.

☐ 16. Remove the scan tool test connector cable from the diagnostic connector.

☐ 17. Detach the scan tool power lead from the cigarette lighter or battery terminals, if applicable.

☐ 18. Clean the work area and return any equipment to storage.

☐ 19. Did you encounter any problems during this procedure? Yes ___ No ___

If Yes, describe the problems: _____

_____

What did you do to correct the problems? _____

_____

☐ 20. Have your instructor check your work and sign this job sheet.

Performance Evaluation—Instructor Use Only

Did the student complete the job in the time allotted? Yes ___ No ___
If No, which steps were not completed?_____
How would you rate this student's overall performance on this job?_____
5–Excellent, 4–Good, 3–Satisfactory, 2–Unsatisfactory, 1–Poor
Comments: _____

_____

_____

INSTRUCTOR'S SIGNATURE _____

Project

Removing and Disassembling an Engine

5

Introduction

While engine design varies greatly between manufacturers, the general steps for removal are similar. The most important part of the job is to make sure that the lifting fixtures are safely and properly installed. Everything that connects the engine to the body, such as hoses, wires, and cables, must be removed before lifting the engine from the vehicle.

In Job 12 you will remove an engine from a vehicle. Once the engine is removed from the vehicle, it can be disassembled for repair or overhaul. In Job 13, you will completely disassemble an engine.

Project 5 Jobs

- Job 12—Remove an Engine
- Job 13—Disassemble an Engine

Tools and Materials

The following list contains the tools and materials that may be needed to complete the jobs in this project. The items used will depend on the make and model of vehicle being serviced.

- Vehicle with an engine in need of removal and overhaul.
- Engine lift.
- Engine lifting hook and chain assembly.
- Engine holding fixture.
- Applicable service information.
- Hand tools as needed.
- Air-powered tools as needed.
- Safety glasses and other protective equipment as needed.

Safety Notice

Before performing this job, review all pertinent safety information in the text and review safety information with your instructor.

Notes

Job 12—Remove an Engine

After completing this job, you will be able to prepare an engine for removal, properly attach a lifting fixture, and remove the engine from the vehicle.

Procedures

> **Note**
>
> Since most engines are removed through the top of the engine compartment, this procedure concentrates on top removal. If the engine is removed from the bottom of the vehicle, substitute the manufacturer's removal procedures at all steps referring to top removal and installation.

☐ 1. Obtain a vehicle to be used in this job. Your instructor may direct you to perform this job on a shop vehicle.

Briefly describe the engine to be used for this job: _____

_____

_____

☐ 2. Gather the tools needed to perform the following steps. Refer to the project's tools and materials list.

> **Warning**
>
> ⚠ Before attempting to do any work on a hybrid vehicle's engine, ensure that the high-voltage system has been properly disabled.

Prepare the Engine for Removal

☐ 1. Scribe around the hood hinges so they can be reinstalled in the same position. Then, remove the hood and set it aside in a place where it will not be damaged.

☐ 2. Disconnect the battery negative cable.

☐ 3. Depressurize the fuel system.

Type of fuel system:
- Multipoint fuel injection ___
- Central fuel injection ___
- Carburetor ___

☐ 4. Drain the engine coolant.

> **Note**
>
> If the engine has an auxiliary oil cooler, drain the oil and remove the cooler lines at this time.

☐ 5. Remove the upper and lower radiator hoses.

☐ 6. Remove the heater hoses.

☐ 7. Remove any engine harness connectors at the firewall or body and any electrical cables or ground straps between the body and engine. Mark them for reinstallation.

☐ 8. Remove any belts as needed.

☐ 9. Remove the mounting bolts from the power steering pump (if used) and tie the pump to the body, away from the engine.

☐ 10. Remove the mounting bolts from the air conditioning compressor (if used) and tie the compressor to the body, away from the engine.

> **Note**
> Usually the alternator and air pump (when used) can be left on the engine during removal. You may, however, want to remove them at this time to prevent possible damage.

☐ 11. Remove the air cleaner assembly or the air inlet ducts to the throttle body.

☐ 12. Remove the throttle cable and the transmission and cruise control cables, if used.

☐ 13. Remove the fuel inlet and return lines.

☐ 14. Remove all vacuum hoses from the intake manifold and mark them for reinstallation.

> **Note**
> To perform steps 15 through 19, the vehicle must be raised. Raise the vehicle in a safe manner to gain access to the underside.

☐ 15. From underneath the vehicle, disconnect the exhaust system.

☐ 16. From underneath the vehicle, remove any brackets holding the engine to the transmission or transaxle.

☐ 17. Remove the flywheel cover.

☐ 18. Remove the starter wiring and remove the starter.

☐ 19. If the vehicle has an automatic transmission or transaxle, mark the relative positions of the engine flywheel and torque converter. Then, remove the flywheel-to-torque converter fasteners.

> **Note**
> If the vehicle has a manual transmission, do not remove the bolts holding the pressure plate to the flywheel. The clutch assembly can be removed with the engine.

Job 12—Remove an Engine (continued)

Remove the Engine

☐ 1. Lower the vehicle and install an engine lifting bracket or other lifting fixture on the engine. **Figure 12-1** shows the use of a bracket originally used to install the engine in the vehicle at the factory.

Describe how you installed the lifting device: _____

_____

_____

_____

_____

> **Warning**
>
> ⚠ The lifting devices must be installed correctly to avoid personal injury or damage to the vehicle. Consult your instructor before using the lifting device to raise the engine. **Figure 12-2** shows the placement of the lifting chain on a 6-cylinder engine.

☐ 2. Connect the lifting fixture to an engine hoist.

☐ 3. Raise the engine to remove the tension on the engine mounts.

☐ 4. Remove the engine mount fasteners and brackets as necessary.

Figure 12-1. This bracket was installed on the engine at the factory and can be used to remove and replace the engine in the field.

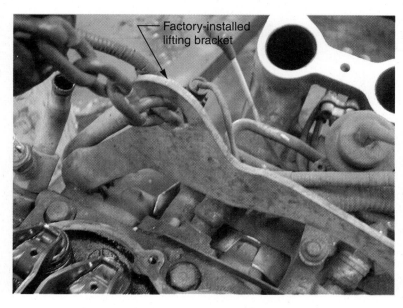

Figure 12-2. The placement of the lifting chain used to remove an inline 6-cylinder engine from a pickup truck is shown here.

☐ 5. Support the transmission or transaxle.

Describe how you supported the transmission or transaxle:_____

_____

_____

_____

_____

☐ 6. Remove the fasteners holding the transmission or transaxle to the engine.

☐ 7. Separate the engine and transmission or transaxle.

☐ 8. Raise the engine and remove it from the vehicle.

> **Caution**
>
> ⚠ While raising the engine, watch carefully for any wires or hoses that you may have forgotten to remove.

☐ 9. Lower the engine as soon as it clears the vehicle and install it on an engine-holding fixture.

☐ 10. Clean the work area and return any equipment to storage.

☐ 11. Did you encounter any problems during this procedure? Yes ___ No ___

If Yes, describe the problems: _____

_____

_____

What did you do to correct the problems? _____

_____

_____

☐ 12. Have your instructor check your work and sign this job sheet.

Job 12—Remove an Engine (continued)

Performance Evaluation—Instructor Use Only

Did the student complete the job in the time allotted? Yes ___ No ___

If No, which steps were not completed?_____

How would you rate this student's overall performance on this job?_____

5–Excellent, 4–Good, 3–Satisfactory, 2–Unsatisfactory, 1–Poor

Comments: _____
_____
_____

INSTRUCTOR'S SIGNATURE _____

Notes

Name _____ Date_____

Instructor _____ Period _____

Job 13—Disassemble an Engine

After completing this job, you will be able to properly disassemble an engine.

Procedures

☐ 1. Obtain an engine to be used in this job. Your instructor may direct you to perform this job on a shop engine.
- Make of engine:_____
- Number of cylinders:_____
- Cylinder arrangement (V-type, inline, etc.):_____
- Cooling system: Liquid ___ Air ___

☐ 2. Gather the tools needed to perform the following tasks. Refer to the tools and materials list for this project.

> **Note**
>
> If the clutch assembly is installed, remove it before proceeding to step 3. Refer to Job 74 for this procedure.

Begin Disassembly

☐ 1. Remove the engine flywheel.

☐ 2. Install the engine on an engine stand.

☐ 3. Drain the oil and remove the oil filter.

☐ 4. Remove all accessory parts, noting their position for reinstallation:
- Engine mount brackets.
- Accessory attaching brackets.
- Belt tensioner bracket, if used.
- Turbocharger or supercharger, if used.
- Dipstick and tube.
- Ignition module, coils, and plug wires.

☐ 5. Remove the distributor, if used.

☐ 6. Remove the valve cover(s). If the ignition coils are installed on the valve cover, remove the coils before removing the cover.

☐ 7. Remove the upper intake manifold or plenum.

☐ 8. Remove the lower intake manifold, if used.

☐ 9. Remove the thermostat housing if it is not part of the intake manifold.

☐ 10. Remove the exhaust crossover pipe, if applicable.

☐ 11. Remove the exhaust manifold(s).

☐ 12. Remove the oil filter adapter, if used.

☐ 13. Mark the rocker arms so they can be installed in their original position, then remove the rocker arms.

☐ 14. Remove the push rods, if applicable.

☐ 15. Remove the valve lifters or lash adjusters as applicable.

☐ 16. If the engine has a mechanical fuel pump, remove it.

☐ 17. Remove the coolant pump.

☐ 18. Remove the vibration damper (crankshaft balancer), any attached pulleys, and the crankshaft key. Inspect the vibration damper and obtain a replacement if needed.

> **Note**
>
> Some vibration dampers will slide from the crankshaft when the bolt is removed from the crankshaft. Others must be removed with a puller. Consult the proper service literature to determine the removal method to be used.

☐ 19. Remove the timing cover.

☐ 20. Rotate the engine on the stand to gain access to the oil pan. Remove the pan.

☐ 21. Remove the oil pump and drive.

☐ 22. Remove the timing gears or sprockets and the related chain or belt. A puller may be needed to remove the timing gear from the crankshaft. See **Figure 13-1**.

☐ 23. If the camshaft is installed in the block, remove it.

Remove the Cylinder Head(s)

☐ 1. Rotate the engine on the stand to gain access to the cylinder head(s).

☐ 2. Loosen and remove the cylinder head bolts. On some overhead cam engines, it may be necessary to remove the camshaft(s) to reach some of the head bolts.

☐ 3. Gently pry between the head(s) and the block to loosen the gasket, then remove the head(s).

Remove the Pistons

☐ 1. Turn the crankshaft with a breaker bar and socket to position the piston to be removed at the bottom of its cylinder.

☐ 2. Check the cylinder for wear or damage. Look for scratches, grooves, and signs of overheating. A shiny surface indicates normal cylinder wear.

Figure 13-1. A puller may be needed to remove the crankshaft gear. (Chrysler)

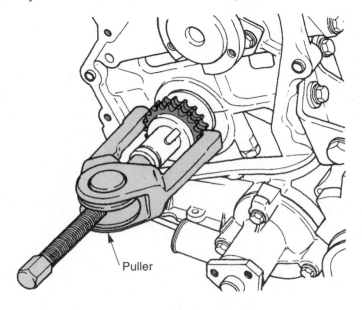

Puller

Job 13—Disassemble an Engine (continued)

☐   3.   Attempt to move your fingernail from the midpoint of the cylinder wall to the very top of the cylinder wall. If your fingernail hangs up solidly on the ridge formed at the top of the piston ring travel, the cylinder is worn excessively.

Does the ring ridge indicate excessive wear? Yes ___ No ___

> **Caution**
>
> The ring ridge must be removed before the piston is removed. Forcing the piston over the ring ridge can break the rings and damage the piston grooves and lands.

☐   4.   Place shop rags in the cylinder to catch metal chips.

☐   5.   Insert the ridge-reaming tool into the cylinder.

☐   6.   Adjust the cutters against the ridge.

☐   7.   Turn the reamer with a hand wrench or a ratchet and socket until the ridge is cut flush with the worn part of the cylinder wall. The new reamed surface must blend smoothly with the existing cylinder. Do not undercut the ridge.

☐   8.   Remove the rags and blow out the cylinder to remove metal shavings.

☐   9.   Rotate the engine on the engine stand to gain access to the connecting rod.

☐ 10.   Check that the rod cap and rod are numbered. If the rods and caps are not numbered, mark them with a punch and hammer or a number punch if available. Also mark the rod and cap so the cap can be reinstalled in the original position.

☐ 11.   Loosen and remove the nuts holding the rod cap on the connecting rod.

☐ 12.   Remove the connecting rod cap.

☐ 13.   Check that the piston head has a mark to indicate which direction it should face when reinstalled. If there is no mark, mark the connecting rod to indicate the front of the engine. The pistons must face in the original direction when reinstalled.

☐ 14.   Push the piston out of the cylinder by pushing the connecting rod toward the top of the block. If the piston is difficult to remove, lightly tap on the connecting rod with a plastic faced hammer or block of wood.

☐ 15.   Repeat steps 1 through 14 to remove all of the pistons. Place all of the pistons in order for reassembly.

☐ 16.   Inspect the pistons for scuffed or scratched skirts, damage to the ring grooves and lands, damage to the piston head, and worn piston (wrist) pin bores. These procedures are covered in Job 18.

Remove the Crankshaft

☐   1.   Mark the crankshaft main bearing caps so that they can be reinstalled in the same position and direction.

☐   2.   Loosen and remove the main bearing cap fasteners.

☐   3.   Remove the main bearing caps from the block.

☐   4.   Lift the crankshaft from the engine block.

> **Note**
> Crankshaft inspection is covered in Job 14.

☐ 5. If necessary, on an aluminum block, remove the cylinder liners using a special removal tool.

> **Note**
> Some aluminum engines have non-removable cylinder liners. Check the applicable service literature before attempting to remove the liners.

☐ 6. If the engine uses one or more balance shafts, remove them from the block.

> **Note**
> Some engine blocks are made in two or more sections. If necessary, remove the fasteners and disassemble the block.

☐ 7. Clean the work area and return any equipment to storage.

☐ 8. Did you encounter any problems during this procedure? Yes ___ No ___

If Yes, describe the problems: _____

_____

What did you do to correct the problems? _____

_____

☐ 9. Have your instructor check your work and sign this job sheet.

Performance Evaluation—Instructor Use Only

Did the student complete the job in the time allotted? Yes ___ No ___

If No, which steps were not completed?_____

How would you rate this student's overall performance on this job?_____

5–Excellent, 4–Good, 3–Satisfactory, 2–Unsatisfactory, 1–Poor

Comments: _____

_____

_____

INSTRUCTOR'S SIGNATURE _____

Project

Performing Bottom End Service

6

Introduction

The engine block is the foundation of the engine. The pistons in the cylinders transform the energy released in combustion into reciprocal motion, and the connecting rods transfer that motion to the crankshaft. The crankshaft converts the up and down motion of the pistons into rotation to drive the wheels. Together, these components are frequently referred to as the engine's bottom end.

While engines last longer than previously, vehicles are staying on the road for many more years than previously. Longer vehicle life means that internal engine parts must operate for longer periods and will eventually wear out. Engine block and crankshaft problems must be diagnosed and corrected. These problems can result in loss of compression, oil burning, and noise. Crankshaft and bearing problems cause noises and low oil pressure.

In this project, you will inspect and service the engine's block, crankshaft, pistons, connecting rods, and other bottom end components. You will inspect the block, crankshaft, and bearings in Job 14. In Job 15, you will learn how to repair a variety of damaged threads. You will measure cylinder wear and hone the cylinders in Job 16. You will inspect the connecting rods in Job 17, and inspect the pistons and replace piston rings in Job 18. In Job 19, you will inspect balance shafts, the vibration damper, and the oil pump.

Project 6 Jobs

- Job 14—Inspect the Block, Crankshaft, and Bearings
- Job 15—Repair Damaged Threads
- Job 16—Measure Cylinder Wear and Hone Cylinders
- Job 17—Inspect Connecting Rods
- Job 18—Inspect Pistons and Replace Piston Rings
- Job 19—Inspect Balance Shafts, Vibration Damper, and Oil Pump

Tools and Materials

The following list contains the tools and materials that may be needed to complete the jobs in this project. The items used will depend on the make and model of vehicle being serviced.

- Disassembled engine.
- Applicable service information.
- Ridge reamer.
- Ring expander.
- Inside micrometer.
- Outside micrometer.
- Set of flat feeler gauges.
- Engine oil.
- Cylinder hone.
- Air or electric drill.
- Crack-detection equipment.
- Thread repair tools.
- Thread repair inserts.
- Straightedge.
- Hand tools as needed.
- Air-powered tools as needed.
- Safety glasses and other protective equipment as needed.

Safety Notice

Before performing this job, review all pertinent safety information in the text and review safety information with your instructor.

Job 14—Inspect the Block, Crankshaft, and Bearings

After completing this job, you will be able to inspect a block, crankshaft, and bearings.

Procedures

> **Note**
>
> This job assumes that the engine block has been disassembled. Refer to Job 13 for disassembly procedures.

☐ 1. Obtain an engine to be used in this job. Your instructor may direct you to perform this job on a shop engine.

☐ 2. Identify the engine by listing the following information.
   - Engine maker:_____
   - Number of cylinders:_____
   - Engine displacement:_____

☐ 3. Gather the tools needed to perform the following job. Refer to the tools and materials list at the beginning of the project.

Check the Engine Block for Defects

☐ 1. Check the engine block for obvious defects, **Figure 14-1**.
   - Obvious cracks.
     Are there any obvious cracks? Yes ___ No ___

> **Note**
>
> If you suspect that the block has a crack that cannot be found by visual inspection, it will be necessary to check the block by magnetic or black light methods. Ask your instructor if further crack testing should be performed. Magnetic detection methods cannot be used to check aluminum blocks.

Figure 14-1. Once the engine is disassembled and cleaned, check the block carefully for cracks and other damage.

- Damaged core (freeze) plugs, **Figure 14-2**. These can usually be spotted by streaks caused by leaking coolant. Most technicians prefer to replace the core plugs whenever the engine is overhauled.

  Are there any damaged core plugs? Yes ___ No ___

- Stripped fastener threads. Some threads can be repaired with special procedures, explained in Job 15.

  Are there any stripped threads? Yes ___ No ___

2. Check the engine block deck (surface of the block) for warping where it seals against the head(s). Place a straightedge across the block deck at various angles and attempt to slide different size feeler gauge blades between the straightedge and the block. Acceptable warpage is around .003″ (0.076 mm) over any 6″ (152 mm) surface.

   What is the largest blade that will fit between the straightedge and head? _____

   If the head is warped more than the specification, what should be done next? _____

   _____

   _____

Inspect the Crankshaft and Bearings

1. Check the condition of the bearings and crankshaft journals.

   Condition observed: Pitting ___ Scratches ___ Ridges ___ Overheating ___
   Damaged keyway ___ Other (describe). _____

   _____

   Describe where the damage was found: _____

2. Take the following measurements using the proper micrometer:

   Connecting rod journal #1   Specified diameter:_____ Measured diameter:_____
   Connecting rod journal #2   Specified diameter:_____ Measured diameter:_____
   Connecting rod journal #3   Specified diameter:_____ Measured diameter:_____
   Connecting rod journal #4   Specified diameter:_____ Measured diameter:_____
   Connecting rod journal #5   Specified diameter:_____ Measured diameter:_____
   Connecting rod journal #6   Specified diameter:_____ Measured diameter:_____

Figure 14-2. Check freeze plugs and pipe fittings for leakage. Change them if there is any sign of leaking.

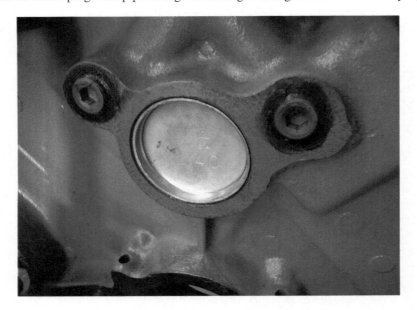

Job 14—Inspect the Block, Crankshaft, and Bearings (continued)

Connecting rod journal #7 Specified diameter:_____ Measured diameter:_____

Connecting rod journal #8 Specified diameter:_____ Measured diameter:_____

Connecting rod journal #9 Specified diameter:_____ Measured diameter:_____

Connecting rod journal #10 Specified diameter:_____ Measured diameter:_____

Main bearing journal #1 Specified diameter:_____ Measured diameter:_____

Main bearing journal #2 Specified diameter:_____ Measured diameter:_____

Main bearing journal #3 Specified diameter:_____ Measured diameter:_____

Main bearing journal #4 Specified diameter:_____ Measured diameter:_____

Main bearing journal #5 Specified diameter:_____ Measured diameter:_____

Main bearing journal #6 Specified diameter_____ Measured diameter:_____

3. Visually check the crankshaft for the following conditions:
   - Damage to the thrust bearing surfaces such as deep scratching or grooving.
     Were any defects found? Yes ___ No ___
   - Deep grooving or other damage at the rear seal surface.
     Were any defects found? Yes ___ No ___
   - Clogged oil passages.
     Were any defects found? Yes ___ No ___
   - Loose or damaged oil gallery plugs.
     Were any defects found? Yes ___ No ___
   - Stripped bolt holes.
     Were any defects found? Yes ___ No ___

4. Clean the work area and return any equipment to storage.

5. Did you encounter any problems during this procedure? Yes ___ No ___

   If Yes, describe the problems: _____

   _____

   What did you do to correct the problems? _____

   _____

6. Have your instructor check your work and sign this job sheet.

Performance Evaluation—Instructor Use Only

Did the student complete the job in the time allotted? Yes ___ No ___

If No, which steps were not completed?_____

How would you rate this student's overall performance on this job?_____

5–Excellent, 4–Good, 3–Satisfactory, 2–Unsatisfactory, 1–Poor

Comments: _____

_____

_____

INSTRUCTOR'S SIGNATURE _____

Notes

Job 15—Repair Damaged Threads

After completing this job, you will be able to repair a variety of damaged threads and chase threads in a bore.

Procedures

☐ 1. Obtain a piece of metal or a part with a threaded bore that has stripped threads.

☐ 2. Gather the tools needed to perform the following job. Refer to the tools and materials list at the beginning of the project.

Repair External Threads

Note

Most stripped external threads occur on capscrews, bolts, or other parts that are usually replaced when damaged. Occasionally, the technician will be asked to repair damaged external threads. The usual reason for repairing an external thread is to remove burrs. External threads may also be repaired if hammering or cross-threading has damaged the beginning thread.

☐ 1. Carefully inspect the external threads and determine whether they can be repaired.
   - Are large sections of the threads missing? Yes ___ No ___
   - Have large sections of the thread been cross-threaded? Yes ___ No ___
   - Are the threads badly damaged throughout their length? Yes ___ No ___
   - Are the threads badly damaged at the spot where they will be under the most tension? Yes ___ No ___

   A Yes answer to any of these questions usually means that the part must be replaced.

   Can the threads be repaired? Yes ___ No ___

Warning

⚠ Do not attempt to repair badly damaged external threads, or damaged threads on critical parts. Part failure and possible injury could result.

☐ 2. Use one of the following methods to repair an external thread.
   Die Method
   a. Install the correct size die in the die stock.
   b. Place the die and die stock over the external threads to be repaired.
   c. Lubricate the external threads.
   d. Begin turning the die stock over the threads to remove burrs or straighten a cross-threaded or damaged beginning thread.
   e. After the damaged threads are repaired, remove the die and die stock by turning it in the opposite direction.
   f. Thoroughly clean and reinspect the threads.

   Can the part be reused? Yes ___ No ___

Thread Chaser Method

> **Note**
>
> Some thread chasers clamp over the damaged thread. Use the die method to operate this type of thread chaser.

   a. Lubricate the external threads.
   b. Select the proper size thread chaser. Most file-type thread chasers will be four-sided, with a different thread on each side of one end, and four other thread sizes on the other end.
   c. Place the thread chaser on the damaged threads. Make sure that the high points of the chaser rest over the low points of the thread.
   d. Slowly draw the thread chaser over the damaged threads, stopping frequently to inspect the work.
   e. Thoroughly clean the threads and reinspect.
   Can the part be reused? Yes ___ No ___

Repair Internal Threads

> **Note**
>
> Stripped threads are usually found in aluminum assemblies, but can occur in any type of metal. Thread repair kits contain the needed drills, taps, thread repair inserts, and special mandrels for installing the inserts.

☐ 1. Determine the size and pitch of the stripped thread.

☐ 2. If necessary, place rags around the stripped thread to catch metal chips.

☐ 3. Select the proper drill bit according to the instructions of the thread-repair kit manufacturer.

☐ 4. Drill out the stripped threads and clean the metal chips from the hole.

☐ 5. Tap the hole with the appropriate tap from the thread-repair kit. Be sure to lubricate the tap before beginning the tapping operation.

☐ 6. Clean the threads to remove any remaining chips.

☐ 7. Thread the repair insert onto the correct mandrel. The insert tang should be engaged with the matching slot on the mandrel.

☐ 8. If installing the insert into an iron or steel part, lubricate it with engine oil. Do not lubricate the insert if it is to be installed in an aluminum part.

☐ 9. Turn the mandrel to advance it into the tapped hole.

☐ 10. Back the mandrel out of the hole when the insert reaches the bottom.

☐ 11. If the tang did not break off when the mandrel was removed, lightly tap it with a drift punch and hammer to remove it.

Remove Broken Fastener

☐ 1. Consult with your instructor to obtain a part with a broken fastener. Remove the fastener using one of the following methods. Indicate the method used.

> **Note**
>
> Sometimes it is necessary to use more than one removal method, depending on fastener tightness and how much of the fastener remains above the part.

☐ 2. Clamp the fastener with locking pliers and turn to remove the fastener.

Job 15—Repair Damaged Threads (continued)

- ☐ 3. Grind flats on each side of the fastener, and use an adjustable wrench to remove the fastener.
- ☐ 4. Cut a slot in the top of the fastener, and use a screwdriver to remove the fastener.
- ☐ 5. Heat the area surrounding the fastener with a torch and remove the fastener.

> **Note**
>
> Work quickly to remove the fastener, before the surrounding area cools.

- ☐ 6. Drill out the center of the fastener and use an extractor to remove the fastener.
- ☐ 7. Weld a nut to the top of the fastener and turn the nut with a wrench.
- ☐ 8. Describe the procedure(s) you used and the results.

  _____

  _____

  _____

- ☐ 9. Were the threads of the part holding the broken fastener damaged? Yes _____ No _____

  If Yes, what did you do to correct the problem? _____

  _____

  _____

- ☐ 10. Clean and return all of the tools, and clean the work area.
- ☐ 11. Did you encounter any problems during this procedure? Yes ___ No ___

  If Yes, describe the problems: _____

  _____

  What did you do to correct the problems? _____

  _____

- ☐ 12. Have your instructor check your work and sign this job sheet.

Performance Evaluation—Instructor Use Only

Did the student complete the job in the time allotted? Yes ___ No ___

If No, which steps were not completed?_____

How would you rate this student's overall performance on this job?_____

5–Excellent, 4–Good, 3–Satisfactory, 2–Unsatisfactory, 1–Poor

Comments: _____

_____

_____

INSTRUCTOR'S SIGNATURE _____

Notes

Name _____ Date_____

Instructor _____ Period _____

Job 16—Measure Cylinder Wear and Hone Cylinders

After completing this job, you will be able to measure cylinder oversize, taper, and out-of-round. You will also be able to hone cylinders.

Procedures

> **Note**
>
> This job assumes that the engine block has been disassembled. Refer to Job 13 for disassembly procedures.

☐ 1. Obtain an engine to be used in this job. Your instructor may direct you to perform this job on a shop engine.

☐ 2. Identify the engine by listing the following information.
   - Engine maker:_____
   - Number of cylinders:_____
   - Engine displacement:_____

☐ 3. Gather the tools needed to perform the following job. Refer to the tools and materials list at the beginning of the project.

Measure Cylinder Wear

☐ 1. Refer to the service literature to find the standard bore size.

   Standard bore size:_____

☐ 2. Obtain an inside micrometer or a telescoping gauge and an outside micrometer.

☐ 3. Take measurements at the top and bottom of the cylinder.

   Smallest reading:_____

   Largest reading:_____

☐ 4. Check cylinder taper at the top and bottom of ring travel at right angles to the engine centerline.

   Top reading:_____

   Bottom reading:_____

☐ 5. Determine cylinder oversize by subtracting the standard bore size from the measurement made in steps 3 and 4 and recording it in the appropriate spaces in the following chart.

   Largest measured bore size reading:_____

   Standard bore size:_____

   Cylinder oversize:_____

☐ 6. Calculate the cylinder taper by subtracting the bottom reading from the top reading.

   Top of cylinder:_____

   Bottom of cylinder:_____

   Cylinder taper:_____

☐ 7. Take measurements at 90° apart at the top of the cylinder. Then, calculate the cylinder out-of-round by subtracting the smallest reading from the largest reading.

   Largest reading:_____

   Smallest reading:_____

   Out-of-round:_____

> **Note**
>
> If the cylinder is more than a few thousandths oversize, tapered, or out-of round, the cylinder should be rebored to the next largest standard oversize and matching oversize rings and pistons installed. Consult your instructor if any of the measurements indicate that any of the cylinders need reboring.

☐ 8. Repeat steps 1 through 7 for all cylinders.

Hone Cylinders

> **Note**
>
> Honing is the process used to restore the proper finish to the cylinder walls. The following task assumes that an electric drill–operated portable hone will be used. Many shops use rigid stone hones, while others use flexible, or bead hones. Some shops create a final finish with fine grit rigid stones or brush or plateau hone. Consult with your instructor to ensure that you select the proper hone or hones for this task.

☐ 1. Clamp the hone into a low-speed electric drill.

☐ 2. Insert the hone in the first cylinder to be honed.

☐ 3. Squirt a moderate amount of hone oil onto the cylinder wall.

☐ 4. Start the drill and move it up and down the full length of the cylinder. Move the hone up and down at a rate that will produce a 50°-crosshatch pattern.

> **Caution**
>
> Be careful not to pull the hone too far out of the bore or hone damage may result.

☐ 5. Turn the drill off and remove the hone. Before removing the hone, hand squeeze or adjust the stones together to prevent vertical scratches in the cylinder wall.

☐ 6. Repeat steps 2 through 5 for all cylinders.

☐ 7. Clean all cylinders with soap and hot water and rinse with clean hot water.

☐ 8. Wipe the cylinders dry with clean rags.

☐ 9. Wipe the cylinders with clean oil-soaked rags until all of the grit is removed.

☐ 10. Clean the work area and return any equipment to storage.

☐ 11. Did you encounter any problems during this procedure? Yes ___ No ___

If Yes, describe the problems: _____

_____

What did you do to correct the problems? _____

_____

☐ 12. Have your instructor check your work and sign this job sheet.

Job 16—Measure Cylinder Wear and Hone Cylinders (continued)

Performance Evaluation—Instructor Use Only

Did the student complete the job in the time allotted? Yes ___ No ___

If No, which steps were not completed?_____

How would you rate this student's overall performance on this job?_____

5–Excellent, 4–Good, 3–Satisfactory, 2–Unsatisfactory, 1–Poor

Comments: _____

_____

_____

INSTRUCTOR'S SIGNATURE _____

Notes

Job 17—Inspect Connecting Rods

After completing this job, you will be able to inspect connecting rods for bending or twisting. You will also be able to check the fit of the piston (wrist) pin.

Procedures

> **Note**
>
> This job assumes that the engine block has been disassembled. Refer to Job 13 for disassembly procedures.

☐ 1. Obtain an engine to be used in this job. Your instructor may direct you to perform this job on a shop engine.

☐ 2. Identify the engine by listing the following information.
- Engine maker:_____
- Number of cylinders:_____
- Engine displacement:_____

☐ 3. Gather the tools needed to perform the following job. Refer to the tools and materials list at the beginning of the project.

Check Piston Pin Fit

☐ 1. Lightly clamp the first connecting rod and piston assembly in a vise and try to rock the piston sideways. Any movement indicates piston pin looseness, which must be corrected.

Does the piston pin show any looseness? Yes ___ No ___

If Yes, the piston pin looseness must be corrected. A loose pin will make a rapping noise as the engine runs. It will eventually cause the piston to break. If the piston and connecting rods have bushings, the bushings can be honed oversize and an oversize pin installed. However, on most modern vehicles, the piston and pin are replaced.

☐ 2. Repeat this operation for all pistons.

Check Connecting Rod for Twisting and Bending

☐ 1. Remove the connecting rod from the first piston to be serviced. If the piston pin is held in place by snap rings, remove one snap ring and slide the pin from the piston and connecting rod. If the pin has a press fit, carefully press it from the connecting rod using a hydraulic press and suitable fixtures. **Figure 17-1** shows the parts of a piston and rod assembly that uses snap rings to retain the piston pin.

☐ 2. Install the measuring fixture on the connecting rod. Some measuring fixtures require that the piston pin be reinserted into the connecting rod.

☐ 3. Replace the rod cap and slide the connecting rod over the measuring stand mounting rod. If necessary, adjust the mounting rod until the connecting rod fits snugly.

☐ 4. Measure the gap at several places on the measuring fixture. If you can insert a feeler gauge larger than .003″ (.075 mm) between the rod and the flat plate on the measuring fixture, the rod is bent or twisted and should be replaced.

☐ 5. After checking the connecting rod, reassemble the connecting rod and piston.

☐ 6. Repeat steps 1 through 5 for all connecting rods.

☐ 7. Clean the work area and return any equipment to storage.

Figure 17-1. Piston and connecting rod components.

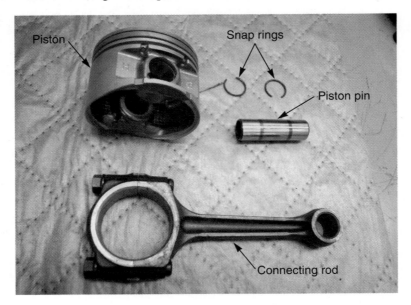

8. Did you encounter any problems during this procedure? Yes ___ No ___

   If Yes, describe the problems: _____

   _____

   What did you do to correct the problems? _____

   _____

9. Have your instructor check your work and sign this job sheet.

Performance Evaluation—Instructor Use Only

Did the student complete the job in the time allotted? Yes ___ No ___
If No, which steps were not completed?_____
How would you rate this student's overall performance on this job?_____
5–Excellent, 4–Good, 3–Satisfactory, 2–Unsatisfactory, 1–Poor
Comments: _____

_____

_____

INSTRUCTOR'S SIGNATURE _____

Job 18—Inspect Pistons and Replace Piston Rings

After completing this job, your will be able to inspect pistons for excessive wear and replace piston rings.

Procedures

> **Note**
>
> This job assumes the engine block has been disassembled. Refer to Job 13 for disassembly procedures.

☐ 1. Obtain an engine to be used in this job. Your instructor may direct you to perform this job on a shop engine.

☐ 2. Identify the engine by listing the following information.
- Engine maker:_____
- Number of cylinders:_____
- Engine displacement:_____

☐ 3. Gather the tools needed to perform the following job. Refer to the tools and materials list at the beginning of the project.

Visually Inspect Pistons

☐ 1. If necessary, clean the piston thoroughly.

☐ 2. Visually inspect the pistons for severe scuffing on the skirts and ring land areas, cracks at the piston pin bosses, erosion or indentations of the piston head, and other obvious damage.

Is there any obvious damage? Yes ___ No ___

If Yes, explain:_____

_____

Measure Piston Clearance

☐ 1. Measure the diameter of the piston across the skirts.

Piston diameter:_____

☐ 2. Find the cylinder diameter, the measurement taken at the bottom of the cylinder. Refer to Job 16 as needed.

Cylinder diameter:_____

☐ 3. Subtract the piston diameter from cylinder diameter to determine piston clearance.

Cylinder diameter:_____

Piston diameter:_____

Piston clearance:_____

☐ 4. Repeat steps 1 through 3 for all other cylinders and pistons.

> **Caution**
>
> If the piston clearance is excessive, do not install the piston, as it will be noisy and wear the cylinder and rings. The piston can be knurled to increase the skirt diameter, or a larger-diameter piston can be substituted.

Remove Piston Rings and Clean Ring Grooves

☐ 1. Place the piston and rod assembly in a vise so that the piston skirt is resting on top of the vise jaws so it cannot swivel. Clamp the vise jaws around the connecting rod.

☐ 2. Use a ring expander to remove the piston rings from the piston, **Figure 18-1**. Open the rings only enough to clear the piston lands.

☐ 3. Remove carbon from the inside of the ring grooves with a ring groove cleaner. Select the correct-width scraper for each groove and be careful not to remove any metal.

> **Note**
>
> If a groove cleaner is not available, a broken piston ring can be used to clean the ring grooves.

☐ 4. Repeat steps 1 through 3 for all other pistons.

Check Ring Groove Wear

☐ 1. Measure the ring side clearance as demonstrated in **Figure 18-2**. Fit a ring into the top and middle ring grooves. Determine the largest size feeler gauge that will fit between the side of each groove and ring. The size of that gauge is the ring side clearance for that ring. The top piston groove will usually be the most worn.

Specified side clearance for the top ring:_____

Specified side clearance for the middle ring:_____

Measured side clearance of the top ring:_____

Measured side clearance of the middle ring:_____

☐ 2. Repeat this operation for all pistons. Replace any pistons that have excessive ring groove wear.

Check Piston-to-Bore Clearance

> **Note**
>
> The piston-to-bore clearance of a cylinder and piston must be measured using the same piston that will be installed in the cylinder during reassembly. Measuring piston-to-bore clearance using a different piston will result in an inaccurate reading.

Figure 18-1. Use a ring expander to remove or install piston rings. (DaimlerChrysler)

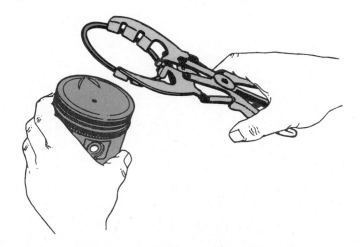

Job 18—Inspect Pistons and Replace Piston Rings (continued)

Figure 18-2. Use a feeler gauge to check piston ring to groove clearance. (DaimlerChrysler)

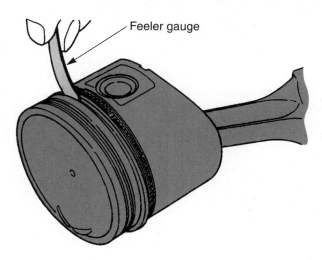

1. Select the piston that will be installed in the cylinder during reassembly.
2. Remove the piston rings if not already removed.
3. Install a spring scale on the piston.
4. Place the piston in the cylinder bore with the piston skirts in the same relative position that they would have with the piston installed.
5. Insert a feeler gauge between piston skirt and bore. Special narrow feeler gauges are available for this procedure.
6. Pull the piston out of the cylinder using the spring scale.
7. Record the spring scale reading.

   Reading _____

   Specified reading _____
8. Change feeler gauges until proper spring gauge reading is obtained.
9. If the gauge reading is too high, substitute a thinner feeler gauge and repeat steps 4 through 7.
10. If the gauge reading is too low, substitute a thicker feeler gauge and repeat steps 4 through 7.
11. When correct spring gauge reading is obtained, the feeler gauge size used at the time is the piston-to-bore clearance.

> **Note**
>
> If piston-to-bore clearance is excessive, consult your instructor. Excessive piston-to-bore clearance must be corrected to avoid piston slap and resulting piston and cylinder wear. Possible corrections include knurling the piston skirts or boring the block oversize and installing oversize pistons.

Check Ring Gap

1. Place a compression ring into the cylinder squarely. Push it to the bottom of ring travel with the head of the piston.

2. Determine the largest size feeler gauge that will fit in the gap between the ends of the ring. This is the ring gap.

Specified upper compression ring gap:_____

Specified lower compression ring gap:_____

Measured upper compression ring gap:_____

Measured lower compression ring gap:_____

Note

If all cylinders are the same size, it is usually not necessary to check the ring gap on every cylinder.

3. If the ring gap is too small, the ends of the ring can be carefully filed to increase the gap. If the gap is too large, replace the ring.

Install Piston Rings

1. Install the oil rings on the piston.
2. Check the ring markings to ensure that they are installed facing up.
3. Install the lower compression ring by hand or with a ring expander.
4. Install the top compression ring by hand or with a ring expander.
5. Stagger the ring gaps so that they do not line up. See **Figure 18-3**.
6. Repeat this operation for all pistons.
7. Clean the work area and return any equipment to storage.

Figure 18-3. Install the piston rings so the gaps are staggered. (DaimlerChrysler)

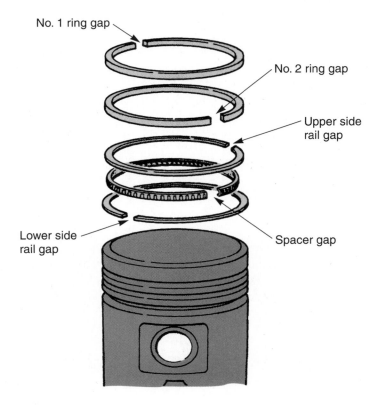

Job 18—Inspect Pistons and Replace Piston Rings (continued)

☐ 8. Did you encounter any problems during this procedure? Yes ___ No ___

    If Yes, describe the problems: _____

    _____

    What did you do to correct the problems? _____

    _____

☐ 9. Have your instructor check your work and sign this job sheet.

Performance Evaluation—Instructor Use Only

Did the student complete the job in the time allotted? Yes ___ No ___

If No, which steps were not completed?_____

How would you rate this student's overall performance on this job?_____

5–Excellent, 4–Good, 3–Satisfactory, 2–Unsatisfactory, 1–Poor

Comments: _____

_____

_____

INSTRUCTOR'S SIGNATURE _____

Notes

Name _____   Date_____

Instructor _____   Period _____

Job 19—Inspect Balance Shafts, Vibration Damper, and Oil Pump

After completing this job, you will be able to inspect balance shafts, vibration dampers, and oil pumps.

Procedures

> **Note**
>
> This job assumes the engine block has been disassembled. Refer to Job 13 for disassembly procedures.

☐ 1. Obtain an engine to be used in this job. Your instructor may direct you to perform this job on a shop engine.

☐ 2. Identify the engine by listing the following information.
   - Engine maker:_____
   - Number of cylinders:_____
   - Engine displacement:_____

☐ 3. Gather the tools needed to perform the following job. Refer to the tools and materials list at the beginning of the project.

Inspect Balance Shafts

☐ 1. Check the condition of the balance shaft journals and any bearings.

   Condition Observed

   Pitting:___ Scratches:___ Ridges:___ Overheating:___ Damaged keyway:___
   Other (describe): _____
   _____
   _____

☐ 2. Measure the balance shaft journals with a micrometer.

   Bearing journal #1 Specified diameter:_____ Measured diameter:_____
   Bearing journal #2 Specified diameter:_____ Measured diameter:_____
   Bearing journal #3 Specified diameter:_____ Measured diameter:_____
   Bearing journal #4 Specified diameter:_____ Measured diameter:_____

☐ 3. If the balance is damaged or excessively worn, replace it.

Check the Vibration Damper

☐ 1. Visually check the vibration damper (sometimes called a balancer) for the following:
   - Missing or extruded (pushed out) rubber between the damper sections.
   - A damaged keyway.

   Were any defects found? Yes ___ No ___

☐ 2. Place the vibration damper on a flat surface and look for misalignment between the damper sections.

   Was any misalignment found? Yes ___ No ___

☐ 3. If the vibration damper has any obvious problems, such as a separated rubber insert or misalignment, replace it.

Inspect the Oil Pump

☐ 1. Disassemble the oil pump, if it is not already disassembled.

☐ 2. Check the pump for obvious wear and scoring.

☐ 3. Using feeler gauges, check clearances between the pump gears and the pump cover and pump body.

Maximum allowable clearance:_____

Actual reading:_____

Is the actual reading within specifications? Yes ___ No ___

☐ 4. If the oil pump has scored or excessively worn parts, replace it.

> **Note**
>
> Some oil pump gears and endplates can be replaced without replacing the pump housing, but this should only be done when the other oil pump parts are in excellent condition.

☐ 5. Clean the work area and return any equipment to storage.

☐ 6. Did you encounter any problems during this procedure? Yes ___ No ___

If Yes, describe the problems: _____

_____

What did you do to correct the problems? _____

_____

☐ 7. Have your instructor check your work and sign this job sheet.

Performance Evaluation—Instructor Use Only

Did the student complete the job in the time allotted? Yes ___ No ___

If No, which steps were not completed?_____

How would you rate this student's overall performance on this job?_____

5–Excellent, 4–Good, 3–Satisfactory, 2–Unsatisfactory, 1–Poor

Comments: _____

_____

_____

INSTRUCTOR'S SIGNATURE _____

Project

Servicing the Cylinder Head(s) and Valve Train

7

Introduction

As a technician, you will likely be required to perform cylinder head service, usually called a "valve job." Cylinder heads can be made of cast iron, aluminum, or magnesium, with two, three, or four valves per cylinder. A variety of camshaft placements and valve train designs can be used to open and close the valves. All engines have at least one camshaft, and most engines have valve lifters.

When the camshaft is located in the block, pushrods are used. This arrangement is known as cam-in-block engine. When the cylinder head contains the camshaft, the arrangement is known as an overhead camshaft (or simply overhead cam) engine. Some overhead camshaft engines have hydraulic lash adjusters that take up the valve train clearance. The service procedures described in this project apply generally to all cylinder heads and will provide you with a guideline for complete engine top-end and front-end service.

In Job 20, you will disassemble and inspect a cylinder head. In Job 21, with the cylinder head disassembled, you will inspect valve train components for wear and damage by observation and by the use of measuring tools. In Job 22, you will learn how to resurface (grind) valves and valve seats. You will reassemble the cylinder head in Job 23. In Job 24, you will inspect camshaft drives and other valve timing components. You will adjust the engine's valve in Job 25. In Job 26, you will learn how to replace a valve seal without removing the cylinder head, thereby saving time and expense.

Project 7 Jobs

- Job 20—Disassemble and Inspect a Cylinder Head
- Job 21—Inspect and Service Valve Train Parts
- Job 22—Resurface and Inspect Valves and Valve Seats
- Job 23—Assemble a Cylinder Head
- Job 24—Inspect Valve Timing Components
- Job 25—Adjust Valves
- Job 26—Replace Valve Seals with the Cylinder Head Installed

Tools and Materials

The following list contains the tools and materials that may be needed to complete the jobs in this project. The items used will depend on the make and model of vehicle being serviced.

- Vehicle in need of valve seal replacement.
- Cylinder head to be serviced.
- Service information.
- Valve spring compressor.
- Air pressure adapter that can be installed in a spark plug opening.
- Air pressure hose.
- Brass hammer.
- Straightedge.
- Feeler gauges.
- Inside micrometers.
- Outside micrometers.
- Telescoping gauges.
- Valve grinding machine.
- Valve seat grinder or cutter set.
- Air or electric drill.
- Rotary wire brush.
- Hand tools as needed.
- Air-powered tools as needed.
- Safety glasses and other protective equipment as needed.

Safety Notice

Before performing this job, review all pertinent safety information in the text and review safety information with your instructor.

Job 20—Disassemble and Inspect a Cylinder Head

After completing this job, you will be able to inspect, disassemble, and clean a cylinder head.

Procedures

- [ ] 1. Obtain a cylinder head assembly to be used in this job.
- [ ] 2. Gather the tools needed to perform the following job. Refer to the project's tools and materials list.

Inspect the Cylinder Head

- [ ] 1. Place the head on a bench or suitable head repair fixture.
- [ ] 2. Identify the cylinder head by listing the following information.
  - Engine maker:_____
  - Engine displacement:_____
  - Number of valves per cylinder:_____
- [ ] 3. Visually inspect the head for damage. **Figure 20-1** shows an obviously burned valve.
- [ ] 4. Place a straightedge across the surface of the head at various angles and attempt to slide different size feeler gauge blades between the straightedge and the head. This checks the cylinder head for a warped sealing surface. See **Figure 20-2**. Acceptable warpage is around .003″ (0.08 mm) over any 6″ (152 mm) surface.

  What is the largest blade that will fit between the straightedge and head? _____

  If the head is warped more than the specification, what should be done next? _____

  _____

  _____

Figure 20-1. The exhaust valve in this photograph is obviously burned. Sometimes a burned valve will be harder to identify, especially if it has just started to fail.

Figure 20-2. Use a feeler gauge and straightedge to check the head for warping.

Disassemble the Cylinder Head

☐ 1. Strike the valve spring retainers to loosen them. Use a brass hammer to avoid damaging the valve stems.

☐ 2. Place a valve spring compressor over the first valve to be removed.

☐ 3. Depress the spring compressor and remove the valve spring keepers.

☐ 4. Open the spring compressor and remove the retainer and spring. Do *not* remove the valves and valves seals at this time.

☐ 5. Place the parts together in a container or on a clean area of the bench.

☐ 6. Repeat steps 2 through 5 for the other valves in the head. Place the parts in order so they can be reinstalled in the same position.

☐ 7. Pull each valve partially out of the head and wiggle it in the valve guide, **Figure 20-3**. Measure the movement with a dial indicator for maximum accuracy. If the valve moves more than 1/32″ (.8 mm), the guide is worn excessively. List the guide wear measurements in the spaces provided:

Valve 1:_____   Valve 2:_____
Valve 3:_____   Valve 4:_____
Valve 5:_____   Valve 6:_____
Valve 7:_____   Valve 8:_____
Valve 9:_____   Valve 10:_____
Valve 11:_____   Valve 12:_____
Valve 13:_____   Valve 14:_____
Valve 15:_____   Valve 16:_____

> **Note**
>
> Worn valve guides must be repaired or replaced to avoid excessive oil consumption, valve noise, and valve and seat damage. If any valves are excessively worn, consult your instructor before proceeding. Some manufacturers require that the internal diameter of the guides be checked with a small inside micrometer or telescoping gauge.

Job 20—Disassemble and Inspect a Cylinder Head (continued)

Figure 20-3. If the valve moves too much, the guide, and possibly the valve stem, is worn excessively.

☐ 8. Remove the valve stem seals if they are not already removed.

☐ 9. Remove the valves, one at a time, keeping them in order.

Caution

 If the valve does not slide easily from the head, the stem may have widened where it contacts the rocker arm. This is called "mushrooming." A mushroomed valve stem must be filed down to allow it to slide through the valve guide. Do not hammer a mushroomed valve out of the head.

☐ 10. Inspect each valve for wear or damage by comparing it to the valves shown in **Figure 20-4**.
Were any valves too badly worn or damaged to reuse? Yes ___ No ___
Explain your answer:_____
_____
_____
_____
_____

☐ 11. Scrape all gasket material from the head sealing surface and use a wire wheel and drill to remove carbon from the combustion chamber. Thoroughly wash the head after removing all debris.

Warning

 Wear eye protection while using a wire brush.

Figure 20-4. Inspect all valves for signs of damage and wear. Valve A is burned, cracked, and has mechanical damage. Note the common areas of wear and the closeup of damage on valve A. Valve B is in good used condition.

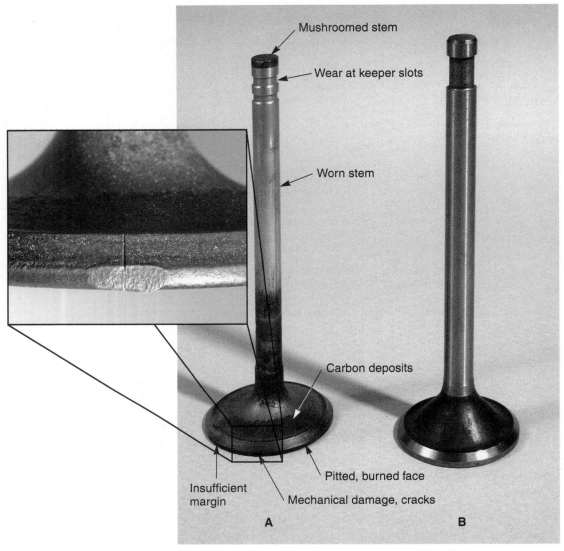

12. Check the cylinder head for obvious cracks.

Are any cracks present? Yes ___ No ___

If Yes, what should you do next? _____

_____

_____

_____

Note

If you suspect that the head has a crack that cannot be found by visual inspection, it will be necessary to check the head by magnetic or black light methods. Ask your instructor if further crack testing should be performed. Magnetic crack detection will not work on aluminum cylinder heads.

Job 20—Disassemble and Inspect a Cylinder Head (continued)

☐ 13. Clean the work area and return any equipment to storage.

☐ 14. Did you encounter any problems during this procedure? Yes ___ No ___

If Yes, describe the problems: _____

_____

What did you do to correct the problems? _____

_____

☐ 15. Have your instructor check your work and sign this job sheet.

Performance Evaluation—Instructor Use Only

Did the student complete the job in the time allotted? Yes ___ No ___

If No, which steps were not completed?_____

How would you rate this student's overall performance on this job?_____

5–Excellent, 4–Good, 3–Satisfactory, 2–Unsatisfactory, 1–Poor

Comments: _____

_____

_____

INSTRUCTOR'S SIGNATURE _____

Notes

Job 21—Inspect and Service Valve Train Parts

After completing this job, you will be able to check valve train components for wear and damage.

Procedures

☐ 1. Obtain an engine or cylinder head assembly to be used in this job.

☐ 2. Gather the tools needed to perform the following job. Refer to the project's tools and materials list.

☐ 3. Identify the valve train parts and the type of valve train being serviced.

Describe the valve train being serviced, including the location of the cam, the type of lifters or lash adjusters, and the number of valves per cylinder:_____

_____

_____

_____

> **Note**
>
> Valve lifters can be divided into two general types, mechanical and hydraulic. Use the procedure that applies to the lifter being checked. A hydraulic lifter will have oil holes and a plunger that allows its overall length to vary with hydraulic pressure.

Inspect Hydraulic Lifters

☐ 1. Examine the lower part of the lifter where it contacts the camshaft, **Figure 21-1**.

Did you find any wear? Yes ___ No ___

☐ 2. Check for flat spots on the roller (for roller lifters), **Figure 21-2**.

Did you find any flat spots? Yes ___ No ___

☐ 3. Check for a loose roller pin.

Was the pin loose? Yes ___ No ___

Figure 21-1. Check the lifter body for damage to the camshaft contact surface, wear and scoring where the body contacts the engine block, and sludge buildup in the oil holes. The camshaft contact surface of this hydraulic lifter is badly pitted.

Figure 21-2. Roller lifters should be checked for flat spots on the rollers. If a roller has a flat spot, the camshaft lobe is probably damaged. This roller lifter appears to be in good condition.

4. Examine the push rod contact surface.

   Was any wear visible? Yes ___ No ___

5. Check for wear or damage to the clip holding the internal parts to the body.

   Was any wear or damage noted? Yes ___ No ___

6. Inspect the oil passages for sludge deposits.

   Was sludge noted? Yes ___ No ___

7. If instructed to do so, check the leakdown rate of the valve lifters.

   What were the results of the leakdown test?_____

   _____

   _____

8. Replace any lifters that are excessively worn.

Inspect Mechanical Lifters

1. Examine the lower part of the lifter where it contacts the camshaft.

   Did you find any wear? Yes ___ No ___

2. Check for flat spots on the roller (for roller lifters).

   Did you find any flat spots? Yes ___ No ___

3. Check for a loose roller pin.

   Was the pin loose? Yes ___ No ___

4. Check the push rod contact surface.

   Was wear evident? Yes ___ No ___

5. Replace any lifters that are excessively worn.

Inspect Valve Lash Adjusters

Note

Lash adjusters are installed in the head and control the clearance of the drive train parts in an overhead cam engine. **Figure 21-3** shows the common types of hydraulic lash adjusters and compares them to a lifter in a push rod—type valve train.

Job 21—Inspect and Service Valve Train Parts (continued)

Figure 21-3. Like hydraulic lifters, valve lash adjusters remove clearance from the drive train. A—Valve lash adjusters may be placed opposite the valve. B—Valve lash adjusters may be part of the rocker arm. C—Compare the placement and function of the lash adjusters in A and B to the lifter shown here.

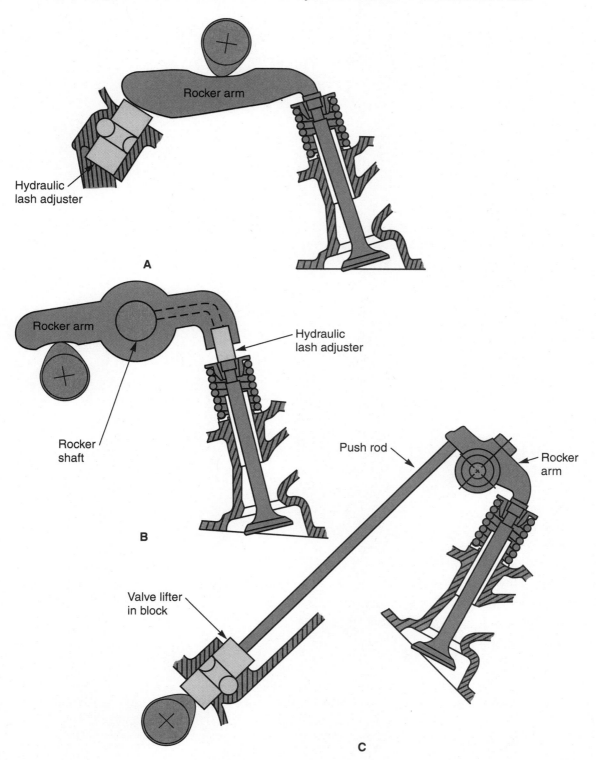

1. Check for wear where the lash adjuster contacts the other moving parts of the drive train and where it contacts the cylinder head.

   Was any wear noted? Yes ___ No ___

2. Check for wear or damage to the clip, pin, or snap ring holding the internal parts in the lash adjuster body.

   Was any wear or damage noted? Yes ___ No ___

3. Inspect the oil passages for sludge deposits.

   Was any sludge noted? Yes ___ No ___

4. Check lash adjuster leakdown if possible and if instructed to do so.

   Describe the results of the leakdown test:_____

   _____

Inspect the Camshaft

1. Remove the camshaft from the engine, if necessary.

2. Visually check the condition of the camshaft lobes. **Figure 21-4** shows some badly worn lobes.

   Are the lobes on your camshaft worn? Yes ___ No ___

3. Check lobe lift using a micrometer. **Figure 21-5** shows a typical procedure. Record the results of the micrometer tests in the following chart.

   Specification:_____   Reading:_____
   Specification:_____   Reading:_____
   Specification:_____   Reading:_____
   Specification:_____   Reading:_____
   Specification:_____   Reading:_____
   Specification:_____   Reading:_____
   Specification:_____   Reading:_____
   Specification:_____   Reading:_____
   Specification:_____   Reading:_____
   Specification:_____   Reading:_____
   Specification:_____   Reading:_____
   Specification:_____   Reading:_____
   Specification:_____   Reading:_____
   Specification:_____   Reading:_____
   Specification:_____   Reading:_____
   Specification:_____   Reading:_____

4. Check camshaft straightness using V-blocks. The camshaft should turn between the blocks with no noticeable wobble. For extreme accuracy, a dial indicator can be used.

5. Check the bearing surfaces for wear and scoring, **Figure 21-6**.

   Is wear or scoring present? Yes ___ No ___

> **Note**
>
> If the camshaft bearings are installed in the engine block, you can use an inside micrometer or telescoping gauge to measure the bearing diameter. The bearing clearance on some overhead camshafts can be checked with Plastigage. Consult your instructor to determine whether you should perform these checks.

Job 21—Inspect and Service Valve Train Parts (continued)

Figure 21-4. The lobes on this camshaft are excessively worn.

Figure 21-5. Measure camshaft lobe lift at the wear zone. To determine camshaft lobe wear, measure the camshaft lobe at the unworn area and subtract the measurement taken at the wear zone. (DaimlerChrysler)

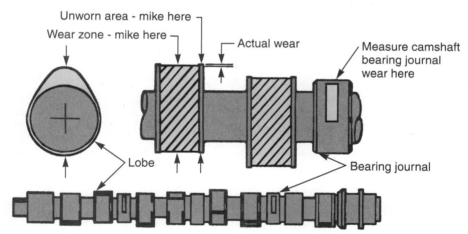

☐ 6. Check the thrust surfaces for wear and scoring.

   Is wear or scoring present? Yes ___ No ___

☐ 7. Check the camshaft bearing journal size with a micrometer. See Figure 21-5. Record the results of the micrometer tests in the following chart.

   Specification: _____ Reading: _____
   Specification: _____ Reading: _____
   Specification: _____ Reading: _____
   Specification: _____ Reading: _____
   Specification: _____ Reading: _____
   Specification: _____ Reading: _____

Figure 21-6. A visual inspection of the camshaft bearing surfaces will reveal many problems. Note that this cylinder head does not have separate bearing inserts. This is true of most overhead cam engines.

Camshaft bearing surfaces

> **Note**
>
> If the camshaft is worn or bent, it must be replaced. Worn or damaged bearing inserts can be replaced without replacing the block or head. If a cylinder head does not have bearing inserts, the head must be replaced if the bearing surfaces are damaged.

Inspect Rocker Arms and Rocker Arm Shafts

 1. Refer to **Figure 21-7** and check the rocker arms visually. Check each rocker arm for the following conditions:
- Wear at the points where the rocker arm contacts other moving parts.
- Wear where the rocker arm oscillates on the rocker shaft.
- Clogged oil passages in the rocker arm.
- Bending, cracks, or other damage to the rocker arm.

Figure 21-7. Check the rocker arm for wear at the places where it contacts other parts and for sludge buildup.

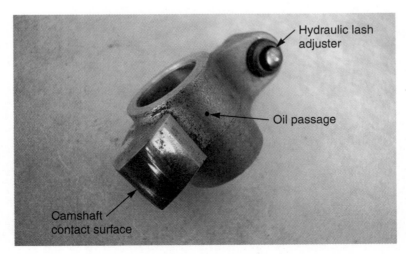

Hydraulic lash adjuster

Oil passage

Camshaft contact surface

Job 21—Inspect and Service Valve Train Parts (continued)

Is any damage present? Yes ____ No ____

If Yes, describe the damage: _____

_____

_____

_____

> **Note**
>
> Damaged rocker arms and related parts should be replaced. If the camshaft directly moves the rocker arm, it should also be replaced. If the rocker arm is worn where it contacts the rocker arm shaft, the shaft should also be replaced.

Inspect Push Rods

☐ 1. Visually check the push rods for wear and bending. If the push rods are hollow, check them for sludge buildup.

☐ 2. Roll each push rod on a flat surface to check for bends.

   Were any of the push rods bent? Yes ____ No ____

☐ 3. Replace any bent or otherwise damaged push rods.

> **Note**
>
> It is usually easier to replace badly clogged push rods rather than attempt to clean them.

Inspect Springs, Valve Locks, and Keepers

☐ 1. Inspect the valve springs.

   Are any coils damaged or broken? Yes _____ No _____

   Are the springs the correct spring height? Yes _____ No _____

   Is the spring assembly straight? Yes _____ No _____

   Are there wear marks between inner and outer springs (when two springs are used)?
   Yes _____ No _____

   Are the individual coils the same size? Yes _____ No _____

> **Note**
>
> Some springs are designed with the coils closer together at one end of the spring. If all springs show this difference in coil size, the springs are good.

☐ 2. Inspect the valve locks and keepers. Do any valve locks and/or keepers show the following?

   Do any have excessive wear? Yes _____ No _____

   Is there bending or other deformation? Yes _____ No _____

   Are there burrs? Yes _____ No _____

☐ 3. Replace any damaged springs, locks, and keepers.

☐ 4. Clean the work area and return any equipment to storage.

☐ 5. Did you encounter any problems during this procedure? Yes ___ No ___

If Yes, describe the problems: _____

_____

What did you do to correct the problems? _____

_____

☐ 6. Have your instructor check your work and sign this job sheet.

Performance Evaluation—Instructor Use Only

Did the student complete the job in the time allotted? Yes ___ No ___

If No, which steps were not completed?_____

How would you rate this student's overall performance on this job?_____

5–Excellent, 4–Good, 3–Satisfactory, 2–Unsatisfactory, 1–Poor

Comments: _____

_____

_____

INSTRUCTOR'S SIGNATURE _____

Job 22—Resurface and Inspect Valves and Valve Seats

After completing this job, you will be able to resurface valves and valve seats.

Procedures

☐ 1. Obtain a disassembled cylinder head assembly to be used in this job.

☐ 2. Gather the tools needed to perform the following job. Refer to the project's tools and materials list.

Resurface (Grind) Valves

> **Note**
>
> Obtain your instructor's permission before using the valve grinding machine.

☐ 1. Prepare the valve grinding machine by checking condition of the grinding wheel, the cooling fluid level, and all related hardware. Be sure that you are familiar with all machine controls.

☐ 2. Determine the valve face angle and set the valve grinding machine to produce this angle. Most modern vehicles will use either a 45° or 30° angle, but this should be confirmed by checking the manufacturer's service literature.

☐ 3. Adjust the grinder to the appropriate angle by loosening the chuck hold-down nut and swiveling the chuck mechanism until its degree marks line up. Some makers call for a 1° interference angle. In these cases, you would set the chuck to cut an angle 1° less than the desired angle. For example, for a 45° valve, you would adjust the valve grinding machine so it produces a 44° cut.

☐ 4. Insert the valve into the valve grinder. Check that the chuck jaws grasp the valve stem on the machined portion of the stem nearest the valve head.

☐ 5. Turn the grinder on and ensure that the valve is not wobbling in the chuck. If the valve wobbles, remount it and recheck.

> **Note**
>
> If the wobble cannot be corrected, the valve is bent and must be replaced.

☐ 6. Make sure cooling fluid is flowing over the valve head. The valve should be positioned so the valve face is parallel with the cutting surface of the stone. Turn the depth wheel a little at a time to move the stone toward the valve so parallelism can be checked and adjusted.

☐ 7. Slowly advance the stone toward the valve. As soon as the stone touches the valve, begin moving the valve back and forth across the grinding surface of the stone.

☐ 8. Watch the face of the valve carefully as you move it across the surface of the grinding stone. As soon as the valve face has been refinished, back off the depth wheel.

☐ 9. Turn off the grinder and check the valve margin, **Figure 22-1**. If the margin is less than 1/32″ (0.79 mm), the valve should be replaced. A thin valve margin will cause the valve to burn.

How wide is the valve margin?_____

Figure 22-1. The valve margin must be as specified after the valve is refaced. (DaimlerChrysler)

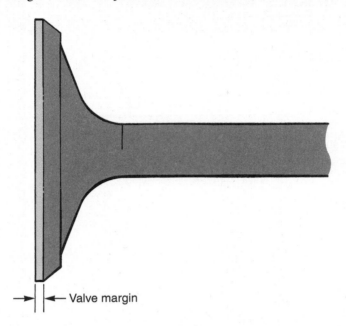

→|←— Valve margin

> **Caution**
>
> Minimum margins will vary by manufacturer. Check the appropriate service information before proceeding.

10. Dress the end of the valve stem by mounting the valve in the V-block on the opposite end of the grinder chuck. Remove as little metal as possible (generally less than .010″, or 0.25 mm).

    How much material had to be removed from the valve stem?_____

11. Repeat steps 4 through 10 for all valves.

Resurface (Grind) Valve Seats

> **Caution**
>
> Some valve seats are actually hardened inserts. They should be replaced instead of ground.

1. Determine and locate the correct size stone or cutter. The cutter should be the proper angle and slightly larger than the head of the valve.

    Cutter angle:_____

    Cutter size:_____

2. Clean the valve guide and install the cutter assembly.

3. Check the fit of the stone or cutter carefully to confirm that it is the correct one.

4. Lubricate the cutter assembly then begin the cutting process. Remove only enough metal to remove pitting and ridges.

5. Inspect the valve seat to ensure that it has been cleaned up sufficiently.

6. Use 60° and 15° cutters or stones to narrow the valve seat as necessary.

7. Repeat steps 2 through 6 for all valve seats.

Job 22—Resurface and Inspect Valves and Valve Seats (continued)

Check the Valve Face-to-Seat Contact

☐ 1. Make a series of pencil marks on the valve seat face.

> **Note**
> This procedure can also be performed using a compound called Prussian Blue.

☐ 2. Place the valve in the cylinder head.

☐ 3. Press down on the valve and turn it about one-fourth turn.

☐ 4. Remove the valve and inspect the pencil marks on the face of the valve seat face. Spots where the pencil marks are rubbed off indicate the contact point between the valve face and the seat. The contact point should be in the middle of the valve face. It should be about 1/16″ (1.59 mm) wide and should extend all the way around the face.

Were the pencil marks wiped off all the way around the valve? Yes ___ No ___

Explain: _____

_____

_____

☐ 5. Repeat steps 1 through 4 for all valve seats.

☐ 6. Did you encounter any problems during this procedure? Yes ___ No ___

If Yes, describe the problems: _____

_____

What did you do to correct the problems? _____

_____

☐ 7. Have your instructor check your work and sign this job sheet.

Performance Evaluation—Instructor Use Only

Did the student complete the job in the time allotted? Yes ___ No ___

If No, which steps were not completed?_____

How would you rate this student's overall performance on this job?_____

5–Excellent, 4–Good, 3–Satisfactory, 2–Unsatisfactory, 1–Poor

Comments: _____

_____

_____

INSTRUCTOR'S SIGNATURE _____

Notes

Job 23—Assemble a Cylinder Head

After completing this job, you will be able to properly assemble a cylinder head.

Procedures

1. Obtain a disassembled cylinder head assembly to be used in this job.
2. Gather the tools needed to perform the following job. Refer to the project's tools and materials list.
3. Oil the valve stem and install the valve in its original position.
4. If the valve stem uses an umbrella seal, install the seal now.
5. Assemble the valve spring(s) and retainer over the valve stem.
6. Use the valve spring compressor to compress the spring.
7. If the valve stem uses an O-ring seal, install the seal now.
8. Install the valve keepers. Keepers can be held in place with heavy grease.
9. Release the spring compressor and ensure that all parts are in their proper place.
10. Check the valve spring assembled height with a small ruler, as in **Figure 23-1**. If the assembled height is correct, proceed to the next step. If the assembled height is not correct, ask your instructor what steps should be taken next.
11. Repeat steps 3 through 10 for all valves.
12. Clean the work area and return any equipment to storage.
13. Did you encounter any problems during this procedure? Yes ___ No ___
    If Yes, describe the problems: _____
    _____
    What did you do to correct the problems? _____
    _____
14. Have your instructor check your work and sign this job sheet.

Figure 23-1. Check the valve spring assembled height using a small ruler.

Performance Evaluation—Instructor Use Only

Did the student complete the job in the time allotted? Yes ___ No ___

If No, which steps were not completed?_____

How would you rate this student's overall performance on this job?_____

5–Excellent, 4–Good, 3–Satisfactory, 2–Unsatisfactory, 1–Poor

Comments: _____

_____

_____

INSTRUCTOR'S SIGNATURE _____

Job 24—Inspect Valve Timing Components

After completing this job, you will be able to inspect valve timing components and set valve timing.

Procedures

1. Obtain a vehicle to be used in this job. Your instructor may direct you to perform this job on a shop vehicle or engine.

2. Gather the tools needed to perform the following job. Refer to the project's tools and materials list.

Check Camshaft Drive Parts

1. Remove the camshaft drive parts from the engine as necessary.

2. Visually inspect the condition of gears and drive sprockets. See **Figure 24-1**.

3. Check the condition of the timing chain or belt. Most checks can be made visually. A worn chain will have excessive slack. A worn belt will have cracks and possible missing teeth.

> **Note**
>
> Refer to Job 28 for procedures for replacing camshaft drive parts. Drive parts should be replaced as a set.

Check Variable Valve Timing Components

1. Inspect the variable valve timing mechanism.

   Is there leakage? Yes _____ No _____

   Are there cracks? Yes _____ No _____

   Is there damaged mounting hardware? Yes _____No _____

Figure 24-1. This kind of wear on a timing gear indicates that the gear should be replaced.

> **Note**
>
> The following procedure can be performed on an assembled and operating engine only. Testing of the variable valve mechanism on a disassembled engine is limited to visual inspection of the component parts.

Check Operation of Variable Valve Timing Mechanism

> **Caution**
>
> Perform the following test only if it is specifically mentioned in the manufacturer's service literature. Do not ground any computer control system wire without checking the service literature.

☐ 1. Locate the solenoid that controls the flow of oil to the variable valve timing mechanism. The solenoid may be installed on the valve cover, on the side of the engine, or in the head under the valve cover.

☐ 2. Determine which solenoid wire receives a ground signal from the ECM by consulting the appropriate wiring diagram.

☐ 3. Start the engine and raise the engine speed to about 2500 rpm.

☐ 4. Ground the solenoid signal wire and observe engine operation. There should be a noticeable change in engine speed and sound as the valve timing changes.

Did the engine speed change? Yes ___ No ___

Did the sound of the engine change? Yes ___ No ___

If the engine speed or sound did not change, the solenoid is defective, the internal advance mechanism is stuck or leaking, or the engine's oil pressure is low. Make further checks according to the manufacturer's service information.

☐ 5. Clean the work area and return any equipment to storage.

☐ 6. Did you encounter any problems during this procedure? Yes ___ No ___

If Yes, describe the problems: _____

_____

What did you do to correct the problems? _____

_____

☐ 7. Have your instructor check your work and sign this job sheet.

Performance Evaluation—Instructor Use Only

Did the student complete the job in the time allotted? Yes ___ No ___

If No, which steps were not completed?_____

How would you rate this student's overall performance on this job?_____

5–Excellent, 4–Good, 3–Satisfactory, 2–Unsatisfactory, 1–Poor

Comments: _____

_____

_____

INSTRUCTOR'S SIGNATURE _____

Job 25—Adjust Valves

After completing this job, you will be able to adjust mechanical and hydraulic valves.

Procedures

☐ 1. Obtain a vehicle to be used in this job. Your instructor may direct you to perform this job on a shop vehicle or engine.

☐ 2. Gather the tools needed to perform the following job. Refer to the project's tools and materials list.

☐ 3. Remove the engine valve cover(s) to reveal the valve adjustment device.

> **Note**
>
> On some overhead camshaft engines clearance is adjusted by changing lifters or adding shims at the lifters. Consult the manufacturer's service information for these procedures.

☐ 4. Determine whether the engine uses mechanical or hydraulic valve lifters.
   Mechanical ____ Hydraulic ____

> **Note**
>
> The following sections provide instructions for adjusting valves with mechanical lifters and adjusting valves with hydraulic lifters. Use the procedure that is appropriate for the type of lifters used in your vehicle. Skip the procedure that does not apply.

Adjust Valves—Mechanical Valve Lifters

☐ 1. Obtain the proper valve-clearance specifications.
   Specifications for intake valve:_____ inches or millimeters (circle one).
   Does the specification call for the engine to be warmed up? Yes ____ No ____
   Specifications for exhaust valve:_____ inches or millimeters (circle one).
   Does the specification call for the engine to be warmed up? Yes ____ No ____

☐ 2. Turn the engine until the camshaft lobe for the valve to be adjusted points away from the lifter.

☐ 3. Loosen the adjuster locknut.

☐ 4. Insert the proper thickness feeler gauge between the valve and the rocker arm. See **Figure 25-1**.

☐ 5. Turn the adjuster while moving the feeler gauge between the valve and rocker arm. When there is a light drag on the feeler gauge, clearance is correct.

☐ 6. Tighten the adjuster locknut while holding the adjuster in position.

☐ 7. Recheck adjustment to ensure that the adjuster did not move when the locknut was tightened.

☐ 8. Repeat steps 2 through 7 for the other engine valves.

Figure 25-1. Insert the feeler gauge between the rocker arm and the valve stem. The cam lobe must be at its lowest point (putting no pressure on the valve train).

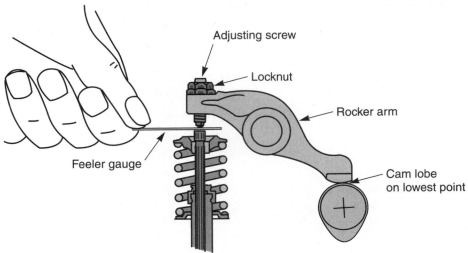

9. Start the engine and recheck the adjustment by inserting a feeler gauge between the valve and rocker arm.

Does the feeler gauge fit between the valve and rocker arm? Yes ___ No ___

Are any valves noisy (clattering or tapping sounds)? Yes ___ No ___

Does the noise stop when the feeler gauge is inserted between the valve and rocker arm? Yes ___ No ___

Adjust Valves—Hydraulic Valve Lifters

1. Place shielding around the valve assemblies to reduce oil spray.
2. Start the engine.
3. Loosen the valve adjuster locknut on the first valve to be adjusted. Note that not all rocker assemblies are equipped with valve adjuster locknuts.
4. Back off the valve adjuster until the valve begins to make a clattering noise. See **Figure 25-2**.
5. Tighten the adjuster until the noise just stops.
6. Tighten the adjuster an additional number of turns as specified by the service literature. If the rocker assembly is equipped with a valve adjuster locknut, retighten it. Be careful not to allow the valve adjuster to rotate.
7. Repeat steps 3 through 6 for all other engine valves.
8. Replace the valve cover using a new gasket as necessary.
9. Clean the work area and return any equipment to storage.
10. Did you encounter any problems during this procedure? Yes ___ No ___

If Yes, describe the problems: _____

_____

What did you do to correct the problems? _____

_____

11. Have your instructor check your work and sign this job sheet.

Job 25—Adjust Valves (continued)

Figure 25-2. Adjust the valve until it just stops clattering, then turn the adjuster in an additional number of turns. This adjusting nut is an interference fit and does use a locknut.

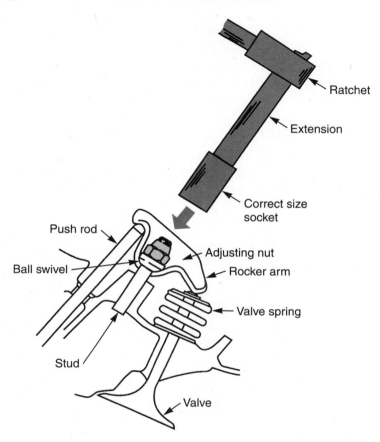

Performance Evaluation—Instructor Use Only

Did the student complete the job in the time allotted? Yes ___ No ___

If No, which steps were not completed?_____

How would you rate this student's overall performance on this job?_____

5–Excellent, 4–Good, 3–Satisfactory, 2–Unsatisfactory, 1–Poor

Comments: _____

_____

_____

INSTRUCTOR'S SIGNATURE _____

Notes

Job 26—Replace Valve Seals with the Cylinder Head Installed

After completing this job, you will be able to replace valve seals without removing the cylinder head from the vehicle.

Procedures

☐ 1. Obtain a vehicle to be used in this job. Your instructor may direct you to perform this job on a shop vehicle or engine.

☐ 2. Gather the tools needed to perform the following job. Refer to the project's tools and materials list.

☐ 3. Remove the valve cover over the valve needing the seal replacement.

Record the number of the cylinder requiring valve seal replacement:_____.

☐ 4. Remove the rocker arm from the valve requiring seal replacement.

Is the seal on an exhaust valve or an intake valve?_____

> **Note**
>
> All valves in this cylinder must be closed or the procedure will not work. Bring the piston to the top of its compression stroke or back off all other cylinder valve rocker arms.

☐ 5. Remove the spark plug on the cylinder of the valve to have its seal replaced.

☐ 6. Install an air pressure adapter in the plug opening.

☐ 7. Introduce air pressure into cylinder. This will hold the valve up as the spring is compressed.

☐ 8. Install a special spring compressor tool on the cylinder head. **Figure 26-1** shows a valve spring compressor used on an overhead camshaft engine.

☐ 9. Compress the valve spring with the valve spring compressor.

Figure 26-1. This valve spring compressor hooks around the camshaft to remove the valve. (DaimlerChrysler)

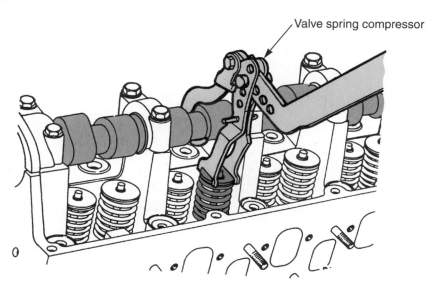

Valve spring compressor

☐ 10. Remove the valve spring keepers.

☐ 11. Release the pressure on the spring and remove the spring and retainer from the head.

☐ 12. Remove the valve seal.

What type of seal is used?
- Umbrella seal. ___
- Positive seal. ___
- O-ring seal. ___

☐ 13. If the seal is an umbrella or positive seal type, install the replacement seal over the valve stem. Use a seal protector to prevent seal damage. See **Figure 26-2**.

☐ 14. Place the spring and retainer over the valve stem.

☐ 15. Compress the valve spring with the valve spring compressor.

☐ 16. If the seal is an O-ring type, lubricate it and place it in its groove on the valve stem. See **Figure 26-3**.

> **Caution**
>
> Be sure that the O-ring is installed in the seal groove, not the keeper groove.

☐ 17. Install the spring keepers.

☐ 18. Relieve pressure from the spring and allow the spring to rise to its normal position.

☐ 19. Check that the keepers are properly in place.

☐ 20. Release the air pressure from the cylinder.

☐ 21. Remove the air pressure adapter from the spark plug opening.

☐ 22. Reinstall the spark plug.

☐ 23. Reinstall the rocker arm and valve cover. Use a new valve cover gasket as necessary.

☐ 24. Clean the work area and return any equipment to storage.

Figure 26-2. Umbrella and positive seals are installed between the valve stem and cylinder head. (DaimlerChrysler)

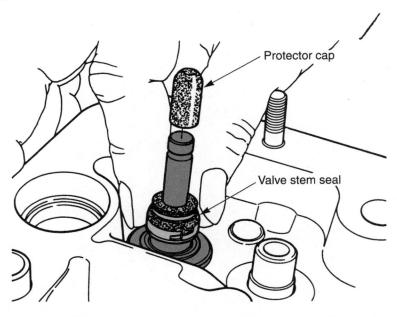

Protector cap

Valve stem seal

Job 26—Replace Valve Seals with the Cylinder Head Installed (continued)

Figure 26-3. The O-ring should be lubricated and placed in the proper groove. Make sure that the O-ring is not twisted.

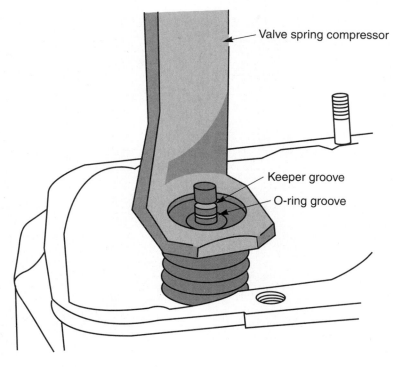

Valve spring compressor

Keeper groove

O-ring groove

25. Did you encounter any problems during this procedure? Yes ___ No ___

    If Yes, describe the problems: _____

    _____

    What did you do to correct the problems? _____

    _____

26. Have your instructor check your work and sign this job sheet.

Performance Evaluation—Instructor Use Only

Did the student complete the job in the time allotted? Yes ___ No ___

If No, which steps were not completed?_____

How would you rate this student's overall performance on this job?_____

5–Excellent, 4–Good, 3–Satisfactory, 2–Unsatisfactory, 1–Poor

Comments: _____

_____

INSTRUCTOR'S SIGNATURE _____

Notes

Project

Reassembling and Reinstalling an Engine

8

Introduction

Although there are many types of engines, the general steps for disassembly and reassembly are similar. The most important parts of the job are to make sure that parts are installed in order, thereby preventing unnecessary removal and reinstallation, and to make sure that all fasteners are tightened to the proper torque values. In Job 27, you will install the crankshaft, piston assemblies, and other bottom end components to create a properly assembled short block. In Job 28, you will assemble the engine to the long block stage by adding the heads, valve train components, and valve timing components to the short block. In Job 29, you will complete the reassembly of the engine. You will reinstall the engine into the vehicle in Job 30.

Project 8 Jobs

- Job 27—Reassemble the Engine's Bottom End
- Job 28—Install the Cylinder Head(s) and Valve Train
- Job 29—Complete the Engine Reassembly
- Job 30—Reinstall the Engine

Tools and Materials

The following list contains the tools and materials that may be needed to complete the jobs in this project. The items used will depend on the make and model of vehicle being serviced.

- Engine to be reassembled.
- Applicable service information.
- Ring compressor.
- Torque wrench.
- Engine lift.
- Engine lifting hook and chain assembly.
- Engine holding fixture.
- Dial indicator.
- Engine oil.
- Hand tools as needed.
- Air-powered tools as needed.
- Safety glasses and other protective equipment as needed.

Safety Notice

Before performing this job, review all pertinent safety information in the text and review safety information with your instructor.

Notes

Job 27—Reassemble the Engine's Bottom End

After completing this job, you will be able to reassemble an engine to the short block stage.

Procedures

1. Obtain an engine to be used in this job. Your instructor may direct you to perform this job on a shop engine.
   - Make of engine:_____
   - Number of cylinders:_____
   - Cylinder arrangement (V-type, inline, etc.):_____
   - Cooling system: Liquid ___ Air ___

2. Gather the tools needed to perform the following tasks. Refer to the tools and materials list for this project.

3. Before starting engine assembly, look up the following torque specifications in the shop manual. Record the information in the appropriate spaces.

   Oil pump fasteners:_____ ft lb _____ N•m

   Main bearing cap fasteners:_____ ft lb _____ N•m

   Connecting rod cap fasteners:_____ ft lb _____ N•m

Install the Crankshaft

1. Install the upper halves of the main bearings in the engine block. Be sure to check for dirt or metal particles between the back of the bearing caps and the block before installation.

2. Heavily lubricate the bearing faces. Do *not* lubricate the backs of the bearings.

3. Carefully install the crankshaft into the engine block.

4. Install bearing inserts into the main bearing caps. Then, install the main bearing caps, being sure to install them according to the marks made during removal.

5. Check the main bearing clearance with Plastigage, using the following method:
   a. Thoroughly dry the cap bearing insert.
   b. Lay a strip of Plastigage across the bearing.
   c. Install and torque the main cap to specifications. Do not turn the engine.
   d. Remove the cap and compare the flattened Plastigage to the scale provided. See **Figure 27-1**.

   Main bearing clearance:_____

   Is the main bearing clearance acceptable? Yes ___ No ___

   If No, select the correct oversize or undersize main bearings as follows:
   a. Add or subtract the Plastigage reading from the specified clearance.
   Result:_____
   b. Select the correct oversize or undersize bearing to bring the clearance to the proper specification.
   c. Replace the original bearing with the replacement bearing and recheck clearance.

6. Install and tighten the main bearing cap fasteners to the proper torque.

7. Ensure that the crankshaft turns easily with the main bearing caps properly torqued.

Figure 27-1. The width of the flattened Plastigage indicates the bearing clearance. (DaimlerChrysler)

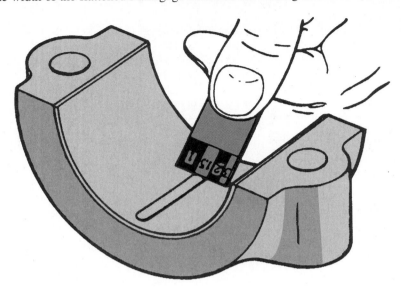

8. Check crankshaft endplay using a dial indicator:
    a. Install the dial indicator on one end of the crankshaft. The dial indicator plunger should be parallel with the crankshaft. **Figure 27-2** shows a typical dial indicator setup.
    b. Push the crankshaft forward.
    c. Zero the dial indicator.
    d. Push the crankshaft backward.
    e. Note the dial indicator reading:_____
    Is the reading within specifications? Yes ___ No ___
    If No, an oversize or undersize thrust bearing must be installed.

Figure 27-2. Set up the dial indicator to record crankshaft endplay. (DaimlerChrysler)

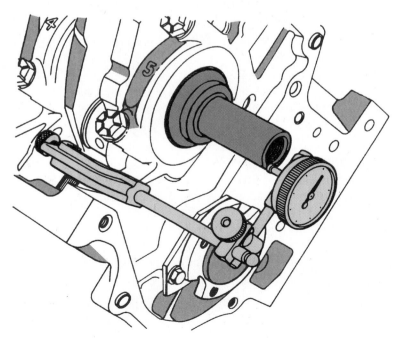

Job 27—Reassemble the Engine's Bottom End (continued)

> **Note**
>
> Some manufacturers call for checking endplay with a feeler gauge between the crankshaft and the thrust bearing.

☐ 9. If necessary, install an oversize or undersize thrust bearing as follows:
   a. Add or subtract the dial indicator reading from the specified endplay.
   Result:_____
   b. Select the correct oversize or undersize bearing to bring the endplay to the proper specification.
   c. Replace the original bearing with the replacement bearing and recheck the endplay.

> **Note**
>
> Some engines have a separate thrust bearing, while on others the thrust bearing is attached to one of the main bearings.

Install Pistons

☐ 1. Decide which piston to install first, and heavily oil the piston, pin, and rings.

☐ 2. Place the ring compressor around the piston, ensuring that the ring compressor is facing the proper direction.

☐ 3. Tighten the ring compressor around the piston rings.

☐ 4. Remove the rod cap and install the bearing insert in the rod. Oil the exposed surface of the bearing. Install the bearing insert in the cap, but do not oil it at this time.

☐ 5. Place protective caps over the connecting rod studs of the piston.

☐ 6. Install the piston by placing the piston and rod assembly into the cylinder and gently tapping the piston into the cylinder with a hammer handle, **Figure 27-3**. Be sure that the rod studs slip over the crankshaft journal.

☐ 7. Place the connecting rod cap over the rod studs, being sure that it faces in the proper direction.

☐ 8. Check the rod bearing clearance with Plastigage, using the following method:
   a. Thoroughly dry the cap bearing insert.
   b. Lay a strip of Plastigage across the bearing.
   c. Install and torque the rod cap to specifications. Do not turn the engine.
   d. Remove the cap and compare the flattened Plastigage to the scale provided.
   Rod bearing clearance:_____
   Is the rod bearing clearance acceptable? Yes ___ No ___
   If Yes, go to step 9.
   If No, oversize or undersize bearings must be installed. Use the following procedure to install oversize or undersize bearings as needed:
   a. Add or subtract the Plastigage reading from the desired clearance.
   Result:_____
   b. Select the correct bearing inserts to bring the clearance to the proper specification.
   c. Replace the original rod and cap inserts with the replacement inserts and recheck the clearance.

Figure 27-3. Carefully install the piston so there is no damage to the rings or crankshaft journals. (DaimlerChrysler)

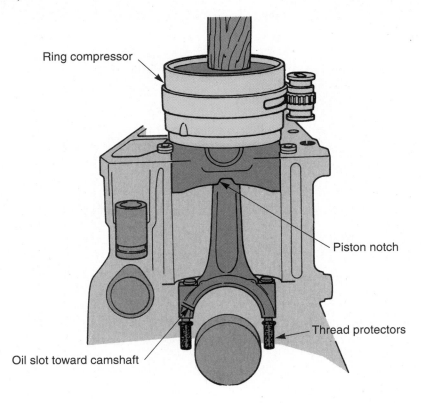

9. Remove the connecting rod cap. Remove the Plastigage. Oil the exposed surface of the rod cap bearing insert. Place the connecting rod cap over the rod studs, being sure that it faces in the proper direction.

10. Install and tighten the rod nuts to the proper torque.

Torque:_____

11. Install long bolts in two of the bores in the crankshaft's flywheel flange. Place a large screwdriver between the bolts and use it as a lever to turn the crankshaft.

Does the crankshaft turn easily? Yes ___ No ___

If No, remove the piston and rod assembly, identify and correct the problem, and reinstall the assembly.

12. Repeat steps 1 through 11 for all piston and rod assemblies.

Install Balance Shafts

1. Install the balance shaft drive chain if used.

2. Lubricate the balance shaft bearings as necessary.

3. Place the balance shaft in position on the engine.

4. Install the balance shaft drive chain on the balance shaft.

> **Caution**
>
> Be sure to line up all balance shaft, crankshaft, and chain timing marks. If this is not done, the engine will vibrate severely.

5. Install all bearing or housing fasteners and tighten them to the proper torque.

Job 27—Reassemble the Engine's Bottom End (continued)

Install the Oil Pump and Oil Pan

☐ 1. Rotate the engine block until the bottom of the engine faces up.

☐ 2. Place the oil pump in position on the engine block and install and tighten the oil pump bolts.

> **Note**
>
> Some oil pumps are driven by a shaft from the distributor or distributor adapter gear. If the pump shaft can only be installed from the bottom, install it as part of the oil pump assembly.
>
> Other oil pumps are installed in the engine front cover and driven directly by the crankshaft. Refer to the manufacturer's service literature for information on how to assemble and install these types of pumps.

☐ 3. If your instructor directs, pre-lubricate the bearings by following the directions provided by the manufacturer of the pre-lubrication equipment.

☐ 4. Scrape the oil pan and block mating surfaces to remove old gasket material.

☐ 5. Place the oil pan gaskets and seals on the engine block. Use the proper type of sealer if required.

☐ 6. Place the oil pan on the block.

☐ 7. Install and tighten the oil pan fasteners. Use the correct sequence and tighten the fasteners to the proper torque.

Install In-Block Camshafts

> **Note**
>
> This procedure is used only if the camshaft is installed in the block. This procedure assumes that the camshaft bearings have been inspected and/or replaced.

☐ 1. Lubricate the camshaft bearings and bearing journals.

☐ 2. Slowly insert the camshaft into the block, making sure that the cam lobes do not damage the cam bearings.

☐ 3. When the camshaft is fully installed, ensure that it turns freely in the block.

☐ 4. Install the thrust plate or other device that holds the cam in the block.

☐ 5. Install the block's rear camshaft plug if necessary.

☐ 6. Check camshaft endplay using the following procedure:
   a. Install a dial indicator in the camshaft at the thrust plate.
   b. Pull the camshaft forward.
   c. Zero the dial indicator.
   d. Push the camshaft rearward.
   e. Read and record the endplay.
   Endplay:_____
   Is the endplay within specifications? Yes ___ No ___
   If No, consult your instructor for the steps needed to correct the endplay.

> **Note**
>
> On many engines, the camshaft gear must be temporarily installed to take endplay readings.

☐ 7. Clean and return all of the tools and clean the work area.

☐ 8. Did you encounter any problems during this procedure? Yes ___ No ___

If Yes, describe the problems: _____

_____

What did you do to correct the problems? _____

_____

☐ 9. Have your instructor check your work and sign this job sheet.

Performance Evaluation—Instructor Use Only

Did the student complete the job in the time allotted? Yes ___ No ___

If No, which steps were not completed? _____

How would you rate this student's overall performance on this job? _____

5–Excellent, 4–Good, 3–Satisfactory, 2–Unsatisfactory, 1–Poor

Comments: _____

_____

_____

INSTRUCTOR'S SIGNATURE _____

Job 28—Install the Cylinder Head(s) and Valve Train

After completing this job, you will be able to install the cylinder heads and valve train components on a short block.

Procedures

> **Note**
>
> This job assumes that crankshaft, pistons, balance shafts, oil pump, and oil pan have been installed on the block. Refer to Job 27 for procedures for installing these components.

1. Obtain an engine to be used in this job. Your instructor may direct you to perform this job on a shop engine.
   - Make of engine:_____
   - Number of cylinders:_____
   - Cylinder arrangement (V-type, inline, etc.):_____
   - Cooling system: Liquid ____ Air ____

2. Gather the tools needed to perform the following tasks. Refer to the tools and materials list for this project.

3. Before starting engine assembly, look up the following torque specifications in the shop manual. Record the information in the appropriate spaces.

   Cylinder head fasteners: _____ ft lb _____ N•m

   Intake manifold/plenum fasteners: _____ ft lb _____ N•m

   Rocker arm fasteners (if applicable): _____ ft lb _____ N•m

   Valve cover fasteners: _____ ft lb _____ N•m

   Timing mechanism fasteners: _____ ft lb _____ N•m

   Front cover fasteners: _____ ft lb _____ N•m

Install the Cylinder Heads

1. Rotate the engine until the top of the engine faces up.

2. Scrape the block and head mating surfaces clean. Remove all traces of old gasket material from the engine. The mating surfaces should be perfectly clean, as shown in **Figure 28-1**.

3. Place the head gasket on the block. Gasket direction is usually stamped on the gasket material. Use gasket sealer only if recommended by manufacturer.

4. Carefully position the cylinder head on the block. Aligning pins will help in installation and reduce the chance of damaging the head gasket.

5. Squirt a small amount of oil on the head bolt threads and thread them into the block by hand. Use a speed handle or ratchet (not an impact wrench) to turn them until they seat in the block. Do not tighten the bolts at this time.

> **Note**
>
> Many modern engines use a torque plus angle tightening method. The bolts are torqued to a relatively low value, then tightened by turning the bolt an additional fraction of a turn, such as 1/4 turn (an angle of 90 degrees). If this engine must be reassembled using this method, consult the manufacturer's service literature for the proper torque values.

149

Figure 28-1. The block and head mating surfaces should be perfectly clean and straight.

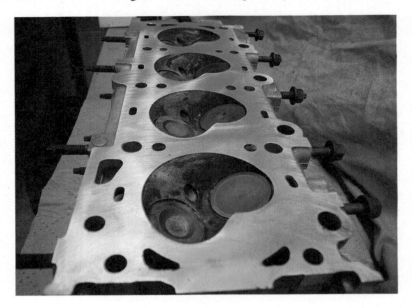

6. Using a torque wrench, tighten the head bolts to one-half of their torque specification. Follow the order or sequence given in the service literature.

   What is the location of the first bolt to be tightened?_____

   What is one-half of the head bolt torque specification? _____

7. Torque the head bolts to three-fourths of the torque specification. Follow the correct sequence.

8. Tighten the head bolts to their full torque value. Following the tightening sequence, recheck each bolt to ensure that all are at full torque.

   Did any of the bolts turn when you tightened them for the final time? Yes ___ No ___

   If Yes, can you think of a reason that some of the bolts could be turned more? _____

   _____

Note

This completes head installation. Some manufacturers recommend that the head bolts be retorqued after the engine has been started and has reached full operating temperature. The instructions with the new head gaskets will normally provide this information.

---

9. If the engine is a V-type, repeat steps 2 through 8 for the other head.

Install Push Rods and Valve Timing Components

Note

Not all of the following steps will apply to every engine. This procedure assumes that, if the engine has an in-block camshaft, the camshaft has been previously installed.

---

1. Install the push rods, lifters, and rocker arms in their original locations. If a rocker shaft is used, start the rocker shaft fasteners by hand. Then, tighten the fasteners, starting with the middle bolts and turning each bolt one-half turn at a time. This will prevent a strain on any one bolt. Remember, the pressure of the compressed valve springs will be pushing against the rocker shaft assembly.

Job 28—Install the Cylinder Head(s) and Valve Train (continued)

> **Note**
>
> On some engines, the valves may need adjusting at this time. See Job 25.

☐ 2. Install the valve cover(s). Use new gaskets and torque the fasteners to specification.

> **Note**
> On some belt-driven overhead valve engines, the timing gear(s) must be held in position by a special tool as the belt is installed. Consult the engine maker's service information for exact procedures.

☐ 3. Loosely place the timing gears or sprockets on the crankshaft and camshaft and rotate the crankshaft and camshaft until the timing marks are aligned. **Figure 28-2** illustrates the three types of timing mechanisms.

Figure 28-2. Three types of timing mechanisms are shown here. A—A gear-driven timing mechanism. B—A chain-driven timing mechanism. C—A belt-driven timing mechanism. (DaimlerChrysler, Ford)

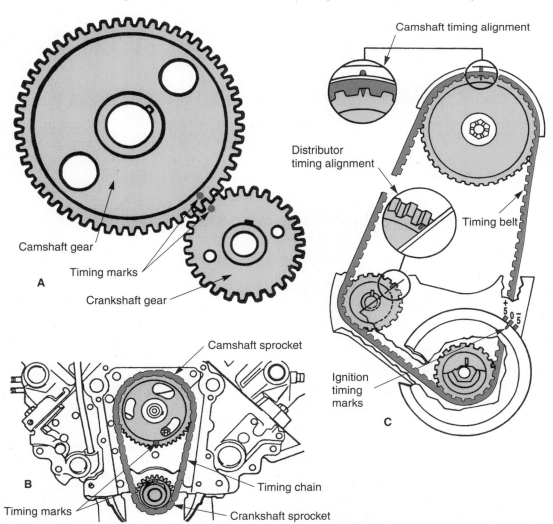

151

☐ 4. Carefully remove the sprockets or gears, being careful not to allow the crankshaft or camshaft to turn.

☐ 5. Place the timing chain or belt in position over the sprockets. Be careful to maintain the alignment of the timing marks.

☐ 6. Install the sprockets and chain or belt into position on the camshaft and crankshaft. Be careful not to allow the crankshaft or camshaft to rotate.

☐ 7. Install and tighten the timing device fasteners.

☐ 8. After all parts are installed, recheck the timing marks to ensure that they are in position.

☐ 9. Index the camshaft position sensor using manufacturer's procedures. On most engines, the camshaft and crankshaft position sensors are indexed by the following procedure:
   a. Place the sensor in position on the engine block or timing cover and lightly attach the fasteners.
   b. Ensure that the marks on the camshaft position sensor are aligned with the marks on the engine block or timing cover.
   c. Tighten the fasteners, being sure that the marks remain in alignment.

☐ 10. Scrape the timing cover, coolant pump, and block mating surfaces to remove old gasket material.

☐ 11. Place the timing cover gaskets and seals on the engine block. Use the proper type of sealant if required.

☐ 12. Place the timing cover on the block.

☐ 13. Install and tighten the timing cover fasteners. Tighten the fasteners to the proper torque in the correct sequence.

☐ 14. Clean and return all of the tools and clean the work area.

☐ 15. Did you encounter any problems during this procedure? Yes ___ No ___

   If Yes, describe the problems: _____

   _____

   What did you do to correct the problems? _____

   _____

☐ 16. Have your instructor check your work and sign this job sheet.

Performance Evaluation—Instructor Use Only

Did the student complete the job in the time allotted? Yes ___ No ___
If No, which steps were not completed?_____
How would you rate this student's overall performance on this job?_____
5–Excellent, 4–Good, 3–Satisfactory, 2–Unsatisfactory, 1–Poor
Comments: _____

_____

_____

INSTRUCTOR'S SIGNATURE _____

Job 29—Complete the Engine Reassembly

After completing this job, you will be able to install the remaining engine components onto a long block to prepare it for reinstallation into the vehicle.

Procedures

> **Note**
>
> This job assumes that the engine has been reassembled to the long block stage. Refer to Job 27 for instructions for assembling an engine to the short block stage and Job 28 for procedures for assembling an engine to the long block stage.

☐ 1. Obtain an engine to be used in this job. Your instructor may direct you to perform this job on a shop engine.
- Make of engine:_____
- Number of cylinders:_____
- Cylinder arrangement (V-type, inline, etc.):_____
- Cooling system: Liquid ____ Air ____

☐ 2. Gather the tools needed to perform the following tasks. Refer to the tools and materials list for this project.

Install the Intake Manifold

☐ 1. Install the intake manifold gaskets.

> **Note**
>
> On some inline engines, the intake and exhaust manifolds are installed together.

☐ 2. Place the intake manifold in position on the engine.

☐ 3. Install all intake manifold fasteners hand tight.

☐ 4. Determine whether there is a tightening sequence for the intake manifold bolts.
Did you find a tightening sequence for the manifold bolts? Yes ____ No ____

☐ 5. Tighten the intake manifold bolts, in the prescribed sequence, to one-half of their full torque specification.
Torque:_____

☐ 6. Tighten the intake manifold bolts to three-fourths of the full torque specification.

☐ 7. Torque the intake manifold bolts to their full specification following the correct sequence.

☐ 8. Following the tightening sequence, recheck each manifold fastener to ensure that all are at full torque.
Did any of the fasteners turn when you tightened them for the final time? Yes ____ No ____
Can you think of a reason that some of the fasteners could be turned more? _____
_____
_____

> **Note**
>
> Some manufacturers recommend retightening the bolts after the engine is warmed to operating temperature and allowed to cool.

Install the Exhaust Manifold(s)

☐ 1. Position the exhaust manifold(s) on the head, using gaskets where applicable.

☐ 2. Start all of the exhaust manifold fasteners by hand.

☐ 3. Tighten the bolts to the proper torque specification, beginning at the center and alternating outward.

   Torque:_____

Complete the Engine Reassembly

☐ 1. Replace the coolant pump gasket and place the coolant pump on the block or timing cover as applicable.

☐ 2. Install and tighten the pump fasteners in the proper sequence and to the proper torque.

☐ 3. Install the vibration damper. If the vibration damper has a rotor for use with a crankshaft position sensor, align the rotor and sensor with the necessary special tool.

☐ 4. Place the flywheel on the crankshaft flange.

> **Note**
>
> Flywheel installation may have to wait until the engine is removed from the engine stand.

☐ 5. Install and tighten the flywheel bolts. These bolts *must* be torqued correctly.

   Torque:_____

> **Note**
>
> If the vehicle has a manual transmission or transaxle, install the clutch and pressure plate, using a pilot shaft to ensure that the clutch disc is properly positioned.

☐ 6. If the valves do not have to be readjusted after the engine is started, install the valve covers.

☐ 7. Install the turbocharger or supercharger if used.

☐ 8. Install any remaining engine parts. If the engine has a distributor, time it to the engine during installation.

> **Caution**
>
> Some engines have a camshaft position sensor and a distributor-mounted crankshaft position sensor. The relationship of these sensors must be checked and adjusted. Failure to make this adjustment may cause random misfires, MIL lamp illumination, or failure to start. The distributor housing may have an indexing mark for proper positioning. On other engines, a scan tool or a dual trace lab scope may be needed to determine the relative sensor positions with the engine running or being cranked. Adjustment is made by turning the distributor housing.

Job 29—Complete the Engine Reassembly (continued)

> **Note**
> At this point, the engine is ready to be reinstalled in the vehicle, as described in Job 30.

☐ 9. Clean and return all of the tools and clean the work area.

☐ 10. Did you encounter any problems during this procedure? Yes ___ No ___

    If Yes, describe the problems: _____

    _____

    What did you do to correct the problems? _____

    _____

☐ 11. Have your instructor check your work and sign this job sheet.

Performance Evaluation—Instructor Use Only

Did the student complete the job in the time allotted? Yes ___ No ___

If No, which steps were not completed?_____

How would you rate this student's overall performance on this job?_____

5–Excellent, 4–Good, 3–Satisfactory, 2–Unsatisfactory, 1–Poor

Comments: _____

_____

_____

INSTRUCTOR'S SIGNATURE _____

Notes

Job 30—Reinstall the Engine

After completing this job, you will be able to install and properly reconnect a vehicle's engine.

Procedures

☐ 1. Obtain a vehicle and engine to be used in this job. Your instructor may direct you to perform this job on a shop vehicle and engine.

☐ 2. Gather the tools needed to perform the following job. Refer to the project's tools and materials list.

Install an Engine

☐ 1. Attach the lifting chain to the engine as was done to remove it. Be sure that the chain is securely attached. Lift the engine enough to remove tension on the holding fixture. Remove any bolts securing the engine to the holding fixture and then carefully lift the engine.

☐ 2. Lower the engine into position in the engine compartment.

☐ 3. Align the engine with the transmission or transaxle.

> **Caution**
>
> ⚠ If the vehicle has a manual transmission or transaxle, be extremely careful to align the clutch disc with the input shaft. Misalignment could result in serious clutch damage.

☐ 4. If the vehicle has an automatic transmission or transaxle, ensure that the flywheel and torque converter are in the same relative position, and that the converter is not binding against the flywheel. If the converter is attached to the flywheel through studs, make sure that the studs pass through the flywheel before proceeding to the next steps. See **Figure 30-1**.

☐ 5. Install the fasteners holding the transmission or transaxle to the engine.

☐ 6. Install and tighten the engine mount fasteners and remove the lifting fixture.

☐ 7. Install and tighten the flywheel-to-torque converter fasteners.

Figure 30-1. Be sure that the converter studs pass through the correct holes in the flywheel before continuing the installation.

Converter stud passing through the flywheel

☐ 8. Install the starter and reattach the starter wires.

☐ 9. Install the flywheel cover and any engine-to-transmission brackets.

☐ 10. Reattach the exhaust system.

☐ 11. Install the vacuum lines on the intake manifold.

☐ 12. Install the fuel lines.

☐ 13. Install the throttle linkage and any related linkages.

☐ 14. Install the air cleaner assembly or the air inlet ducts.

☐ 15. Install the power steering pump if used.

☐ 16. Install the air conditioner compressor if used.

☐ 17. Install any belts and tighten them to specifications.

☐ 18. Install all electrical connectors and ground straps.

☐ 19. Install the radiator and heater hoses.

☐ 20. Refill the cooling system with the proper type of coolant.

☐ 21. Add the proper type of engine oil until the level reaches the full mark on the dipstick.

☐ 22. Reconnect the battery negative cable.

☐ 23. If the vehicle has an electric fuel pump, turn the ignition switch to the on position and allow the fuel system to refill.

☐ 24. Check the fluid level in the power steering and transmission or transaxle. Add the proper type of fluid as needed.

☐ 25. Start the engine.

☐ 26. Check for proper engine operation and leaks as the engine warms up. Make any necessary ignition or fuel system adjustments.

☐ 27. Reinstall the hood.

☐ 28. Road test the vehicle to ensure that all repairs are satisfactory.

☐ 29. Clean the work area and return any equipment to storage.

☐ 30. Did you encounter any problems during this procedure? Yes ___ No ___

If Yes, describe the problems: _____

_____

What did you do to correct the problems? _____

_____

☐ 31. Have your instructor check your work and sign this job sheet.

Performance Evaluation—Instructor Use Only

Did the student complete the job in the time allotted? Yes ___ No ___
If No, which steps were not completed?_____
How would you rate this student's overall performance on this job?_____
5–Excellent, 4–Good, 3–Satisfactory, 2–Unsatisfactory, 1–Poor
Comments: _____

_____

_____

INSTRUCTOR'S SIGNATURE _____

Project
Servicing a Cooling System

9

Introduction

The cooling system consists of many components, all of which can fail. While modern vehicles are designed to require less maintenance, they also last for many more years than they did previously, so the chance of encountering a defective cooling system part remains the same. If defective cooling system parts are not replaced, the engine can overheat and be destroyed.

As you work through the jobs in this project, you will perform periodic maintenance and replace faulty or worn parts in a cooling system. In Job 31, you will inspect and perform diagnostic tests on a cooling system. You will flush and bleed the cooling system in Job 32. In Job 33, you will replace cooling system belts and hoses. You will replace a vehicle's coolant pump in Job 34, and a vehicle's radiator in Job 35. In Job 36, you will remove and replace a vehicle's thermostat.

Project 9 Jobs

- Job 31—Inspect and Test a Cooling System; Identify Causes of Engine Overheating
- Job 32—Flush and Bleed a Cooling System
- Job 33—Replace Belts and Hoses
- Job 34—Replace a Coolant Pump
- Job 35—Replace a Radiator
- Job 36—Remove and Replace a Thermostat

Tools and Materials

The following list contains the tools and materials that may be needed to complete the jobs in this project. The items used will depend on the make and model of vehicle being serviced.

- Vehicle in need of cooling system service.
- Applicable service information.
- Cooling system pressure tester.
- Hydrometer or other type of antifreeze tester.
- Temperature tester.
- Drain pan(s).
- Hand tools as needed.
- Air-powered tools as needed.
- Safety glasses and other protective equipment as needed.

Safety Notice

Before performing this job, review all pertinent safety information in the text and review safety information with your instructor.

Job 31—Inspect and Test a Cooling System; Identify Causes of Engine Overheating

After completing this job, you will be able to inspect a cooling system and diagnose cooling system problems.

Procedures

- [ ] 1. Obtain a vehicle on which cooling system service can be performed. Your instructor may direct you to perform this job on a shop vehicle.

- [ ] 2. Gather the tools needed to perform the following job. Refer to the project's tools and materials list.

- [ ] 3. Determine the type of antifreeze to be used in the vehicle cooling system.

   Antifreeze type: _____

Inspect Belts and Hoses

- [ ] 1. Open the hood and visually inspect the condition of the cooling system hoses. Examine the hoses for bulging, leaks, or fraying. Ensure that the hoses do not contact any moving parts. Squeeze the hoses to check for hardness, cracks, or softness, **Figure 31-1**.

   Were any problems found? Yes ___ No ___

   If Yes, describe the problems: _____

   _____

- [ ] 2. Determine the condition of the vehicle's belts. Examine the belts for glazing, fraying, or oil contamination. Note whether the belts are loose by attempting to turn the pulley on one of the accessories. If applicable, note the condition of the belt tensioner.

   Were any problems found? Yes ___ No ___

   If Yes, describe the problems: _____

   _____

Figure 31-1. Squeeze the hose to determine whether it is excessively soft. Be careful when doing this on a warmed up engine as the hose will be hot.

3. Inspect the belt tensioner for the following defects:
   - Weak tensioner spring resistance.
   - Worn pulley grooves.
   - Worn bearings (roughness when turning the tensioner).

   Can the tensioner be reused? Yes ___ No ___

   If No, explain: _____

   _____

4. Check the pulley alignment of the belt system(s). Sight along the pulleys to ensure that they are properly aligned, as in **Figure 31-2**.

   Are the pulleys and belts aligned? Yes ___ No ___

   If No, describe the relative positions of the pulleys: _____

   _____

Inspect Other Cooling System Components

1. Visually check the radiator's condition. Make sure the radiator fins are free of leaves and other debris, the radiator's cooling fins are not damaged, and there are no obvious leaks.

   Were any problems found? Yes ___ No ___

   If Yes, describe the problems: _____

   _____

2. Check the condition of the coolant recovery system. Inspect the coolant recovery tank, cap, and hoses for damage. Inspect the coolant in the tank for excessive rust and discoloration. This can be done visually, as shown in **Figure 31-3**.

   Were any problems found? Yes ___ No ___

   If Yes, describe the problems: _____

   _____

3. Ensure that the engine and cooling system are sufficiently cool to allow the radiator cap to be safely removed.

4. Remove the radiator cap and check its condition. Replace the cap if it is corroded or if the rubber seal is hard, cracked, or otherwise damaged.

   Is the cap in good condition? Yes ___ No ___

Figure 31-2. The easiest way to determine whether the pulleys and belts are aligned is to make a visual check. Note the alignment of the various pulleys and the belt in this figure.

Job 31—Inspect and Test a Cooling System; Identify Causes of Engine Overheating (continued)

Figure 31-3. Visually check the reservoir for discoloration. Rust will stain the reservoir.

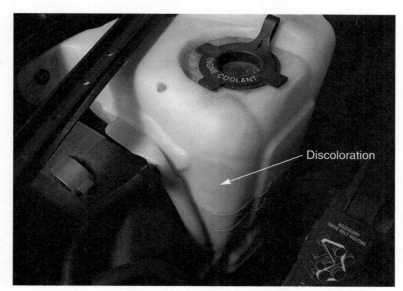

Discoloration

☐ 5. Check the coolant level. If the cooling system has a coolant recovery system, the coolant level should be at the top of the filler neck. If the vehicle does not have a coolant recovery system the coolant level should be about 3″ (76 mm) below the filler neck.

   Is the coolant level correct? Yes ___ No ___

☐ 6. Visually check the coolant for rust.

   Is the coolant rusty? Yes ___ No ___

☐ 7. Check degree of freeze protection (antifreeze percentage) with an antifreeze tester. Follow the directions provided by the tester manufacturer.

   Antifreeze will protect to _____ °F or °C (circle one).

Pressure Test the Pressure Cap

☐ 1. Obtain a cooling system pressure tester.

☐ 2. Check the condition of the radiator cap by installing it on the tester, **Figure 31-4**. Adapters may be needed to properly test the cap. Pump up the tester pressure until it levels off. A pressure specification should be printed somewhere on the cap. The radiator should hold this pressure without allowing it to drop.

   Did the radiator cap pass the pressure test? Yes ___ No ___
   If No, replace the radiator cap.

Vacuum Test the Pressure Cap

☐ 1. Obtain a vacuum pump and radiator cap vacuum test adapter.

☐ 2. Install the pressure cap on the adapter and connect the adapter to the vacuum pump.

☐ 3. Apply vacuum to the radiator cap. The cap should allow vacuum to build up to 2″–5″ Hg, then release and allow air to flow through the cap.

Figure 31-4. Install the pressure cap on the pressure tester. Adapters may be needed. (Jack Klasey)

> **Note**
>
> Release vacuum varies between manufacturers. Check the specifications when available to determine the proper release vacuum for a particular vehicle.

Does the cap allow vacuum to build up, then release? Yes _____ No_____
If No, replace the radiator cap.

Pressure Test the Cooling System

☐ 1. Remove the radiator cap from the pressure tester and install the pressure tester on the radiator filler neck. See **Figure 31-5**.

☐ 2. Pump the pressure tester handle until the cooling system has been pressurized to the rating on the radiator cap. If the system is very low on coolant, add water to make the job of pressurizing easier. Carefully observe the cooling system for leaks. If no leaks are visible and the pressure holds for two minutes, the system is not leaking.

Did the cooling system pass the pressure test? Yes ___ No ___
If No, explain why the cooling system failed: _____
_____
_____
_____

Figure 31-5. Install the pressure tester on the radiator filler neck. (Jack Klasey)

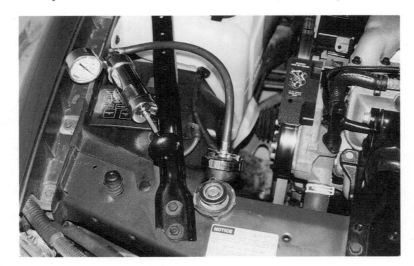

Job 31—Inspect and Test a Cooling System; Identify Causes of Engine Overheating (continued)

> **Note**
>
> If you add water to the cooling system for diagnostic testing, make sure you drain and refill the system with the proper antifreeze mixture before releasing the vehicle to its owner.

Make Visual Checks for Leaks

☐ 1. Check all visible gaskets, fittings, and freeze plugs for the presence of coolant.

Describe any leaks or suspicious areas: _____

_____

_____

☐ 2. Check all engine surfaces for pools of coolant.

Did you find pools of coolant on any of the engine surfaces? Yes ___ No ___

If Yes, describe the location: _____

_____

Based on your visual inspection, does the engine have leaks that must be corrected?
Yes ___ No ___

If Yes, consult with your instructor about the needed repairs.

> **Note**
>
> If the cooling system is leaking, you can make the leak easier to find by adding fluorescent dye to the cooling system, allowing it to circulate with the coolant, and then inspecting the system with a black light.

Check the Coolant Pump

☐ 1. Check the coolant pump for leaks at the gaskets and weep hole. The weep hole is usually located on the housing between the pulley flange and the rear of the pump.

Did you find any leaks? Yes ___ No ___

If Yes, consult with your instructor about the needed repairs.

> **Note**
>
> It is normal for a slight amount of coolant to be present at the weep hole.

Belt-Driven Pump

☐ 1. Check the pump bearings for wear by attempting to move the pump up and down and back and forth. There should be almost no movement.

Did you detect excessive movement? Yes ___ No ___

If Yes, consult your instructor about the needed repairs.

Electric Motor–Driven Pump

☐ 1. Ensure that the pump motor runs whenever the ignition switch is in the *on* position.

Does the pump motor operate with the ignition switch in the on position? Yes ___ No ___

If No, consult your instructor about needed repairs.

Check Coolant Temperature and Operation of a Cooling Fan

☐ 1. Ensure that the vehicle cooling system is full.

☐ 2. Replace the radiator cap if it was removed.

☐ 3. Start the vehicle and allow it to warm up. If the vehicle has an air conditioner, turn it on to shorten the warm-up time.

☐ 4. As the engine warms up, periodically check the engine temperature and record it below. Circle the temperature scale used, °F or °C.

One minute after startup: _____°F or °C

Three minutes after startup: _____°F or °C

Five minutes after startup: _____°F or °C

Ten minutes after startup (if necessary): _____°F or °C

Electric Cooling Fan

☐ 1. Observe the cooling fan. It should come on shortly after the engine reaches operating temperature. On some vehicles, the fan will come on when the air conditioner high-side pressures reach a certain value, or it may run whenever the air conditioner is on. Many vehicles have two cooling fans. One or both fans will come on as the engine warms up.

Does the fan come on when needed? Yes ___ No ___

If No, consult your instructor for further instructions.

Belt-Driven Fan and Fan Clutch

☐ 1. Observe the cooling fan airflow. It should be pulling air through the radiator when the engine is at operating temperature.

> **Note**
>
> Consult your instructor if you are not sure whether the fan is pulling enough air through the radiator.

☐ 2. Stop the engine while observing the fan. The fan should stop turning within one revolution.

Does the fan stop within one revolution? Yes ___ No ___

If No, consult your instructor for further instructions.

Check Fan Shroud and Air Dams

☐ 1. Visually inspect the fan shroud for:
- Cracks.
- Missing parts.
- Loose or missing shroud fasteners.
- Evidence that the fan is hitting the shroud.

☐ 2. Visually inspect the air dams for:
- Tears or cracks.
- Missing dams.
- Loose or missing dam fasteners.

Did you observe any of the above problems? Yes ___ No ___

If Yes, consult your instructor for further instructions.

Check for Combustion Leaks

> **Note**
>
> Due to the variation in types of combustion testers, the following steps outline only a general method of combustion testing.

Job 31—Inspect and Test a Cooling System; Identify Causes of Engine Overheating (continued)

- [ ] 1. Ensure that the vehicle's cooling system is full.
- [ ] 2. Remove the radiator cap.
- [ ] 3. Install the combustion tester on the filler neck.
- [ ] 4. Start the engine and follow the manufacturer's directions to test for the presence of combustion gases in the cooling system.

    Were combustion gases detected in the cooling system? Yes ___ No ___

    If Yes, consult your instructor for further instructions before continuing.
- [ ] 5. Remove the tester and replace the radiator cap.

Inspect the Engine Oil Cooler

- [ ] 1. If the vehicle has an auxiliary engine oil cooler, check it for the following:
    - Visible leaks.
    - Clogged fins.
    - Cracked or damaged hoses or lines.

    Were any problems found? Yes ___ No ___

    If Yes, describe the problems: _____

    _____

    _____

Inspect for Non-Cooling System Causes of Overheating

- [ ] 1. Inspect the vehicle and operating conditions for the following.

    Is the radiator blocked by debris or front body modifications? Yes _____ No _____

    Is there a missing air dam below front bumper? Yes _____ No _____

    Is there a restricted exhaust system (Refer to Job 8 for testing procedures)?
    Yes _____ No _____

    Is there a binding or resistance in drivetrain? Yes _____ No _____

    Are there dragging brakes or other resistance in brake system? Yes _____ No _____

    Is there excessive idling, especially in gear and/or with the air conditioning operating?
    Yes _____ No _____

    Does it operate at extreme high speed? Yes _____ No _____

    Is there a towing or other extra weight situation? Yes _____ No _____

    Is there extreme ambient temperatures? Yes _____ No _____

    If Yes, describe the problem(s): _____

    _____

    _____

- [ ] 2. Clean up any spilled coolant.
- [ ] 3. Clean the work area and return any equipment to storage.
- [ ] 4. Did you encounter any problems during this procedure? Yes ___ No ___

    If Yes, describe the problems: _____

    _____

    What did you do to correct the problems? _____

    _____

- [ ] 5. Have your instructor check your work and sign this job sheet.

Performance Evaluation—Instructor Use Only

Did the student complete the job in the time allotted? Yes ___ No ___

If No, which steps were not completed?_____

How would you rate this student's overall performance on this job?_____

5–Excellent, 4–Good, 3–Satisfactory, 2–Unsatisfactory, 1–Poor

Comments: _____

_____

_____

INSTRUCTOR'S SIGNATURE _____

Job 32—Flush and Bleed a Cooling System

After completing this job, you will be able to flush and bleed a cooling system.

Procedures

☐ 1. Obtain a vehicle on which cooling system service can be performed. Your instructor may direct you to perform this job on a shop vehicle.

☐ 2. Gather the tools needed to perform the following job. Refer to the project's tools and materials list.

☐ 3. Determine the type of antifreeze to be used in the vehicle's cooling system.

Antifreeze type: _____

Flush and Bleed the Cooling System

> **Note**
>
> Many shops have pressure flushing and refilling machines. To use such a machine, use the manufacturer's procedures rather than the steps given here.

☐ 1. Ensure that the engine and cooling system are sufficiently cool to allow the radiator cap to be safely removed.

☐ 2. Remove the radiator cap, then locate the radiator drain plug. Locate any coolant drain plugs on the engine.

☐ 3. Place a suitable container under the drain plug(s).

☐ 4. Open the radiator drain plug and any coolant drain plugs on the engine. Make sure that coolant drains into the containers.

> **Warning**
>
> ⚠ Clean up any coolant spills. Antifreeze is poisonous to people and animals.

☐ 5. Allow the coolant to drain into the container(s) until coolant stops coming from the drain plugs.

☐ 6. Close the drain plugs.

☐ 7. Refill the cooling system with water and flushing agent. Carefully follow the directions provided by the manufacturer of the flushing agent.

☐ 8. Start the engine and allow it to idle for the recommended time. Turn the vehicle heater to high and closely monitor the level in the radiator.

☐ 9. After the recommended time has elapsed, turn the engine off.

☐ 10. Repeat steps 4 through 9 until the water from the drain plugs is clear. The length of this procedure varies depending on the condition of the cooling system and the number of times the cooling system must be flushed to remove all of the cleaning agent.

☐ 11. Refill the cooling system with a 50-50 mixture of the proper type of antifreeze and clean water.

What type of antifreeze does this engine require? _____

_____

> **Caution**
>
> Many modern vehicles take special long-life coolant. Always install the proper type of coolant. Do not mix coolant types.

☐ 12. Bleed the system using one of the following procedures, depending on the relative placement of the engine and radiator:
- **Radiator higher than engine:** Fill the system to about 3″ (76 mm) below the top of the filler neck. Allow the engine to reach operating temperature. When the thermostat opens, the radiator level will drop. Add more coolant until the level is at the top of the filler neck (coolant recovery system) or 3″ (76 mm) below the filler neck (no coolant recovery system).
- **Engine higher than radiator:** Fill the system to about 3″ (76 mm) below the top of the filler neck. Allow the engine to reach operating temperature. When the thermostat opens, the radiator level will drop. Open the bleed valve(s) on the engine and add more coolant until the coolant begins to exit from the bleeder valve. Close the bleeder valve and add coolant to fill the radiator. On some vehicles the bleeder may need to be opened and closed several times to remove all air from the system.

☐ 13. Install the radiator cap.

☐ 14. Add coolant to the cold level on the coolant recovery reservoir.

☐ 15. Clean up any spilled coolant.

☐ 16. Clean the work area and return any equipment to storage.

☐ 17. Did you encounter any problems during this procedure? Yes ___ No ___

If Yes, describe the problems: _____

_____

What did you do to correct the problems? _____

_____

☐ 18. Have your instructor check your work and sign this job sheet.

Performance Evaluation—Instructor Use Only

Did the student complete the job in the time allotted? Yes ___ No ___
If No, which steps were not completed?_____
How would you rate this student's overall performance on this job?_____
5–Excellent, 4–Good, 3–Satisfactory, 2–Unsatisfactory, 1–Poor
Comments: _____

_____

_____

INSTRUCTOR'S SIGNATURE _____

Job 33—Replace Belts and Hoses

After completing this job, you will be able to replace V-belts and serpentine belts.

Procedures

☐ 1. Obtain a vehicle to be used in this job. Your instructor may direct you to perform this job on a shop vehicle.

☐ 2. Gather the tools needed to perform the following job. Refer to the tools and materials list at the beginning of the project.

Replace V-Belt

☐ 1. Determine which belt must be replaced. V-belts are used to drive combinations of accessories including the coolant pump, alternator, air conditioning compressor, power steering pump, and air pump. Some or all of these accessories must be loosened to remove a particular belt. Sometimes one or more serviceable belts must be removed to remove the defective belt.

What accessories are driven by the belt that will be replaced? _____

_____

☐ 2. Loosen the fasteners holding the accessory driven by the belt. You may need a droplight so that you can find all of the fasteners.

☐ 3. Push the accessory toward the engine to remove tension on the belt.

☐ 4. Remove the belt from the pulleys.

☐ 5. Compare the old and new belts.

Is the new belt the correct replacement? Yes ___ No ___

> **Note**
> Even a slight difference in size means that the new belt is unusable.

☐ 6. Slip the new belt over the pulleys, making sure that it is fully seated in the pulley grooves.

☐ 7. Pull the accessories away from the engine to place slight tension on the belt.

☐ 8. Adjust belt tension. Use a belt tension gauge to measure exact tension. If a gauge is not available, use the belt deflection method to adjust the belt. The belt should deflect about 1/2″ to 5/8″ (13 mm to 9.5 mm) with 25 pounds (11 kg) of push applied. The belts should only be tight enough to prevent slipping or squealing. Excessive tension on the accessory bearings will quickly wear them out.

☐ 9. Start the vehicle and check belt operation.

Replace Serpentine Belt

> **Note**
> This procedure assumes that the vehicle has one serpentine belt. If you must remove other belts in order to replace the serpentine belt, use this procedure (for serpentine belts) or the previous procedure (for V-belts) to remove them.

☐ 1. Open the vehicle hood and determine whether the engine contains a label showing serpentine belt routing, or obtain service literature showing serpentine belt routing for the vehicle that you are working on.

☐ 2. Locate the belt-tensioning device and determine what tool(s) are needed to remove tension. This may be simple hand tools or a special tool, **Figure 33-1**.

Tool(s) needed to remove tension: _____

☐ 3. Remove tension from the belt using the proper tool(s).

☐ 4. Remove the serpentine belt.

☐ 5. Release the belt-tensioning device.

☐ 6. Compare the old and new belts.

Are the belts the same size? Yes ___ No ___
If No, obtain the correct belt before proceeding.

☐ 7. Place the new belt in position, using the proper belt routing diagram.

☐ 8. Turn the pulley of the belt-tensioning device.

Is any roughness noted? Yes ___ No ___
If Yes, the pulley should be replaced. Consult your instructor before proceeding.

☐ 9. Pull the belt-tensioning device away from the belt to allow the belt to be installed.

☐ 10. Finish installing the new belt, making sure that the belt is properly positioned over the pulleys.

☐ 11. Release the belt-tensioning device.

☐ 12. Start the engine and check belt operation.

Remove and Replace Cooling System and Heater Hoses

☐ 1. Determine which hose must be replaced.

Which hose must be replaced? _____

☐ 2. Ensure that the engine and cooling system are sufficiently cool to allow the radiator cap to be safely removed.

Figure 33-1. Serpentine belt tension can usually be removed by inserting a 3/8″ or 1/2″ ratchet or flex handle into a square hole in the tensioner and pulling away from the direction of tension. Sometimes a special tool is needed to remove belt tension.

Job 33—Replace Belts and Hoses (continued)

☐ 3. Remove the radiator cap, then locate the radiator drain plug.

☐ 4. Place a suitable container under the drain plug.

☐ 5. Open the radiator drain plug. Ensure that coolant drains into the container.

Warning

⚠ Clean up any coolant spills. Antifreeze is poisonous to people and animals.

☐ 6. Allow the coolant to drain out so that the level of coolant in the engine is below the level of the hose nipples. It is not necessary to remove all of the coolant.

☐ 7. Loosen and remove the hose clamps. Most modern vehicles use spring (sometimes called "hog ring") clamps, which can be removed by compressing the spring tabs with a pair of pliers, **Figure 33-2**.

☐ 8. Remove the hose. It may be necessary to split the hose at the hose nipple to make removal easier.

☐ 9. Compare the old hose to the replacement hose.

   Is the replacement hose correct? Yes ____ No ____

☐ 10. Lightly coat the hose nipples with nonhardening sealer.

☐ 11. Place new hose clamps over the replacement hose.

☐ 12. Install the hose on the hose nipples.

☐ 13. Position and tighten the hose clamps.

☐ 14. Refill the system with the proper kind of coolant, following the procedures in Job 32. Many modern vehicles use special long-life coolant. Do not mix coolant types.

☐ 15. Start the engine and check for leaks.

☐ 16. Clean the work area, including any coolant spills, and return any equipment to storage.

Figure 33-2. Compress the tabs on the hose clamp to loosen the clamp.

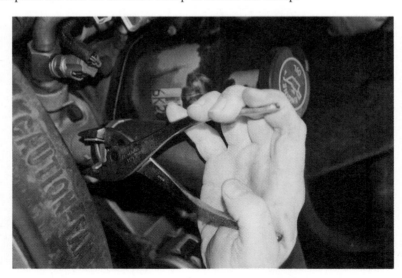

☐ 17. Did you encounter any problems during this procedure? Yes ___ No ___

If Yes, describe the problems: _____

_____

What did you do to correct the problems? _____

_____

☐ 18. Have your instructor check your work and sign this job sheet.

Performance Evaluation—Instructor Use Only

Did the student complete the job in the time allotted? Yes ___ No ___

If No, which steps were not completed?_____

How would you rate this student's overall performance on this job?_____

5–Excellent, 4–Good, 3–Satisfactory, 2–Unsatisfactory, 1–Poor

Comments: _____

_____

_____

INSTRUCTOR'S SIGNATURE _____

Job 34—Replace a Coolant Pump

After completing this job, you will be able to remove and replace a coolant pump.

Procedures

☐ 1. Obtain a vehicle to be used in this job. Your instructor may direct you to perform this job on a shop vehicle.

☐ 2. Gather the tools needed to perform the following job. Refer to the tools and materials list at the beginning of the project.

Remove and Replace the Coolant Pump

☐ 1. Ensure that the engine and cooling system are sufficiently cool to allow the radiator cap to be safely removed.

☐ 2. Remove the radiator cap, then locate the radiator drain plug.

☐ 3. Place a suitable container under the drain plug.

☐ 4. Open the radiator drain plug. Ensure that coolant drains into the container.

> **Warning**
>
> ⚠ Clean up any coolant spills. Antifreeze is poisonous to people and animals.

☐ 5. Allow the coolant to drain out so that the level of coolant in the engine is below the level of the coolant pump. It is not necessary to remove all of the coolant.

☐ 6. Loosen and detach the cooling system hoses at the coolant pump.

☐ 7. Remove any accessories or brackets blocking the coolant pump.

☐ 8. Loosen and remove the pump drive belt.

☐ 9. Loosen and remove the pump pulley fasteners. Then, remove the pulley, and fan if necessary.

☐ 10. Remove the fasteners holding the coolant pump to the engine.

☐ 11. Remove the coolant pump from the engine. It may be necessary to carefully pry on the pump to loosen it.

☐ 12. Remove any gasket material from the engine-to-pump sealing areas. Be careful not to gouge or nick the metal surfaces.

☐ 13. Compare the old and replacement pumps. Check the following:

Do the pump sizes match? Yes ____ No ____

Do the pump bolt patterns match? Yes ____ No ____

Do the direction of pump impellers match? Yes ____ No ____

Do the pump shaft lengths match? Yes ____ No ____

If the pumps do not match, explain why: _____

_____

_____

> **Caution**
>
> If the pump shows heavy rust deposits, such as those in **Figure 34-1**, be sure to flush the cooling system before releasing the vehicle. If you do not flush the system, rust deposits will ruin the new pump.

☐ 14. If specified, coat the coolant pump and engine sealing surfaces with nonhardening sealer.

☐ 15. Install the new coolant pump gasket on the engine.

☐ 16. Place the coolant pump on the engine and install the pump fasteners.

☐ 17. Tighten the pump fasteners in an alternating pattern to evenly draw the pump to the engine. Be careful not to overtighten the fasteners.

☐ 18. Install the pump pulley, and fan if necessary, and loosely install the pulley fasteners.

☐ 19. Install the pump drive belt.

☐ 20. Tighten the pump pulley fasteners.

☐ 21. Reinstall any accessories or brackets that had to be removed to access the coolant pump. Make sure the belts are properly adjusted.

☐ 22. Refill the system with the proper kind of coolant, following the procedures in Job 32. Many modern vehicles use special long-life coolant. Do not mix coolant types.

☐ 23. Start the engine and check for leaks and proper pump operation.

Replace a Timing Gear–Driven Coolant Pump

☐ 1. Drain the engine coolant.

☐ 2. Remove any exhaust system and wiring blocking access to the timing cover.

☐ 3. Remove the timing cover fasteners and the cover.

☐ 4. Relieve tension on the timing chain tensioner.

☐ 5. Remove the lower radiator hose from the coolant pump or coolant pump housing as necessary.

☐ 6. Remove the coolant pump housing if one is used.

Figure 34-1. Rust deposits on this pump impeller indicate that the cooling system has not been maintained.

Job 34—Replace a Coolant Pump (continued)

☐ 7. Remove the fasteners holding the coolant pump to the engine.

☐ 8. Remove the coolant pump by maneuvering the driven gear out of engagement with the timing chain and sliding the pump out of the engine.

☐ 9. Scrape all mounting surfaces as needed.

☐ 10. Compare the old and new pumps.

Do the pumps match? Yes ___ No ___

If No, consult your instructor.

☐ 11. Place the new gaskets in position, using sealer as recommended.

☐ 12. Place the coolant pump in position on the engine, making sure that the driven gear is lined up with the timing chain.

☐ 13. Install and tighten the coolant pump fasteners.

☐ 14. Install the coolant pump housing gaskets, using sealer as necessary.

☐ 15. Reinstall the coolant pump housing.

☐ 16. Reinstall the lower radiator hose on the coolant pump housing.

☐ 17. Ensure that the timing chain is properly aligned with the pump, then allow the timing chain tensioner to assume its original position.

☐ 18. Using gaskets and sealer as needed, reinstall the timing chain cover.

☐ 19. Reinstall exhaust system components and wiring as necessary.

> **Caution**
>
> ⚠ If any coolant has entered the engine oil pan, drain and replace the oil before proceeding. Antifreeze in the oil can badly damage an engine.

☐ 20. Replace the coolant and start the engine.

☐ 21. With the engine running, bleed the cooling system and check the pump for leaks.

Replace an Electrically Driven Coolant Pump

☐ 1. Disconnect the vehicle battery.

☐ 2. Drain the engine coolant.

☐ 3. Remove any engine or system components blocking access to the coolant pump.

☐ 4. Remove hoses from the coolant pump as needed.

☐ 5. Disconnect the pump motor's electrical connectors.

☐ 6. Remove the fasteners holding the coolant pump to the engine.

☐ 7. Remove the coolant pump from the engine.

☐ 8. Scrape all mounting surfaces as needed.

☐ 9. Compare the new and old pumps.

Do the pumps match? Yes ___ No ___

If No, consult your instructor before proceeding.

☐ 10. Place new gaskets in position, using sealer as recommended.

☐ 11. Place the coolant pump in position on the engine.

- [ ] 12. Install and tighten the coolant pump fasteners.
- [ ] 13. Reconnect the pump motor's electrical connectors.
- [ ] 14. Reinstall any other components that were removed.
- [ ] 15. Reinstall hoses as needed.
- [ ] 16. Replace the coolant and reconnect the battery negative cable.
- [ ] 17. Start the engine and bleed the cooling system with the engine running.
- [ ] 18. Check the coolant pump for leaks.
- [ ] 19. Clean up any spilled coolant.
- [ ] 20. Clean the work area and return any equipment to storage.
- [ ] 21. Did you encounter any problems during this procedure? Yes ___ No ___

    If Yes, describe the problems: _____

    _____

    What did you do to correct the problems? _____

    _____

- [ ] 22. Have your instructor check your work and sign this job sheet.

Performance Evaluation—Instructor Use Only

Did the student complete the job in the time allotted? Yes ___ No ___

If No, which steps were not completed?_____

How would you rate this student's overall performance on this job?_____

5–Excellent, 4–Good, 3–Satisfactory, 2–Unsatisfactory, 1–Poor

Comments: _____

_____

_____

INSTRUCTOR'S SIGNATURE _____

Job 35—Replace a Radiator

After completing this job, you will be able to remove and replace a radiator.

Procedures

☐ 1. Obtain a vehicle to be used in this job. Your instructor may direct you to perform this job on a shop vehicle.

☐ 2. Gather the tools needed to perform the following job. Refer to the tools and materials list at the beginning of the project.

☐ 3. Ensure that the engine and cooling system are sufficiently cool to allow the radiator cap to be safely removed.

☐ 4. Remove the radiator cap, then locate the radiator drain plug.

☐ 5. Place a suitable container under the drain plug.

☐ 6. Open the radiator drain plug. Ensure that coolant drains into the container.

> **Warning**
>
> ⚠ Clean up any coolant spills. Antifreeze is poisonous to people and animals.

☐ 7. Allow as much coolant as possible to drain out.

☐ 8. Remove the upper and lower radiator hoses, any heater hoses attached to the radiator, and the coolant recovery system hose.

☐ 9. Remove any electrical connectors at the radiator tanks.

☐ 10. Remove the transmission cooler lines if used.

☐ 11. Remove any shrouds that prevent access to the radiator.

☐ 12. Where applicable, remove the fasteners holding the electric fan assembly to the radiator, and then remove the fan assembly.

☐ 13. Remove the fasteners and brackets holding the radiator to the vehicle.

☐ 14. Lift the radiator from the vehicle. Compare the original and replacement radiators.

Do the sizes match? Yes ___ No ___

Do the shapes match? Yes ___ No ___

Are the hose nipples in the same positions? Yes ___ No ___

Are the cooler lines and electrical fittings in the same positions? Yes ___ No ___

Is the drain plug/petcock in the same location? Yes ___ No ___

If the radiators do not match, describe how they differ:_____

_____

_____

☐ 15. Transfer any parts, such as coolant level switches, between the original and replacement radiators.

> **Note**
>
> 📄 Many new radiators require that you transfer the drain petcock or plug from the old radiator.

☐ 16. Place the radiator in position in the vehicle, making sure that the radiator is properly positioned and seated.

☐ 17. Install the attaching brackets.

☐ 18. Install the electric fan assembly, if necessary.

☐ 19. Install the transmission oil cooler lines and electrical connectors if necessary.

☐ 20. Install the upper and lower radiator hoses, the coolant recovery system hose, and any heater hoses.

☐ 21. Refill the system with the proper kind of coolant, following the procedures in Job 32. Many modern vehicles use special long-life coolant. Do not mix coolant types.

☐ 22. Start the engine and check for leaks. Bleed the system as necessary. See Job 32 for bleeding procedures.

☐ 23. Clean the work area, including any coolant spills, and return any equipment to storage.

☐ 24. Did you encounter any problems during this procedure? Yes ___ No ___

If Yes, describe the problems: _____

_____

What did you do to correct the problems? _____

_____

☐ 25. Have your instructor check your work and sign this job sheet.

Performance Evaluation—Instructor Use Only

Did the student complete the job in the time allotted? Yes ___ No ___

If No, which steps were not completed?_____

How would you rate this student's overall performance on this job?_____

5–Excellent, 4–Good, 3–Satisfactory, 2–Unsatisfactory, 1–Poor

Comments: _____

_____

_____

INSTRUCTOR'S SIGNATURE _____

Job 36—Remove and Replace a Thermostat

After completing this job, you will be able to remove and replace a thermostat.

Procedures

☐ 1. Obtain a vehicle to be used in this job. Your instructor may direct you to perform this job on a shop vehicle.

☐ 2. Gather the tools needed to perform the following job. Refer to the tools and materials list at the beginning of the project.

☐ 3. Ensure that the engine and cooling system are sufficiently cool to allow the radiator cap to be safely removed.

☐ 4. Remove the radiator cap, then locate the radiator drain plug.

☐ 5. Place a suitable container under the drain plug.

☐ 6. Open the radiator drain plug. Ensure that coolant drains into the container.

Warning

⚠ Clean up any coolant spills. Antifreeze is poisonous to people and animals.

☐ 7. Allow coolant to drain out so that the level of coolant in the engine is below the level of the thermostat. It is not necessary to remove all of the coolant.

☐ 8. Loosen the upper radiator hose clamp and remove the hose from the thermostat housing.

☐ 9. Remove the fasteners holding the thermostat housing to the engine and remove the housing. The thermostat should now be visible. See **Figure 36-1**.

☐ 10. Remove the thermostat from the engine.

Figure 36-1. On V-type engines, the thermostat is usually located between the cylinder banks at the front of the engine. On inline engines, thermostats are usually near the upper radiator hose at the front of the engine.

☐ 11. Remove any gasket material from the engine and the thermostat housing sealing areas. If seals are used instead of gaskets, remove them and make sure the seal grooves are clean. You may wish to stuff a rag in the thermostat housing to prevent any debris from entering.

> **Note**
>
> The following step is optional and may be assigned by your instructor.

☐ 12. Test the thermostat by suspending it in a water-filled container over a burner or other source of heat. The thermostat should not touch the sides of the container. Place a thermometer in the water and observe the thermostat as the water is heated. If the thermostat does not begin to open at the rated temperature or is not fully open before the water begins boiling, it is defective.

☐ 13. If the thermostat is being replaced, compare the old and new thermostats. The new thermostat should be the same size as the original thermostat. The thermostat opening temperature should match the manufacturer's specifications. The thermostat opening temperature is usually stamped on the thermostatic element, as shown in **Figure 36-2**.

Do the sizes match? Yes ___ No ___

Does the opening temperature match? Yes ___ No ___

> **Caution**
>
> Do not install a thermostat with a lower operating temperature than the original thermostat. This will affect emissions control operations and may set a trouble code on OBD II-equipped vehicles.

☐ 14. Place the thermostat in position. The heat-sensing element should face the interior of the engine.

Is the thermostat installed properly? Yes ___ No ___

☐ 15. If specified, coat the thermostat housing and engine sealing surfaces with nonhardening sealer.

Figure 36-2. The opening temperature is usually stamped on the thermostatic element of the thermostat. Most thermostats will have both Fahrenheit and Celsius figures.

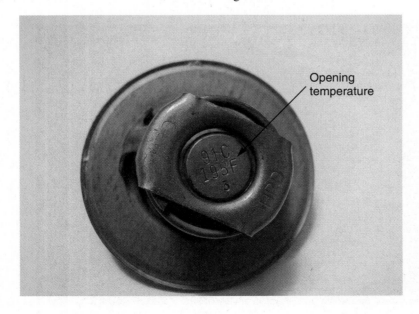

Job 36—Remove and Replace a Thermostat (continued)

16. Install the thermostat gasket on the engine.

17. Place the thermostat housing on the engine and install the housing fasteners.

18. Tighten the thermostat housing fasteners in an alternating pattern to prevent distortion. Be careful not to overtighten the fasteners. Overtightening the housing can cause distortion and leaks, and may break the housing.

19. Refill the system with the proper kind of coolant, following the procedures in Job 32. Many modern vehicles use special long-life coolant. Do not mix coolant types.

20. Start the engine and check for leaks and proper thermostat operation.

21. Clean up any coolant spills.

22. Clean the work area, including coolant spills, and return any equipment to storage.

23. Did you encounter any problems during this procedure? Yes ____ No ____

    If Yes, describe the problems: _____

    _____

    What did you do to correct the problems? _____

    _____

24. Have your instructor check your work and sign this job sheet.

Performance Evaluation—Instructor Use Only

Did the student complete the job in the time allotted? Yes ____ No ____

If No, which steps were not completed?_____

How would you rate this student's overall performance on this job?_____

5–Excellent, 4–Good, 3–Satisfactory, 2–Unsatisfactory, 1–Poor

Comments: _____

_____

_____

INSTRUCTOR'S SIGNATURE _____

Notes

Project
10
Inspecting, Replacing, and Aligning Powertrain Mounts

Introduction

A broken or misaligned powertrain mount can cause several problems. Misaligned powertrain mounts can cause noises, vibration or pulling on acceleration. A broken or misaligned powertrain mount can cause an engine or drive train part to strike the body. You will inspect powertrain mounts in Job 37, and then replace and align them in Job 38.

Project 10 Jobs

- Job 37—Inspect Powertrain Mounts
- Job 38—Replace and Align Powertrain Mounts

Tools and Materials

The following list contains the tools and materials that may be needed to complete the jobs in this project. The items used will depend on the make and model of vehicle being serviced.

- One or more vehicles needing powertrain mount service.
- Applicable service information.
- Pry bars.
- Engine lifting fixture (not always needed).
- Droplight or other source of illumination.
- Hand tools as needed.
- Safety glasses and other protective equipment as needed.

Safety Notice

Before performing this job, review all pertinent safety information in the text, and review safety information with your instructor.

Notes

186

Job 37—Inspect Powertrain Mounts

After completing this job, you will be able to check powertrain mounts on front-wheel and rear-wheel drive vehicles.

Procedures

☐ 1. Obtain a vehicle to be used in this job. Your instructor may direct you to perform this job on a shop vehicle.

☐ 2. Gather the tools needed to perform the following job. Refer to the project's tools and materials list.

Preliminary Check

☐ 1. Open the hood of the vehicle.

☐ 2. Apply the parking brake.

☐ 3. Have an assistant start the engine and tightly press on the brake pedal while placing the engine in drive.

☐ 4. Have the assistant slightly open the throttle.

> **Note**
> If the engine rises more than 1″ (25.4 mm) in the following steps, a motor mount may be broken or soft. If this occurs, make a visual inspection of the mounts using the procedure presented later in this job.

☐ 5. Observe the engine as the throttle is opened with the brakes applied.
   Did the engine move excessively? Yes ___ No ___
   If Yes, perform a visual inspection of the motor mounts.

☐ 6. Continue to observe the engine as you have your assistant place the transmission in reverse and slightly open the throttle while keeping the brakes engaged.
   Did the engine move excessively? Yes ___ No ___
   If Yes, perform a visual inspection of the motor mounts.

Visual Inspection

☐ 1. Make sure that there is sufficient illumination to see the mount. If necessary, have an assistant hold a light in position to illuminate the mount. Mount locations vary. **Figure 37-1** and **Figure 37-2** show the general location of motor and drive train mounts on front-wheel and rear-wheel drive vehicles.

☐ 2. Carefully pry on the engine to raise it in the area of a solid mount. Check for cracked rubber as you move the mount.
   Did you find any cracked rubber in the mount? Yes ___ No ___
   If Yes, replace the mount using the procedures outlined in Job 38.

☐ 3. Check fluid-filled mounts for exterior oil leakage.
   Did you detect any oil leakage? Yes ___ No ___
   If Yes, replace the mount using the procedures outlined in Job 38.

Figure 37-1. Location of mounts on a rear-wheel drive vehicle. Most rear-wheel drive vehicles will have mounts in these general locations.

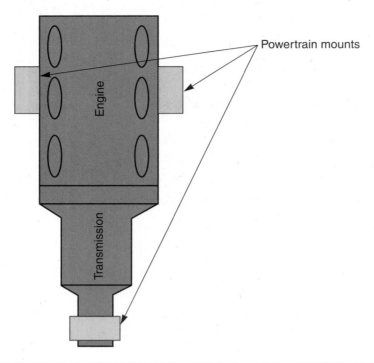

Figure 37-2. Location of mounts on a front-wheel drive vehicle. Mount locations on front-wheel drive vehicles vary between makers, and even between engines from the same maker. Always check the service literature for the exact mount locations.

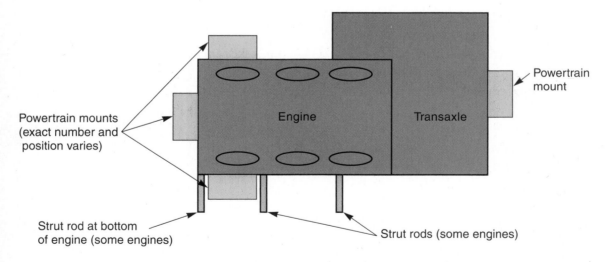

4. Check the mounts for loose fasteners.

   Did you find any loose fasteners? Yes ___ No ___

   If Yes, tighten the fasteners to the proper torque.

5. Clean the work area and return any equipment to storage.

Job 37—Inspect Powertrain Mounts (continued)

☐   6.   Did you encounter any problems during this procedure? Yes ___ No ___

       If Yes, describe the problems: _____

       _____

       What did you do to correct the problems? _____

       _____

☐   7.   Have your instructor check your work and sign this job sheet.

Performance Evaluation—Instructor Use Only

Did the student complete the job in the time allotted? Yes ___ No ___

If No, which steps were not completed?_____

How would you rate this student's overall performance on this job?_____

5–Excellent, 4–Good, 3–Satisfactory, 2–Unsatisfactory, 1–Poor

Comments: _____

_____

_____

INSTRUCTOR'S SIGNATURE _____

Notes

_____

_____

_____

_____

_____

_____

_____

_____

_____

_____

_____

_____

_____

_____

_____

_____

_____

_____

_____

_____

_____

_____

_____

_____

_____

_____

_____

_____

_____

_____

_____

_____

_____

_____

_____

_____

_____

_____

_____

_____

_____

_____

Job 38—Replace and Align Powertrain Mounts

After completing this job, you will be able to replace and align powertrain mounts on front-wheel drive and rear-wheel drive vehicles.

Procedures

☐ 1. Obtain a vehicle to be used in this job. Your instructor may direct you to perform this job on a shop vehicle.

☐ 2. Gather the tools needed to perform the following job. Refer to the project's tools and materials list.

Replace Powertrain Mounts

☐ 1. Raise the vehicle as necessary to gain access to the powertrain mount to be replaced.

> **Warning**
> ⚠ The vehicle must be raised and supported in a safe manner. Always use approved lifts or jacks and jack stands.

☐ 2. Place a jack stand under the engine or transmission/transaxle near the powertrain mount to be replaced.

> **Caution**
> ◇ Place the jack stand so that it will not damage the underside of the engine or transmission/transaxle. If necessary place a wood block or other protective device on the jack stand.

> **Note**
> 📄 On some front-wheel drive vehicles, an engine-lifting device can be used to raise the engine and remove pressure on the mounts.

☐ 3. Lower the vehicle just enough to remove engine and drive train weight from the mount to be replaced.

☐ 4. Lightly shake the vehicle to ensure that the jack stand is solidly placed.

☐ 5. Remove the fasteners holding the mount to the engine and vehicle.

☐ 6. Compare the old and new mounts to ensure that the new mount is correct.

☐ 7. Place the new mount in position and install the fasteners loosely.

☐ 8. Once all fasteners are in place, tighten them to the proper torque.

☐ 9. Raise the vehicle and remove the jack stand.

> **Note**
> 📄 If necessary, align the new powertrain mount using the following procedure.

Adjust Powertrain Mounts on Rear-Wheel Drive Vehicles

> **Note**
> Many rear-wheel drive powertrain mounts are not adjustable. Some rear mounts can be shimmed to correct drive shaft angle problems. Refer to Job 62.

☐ 1. Loosen the mounting bolts.

☐ 2. Move the tailshaft until it is centered in the drive shaft tunnel.

☐ 3. Retighten the mounting bolts.

Adjust Powertrain Mounts on Front-Wheel Drive Vehicles

☐ 1. Observe the marks made by the cradle mounting bolts and washers. If the original marks made by the washers are not completely covered, the cradle is incorrectly positioned.

Is the cradle properly positioned? Yes ___ No ___

> **Note**
> If the cradle is correctly positioned, skip steps 2 through 4.

☐ 2. Loosen the cradle fasteners.

☐ 3. Move the cradle to its original position.

☐ 4. Tighten the cradle fasteners.

☐ 5. Measure the mount adjustment. On many vehicles mount adjustment is checked by measuring the length of the CV axle on the same side as the adjustable mount. If the length of the CV axle is correct, the mount is properly adjusted. On other vehicles the difference between the mount and a stationary part of the vehicle frame must be measured. Sometimes the only procedure is to loosen the fasteners and allow the mount to assume its normal unloaded position.

Briefly describe the mount adjustment method: _____

_____

_____

_____

Is the mount adjustment correct? Yes ___ No ___.

If Yes, skip steps 6 through 10, if your instructor approves.

☐ 6. Remove the load on the mount by raising the engine and transaxle assembly with a floor jack.

> **Caution**
> Raise the assembly only enough to unload the mount. Be careful not to damage the oil pan.

☐ 7. Loosen the adjustable mount fasteners.

☐ 8. Reposition the mount as needed.

☐ 9. Measure the mount using the same method used in step 5.

Is the mount adjustment correct? Yes ___ No ___

If No, repeat steps 8 and 9.

Job 38—Replace and Align Powertrain Mounts (continued)

☐ 10. Tighten the mount fasteners.

What is the proper fastener torque? _____

☐ 11. Remove the jack from under the engine and transaxle.

☐ 12. Clean the work area and return any equipment to storage.

☐ 13. Did you encounter any problems during this procedure? Yes ___ No ___

If Yes, describe the problems: _____

_____

What did you do to correct the problems? _____

_____

☐ 14. Have your instructor check your work and sign this job sheet.

Performance Evaluation—Instructor Use Only

Did the student complete the job in the time allotted? Yes ___ No ___

If No, which steps were not completed?_____

How would you rate this student's overall performance on this job?_____

5–Excellent, 4–Good, 3–Satisfactory, 2–Unsatisfactory, 1–Poor

Comments: _____

_____

_____

INSTRUCTOR'S SIGNATURE _____

Notes

Project

11

Diagnosing Transmission and Transaxle Problems

Introduction

Automatic transmissions and transaxles are complex units and it may seem that they are impossible to diagnose. Automatic transmissions and transaxles, however, operate according to well-defined mechanical, hydraulic, and electrical/electronic principles. If these principles are understood, that knowledge can be put to use to determine the problem. The key to successfully diagnosing automatic transmission and transaxle problems is to use logical troubleshooting processes and avoid guesswork.

Pressure tests are often used to determine whether an automatic transmission or transaxle has an internal problem. Pressures can pinpoint internal leaks, a defective pump, or sticking valves. The circuit pressures that can be measured depend on the type and number of pressure taps provided by the manufacturer.

In Job 39, you will diagnose transmission and transaxle problems using the seven-step diagnostic process. In Job 40, you will measure transmission/transaxle pressures and determine what these pressures mean.

Project 11 Jobs

- Job 39—Use the Seven-Step Diagnostic Process to Identify Transmission/Transaxle Problems
- Job 40—Perform Transmission/Transaxle Pressure Tests

Tools and Materials

The following list contains the tools and materials that may be needed to complete the jobs in this project. The items used will depend on the make and model of vehicle being serviced.

- Vehicle with a potential automatic transmission or transaxle problem.
- Applicable transmission or transaxle service information.
- One or more transmission pressure gauges.
- Transmission fluid of the correct type.
- Hand tools as needed.
- Safety glasses and other protective equipment as necessary.

195

Safety Notice

Before performing this job, review all pertinent safety information in the text, and review safety information with your instructor.

Job 39—Use the Seven-Step Diagnostic Process to Identify Transmission/Transaxle Problems

After completing this job, you will be able to apply the seven-step diagnostic process to diagnose transmission and transaxle problems.

Procedures

☐ 1. Obtain a vehicle to be used in this job. Your instructor may direct you to perform this job on a shop vehicle.

☐ 2. Gather the tools needed to perform the following job. Refer to the project's tools and materials list.

☐ 3. Identify the type of transmission/transaxle being serviced.

Is it a conventional transmission? Yes _____ No _____

Is it a transaxle? Yes _____ No _____

Is it a CVT? Yes _____ No _____

Is it a hybrid vehicle transmission? Yes _____ No _____

☐ 4. Briefly describe the operational characteristics of the transmission or transaxle being serviced.

_____

_____

_____

_____

_____

_____

Applying the Seven-Step Diagnostic Process

Note

The following procedures make use of a seven-step diagnostic process. The steps of the process are shown in *italics*.

Step 1—Determine the Exact Problem

☐ 1. Identify the apparent problem by interviewing the vehicle driver.

Note

Step 1 can be completed by interviewing a fellow student. Your instructor may present you with a possible problem as part of his instructions for this job.

Step 2—Check for Obvious Problems

☐ 2. Observe the transmission/transaxle area and all other vehicle components for the following:
- Disconnected, grounded, or shorted wires.
- Corroded or overheated electrical connectors.
- Loose or missing ground wires.
- Blown fuses or melted fusible links.
- Disconnected throttle linkage.
- Loose, damaged, or disconnected vacuum lines.
- Fluid leaks. Refer to Job 65.

Were any obvious problems found? Yes ___ No ___

If Yes, could the observed problems be the cause of the transmission/transaxle problem? Yes ___ No ___

☐ 3. Start the engine and allow it to reach operating temperature.

☐ 4. Shift the transmission/transaxle through all ranges.

☐ 5. Remove the transmission/transaxle dipstick and note the following:

Fluid level: Normal ___ High ___ Low ___

Fluid condition: Normal color and smell ___ Overheated ___ Burned ___ Debris in fluid ___ Other (explain): _____

Does fluid level and condition indicate a problem? Yes ___ No ___

If Yes, could this be the cause of the transmission/transaxle problem? Yes ___ No ___

Note

If the transmission/transaxle is not equipped with a dipstick, follow the manufacturer's instructions for checking the fluid level.

☐ 6. Add fluid, if needed, before proceeding to the next step.

Step 3—Determine Which Component or System Is Causing the Problem

☐ 7. Obtain the proper service literature and record the specified upshift and downshift ranges in the chart for step 8.

☐ 8. Road test the vehicle and note shift points. List the shift points in the following chart.

Shift Chart

Shift	Shift range per service manual	Actual shift point
1–2	_____	_____
2–3	_____	_____
3–4	_____	_____
4–3	_____	_____
3–2	_____	_____
2–1	_____	_____

Is the transmission/transaxle shifting within the proper ranges? Yes ___ No ___

How do the shifts feel? Firm ___ Harsh ___ Soft and slipping ___

If any problems are noted, what could be the cause? _____

_____

Job 39—Use the Seven-Step Diagnostic Process to Identify Transmission/Transaxle Problems (continued)

☐ 9. Test the lockup converter application by driving the vehicle with a scan tool or tachometer attached. When the vehicle reaches the speed at which the lockup clutch would apply, check that the engine speed (rpm) drops slightly. Do not confuse the rpm drop of a lockup clutch with the rpm drop caused by upshifts. When the throttle opening is the same, upshifts cause a larger rpm drop than that caused by lockup clutch application.

> **Caution**
>
> ⚠️ A scan tool may indicate lockup clutch application when the clutch members are worn and unable to hold. Always back up the scan tool indication by checking for an rpm drop.

What gear and speed were specified in the service manual? _____

In what gear and at what speed did the lockup clutch actually apply? _____

Does the lockup converter apply at the right time? Yes ___ No ___

If any problems are noted, what could be the cause? _____

_____

☐ 10. Obtain the band and clutch application charts corresponding to the transmission/transaxle being diagnosed.

☐ 11. Use the band and clutch application chart to determine which holding members are applied when the problem occurs.

List the holding members that are applied when the problem occurs:_____

_____

_____

Could a slipping holding member be the cause of the problem? Yes ___ No ___

☐ 12. Attach a scan tool to the data link (diagnostic) connector. Refer to Job 11 as needed.

☐ 13. Retrieve any trouble codes.

☐ 14. Road test the vehicle and note:
- The status of the transmission/transaxle electronic components at different speeds.
- Whether shifts occur at the proper times and whether related electronic components activate or deactivate properly.
- Whether the converter clutch applies at the proper time and in the proper gear(s).

Are the electronic components operating properly? Yes ___ No ___

Could a defective electronic control system component be the cause of the problem? Yes ___ No ___

Step 4—Eliminate Causes of the Problem

☐ 15. Make throttle linkage checks and adjustments. Follow the procedures outlined in Job 45.

☐ 16. Make band adjustments where possible, following the procedures in Job 45.

☐ 17. Road test the vehicle and check transmission/transaxle shift points and shift quality.

Did the linkage/band adjustments correct the problem? Yes ___ No ___

☐ 18. Check transmission/transaxle hydraulic components by performing a pressure test. Refer to Job 40.

Are the pressures satisfactory? Yes ___ No ___

Could a hydraulic system defect be the cause of the problem? Yes ___ No ___

☐ 19. If specified by the manufacturer, perform a stall test. Follow manufacturer's specifications exactly and *do not* exceed the time allowed to make the stall test.

Were the results of the stall test satisfactory? Yes ___ No ___

If No, what did the results of a stall test reveal?_____

_____

_____

_____

☐ 20. Make sure that the problem is not caused by the engine or other parts of the drivetrain. Does the engine have any of the following problems?

Is there restricted exhaust (Refer to Job 8 for testing procedures)? Yes _____ No _____

Is there an ignition defect causing low power or engine miss? Yes _____ No _____

Is there a fuel system defect causing excessively rich or lean mixture? Yes _____ No _____

Is low engine compression on one or more cylinders proper? Yes _____ No _____

Is there binding in other part of drivetrain? Yes _____ No _____

Is a defective cooling system causing transmission or transaxle to overheat?
Yes _____ No _____

Is there a battery or charging system defect affecting electronic controls?
Yes _____ No _____

Does it operate at extreme high speed? Yes _____ No _____

Is there a towing or other excessive weight vehicle? Yes _____ No _____

If Yes, describe the problem(s):_____

_____

_____

Step 5—Isolate and Recheck the Cause of the Problem

☐ 21. Repeat any (or all) of the above steps until you are certain that you have isolated the cause of the problem.

Step 6—Correct the Defect

☐ 22. After consulting with your instructor, make transmission/transaxle repairs as necessary.

What repairs were made? _____

_____

_____

Step 7—Recheck System Operation

☐ 23. Road test the vehicle to ensure that the transmission/transaxle problem has been corrected.

☐ 24. Clean the work area and return any equipment to storage.

Job 39—Use the Seven-Step Diagnostic Process to Identify Transmission/ Transaxle Problems (continued)

☐ 25. Did you encounter any problems during this procedure? Yes ___ No ___

If Yes, describe the problems: _____

_____

What did you do to correct the problems? _____

_____

☐ 26. Have your instructor check your work and sign this job sheet.

Performance Evaluation—Instructor Use Only

Did the student complete the job in the time allotted? Yes ___ No ___

If No, which steps were not completed?_____

How would you rate this student's overall performance on this job?_____

5–Excellent, 4–Good, 3–Satisfactory, 2–Unsatisfactory, 1–Poor

Comments: _____

_____

_____

INSTRUCTOR'S SIGNATURE _____

Notes

_____

_____

_____

_____

_____

_____

_____

_____

_____

_____

_____

_____

_____

_____

_____

_____

_____

_____

_____

_____

_____

_____

_____

_____

_____

_____

_____

_____

_____

_____

_____

_____

_____

_____

_____

_____

_____

_____

_____

_____

_____

Job 40—Perform Transmission/Transaxle Pressure Tests

After completing this job, you will be able to use pressure gauges and service literature to determine whether transmission or transaxle hydraulic pressures are correct.

Procedures

☐ 1. Obtain a vehicle to be used in this job. Your instructor may direct you to perform this job on a shop vehicle.

☐ 2. Gather the tools needed to perform the following job. Refer to the project's tools and materials list.

☐ 3. Obtain service information for the transmission/transaxle in the vehicle selected in step 1.

   Title of literature:_____

☐ 4. Using the table of contents or the index, locate the chapter or section listing transmission or transaxle hydraulic pressures. Use the spaces provided to list all of the types of pressures that can be tested, conditions under which the pressures are measured, and the acceptable ranges for those pressures.

Type of Pressure (line, governor, other)	Selector Lever Position	Engine RPM (when applicable)	Modulator Vacuum	Pressure Range (indicate psi or kPa)
_____	_____	_____	_____	_____
_____	_____	_____	_____	_____
_____	_____	_____	_____	_____
_____	_____	_____	_____	_____
_____	_____	_____	_____	_____
_____	_____	_____	_____	_____
_____	_____	_____	_____	_____
_____	_____	_____	_____	_____

☐ 5. Safely raise and secure the vehicle.

Note

If shift solenoid operation or governor pressure will be checked, make sure that the drive wheels are free to turn and that all safety precautions for operating a vehicle on the lift are observed.

☐ 6. Locate the transmission or transaxle hydraulic pressure taps. See **Figure 40-1**.

☐ 7. Obtain the needed hydraulic pressure gauge or gauges.

☐ 8. Place a drip pan under the transmission or transaxle and remove the line pressure tap plug.

☐ 9. Attach the hydraulic pressure gauges to the transmission/transaxle line pressure tap. Use the appropriate service information to locate the line pressure tap.

Caution

Position the pressure gauge hoses away from moving parts and the exhaust system.

Figure 40-1. Pressure taps on one kind of transaxle. Most automatic transmissions and transaxles do not have this many pressure taps. (DaimlerChrysler)

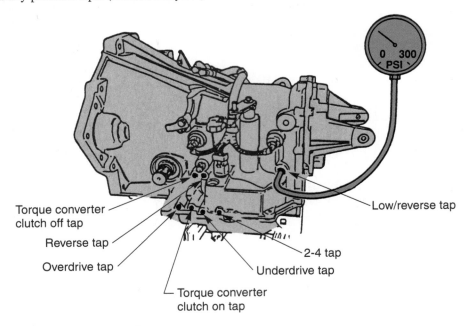

10. Start the engine and observe the line pressure in various gears. If applicable, use a vacuum pump to apply varying pressure to the vacuum modulator and record the line pressure at various vacuum levels. Record the pressures and testing conditions in the following chart.

> **Note**
>
> It is usually not necessary to drive the vehicle in order to measure line pressures. However, on older vehicles with governors, the vehicle may need to be driven. If so, record the speed in the following chart.

	Selector Lever Position	Engine RPM	Vehicle Speed (when applicable)	Modulator Vacuum Line Pressure (indicate psi or kPa)
Park	_____	_____	_____	_____
Reverse	_____	_____	_____	_____
Neutral	_____	_____	_____	_____
Drive	_____	_____	_____	_____
Manual Fourth*	_____	_____	_____	_____
Manual Third*	_____	_____	_____	_____
Manual Second	_____	_____	_____	_____
Manual First	_____	_____	_____	_____

* If applicable

11. Stop the engine.

12. Remove the pressure gauge from the line pressure tap and reinstall the pressure tap plug.

13. Repeat steps 8 through 12 by attaching the hydraulic pressure gauges to other available transmission/transaxle pressure taps.

14. Observe and record pressures as indicated. Vary the engine speed, gear selector position, vacuum, and other factors as needed. Refer to the service information if necessary. Record the pressures and other readings in the following chart.

Job 40—Perform Transmission/Transaxle Pressure Tests (continued)

> **Caution**
> Always stop the engine before removing the pressure gauge or any pressure taps.

Type of Pressure (as described in service literature)	Selector Lever Position	Engine RPM	Modulator Vacuum or Vehicle Speed (when applicable)	Pressure Range (indicate psi or kPa)
_____	_____	_____	_____	_____
_____	_____	_____	_____	_____
_____	_____	_____	_____	_____
_____	_____	_____	_____	_____
_____	_____	_____	_____	_____
_____	_____	_____	_____	_____
_____	_____	_____	_____	_____

☐ 15. Replace all pressure taps and ensure that they do not leak.

☐ 16. Lower the vehicle and remove it from the shop.

☐ 17. Compare the readings with the specifications listed in step 4.

Do all actual pressures agree with the published pressure specifications? Yes ___ No ___

If Yes, skip to step 18. If No, list the differences in the spaces provided.

Type of pressure:_____ Too high ___ Too low ___

Type of pressure:_____ Too high ___ Too low ___

Type of pressure:_____ Too high ___ Too low ___

Type of pressure:_____ Too high ___ Too low ___

State your opinion as to the possible causes of incorrect pressures. _____

_____

_____

_____

_____

_____

_____

_____

_____

_____

☐ 18. Return the service manual to your instructor.

☐ 19. Return the pressure gauge(s) to storage.

☐ 20. Clean the work area and return any equipment to storage.

☐ 21. Did you encounter any problems during this procedure? Yes ___ No ___

If Yes, describe the problems: _____

_____

What did you do to correct the problems? _____

_____

☐ 22. Have your instructor check your work and sign this job sheet.

Performance Evaluation—Instructor Use Only

Did the student complete the job in the time allotted? Yes ____ No ____

If No, which steps were not completed?_____

How would you rate this student's overall performance on this job?_____

5–Excellent, 4–Good, 3–Satisfactory, 2–Unsatisfactory, 1–Poor

Comments: _____

_____

_____

INSTRUCTOR'S SIGNATURE _____

Project

12

Servicing a Transaxle Final Drive

Introduction

The final drive is an important part of every transaxle. The final drive serves the same function as the rear axle assembly installed on a rear-wheel drive vehicle. There are three types of transaxle final drives: planetary gear, helical gear, and hypoid gear. You will service a planetary-type final drive in Job 41. In Job 42, you will service a helical gear final drive. You will service the remaining type of final drive, a hypoid gear final drive, in Job 43.

Project 12 Jobs

- Job 41—Service a Planetary-Type Final Drive
- Job 42—Service a Helical Gear Final Drive
- Job 43—Service a Hypoid Gear Final Drive

Tools and Materials

The following list contains the tools and materials that may be needed to complete the jobs in this project. The items used will depend on the make and model of vehicle being serviced.

- One or more transaxle final drives.
- Applicable service information.
- Dial indicator.
- Feeler gauges.
- Correct final drive lubricant.
- Drain pan.
- Hand tools as needed.
- Air-powered tools as needed.
- Safety glasses and other protective equipment as needed.

Safety Notice

Before performing this job, review all pertinent safety information in the text, and review safety information with your instructor.

Notes

Job 41—Service a Planetary-Type Final Drive

After completing this job, you will be able to service a planetary-type transaxle final drive.

Procedures

> **Caution**
>
> It is very important that all metal and sludge be removed from the final drive parts and the case. Movement of new lubricant will throw contaminants onto the rebuilt drive components.

☐ 1. Obtain a vehicle to be used in this job. Your instructor may direct you to perform this job on a shop vehicle.

☐ 2. Gather the tools needed to perform the following job. Refer to the project's tools and materials list.

> **Note**
>
> If the final drive can only be reached by removing and disassembling the transaxle or transmission, refer to Jobs 50 and 52 for automatic transmissions or Jobs 51 and 55 for automatic transaxles.

☐ 3. Remove the CV axle on the final-drive side of the transaxle. Refer to Job 67.

Which CV axle was removed? Right ___ Left ___

☐ 4. Remove the speedometer gear, governor, or speed sensor as necessary.

☐ 5. Remove the bolts holding the extension housing to the transaxle case.

☐ 6. Place a drip pan under the extension housing and remove the housing.

☐ 7. Remove the clip holding the final drive to the transaxle output shaft.

☐ 8. Pull the final drive assembly from the transaxle.

☐ 9. Place the final drive assembly on a clean bench, **Figure 41-1**.

☐ 10. Clean all parts.

☐ 11. Remove the differential side gears and spider gears.

☐ 12. Check the final drive sun, planet, and ring gears for wear and damage.

Do any parts show signs of wear? Yes ___ No ___

If Yes, is the wear serious enough to require replacement? Check with your instructor if you are unsure. Yes ___ No ___

Do any parts show signs of damage? Yes ___ No ___

List any parts requiring replacement: _____

_____

_____

_____

_____

☐ 13. Replace defective planet carrier parts as necessary.

☐ 14. Check the clearance between the planet gears and the final drive carrier and check the pinion gears for smooth operation.

Figure 41-1. A typical planetary-type final drive assembly is shown here. Note the planet gears used to produce the final drive ratio and the differential gears that allow the vehicle to turn.

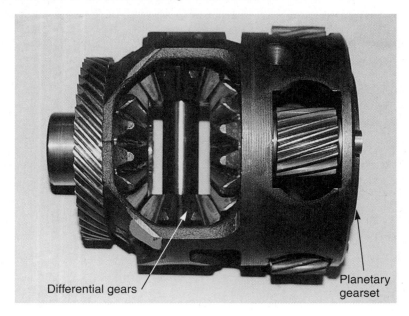

Differential gears

Planetary gearset

15. Check the speedometer drive gear for damage. This is especially important if the speedometer driven gear shows signs of damage.

16. If necessary, remove the speedometer drive gear with a suitable puller.

17. Heat the replacement speedometer drive gear in an oven or by immersing it in hot water.

18. Lightly tap the speedometer drive gear into place.

19. Install the differential side gears, being sure that the washers are in position.

20. Roll the spider gears and their related washers into position.

21. Reinstall the spider gear shaft and install a new shaft-to-case roll pin.

22. Turn the internal parts of the differential assembly to ensure that it operates smoothly.

23. Recheck the clearance between the planet gears and the final drive carrier.

24. If clearance is excessive, replace the thrust washers between the gears and carrier.

Note

Thicker service washers may be available for reducing clearance.

25. Remove all metal and debris from the interior of the extension housing.

Note

If the transaxle does not have a removable extension housing, reinstall the final drive along with the other transaxle internal parts as explained in Jobs 57 and 59 for automatic transaxles or Jobs 79 and 80 for manual transaxles.

26. Install the final drive assembly over the output shaft.

27. Install the retaining clip that holds the final drive to the shaft.

28. Install the extension housing using a new O-ring or gasket.

Job 41—Service a Planetary-Type Final Drive (continued)

☐ 29. Install the speedometer gear, governor, or speed sensor as applicable using new seals.

☐ 30. If necessary, install a new CV axle seal and then reinstall the CV axle.

☐ 31. After all parts have been installed, check the transaxle fluid level and add fluid as necessary.

☐ 32. Clean the work area and return any equipment to storage.

☐ 33. Did you encounter any problems during this procedure? Yes ___ No ___

If Yes, describe the problems: _____

_____

What did you do to correct the problems? _____

_____

☐ 34. Have your instructor check your work and sign this job sheet.

Performance Evaluation—Instructor Use Only

Did the student complete the job in the time allotted? Yes ___ No ___

If No, which steps were not completed?_____

How would you rate this student's overall performance on this job?_____

5–Excellent, 4–Good, 3–Satisfactory, 2–Unsatisfactory, 1–Poor

Comments: _____

_____

_____

INSTRUCTOR'S SIGNATURE _____

Notes

Job 42—Service a Helical Gear Final Drive

After completing this job, you will be able to service a helical gear transaxle final drive.

Procedures

> **Caution**
>
> It is very important that all metal and sludge be removed from the final drive parts and the case. Movement of new lubricant will throw contaminants onto the rebuilt drive components.

☐ 1. Obtain a vehicle to be used in this job. Your instructor may direct you to perform this job on a shop vehicle.

☐ 2. Gather the tools needed to perform the following job. Refer to the project's tools and materials list.

> **Note**
>
> This procedure can usually be performed without removing the transaxle from the vehicle. Consult the proper service manual before beginning disassembly.

☐ 3. Remove the CV axle on the final drive side of the transaxle. Refer to Job 67. On some transaxles, it may be necessary to remove both CV axles.

Which axles were removed? Right ___ Left ___ Both ___

☐ 4. Remove the speedometer gear, governor, or speed sensor as necessary.

☐ 5. Place a drip pan under the final drive cover (differential cover).

☐ 6. Remove the cover fasteners and lightly pry the cover away from the transaxle housing.

☐ 7. Allow oil to drain from the final drive housing.

☐ 8. Remove the differential bearing retainer bolt and bearing retainers.

☐ 9. Remove the extension housing if necessary to remove the final drive and differential assembly.

> **Note**
>
> To remove the final drive pinion gear, it may be necessary to remove the transfer gear that attaches to the pinion gear shaft. Some transfer shaft gears have selective thickness shims that should be saved for reuse.

☐ 10. Remove the final drive and differential assembly. A typical final drive and differential assembly is shown in **Figure 42-1**.

☐ 11. Place the final drive assembly on a clean bench.

☐ 12. Clean all parts thoroughly and clean the interior of the final drive housing.

☐ 13. Remove the pin holding the spider gear shaft in place and remove the shaft, spider gears and side gears.

Figure 42-1. A common helical gear and differential assembly is shown here.

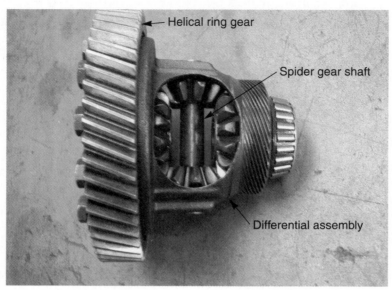

Helical ring gear

Spider gear shaft

Differential assembly

14. Inspect all parts for wear and damage.

Do any of the parts have signs of wear? Yes ___ No ___

If Yes, is the wear serious enough to require replacement? Yes ___ No ___

Do any parts show signs of damage? Yes ___ No ___

List any parts requiring replacement. Check with your instructor if you are unsure. _____

_____

_____

_____

15. Obtain new parts as needed. Clean and lubricate all parts that are to be reused.

> **Note**
>
> Differential gears should be replaced as a set.

16. Install the side gears, and then roll the spider gears into position.

17. Make sure that all spider and side gear washers are in position, then install the spider gear shaft.

18. Install a new shaft-to-case roll pin. A threaded retaining pin can be reused.

19. Once the differential is reassembled, check that it operates smoothly.

20. If the ring gear must be replaced, remove the bolts holding the ring gear to the differential carrier.

21. Remove the ring gear from the carrier. Some ring gears will slide from the carrier, but most will require pressing or light hammering to remove.

> **Caution**
>
> When removing the ring gear with a hammer, hammer evenly around the outside of the ring gear to avoid distorting the carrier.

22. Heat the new ring gear with a heat gun or by immersing it in boiling water.

Job 42—Service a Helical Gear Final Drive (continued)

☐ 23. Lay the ring gear on a flat surface and quickly place the differential carrier over it.

> **Caution**
> ⚠ Be sure to align the differential carrier and ring gear bolt holes.

☐ 24. Install and tighten the ring gear to carrier bolts. Tighten in an alternating pattern to avoid distorting the ring gear.

> **Note**
> 📄 Helical gear final drive service often involves replacing the tapered roller bearings. Refer to Job 5 for bearing replacement procedures.

☐ 25. Install the helical gear and carrier assembly in the case.

> **Caution**
> ⚠ Be sure to properly preload the tapered roller bearings following the manufacturer's bearing preloading procedures.

☐ 26. After the final drive is reinstalled in the case, check endplay. Change shims or adjust the threaded bearing retainers as necessary to obtain the proper endplay reading.

☐ 27. Replace the CV axle seals if necessary and install the CV axles.

☐ 28. Scrape the old gasket material from the differential and final drive assembly cover.

☐ 29. Install the cover using a new gasket.

☐ 30. Fill the final drive with the proper kind of fluid. Some helical final drives are lubricated with the same transmission fluid that lubricates the transmission hydraulic system. Other final drives are lubricated with manual transmission or hypoid axle gear oil. Always check to determine what kind of oil should be used.

What kind of fluid did you use to refill the unit? _____

> **Caution**
> ⚠ If the final drive uses a lubricant other than transmission fluid, the seals separating the transmission and final drive must be replaced to prevent cross contamination of either fluid.

☐ 31. Clean the work area and return any equipment to storage.

☐ 32. Did you encounter any problems during this procedure? Yes ___ No ___

If Yes, describe the problems: _____

_____

What did you do to correct the problems? _____

_____

☐ 33. Have your instructor check your work and sign this job sheet.

Performance Evaluation—Instructor Use Only

Did the student complete the job in the time allotted? Yes ___ No ___

If No, which steps were not completed?_____

How would you rate this student's overall performance on this job?_____

5–Excellent, 4–Good, 3–Satisfactory, 2–Unsatisfactory, 1–Poor

Comments: _____

_____

_____

INSTRUCTOR'S SIGNATURE _____

Job 43—Service a Hypoid Gear Final Drive

After completing this job, you will be able to service a hypoid gear final drive.

Procedures

> **Caution**
>
> It is very important that all metal and sludge be removed from the final drive parts and the case. Movement of new lubricant will throw contaminants onto the rebuilt drive components.

☐ 1. Obtain a vehicle to be used in this job. Your instructor may direct you to perform this job on a shop vehicle.

☐ 2. Gather the tools needed to perform the following job. Refer to the project's tools and materials list.

> **Note**
>
> Hypoid gear final drives are used on only a few transaxles. **Figure 43-1** illustrates the most common type of transaxle that uses a hypoid gear final drive.

☐ 3. Remove the CV axles.

☐ 4. Locate the side cover.

☐ 5. Obtain a drip pan and place it under the side cover.

☐ 6. Loosen the cover-to-transaxle fasteners and slightly pry the cover away from the transaxle.

☐ 7. Allow oil to drain from the final drive housing.

☐ 8. Finish removing the fasteners and remove the cover from the transaxle.

☐ 9. Remove the axle bearing retainers as necessary.

☐ 10. Pull the ring gear and differential assembly from the transaxle case.

Figure 43-1. The most common transaxle using a hypoid gear final drive is shown here.

☐ 11. Place the ring gear and differential assembly on a clean bench.

> **Note**
>
> For hypoid ring and pinion gear disassembly, reassembly, and checking procedures, refer to Jobs 87 and 88. Hypoid ring and pinion gears should always be replaced as a set.

☐ 12. Remove the spider gear shaft locking pin and remove the spider gear shaft.

☐ 13. Remove the spider gears and side gears.

☐ 14. Inspect all parts for wear and damage.

Do any parts show wear? Yes ___ No ___

If Yes, is the wear serious enough to require replacement? Check with your instructor if you are unsure. Yes ___ No ___

Do any parts show signs of damage? Yes ___ No ___

List any parts requiring replacement: _____

_____

_____

☐ 15. Obtain new parts as needed, clean and lubricate all parts to be reused.

☐ 16. Reassemble the side gears, spider gears, and shaft. Make sure that all washers are in position.

☐ 17. Install the locking pin.

> **Note**
>
> Tapered roller bearings are often replaced as part of hypoid gear final drive service. Refer to Job 5.

☐ 18. Check the differential assembly for smooth operation.

☐ 19. Reinstall the differential assembly in the case.

☐ 20. Install the final drive cover using a new gasket.

☐ 21. Install the CV axles.

☐ 22. Fill the final drive with the proper weight of hypoid axle gear oil.

What weight of oil did you use to refill the unit?_____

> **Caution**
>
> All hypoid gear final drives are lubricated with hypoid axle gear oil. Be sure to use the proper weight oil. Do not reuse the old oil.

☐ 23. Clean the work area and return any equipment to storage.

☐ 24. Did you encounter any problems during this procedure? Yes ___ No ___

If Yes, describe the problems: _____

_____

What did you do to correct the problems? _____

_____

☐ 25. Have your instructor check your work and sign this job sheet.

Job 43—Service a Hypoid Gear Final Drive (continued)

Performance Evaluation—Instructor Use Only

Did the student complete the job in the time allotted? Yes ___ No ___

If No, which steps were not completed?_____

How would you rate this student's overall performance on this job?_____

5–Excellent, 4–Good, 3–Satisfactory, 2–Unsatisfactory, 1–Poor

Comments: _____

_____

_____

INSTRUCTOR'S SIGNATURE _____

Notes

Project

13

Servicing an Automatic Transmission or Transaxle

Introduction

As technology advances, automatic transmissions and transaxles require less frequent fluid and filter changes. However, it is still a job that is performed regularly. Vehicles used in heavy-duty service still require frequent fluid and filter changes. A fluid and filter change is relatively simple, but must be done correctly or the unit will be damaged. In Job 44, you will change the fluid and filter on an automatic transmission or transaxle.

Although a vehicle computer controls most operations in a modern automatic transmission or transaxle, there are still some manual adjustment points. The most common is the manual linkage between the shifter and the case. Older vehicles will have throttle linkage that controls the shift points. Most older and a few newer transmissions and transaxles have bands that should be periodically adjusted. These jobs must be done correctly or severe damage can occur. In Job 45, you will adjust linkage and bands on an automatic transmission or transaxle.

Many automatic transmission and transaxle components can be serviced without removing the unit from the vehicle. These include the extension housing bushing and seals, the oil cooler and lines, speedometer drive and driven gears, speed sensors, and the vacuum modulator and governor when used. In the last four jobs of this project, you will service a variety of transmission/transaxle components while the transmission or transaxle is still installed in the vehicle. You will service a transmission/transaxle oil cooler and lines in Job 46, speedometer drive and driven gears in Job 47, speed sensors in Job 48, and a vacuum modulator and governor in Job 49.

Project 13 Jobs

- Job 44—Change Transmission/Transaxle Oil and Filter
- Job 45—Adjust Transmission Linkage and Bands and Neutral Safety Switch/Range Switch
- Job 46—Service an Oil Cooler and Lines
- Job 47—Service Speedometer Drive and Driven Gears
- Job 48—Service Electronic and Electrical Components
- Job 49—Service Vacuum Modulators and Governors

Tools and Materials

The following list contains the tools and materials that may be needed to complete the jobs in this project. The items used will depend on the make and model of vehicle being serviced.

- Vehicle in need of automatic transmission or transaxle service.
- Hydraulic lift (or jacks and jack stands).
- Applicable service information.
- Transmission pressure gauge(s).
- Oil cooler and line flusher (optional).
- Linkage adjustment tools and gauges.
- Band adjustment tools and gauges.
- Drain pan.
- Replacement filter and gasket.
- Replacement gaskets.
- Gasket sealer.
- Specified automatic transmission fluid.
- Hand tools as needed.
- Air-powered tools as needed.
- Safety glasses and other protective equipment as needed.

Safety Notice

Before performing this job, review all pertinent safety information in the text, and review safety information with your instructor.

Job 44—Change Transmission/Transaxle Oil and Filter

After completing this job, you will be able to change the oil and filter on an automatic transmission or transaxle.

Procedures

☐ 1. Obtain a vehicle to be used in this job. Your instructor may direct you to perform this job on a shop vehicle.

☐ 2. Gather the tools needed to perform the following job. Refer to the project's tools and materials list.

Preliminary Checks

☐ 1. Locate the transmission/transaxle dipstick and check the fluid level as specified by the vehicle manufacturer.

> **Note**
>
> Some newer vehicles do not have a dipstick. Check the owner's manual or service literature for checking procedures. Some vehicles without a dipstick require the use of a scan tool to check oil level. Other vehicles are checked by noting whether the fluid wets the bottom of the filler plug's opening threads.

☐ 2. Check the fluid condition by examining the fluid on the dipstick, if used.

☐ 3. Add fluid if needed. Transmissions without a dipstick are usually refilled through a filler plug.

> **Caution**
>
> Be sure to add the transmission fluid specified by the vehicle manufacturer.

☐ 4. Road test the vehicle and determine whether a problem is evident at this point.

☐ 5. Based on the results of the road test, determine if the transmission/transaxle needs further checking or if a fluid and filter change is all that is required.

Change Transmission/Transaxle Oil and Filter

☐ 1. Raise the vehicle.

> **Warning**
>
> The vehicle must be raised and supported in a safe manner. Always use approved lifts or jacks and jack stands.

☐ 2. Position a drain pan at the rear of the transmission/transaxle oil pan.

☐ 3. Remove all but the front four transmission/transaxle oil pan attaching screws.

☐ 4. Pry on the pan to loosen the gasket and allow the pan to drop at the rear.

> **Warning**
>
> Transmission fluid may begin to drain at this point. If the fluid is hot, be careful to avoid contacting it because it may cause burns.

5. Loosen the front four screws to allow the pan to drop further and the fluid to drain.

6. When the fluid stops draining, remove the four front screws and remove the pan.

7. Check the bottom of the pan for excessive amounts of sludge or metal, **Figure 44-1**.

   Inspect the contents of the pan, and place a check mark next to each type of problem found:
   - Sludge buildup greater than 1/4″ (6.4 mm). ___
   - Varnish buildup (baked on fluid resembling household varnish). ___
   - Many metal particles. ___

> **Note**
>
> Consult your instructor if excessive sludge, varnish, or metal is found. The transmission may require further repair.

8. Remove the old transmission oil filter. Some filters are held in place with clips, while others are attached with threaded fasteners.

> **Note**
>
> Allow the old transmission fluid to drain from the valve body before installing the new filter. This prevents the old contaminated fluid from soaking the new filter.

9. If the torque converter has drain plugs, remove them and allow fluid to drain from the converter into the drip pan.

10. Install the new transmission oil filter in the correct position.

11. Reinstall the torque converter drain plugs, if applicable.

12. Check the oil pan for warping or dents. Straighten or replace the pan as needed.

13. Scrape all old gasket material from the transmission or transaxle oil pan and the case.

> **Note**
>
> Some new vehicles have a reusable gasket that does not have to be replaced. Check the service literature to determine whether your unit has this type of gasket.

Figure 44-1. After removing the oil pan, check for sludge, varnish, or metal particles.

Job 44—Change Transmission/Transaxle Oil and Filter (continued)

☐ 14. Install a new gasket on the transmission oil pan, using a light coat of gasket sealer on the pan side of the gasket.

☐ 15. Correctly position the oil pan on the transmission case.

☐ 16. Install the transmission oil pan attaching fasteners and torque them to specifications. If the pan has a tightening sequence, follow it, otherwise start at the middle of each side of the pan and work toward the corners.

☐ 17. Lower the vehicle.

☐ 18. Refill the transmission with the proper type of fluid, being careful not to overfill.

What type of fluid did you use? _____

> **Note**
>
> Transmissions or transaxles without dipsticks may require special filling procedures. Some units have a fluid level sensor in the case, and fluid level can only be read with a scan tool. On other transmissions or transaxles, the fluid level is read by noting whether the fluid wets the bottom threads of the fill plug, similar to the checking procedure for a manual transmission or rear axle. Check the manufacturer's service literature for exact procedures.

☐ 19. Start the vehicle and shift through all gear positions.

☐ 20. Recheck the fluid level according to the manufacturer's instructions.

☐ 21. Check for fluid leaks from the transmission oil pan.

☐ 22. Road test the vehicle, carefully observing transmission operation.

☐ 23. Recheck the fluid level and add fluid if necessary.

☐ 24. Raise the vehicle and inspect the transmission or transaxle for leaks.

> **Note**
>
> See Job 65 for detailed procedures for the different methods of leak detection. The procedures in that job can be used to check automatic transmissions and transaxles for leaks.

☐ 25. Clean the work area and return any equipment to storage.

☐ 26. Did you encounter any problems during this procedure? Yes ___ No ___

If Yes, describe the problems: _____

_____

What did you do to correct the problems? _____

_____

☐ 27. Have your instructor check your work and sign this job sheet.

Performance Evaluation—Instructor Use Only

Did the student complete the job in the time allotted? Yes ___ No ___

If No, which steps were not completed?_____

How would you rate this student's overall performance on this job?_____

5–Excellent, 4–Good, 3–Satisfactory, 2–Unsatisfactory, 1–Poor

Comments: _____

_____

_____

INSTRUCTOR'S SIGNATURE _____

Job 45—Adjust Transmission Linkage and Bands and Neutral Safety Switch/Range Switch

After completing this job, you will be able to perform linkage and band adjustments on an automatic transmission or transaxle.

Procedures

> **Note**
>
> There are many variations in automatic transmission design. The following procedure is to be used as a general guide only. Always consult the manufacturer's service manual for exact linkage and band adjustment procedures and specifications.

☐ 1. Obtain a vehicle to be used in this job. Your instructor may direct you to perform this job on a shop vehicle.

☐ 2. Gather the tools needed to perform the following job. Refer to the project's tools and materials list.

Adjust Manual Shift Linkage

☐ 1. Make sure the transmission shift selector lever is in Park or another position as specified by the manufacturer.

☐ 2. Raise the vehicle in a safe manner.

> **Note**
>
> The shift selector indicator adjuster is located under the hood on some transaxle-equipped vehicles. On other vehicles, the shift selector indicator adjusters are located under the vehicle instrument panel or in the center console. It is unnecessary to raise these vehicles, **Figure 45-1**.

Figure 45-1. This manual linkage adjuster is located near the air cleaner in the engine compartment. Other linkage adjusters can only be reached by raising the vehicle.

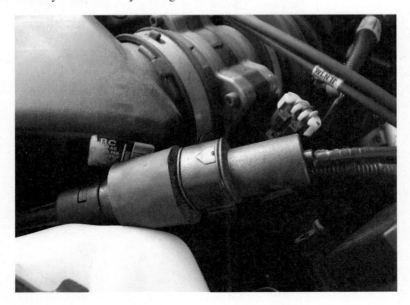

3. Locate the adjuster lock button or lock nut.

Where is the adjuster locking device located? _____

What type of locking device is used? _____

4. Pull the lock button up or loosen the lock nut.

5. Use the outer manual lever on the transmission case to place the transmission or transaxle in the same gear as the shift selector lever.

6. Push the lock button down or tighten the adjuster lock nut.

7. Recheck the shift selector lever operation and indicator position.

8. Lower the vehicle and road test it, carefully observing transmission operation.

Adjust Throttle Linkage and TV Linkage

1. With the engine off, have an assistant press the accelerator completely to the floor.

2. Observe the throttle plate(s).

Are they open completely? Yes ___ No ___

If No, adjust the throttle linkage (linkage between the accelerator pedal and the throttle plates).

> **Note**
>
> Adjustment procedure will vary, depending on the manufacturer. **Figure 45-2** shows the throttle and TV linkage adjusters on one vehicle. Check that carpeting or debris is not keeping the pedal from traveling to the fully open position.

3. Loosen the adjuster locking device of the TV linkage (linkage between the throttle body or carburetor and the transmission), as your assistant continues to hold the accelerator to the floor.

Figure 45-2. On this vehicle, the throttle and TV linkage adjusters are located at the throttle valve and can be reached from under the hood.

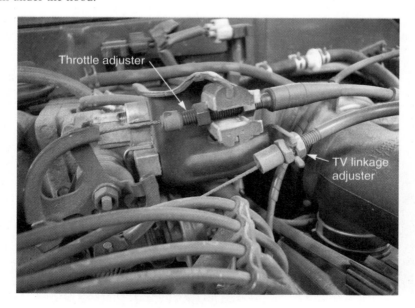

Job 45—Adjust Transmission Linkage and Bands and Neutral Safety Switch/Range Switch (continued)

> **Note**
>
> Most TV cable adjustments can be accomplished without an assistant. Also, some throttle cables are self-adjusting. Pressing the accelerator to the floor completes the adjustment.

☐ 4. Pull the TV linkage to the wide-open throttle position.

☐ 5. Tighten the locking device.

☐ 6. Road test the vehicle and observe transmission/transaxle operation.

Adjust Transmission Bands

> **Note**
>
> Since very few modern transmissions and transaxles have band adjustment procedures, your instructor may want you to skip this part of the job.

☐ 1. Raise the vehicle.

> **Warning**
>
> The vehicle must be raised and supported in a safe manner. Always use approved lifts or jacks and jack stands.

☐ 2. Loosen the locknut holding the band adjustment screw.

> **Note**
>
> Some transmissions have more than one band adjustment screw, as in **Figure 45-3**. Determine which screw adjusts which band. If the adjuster is inside the oil pan, the pan must be removed first. See Job 44.

Figure 45-3. This transmission has two band adjustment screws on the left side. Be sure that you know which band is being adjusted, and whether specifications for the bands are different.

☐ 3. Tighten the band adjustment screw to the torque specified in the service manual.
What is the specified torque value? _____

☐ 4. Back off the band adjustment screw the exact number of turns specified in the service manual.
How many turns did you back off the adjusting screw?_____

☐ 5. Hold the band adjustment screw stationary and tighten the locknut.

☐ 6. Lower and road test the vehicle, carefully observing transmission/transaxle operation.
Did the band adjustment appear to improve shift quality? Yes ___ No ___

Adjust Neutral Safety Switch/Mechanically Operated Range Switch

☐ 1. Place the transmission in neutral, apply the parking brake, and block the drive wheels.

☐ 2. Determine the neutral safety switch/range switch problem:
Does it crank with the shift selector in Park and Neutral? Yes _____ No _____
Does it crank in with the shift selector in any gear other than Park and Neutral?
Yes _____ No _____

☐ 3. Locate the neutral safety switch/range switch on the vehicle. The switch could be installed on the steering column, linkage, or at the transmission manual shaft.

☐ 4. Loosen the switch fasteners.

☐ 5. Move the switch in small amounts until the starter cranks in Park and Neutral and does not crank in any other gear.

☐ 6. Tighten the switch fasteners and recheck operation.

☐ 7. Clean the work area and return any equipment to storage.

☐ 8. Did you encounter any problems during this procedure? Yes ___ No ___
If Yes, describe the problems: _____
_____
What did you do to correct the problems? _____
_____

☐ 9. Have your instructor check your work and sign this job sheet.

Performance Evaluation—Instructor Use Only

Did the student complete the job in the time allotted? Yes ___ No ___
If No, which steps were not completed?_____
How would you rate this student's overall performance on this job?_____
5–Excellent, 4–Good, 3–Satisfactory, 2–Unsatisfactory, 1–Poor
Comments: _____
_____
_____

INSTRUCTOR'S SIGNATURE _____

Job 46—Service an Oil Cooler and Lines

After completing this job, you will be able to service an oil cooler and oil cooler lines.

Procedures

☐ 1. Obtain a vehicle to be used in this job. Your instructor may direct you to perform this job on a shop vehicle.

☐ 2. Gather the tools needed to perform the following job. Refer to the project's tools and materials list.

☐ 3. Visually inspect the oil cooler lines and connections for leaks and kinked or flattened sections.

☐ 4. If the cooler is an external or add-on design, visually inspect it for leaks and accumulations of dirt and other debris.

Check an In-Radiator Transmission Oil Cooler for Leaks

☐ 1. Remove pressure from the vehicle's cooling system.

☐ 2. Remove the radiator cap or coolant recovery system cap.

☐ 3. Visually check for transmission fluid in the cooling system. Transmission fluid will rise to the top of the system and be easily visible. Transmission fluid will also form a ring on the interior of the coolant recovery system reservoir.

☐ 4. Remove the transmission or transaxle dipstick.

☐ 5. Inspect the transmission fluid. If it looks milky, water or coolant has entered the transmission or transaxle.

☐ 6. If the fluid appears milky, confirm that the vehicle has not been driven through deep water.

☐ 7. If transmission fluid is present in the cooling system or if water is present in the transmission or transaxle, the oil cooler is leaking and should be replaced. See Job 35 for radiator replacement procedures.

Flush an Oil Cooler and Cooler Lines Using a Siphon Blowgun

☐ 1. Remove both cooler lines at the transmission or transaxle.

☐ 2. Attach a flexible drain hose to the cooler inlet (transmission/transaxle outlet) line and place the other end of the hose in a used oil container.

☐ 3. Use air pressure at the cooler return line to blow out as much transmission fluid as possible from the cooler lines.

☐ 4. Prepare the flushing solution according to manufacturer's directions.

☐ 5. Place the siphon hose of the blowgun in the container of flushing solution.

☐ 6. Attach the air hose to the blowgun.

☐ 7. Set air pressure as necessary. Do not exceed the recommended air pressure, as this may cause the cooler to rupture.

☐ 8. Place the gun nozzle on the cooler inlet line and operate the gun to allow the entire contents of the solution container to enter the cooler inlet line.

☐ 9. Allow the flushing solution to soak the lines and cooler for the time recommended by the flushing solution manufacturer.

10. After the proper time has elapsed, blow the flushing solution out with compressed air. Try to remove as much of the flushing solution as possible, as excessive amounts could damage the transmission components.

> **Note**
>
> Skip the following steps if the transmission or transaxle has been removed from the vehicle for repairs.

11. Reattach the cooler inlet line to the transmission or transaxle and place the drain hose on the outlet line.
12. Start the engine and allow it to idle for 30 seconds. This will allow fluid to flow through the cooler into the used oil container, removing any remaining flushing solution.
13. Reinstall the cooler outlet line on the transmission or transaxle.
14. Start the engine and check the cooler line fittings for leaks.
15. Check the fluid level and add fluid as necessary.

Flush an Oil Cooler and Cooler Lines Using a Pressurized Container Flushing Device

1. Remove both cooler lines at the transmission or transaxle.
2. Attach a flexible hose to the cooler inlet (transmission/transaxle outlet) line and place the other end of the hose in a container for used oil.
3. Use air pressure to blow as much transmission fluid as possible out of the cooler lines.
4. Prepare the flushing solution to be used according to manufacturer's directions.
5. Pour the flushing solution into the container.
6. Install the container lid being sure that it makes an airtight seal.
7. Attach the air hose to the container.

> **Note**
>
> Some pressurized container flushing devices have a self-contained pump and motor and air hoses are not needed. Plug the motor into a grounded electrical outlet.

8. Set air pressure as necessary. Do not exceed the recommended air pressure as this may cause the cooler to rupture.
9. Attach the container liquid line to the cooler inlet line.
10. Open the container valve and allow the entire contents of the container to enter the lines and cooler.
11. Allow the flushing solution to soak the lines and cooler for the time recommended by the flushing solution manufacturer.
12. After the proper time has elapsed, blow the flushing solution out with compressed air. Try to remove as much of the flushing solution as possible, as excessive amounts could damage the transmission components.

> **Note**
>
> Skip the following steps if the transmission or transaxle has been removed from the vehicle for repairs.

Job 46—Service an Oil Cooler and Lines (continued)

☐ 13. Reattach the cooler inlet line to the transmission or transaxle and place the flexible hose on the outlet line.

☐ 14. Start the engine and allow it to idle for 30 seconds. This will allow fluid to flow through the cooler into the used oil container, removing any remaining flushing solution.

☐ 15. Reattach the cooler inlet line.

☐ 16. Start the engine and check the cooler line fittings for leaks.

☐ 17. Check the fluid level and add fluid as necessary.

Flush an Oil Cooler and Cooler Lines Using an Aerosol Flushing Can

☐ 1. Remove both cooler lines at the transmission or transaxle.

☐ 2. Attach a flexible hose to the cooler inlet (transmission/transaxle outlet) line and place the other end of the hose in a used oil container.

☐ 3. Use air pressure at the cooler outlet line to blow as much transmission fluid as possible out of the cooler lines.

☐ 4. Attach the aerosol flushing can hose to the cooler outlet line.

☐ 5. Depress the can valve and allow the entire contents of the can to enter the lines and cooler.

☐ 6. Allow the flushing solution to soak the lines and cooler for the time recommended on the flushing can.

☐ 7. After the proper time has elapsed, detach the flushing can and blow the flushing solution out with compressed air. Try to remove as much of the flushing solution as possible, as excessive amounts could damage the transmission components.

> **Note**
> Skip the following steps if the transmission or transaxle has been removed from the vehicle for repairs.

☐ 8. Reattach the cooler inlet line to the transmission or transaxle and place the flexible hose on the outlet line.

☐ 9. Start the engine and allow it to idle for 30 seconds. This will allow fluid to flow through the cooler into the used oil container, removing any remaining flushing solution.

☐ 10. Reattach the cooler inlet line.

☐ 11. Start the engine and check the cooler line fittings for leaks.

☐ 12. Check the fluid level and add fluid as necessary.

Replace a Damaged or Leaking Oil Cooler Line

☐ 1. Place a drip pan under the oil cooler line fittings.

☐ 2. Loosen and remove the fittings. Use a line wrench to turn the male fitting at the end of the line while using a backup wrench to hold the fitting that is threaded into the transmission or transaxle. This technique helps prevent twisting and damage of the cooler line.

☐ 3. Remove any clips holding the cooler line to the vehicle frame.

☐ 4. Remove the cooler line.

☐ 5. Place the new cooler line in position. Tight clearances often require that the line be snaked around other vehicle components. Be careful not to damage the new line.

☐ 6. Install and tighten the line fittings.

☐ 7. Install any line clips that were removed.

☐ 8. Start the engine and add fluid as necessary.

☐ 9. Check the new line and fittings for leaks.

Replace a Damaged or Leaking External Oil Cooler

☐ 1. Place a drip pan under the oil cooler fittings.

☐ 2. Loosen and remove the fittings. Use a line wrench to turn the male fitting at the end of the line and a backup wrench to hold the fitting threaded into the cooler. This technique helps avoid twisting the cooler lines.

☐ 3. Remove the fasteners holding the cooler to the front of the vehicle.

☐ 4. Remove the cooler.

☐ 5. Place the new cooler in position.

☐ 6. Loosely install the cooler line fittings.

☐ 7. Install and tighten the cooler fasteners.

☐ 8. Tighten the cooler line fittings.

☐ 9. Start the engine and add fluid as necessary.

☐ 10. Check the cooler fittings for leaks.

☐ 11. Clean the work area and return any equipment to storage.

☐ 12. Did you encounter any problems during this procedure? Yes ___ No ___

If Yes, describe the problems: _____

_____

What did you do to correct the problems? _____

_____

☐ 13. Have your instructor check your work and sign this job sheet.

Performance Evaluation—Instructor Use Only

Did the student complete the job in the time allotted? Yes ___ No ___

If No, which steps were not completed?_____

How would you rate this student's overall performance on this job?_____

5–Excellent, 4–Good, 3–Satisfactory, 2–Unsatisfactory, 1–Poor

Comments: _____

_____

_____

INSTRUCTOR'S SIGNATURE _____

Job 47—Service Speedometer Drive and Driven Gears

After completing this job, you will be able to remove, inspect, and replace speedometer drive gears and driven gears.

Procedures

- ☐ 1. Obtain a vehicle to be used in this job. Your instructor may direct you to perform this job on a shop vehicle.

- ☐ 2. Gather the tools needed to perform the following job. Refer to the project's tools and materials list.

- ☐ 3. Remove the speedometer cable.

- ☐ 4. Remove the speedometer gear retainer hold-down bolt.

- ☐ 5. Pull the speedometer gear from the housing.

- ☐ 6. Inspect the speedometer driven gear for the following:
    - Damaged or worn teeth. ___
    - Damaged bushing. ___
    - Loose cable fit. ___
    - Leaking seal(s). ___
    Is the speedometer driven gear suitable for reuse? Yes ___ No ___

- ☐ 7. Reinstall the speedometer gear if the following steps will not be performed immediately.

- ☐ 8. To remove the speedometer drive gear, begin by removing the extension housing.

> **Note**
>
> On many automatic transaxles, the speedometer drive gear is part of the final drive assembly. Refer to the appropriate job in Project 12 for specific repair procedures.

- ☐ 9. Depress the clip holding the speedometer drive gear to the output shaft.

- ☐ 10. Slide the gear from the output shaft.

- ☐ 11. Inspect the speedometer drive gear for the following defects:
    - Damaged or worn teeth. ___
    - Damaged bushing. ___
    - Loose fit on the output shaft. ___
    Is the speedometer drive gear suitable for reuse? Yes ___ No ___

- ☐ 12. Ensure that the retaining clip is in place in the mating hole of the output shaft.

- ☐ 13. Slide the drive gear over the output shaft until it engages the clip.

- ☐ 14. Reinstall the extension housing.

- ☐ 15. Clean the work area and return any equipment to storage.

- ☐ 16. Did you encounter any problems during this procedure? Yes ___ No ___

    If Yes, describe the problems: _____

    _____

    What did you do to correct the problems? _____

    _____

- ☐ 17. Have your instructor check your work and sign this job sheet.

Performance Evaluation—Instructor Use Only

Did the student complete the job in the time allotted? Yes ____ No ____

If No, which steps were not completed?_____

How would you rate this student's overall performance on this job?_____

5–Excellent, 4–Good, 3–Satisfactory, 2–Unsatisfactory, 1–Poor

Comments: _____

_____

_____

INSTRUCTOR'S SIGNATURE _____

Job 48—Service Electronic and Electrical Components

After completing this job, you will be able to remove, inspect, and install speed sensors, pressure sensors, hydraulic control solenoids, and electronic control modules (ECMs).

Procedures

☐ 1. Obtain a vehicle to be used in this job. Your instructor may direct you to perform this job on a shop vehicle.

☐ 2. Gather the tools needed to perform the following job. Refer to the project's tools and materials list.

Service Speed Sensor

☐ 1. Locate the speed sensor.

> **Note**
>
> There are two general types of speed sensors, **Figure 48-1**. One type extends into the transmission case and obtains a signal from a sensor wheel attached to one of the rotating shafts. The other design is self-contained and is driven by a cable from a gear turned by the output shaft. Some older speed sensors are part of the governor assembly.

☐ 2. Remove the speed sensor electrical connector.

☐ 3. Remove the sensor retainer hold-down bolt.

☐ 4. Pull the sensor from the housing.

☐ 5. Check the sensor for metal particles at the sensor tip.

   Were there metal particles on the sensor tip? Yes ___ No ___

☐ 6. Use an ohmmeter to check the sensor for continuity.

   Does the sensor have continuity? Yes ___ No ___

   Is the sensor suitable for reuse? Yes ___ No ___

Figure 48-1. Two common types of speed sensors are shown here. A—This speed sensor pickup extends into the transmission case or extension housing and picks up a signal from a rotating wheel mounted on an input or output shaft. B—This type of speed sensor contains both the rotating wheel and the pickup in a single unit. The shaft meshes with and is turned by a gear inside of the transmission or transaxle.

Sensor
pickup

A

B

☐ 7. Reinstall the speed sensor, or install a replacement sensor if the original sensor is defective.

Service Pressure Sensor

☐ 1. Remove the vehicle oil pan.

> **Note**
>
> A few transmissions and transaxles have case-mounted pressure sensors, and it may not be necessary to remove the oil pan to service the sensor.

☐ 2. Locate the pressure sensor. There may be more than one sensor, so be sure you have the correct one.

☐ 3. Remove the pressure control solenoid electrical connector. Mark the connector if there is any chance of reinstalling the connector in the wrong position.

☐ 4. Using the correct size wrench, remove the pressure sensor.

☐ 5. Use an ohmmeter to check the sensor for continuity.

> **Note**
>
> The sensor may or may not have continuity, depending on its design. Check the manufacturer's specifications to ensure that the sensor reading is correct.

Does the sensor have continuity? Yes _____ No _____

Is the sensor suitable for reuse? Yes _____ No _____

☐ 6. Reinstall the pressure sensor, or install a replacement sensor if the original sensor is defective.

☐ 7. If necessary, reinstall the oil pan and refill the transmission.

☐ 8. Check oil level as explained in Job 44.

Service Pressure Control Solenoid

☐ 1. Remove the vehicle oil pan.

> **Note**
>
> A few transmissions and transaxles have case-mounted pressure control solenoids, and it may not be necessary to remove the oil pan to service the solenoid.

☐ 2. Locate the pressure control solenoid. There may be more than one solenoid, so be sure you have the correct one.

☐ 3. Remove the pressure control solenoid electrical connector. Mark the connector if there is any chance of reinstalling the connector in the wrong position.

☐ 4. Using the correct size wrench, remove the pressure control solenoid.

☐ 5. Use an ohmmeter to check the solenoid winding resistance.

☐ 6. Compare the resistance against the manufacturer's specifications.

Does the solenoid have the proper resistance? Yes _____ No _____

Is the solenoid suitable for reuse? Yes _____ No _____

☐ 7. Reinstall the pressure control solenoid, or install a replacement solenoid if the original solenoid is defective.

☐ 8. If necessary, reinstall the oil pan and refill the transmission with the proper type of fluid.

☐ 9. Check oil level as explained in Job 44.

Job 48—Service Electronic and Electrical Components (continued)

Service Electronic Control Module

> **Note**
>
> This procedure assumes the electronic control module is defective. Most modules can only be checked with a scan tool while installed on the vehicle.

☐ 1. Locate the electronic control module. Modules may be located in the side kick panel, the engine compartment, under the instrument panel, or on the side of the transmission or transaxle.

> **Caution**
>
> Be sure you have located the proper control module. Modern vehicles often have several modules to control the engine, transmission or transaxle, and body/chassis electrical components.

☐ 2. Remove the fasteners holding the control module to the vehicle.

☐ 3. Remove the control module electrical connectors.

☐ 4. Remove the control module from the vehicle.

> **Caution**
>
> Before installing the new module, be sure any shorted component that could damage the module has been replaced.

☐ 5. Install the electrical connectors on the control module.

☐ 6. Place the control module in position on the vehicle.

☐ 7. Install and tighten the control module fasteners.

☐ 8. Clean the work area and return any equipment to storage.

☐ 9. Did you encounter any problems during this procedure? Yes ___ No ___
   If Yes, describe the problems: _____
   _____
   What did you do to correct the problems? _____
   _____

☐ 10. Have your instructor check your work and sign this job sheet.

Performance Evaluation—Instructor Use Only

Did the student complete the job in the time allotted? Yes ___ No ___
If No, which steps were not completed?_____
How would you rate this student's overall performance on this job?_____
5–Excellent, 4–Good, 3–Satisfactory, 2–Unsatisfactory, 1–Poor
Comments: _____
_____
_____

INSTRUCTOR'S SIGNATURE _____

Notes

Job 49—Service Vacuum Modulators and Governors

After completing this job, you will be able to remove, inspect, and replace governors and vacuum modulators.

Procedures

☐ 1. Obtain a vehicle to be used in this job. Your instructor may direct you to perform this job on a shop vehicle.

☐ 2. Gather the tools needed to perform the following job. Refer to the project's tools and materials list.

Service Vacuum Modulators

☐ 1. Remove the vacuum line and note whether it contains transmission fluid. Transmission fluid in the vacuum line indicates that the vacuum modulator is leaking. Do not confuse transmission fluid with condensed gasoline or water vapor.

Is transmission fluid present? Yes ___ No ___

☐ 2. Remove the vacuum modulator from the case. Most vacuum modulators are held to the case by a retainer and bolt. A few are screwed directly into the case.

☐ 3. Invert the vacuum modulator and note whether transmission fluid drips from the vacuum-hose side. Do not confuse transmission fluid with condensed gasoline or water vapor.

Is transmission fluid present? Yes ___ No ___

☐ 4. Use a vacuum pump to pull a vacuum on the modulator, **Figure 49-1**.

Does the modulator hold vacuum? Yes ___ No ___

☐ 5. Reinstall the modulator into the case. If the modulator is held by a retainer and bolt, use a new O-ring. If the modulator is threaded into the case, use a new gasket.

☐ 6. Reattach the vacuum line.

Figure 49-1. Use a vacuum pump to apply vacuum to the modulator. Vacuum should hold and the valve plunger, if visible, should move.

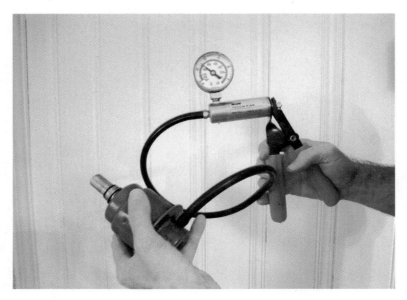

7. If the modulator is adjustable, locate the adjusting screw inside of the vacuum nipple.

8. Locate the shift speed range in the service literature.

9. Adjust the screw until shifts take place within the range specified by the service literature. In most cases, setting the shift on the high side of the range will increase holding member life.

Service Governors

1. Remove the governor cover fasteners or snap ring and slide the governor from the case.

   or

   Remove the extension housing and remove the fasteners holding the governor to the output shaft. Remove the speedometer gear if necessary and slide the governor from the output shaft.

2. Check that the governor is not damaged or worn and that all parts move freely, **Figure 49-2**.

   Are any defects noted? Yes ___ No ___

   If Yes, describe the defects: _____

   _____

   Do all parts move freely? Yes ___ No ___

3. Lightly lubricate the governor and slide it into the case. Replace the governor cover. Install and tighten the fasteners or install the snap ring.

   or

   Slide the governor over the output shaft and install and tighten the fasteners. Reinstall the speedometer gear if necessary and reinstall the extension housing.

4. Clean the work area and return any equipment to storage.

5. Did you encounter any problems during this procedure? Yes ___ No ___

   If Yes, describe the problems: _____

   _____

   What did you do to correct the problems? _____

   _____

6. Have your instructor check your work and sign this job sheet.

Figure 49-2. Check for governor wear or damage and make sure that all moving parts operate easily. Notice that this governor also contains the trigger wheel for a speed sensor.

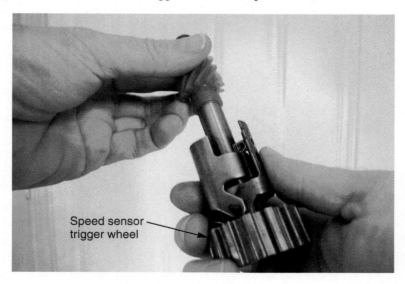

Speed sensor trigger wheel

Job 49—Service Vacuum Modulators and Governors (continued)

Performance Evaluation—Instructor Use Only

Did the student complete the job in the time allotted? Yes ___ No ___

If No, which steps were not completed?_____

How would you rate this student's overall performance on this job?_____

5–Excellent, 4–Good, 3–Satisfactory, 2–Unsatisfactory, 1–Poor

Comments: _____

_____

_____

INSTRUCTOR'S SIGNATURE _____

Notes

Project

Removing an Automatic Transmission or Transaxle

14

Introduction

Automatic transmission removal and installation is relatively simple. Automatic transmissions are always heavy and precautions must be taken to avoid damage and injury. In Job 50, you will remove an automatic transmission from a vehicle.

Automatic transaxle removal is also a relatively simple procedure. Unlike rear-wheel drive transmission removal, transaxle removal requires that the engine be raised slightly to keep it from falling out when the mounts are removed. Precautions must be taken to avoid damage and injury. In Job 51, you will remove an automatic transaxle from a vehicle.

Project 14 Jobs

- Job 50—Remove an Automatic Transmission
- Job 51—Remove an Automatic Transaxle

Tools and Materials

The following list contains the tools and materials that may be needed to complete the jobs in this project. The items used will depend on the make and model of vehicle being serviced.

- Vehicle with an automatic transmission requiring removal.
- Vehicle with an automatic transaxle requiring removal.
- Applicable service information.
- Hydraulic lift (or jacks and jack stands).
- Engine holding fixture.
- Drain pan.
- Flywheel turning tool.
- Transmission jack.
- Torque converter holding tool (or C-clamp).
- Linkage adjustment tools and gauges.
- Specified automatic transmission fluid.
- Hand tools as needed.
- Air-operated tools as needed.
- Safety glasses and other protective equipment as needed.

Safety Notice

Before performing this job, review all pertinent safety information in the text, and review safety information with your instructor.

Notes

Job 50—Remove an Automatic Transmission

After completing this job, you will be able to properly and safely remove an automatic transmission from a vehicle.

Procedures

> **Note**
>
> The following procedure is a general guide only. This procedure assumes that the engine remains installed and only the transmission will be removed. Always refer to the manufacturer's service manual for specific procedures.

☐ 1. Obtain a vehicle to be used in this job. Your instructor may direct you to perform this job on a shop vehicle.

☐ 2. Gather the tools needed to perform the following job. Refer to the project's tools and materials list.

☐ 3. Disconnect the vehicle's negative battery cable.

☐ 4. Remove the upper converter housing–to-engine attaching bolts, if access is easier than from underneath the vehicle.

> **Note**
>
> It may be necessary to install an engine holding fixture before proceeding.

☐ 5. Raise the vehicle.

> **Warning**
>
> ⚠ Raise and support the vehicle in a safe manner. Always use approved lifts or jacks and jack stands.

☐ 6. Inspect the engine mounts to ensure that they are not broken. Refer to Job 37 as needed. Broken engine mounts can cause the engine to fall backward once the rear cross member is removed.

☐ 7. Drain the transmission fluid.

☐ 8. Mark the drive shaft U-joint and rear axle flange to ensure that the drive shaft is reinstalled in the same position as it was in originally.

☐ 9. Remove the drive shaft and cap the rear of the transmission to reduce oil dripping.

☐ 10. Position a drain pan to catch any fluid that spills from the oil cooler lines.

☐ 11. Disconnect the oil cooler lines at the transmission. Use two wrenches to avoid twisting the cooler lines.

☐ 12. If necessary, remove the exhaust pipes to allow transmission removal.

☐ 13. Disconnect the following components at the transmission:
- Manual shift linkage.
- Speedometer cable.
- Oil filler tube.
- Electrical connectors.
- Throttle or kickdown linkage (if used).
- Modulator vacuum line (if used).

☐ 14. Remove the torque converter inspection cover or dust cover.

☐ 15. Remove the converter-to-flywheel attaching bolts or nuts.

> **Note**
>
> The flywheel will have to be turned to gain access to all of the attaching bolts. This can be done with a special turning tool, a screwdriver, or by operating the starter with a remote switch.

☐ 16. If an engine holding fixture is not being used, support the engine with a jack placed under the rear of the engine.

> **Warning**
>
> ⚠ Never let the engine hang, suspended by only the front engine mounts. Failure to support the engine before removing the transmission can result in serious damage or injury.

☐ 17. Place a transmission jack under the transmission. Raise the transmission slightly to remove weight from the cross member.

> **Caution**
>
> Always install the safety chain to ensure that the transmission does not fall from the transmission jack. See **Figure 50-1**.

☐ 18. Remove the parking brake linkage from the transmission cross member.

☐ 19. Remove the fasteners holding the cross member to the vehicle frame and to the transmission mount.

☐ 20. Remove the cross member.

☐ 21. If the starter motor is bolted to the converter housing, remove it and position it out of the way.

Figure 50-1. Always install the safety chain on the transmission to keep the transmission from falling to the floor.

Job 50—Remove an Automatic Transmission (continued)

- [ ] 22. Remove the remaining converter housing attaching bolts holding the transmission to the engine.
- [ ] 23. Slide the transmission straight back, away from the engine.
- [ ] 24. Secure the torque converter with a small C-clamp or holding tool.

> **Warning**
> ⚠ If the torque converter is not secured, it could slide from the transmission and fall on the floor, possibly causing injury or damage.

- [ ] 25. Lower the transmission from the vehicle.

Inspect Engine Core Plugs, Rear Crankshaft Seal, Dowel Pins, Dowel Pin Holes, and Mating Surfaces

1. Once the transmission has been removed, inspect the following parts at the rear of the engine block:

   Do the core plugs have corrosion? Yes _____ No _____

   Do the core plugs have leaks? Yes _____ No _____

   Were the core plugs installed improperly? Yes _____ No _____

   Does the rear seal have leaks? Yes _____ No _____

   Is the rear seal damaged? Yes _____ No _____

   Are there missing dowels? Yes _____ No _____

   Do the dowel pins have burrs? Yes _____ No _____

   Are there bent dowels? Yes _____ No _____

   Are the dowel pin holes elongated? Yes _____ No _____

   Do the dowel pin holes have burrs? Yes _____ No _____

   Does the flywheel/flex plate have cracks? Yes _____ No _____

   Is the flywheel/flex plate damaged or missing teeth? Yes _____ No _____

   Are there loose mounting bolts in the flywheel/flex plate? Yes _____ No _____

   Is the crankshaft converter pilot hole elongated? Yes _____ No _____

   Does the crankshaft converter pilot hole have burrs? Yes _____ No _____

   Does the engine mounting surface have cracks? Yes _____ No _____

   Does the engine mounting surface contain foreign material that would prevent proper engine-transmission mating? Yes _____ No _____

   Were any problems found? Yes _____ No _____

   If Yes, describe the problem(s): _____

   _____

   _____

- [ ] 2. Clean the work area and return any equipment to storage.

- [ ] 3. Did you encounter any problems during this procedure? Yes ___ No ___

   If Yes, describe the problems. _____

   _____

   What did you do to correct the problems? _____

   _____

☐ 4. Have your instructor check your work and sign this job sheet.

Performance Evaluation—Instructor Use Only

Did the student complete the job in the time allotted? Yes ___ No ___

If No, which steps were not completed?_____

How would you rate this student's overall performance on this job?_____

5–Excellent, 4–Good, 3–Satisfactory, 2–Unsatisfactory, 1–Poor

Comments: _____

_____

_____

INSTRUCTOR'S SIGNATURE _____

Job 51—Remove an Automatic Transaxle

After completing this job, you will be able to properly and safely remove an automatic transmission from a vehicle.

Procedures

> **Note**
> The following procedure is a general guide only. This procedure assumes that the engine remains installed and only the transaxle will be removed. Always refer to the manufacturer's service manual for specific procedures.

☐ 1. Obtain a vehicle to be used in this job. Your instructor may direct you to perform this job on a shop vehicle.

☐ 2. Gather the tools needed to perform the following job. Refer to the project's tools and materials list.

☐ 3. Disconnect the battery negative cable.

☐ 4. Open the vehicle hood or engine compartment cover.

☐ 5. Remove the following parts from the transaxle:
- Dipstick.
- Speedometer cable.
- Manual and throttle linkage.
- Electrical connectors.
- Ground straps.
- Vacuum modulator line if used.

☐ 6. Support the engine from the top to take the weight off the bottom engine mounts.

> **Note**
> A special fixture should be used to support the engine. Do not attempt to remove the transaxle without a way to support the engine.

☐ 7. Raise the vehicle.

> **Warning**
> ⚠ The vehicle must be raised and supported in a safe manner. Always use approved lifts or jacks and jack stands.

☐ 8. Inspect the engine mounts to ensure that they are not broken.

> **Note**
> In instances where the engine mounts are accessed from under the hood, it is necessary to inspect the engine mounts before raising the vehicle.

☐ 9. Remove splash shields and any other under-vehicle parts that will be in the way during CV axle removal.

☐ 10. Remove the CV axles. Refer to Job 67.

☐ 11. Drain the transmission oil into a drain pan.

☐ 12. If necessary, remove stabilizer bars, other suspension components, and the exhaust pipes to provide adequate clearance for transaxle removal.

☐ 13. Remove the engine cradle, if applicable.

☐ 14. Disconnect the TV and shift linkages at the transaxle.

☐ 15. Remove the filler tube.

☐ 16. Position a drip pan to catch any fluid that spills from the cooler lines.

☐ 17. Disconnect the oil cooler lines at the transaxle.

> **Caution**
>
> Use two wrenches to avoid twisting the cooler lines.

☐ 18. Remove the torque converter dust cover.

☐ 19. Remove the converter-to-flywheel attaching bolts.

> **Note**
>
> The flywheel will have to be turned to gain access to all the attaching bolts. This can be done using a special turning tool or a screwdriver, or it can be accomplished by operating the starter with a remote switch.

☐ 20. Push the converter back from the flywheel.

☐ 21. Remove the starter if it is bolted to the transaxle converter housing.

☐ 22. Loosen the transaxle converter housing attaching bolts.

> **Note**
>
> Leave at least two bolts holding the transaxle to the engine until you have a transmission jack under the transaxle.

☐ 23. Place a transmission jack under the transaxle oil pan. **Figure 51-1** shows the positioning of the transmission jack.

Figure 51-1. Always install the safety chain before removing the transaxle from the engine. (DaimlerChrysler)

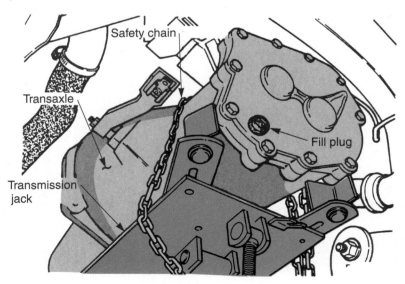

Name _____   Date_____
Instructor _____   Period _____

PRNDL

Job 51—Remove an Automatic Transaxle (continued)

☐ 24. Raise the transaxle slightly to remove transaxle weight from the lower transaxle mounts.

> **Note**
>
> If the engine holding fixture has already removed the weight from the mounts, step 24 is not necessary.

☐ 25. Remove the lower transaxle mount fasteners.

☐ 26. Remove the last bolts holding the transaxle to the engine.

☐ 27. Slide the transaxle away from the engine.

☐ 28. Secure the torque converter with a small C-clamp or a holding tool, or remove the converter from the transaxle.

> **Warning**
>
> ⚠ Failure to secure the torque converter could result in it sliding off and dropping to the floor, possibly causing injury or damage.

☐ 29. Lower the transaxle from the vehicle, **Figure 51-2**.

Inspect Engine Core Plugs, Rear Crankshaft Seal, Dowel Pins, Dowel Pin Holes, and Mating Surfaces

1. Once the transmission has been removed, inspect the following parts at the rear of the engine block:

Do the core plugs have corrosion? Yes _____ No _____

Do the core plugs have leaks? Yes _____ No _____

Were the core plugs installed improperly? Yes _____ No _____

Does the rear seal have leaks? Yes _____ No _____

Figure 51-2. Once the transaxle is removed from the vehicle, lower it and bring it to the workbench.

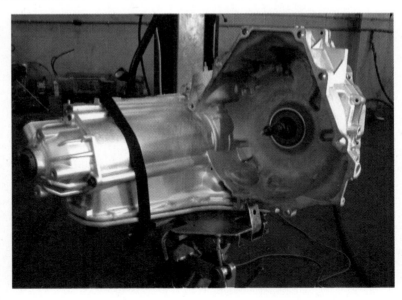

Is the rear seal damaged? Yes _____ No _____

Are there missing dowels? Yes _____ No _____

Do the dowel pins have burrs? Yes _____ No _____

Are there bent dowels? Yes _____ No _____

Are the dowel pin holes elongated? Yes _____ No _____

Do the dowel pin holes have burrs? Yes _____ No _____

Does the flywheel/flex plate have cracks? Yes _____ No _____

Is the flywheel/flex plate damaged or missing teeth? Yes _____ No _____

Are there loose mounting bolts in the flywheel/flex plate? Yes _____ No _____

Is the crankshaft converter pilot hole elongated? Yes _____ No _____

Does the crankshaft converter pilot hole have burrs? Yes _____ No _____

Does the engine mounting surface have cracks? Yes _____ No _____

Does the engine mounting surface contain foreign material that would prevent proper engine-transmission mating? Yes _____ No _____

Were any problems found? Yes _____ No _____

If Yes, describe the problem(s): _____

_____

_____

☐ 2. Clean the work area and return any equipment to storage.

☐ 3. Did you encounter any problems during this procedure? Yes ___ No ___

If Yes, describe the problems: _____

_____

What did you do to correct the problems? _____

_____

☐ 4. Have your instructor check your work and sign this job sheet.

Performance Evaluation—Instructor Use Only

Did the student complete the job in the time allotted? Yes ___ No ___

If No, which steps were not completed?_____

How would you rate this student's overall performance on this job?_____

5–Excellent, 4–Good, 3–Satisfactory, 2–Unsatisfactory, 1–Poor

Comments: _____

_____

_____

INSTRUCTOR'S SIGNATURE _____

Project

Rebuilding an Automatic Transmission

15

Introduction

Automatic transmission overhaul is complex, but not impossible. Technicians rebuild automatic transmissions every day. With practice, you can easily develop the needed skills. The key is to perform a detailed inspection of the transmission, clean all of the parts thoroughly, and reassemble the transmission carefully. Precautions must be taken to avoid damage and injury. In Job 52, you will disassemble and inspect an automatic transmission. You will identify and repair or replace worn or defective internal transmission parts in Job 53 and reassemble an automatic transmission in Job 54.

Project 15 Jobs

- Job 52—Disassemble and Inspect an Automatic Transmission
- Job 53—Service Automatic Transmission Components and Rebuild Subassemblies
- Job 54—Reassemble an Automatic Transmission

Tools and Materials

The following list contains the tools and materials that may be needed to complete the jobs in this project. The items used will depend on the make and model of vehicle being serviced.

- Automatic transmission requiring overhaul.
- Applicable service information.
- Miscellaneous measuring instruments.
- Transmission holding fixture.
- Converter flusher (optional).
- Oil cooler and line flusher (optional).
- Slide hammer.
- Clutch pack spring compressor.
- Servo-cover depressor.

- Clutch/servo seal installers.
- Seal driver.
- Pump alignment tool.
- Band adjustment tools and gauges.
- Specified automatic transmission fluid.
- Hand tools as needed.
- Air-powered tools as needed.
- Safety glasses and other protective equipment as needed.

Safety Notice

Before performing each job, review all pertinent safety information in the text and review safety information with your instructor.

Job 52—Disassemble and Inspect an Automatic Transmission

After completing this job, you will be able to disassemble an automatic transmission and identify worn or defective parts.

> **Note**
>
> There are many variations in automatic transmission design. The following procedures are general guides only. Always consult the manufacturer's service manual for exact service procedures and specifications.

Procedures

☐ 1. Obtain an automatic transmission to be used in this job. Your instructor may direct you to perform this job on a shop transmission.

☐ 2. Gather the tools needed to perform the following job. Refer to the tools and materials list at the beginning of the project.

☐ 3. Mount the transmission in a holding fixture, if available, or secure it on a clean workbench.

☐ 4. Remove the torque converter.

☐ 5. Check the transmission endplay, following the manufacturer's instructions. Compare the measured endplay against specifications. The manufacturer may call for both input and output shaft endplay measurements.

Measured input shaft endplay:_____

Specified input shaft endplay:_____

Measured output shaft endplay:_____

Specified output shaft endplay:_____

Is correction necessary? Yes ___ No ___

If Yes, what did you do to correct the endplay? _____

_____

☐ 6. Remove the transmission oil pan and check it for metal particles, varnish buildup, and sludge.

Remove Case Components

☐ 1. Remove the following parts from the transmission case:
   - Oil filter.
   - Valve body.
   - Spacer plate.
   - Check balls.

> **Note**
>
> Carefully observe the position of all valve parts.

☐ 2. Remove the vacuum modulator, push rod, and throttle valve from the case, if applicable.

☐ 3. Remove the transmission extension housing, if applicable.

☐ 4. Remove the governor, if it is mounted on the output shaft.

☐ 5. Remove the front pump seal.

Remove Internal Components

☐ 1. Remove the front pump. Most oil pumps must be removed with a slide hammer–type puller or a special tool that fits over the input shaft and stator support.

☐ 2. Remove the following internal parts from the front of the transmission, removing snap rings as necessary. The actual sequence of part removal will vary, depending on the transmission's design:
- Input shaft.
- Clutch packs.
- Front planetary gearset.
- Front band(s).

Note

Be sure to save all thrust washers as they come out with these parts. Make a note of the positions of the thrust washers so you can refer to it during reassembly.

☐ 3. Remove the center support, if used. Some center supports are held in place with snap rings, while others are bolted in place.

☐ 4. Remove the following internal parts from the rear of the transmission, removing snap rings as necessary. The actual sequence of part removal will vary, depending on the transmission design:
- Rear planetary gearset.
- Clutch packs.
- Rear bands.
- Output shaft.

☐ 5. Remove the following internal case-mounted accessories by removing snap rings or bolted covers as necessary.
- Pressure switches.
- Solenoids.
- Accumulators.
- Band servos.
- Case-mounted governor, if applicable.
- Case-mounted throttle valve, if applicable.
- Cracked or deformed flex plate.
- Burrs or damage to converter pilot projection.
- Dents or obvious leakage (pressure test may be necessary).
- Stripped or damaged converter attaching bolts.
- Stripped converter attaching bolt holes.

Were any problems found? Yes _____ No _____

If Yes, describe the problem(s): _____

_____

_____

Clean and Inspect Parts

☐ 1. Thoroughly drain the torque converter and inspect it for the following conditions:
- Worn hub at the bushing area.
- Worn or damaged pump drive lugs.
- Stripped threads.
- Proper stator one-way clutch operation.

Job 52—Disassemble and Inspect an Automatic Transmission (continued)

☐ 2. Check the converter endplay as explained in the manufacturer's instructions and compare it to specifications.

Measured endplay:_____

Specified endplay:_____

Can the converter be reused? Yes ___ No ___

Explain your answer:_____

_____

> **Note**
> If the internal transmission parts are very dirty or coated with varnish, replace the converter or clean it with a converter flushing machine.

☐ 3. Inspect all parts for excessive varnish and sludge deposits.

☐ 4. Carefully scrape all old gasket material from the transmission case, oil pan, and other parts.

☐ 5. Clean all metal parts in a safe solvent. Dry the parts with compressed air or allow them to air dry.

> **Warning**
> ⚠ Do not spin antifriction bearings with compressed air.

☐ 6. If there is evidence of fluid contamination, flush the transmission cooler lines. Blow the lines clear with compressed air. If a dedicated cooler and line flushing device is available, use it in accordance with manufacturer's instructions.

> **Caution**
> Do not apply full air pressure to the cooler and lines; they could rupture.

Inspect Transmission Subassemblies

☐ 1. Disassemble one of the clutch packs by removing the snap ring holding the steel plates and friction discs in the clutch drum.

☐ 2. Check the steel plates for discoloration or leopard spots (caused by overheating), scoring, and other damage.

Describe the condition of the steel plates: _____

_____

_____

_____

☐ 3. Check the friction discs for friction linings that are charred, glazed, or heavily pitted and for linings that have separated from the steel backing. See **Figure 52-1A**.

Describe the condition of the friction discs: _____

_____

_____

4. Repeat the previous three steps for all clutch packs.

5. Check the transmission bands (when used) for broken ears and check the friction linings for burned, glazed, or worn friction material. See **Figure 52-1B.**

   Describe the condition of the bands' friction material:_____

   _____

6. Check the drums for scored, overheated, or worn contact surfaces where the bands apply, **Figure 52-1C**. Also, check for wear at the drum bushings and areas where seal rings ride, **Figure 52-1D**.

   Describe the drum's condition: _____

   _____

   _____

7. Check for worn or improperly functioning one-way clutches by visual inspection and by turning the clutches by hand.

   Describe clutch condition:_____

   _____

   _____

Figure 52-1. Four common problems with transmission subassemblies are shown here. A—Burned clutch friction material can be scraped off with a fingernail. The clutch plate must be replaced. B—The friction material has worn away on the edge of this band. C—Check each drum, on the area where a band applies, for wear. D—Check the drum bushings and seal ring surfaces for wear.

A

B

C

D

Job 52—Disassemble and Inspect an Automatic Transmission (continued)

☐ 8. Check the valve body for the following conditions:
- Sticking valves.
- Nicked, scored, or burred valves.
- Cracked accumulator pistons.
- Worn or broken accumulator seals or rings.
- Worn, scored, or scratched accumulator bores.
- Damaged pressure sensors or solenoids.
- Plugged solenoid screens.
- Porous or cracked valve body casting.
- Damaged machined surfaces.
- Damaged spacer plates or excessive wear where check balls operate.

Describe the valve body's condition: _____

_____

_____

_____

☐ 9. Check the sealing rings, ring grooves, and bores for wear and damage.

Describe the sealing ring, groove, and bore conditions:_____

_____

☐ 10. Check the oil pump for the following conditions. See **Figure 52-2**.
- Worn, scored, or pitted gears.
- Worn, scored, or pitted pump body.
- Excessive clearance between the pump gears.
- Excessive clearance between the gears and body.
- Worn pump bushing.

Describe the oil pump's condition: _____

_____

Figure 52-2. A pump body that is this badly scored should be replaced.

261

☐ 11. Inspect the planetary gearsets for the following conditions:
- Worn gears.
- Damaged gear teeth.
- Loose gears on the carrier.
- Worn or damaged bushings and needle bearings.

Describe the condition of the planetary gearsets: _____

_____

☐ 12. Check all thrust washers for wear, scoring, or other damage.

Describe the condition of the thrust washers: _____

_____

☐ 13. Check all snap rings for cracks and distortion.

Describe the condition of the snap rings: _____

_____

☐ 14. Check antifriction bearings for wear, scoring, and roughness.

Describe the condition of the bearings: _____

_____

☐ 15. Check the transmission case for cracks or distortion at the sealing surfaces.

Describe the condition of the transmission case: _____

_____

☐ 16. If a vacuum modulator is used, check for transmission fluid on the vacuum-hose side of the modulator.

Is transmission fluid present? Yes ___ No ___

☐ 17. If a governor is used, check that the governor valve moves freely in the governor body.

Does the valve move freely? Yes ___ No ___

☐ 18. Make ohmmeter checks to isolate any damaged or inoperative pressure or temperature sensors.

Did the sensors pass the ohmmeter checks? Yes ___ No ___

Identify any sensors that failed the ohmmeter checks: _____

_____

☐ 19. Make ohmmeter checks to isolate any damaged or inoperative control solenoids.

Did the solenoid(s) pass the ohmmeter checks? Yes ___ No ___

Identify any solenoids that failed the ohmmeter checks: _____

_____

☐ 20. Report any transmission component defects that you found to your instructor and determine which parts must be replaced. If your instructor directs, list needed parts on a separate sheet of paper.

☐ 21. Obtain the proper replacement parts and compare them with the old parts to ensure the replacements are correct.

☐ 22. Clean the work area and return tools and equipment to storage.

☐ 23. Did you encounter any problems during this procedure? Yes ___ No ___

If Yes, describe the problems: _____

_____

What did you do to correct the problems? _____

_____

☐ 24. Have your instructor check your work and sign this job sheet.

Job 52—Disassemble and Inspect an Automatic Transmission (continued)

Performance Evaluation—Instructor Use Only

Did the student complete the job in the time allotted? Yes ____ No ____

If No, which steps were not completed?_____

How would you rate this student's overall performance on this job?_____

5–Excellent, 4–Good, 3–Satisfactory, 2–Unsatisfactory, 1–Poor

Comments: _____

_____

_____

INSTRUCTOR'S SIGNATURE _____

Notes

Job 53—Service Automatic Transmission Components and Rebuild Subassemblies

After completing this job, you will be able to service clutch packs, servos, accumulators, and other automatic transmission subassemblies.

Procedures

> **Note**
>
> This job begins with a partially disassembled transmission. Refer to Job 52 for procedures for disassembling an automatic transmission.

☐ 1. Obtain an automatic transmission to be used in this job. Your instructor may direct you to perform this job on a shop transmission.

☐ 2. Gather the tools needed to perform the following job. Refer to the tools and materials list at the beginning of the project.

☐ 3. Soak all replacement friction discs, bands, and seals in clean transmission fluid of the proper type.

Service Clutch Packs

☐ 1. Use a clutch-pack spring compressor to compress the clutch apply piston return spring(s), and then remove the snap ring holding the return spring retainer.

☐ 2. Slowly release the spring compressor and remove the spring and retainer.

☐ 3. Remove the clutch apply piston.

☐ 4. Remove the internal piston seals.

☐ 5. Clean all clutch piston parts and the interior of the clutch drum.

> **Note**
>
> Some clutch drums and pistons contain air-bleed check balls. Check balls should be visible when you look straight down into their seat assembly, **Figure 53-1**. If you can see through the seat assembly, the ball is missing. Be sure that check balls are in place and can move freely. A drum or piston with a damaged or missing check ball is usually replaced. A few drums and pistons have replaceable check-ball assemblies.

☐ 6. Acquire new piston seals. Compare the replacement piston seals to the old seals.

   Are the new seals correct? Yes ____ No ____

☐ 7. Install the new seals on the clutch piston (and clutch drum hub, if applicable).

> **Note**
>
> Seals must be lubricated with the appropriate transmission fluid, special assembly lube, or petroleum jelly.

☐ 8. Install the piston in the drum, being careful not to damage the seals.

☐ 9. Use the clutch spring compressor to install the return spring and snap ring.

☐ 10. Install the clutches in the clutch drum, alternating friction discs and steel plates. Install any reaction plates at the top and bottom as required.

Figure 53-1. Make sure that any drum air bleed check balls are in place and free to move. Most check balls can be heard to rattle when the drum is shaken.

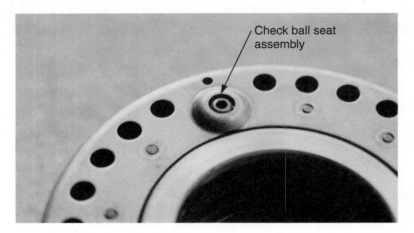

11. Check the clutch pack clearance. **Figure 53-2** shows clutch pack clearance being checked with a feeler gauge, which is the usual procedure. However, some clutch pack clearances are checked with a dial indicator. Refer to the manufacturer's instructions.

Specified clearance:_____

Measured clearance:_____

Is the clearance correct? Yes ____ No ____

If the clearance is incorrect, correct it by installing a different thickness spacer plate or snap ring.

If the endplay was incorrect, what did you do to correct it? _____

_____

12. Repeat steps 1 through 11 for all clutch packs.

Figure 53-2. Check and correct clutch pack clearance. Improper clearances will cause poor shift feel and early clutch failure. (DaimlerChrysler)

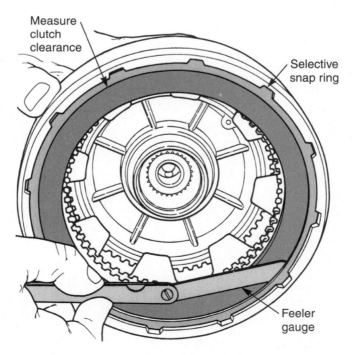

Job 53—Service Automatic Transmission Components and Rebuild Subassemblies (continued)

Service Servos and Accumulators

☐ 1. Remove the snap ring or bolts securing the servo or accumulator cover.

> **Note**
>
> A special tool may be needed to depress the cover.

☐ 2. Remove the servo or accumulator cover.

☐ 3. Remove the servo or accumulator piston and return spring.

☐ 4. Remove the seals from the servo or accumulator piston.

☐ 5. Clean and inspect all servo or accumulator parts. Pay particular attention to the condition of the pistons and bores. Replace any cracked pistons, broken springs, or parts with scored bores. Some common transmission and transaxle cases can be repaired with special bore sleeves. Smaller parts with damaged bores are usually replaced.

☐ 6. Acquire new seals. Compare the old and new seals.

Are the new seals correct? Yes ___ No ___

If No, acquire the correct seals.

☐ 7. Install new piston seals on the servo or accumulator piston.

> **Note**
>
> Seals must be lubricated with the appropriate transmission fluid, special assembly lube, or petroleum jelly.

☐ 8. Install the servo or accumulator piston.

☐ 9. Install the servo or accumulator cover, using the proper new O-ring or gasket.

☐ 10. Repeat steps 1 through 8 for all servos and accumulators.

Service Other Components

☐ 1. Clean, inspect, and rebuild other subassemblies in the transmission according to the manufacturer's instructions.

- **Valve body:** Ensure that all valves and springs are correctly installed and all fasteners are correctly tightened. Check that solenoids are installed correctly and clean or replace solenoid and other screens. Make a final check to ensure that the valves move freely.
- **Oil pump:** Replace worn gears and the front pump bushing. Lightly oil the pump before reassembling. Tighten the pump fasteners to the exact torque, following the proper sequence. Install a new pump seal.
- **Seal rings:** Always install new seal rings. Make sure that the rings are correct, and that the proper fit exists between the rings and ring grooves and the ring bore. Do not expand the rings any more than necessary to install them into the groove.
- **Planetary gears:** Replace worn gears, bushings, thrust washers, and needle bearings. Sometimes it is possible to replace one gear of a planetary gearset instead of replacing the entire gearset. Consult your instructor to determine whether individual planetary gears can be replaced.
- **Bushings:** Carefully press out worn bushings and press in new bushings using a hydraulic press and the proper bushing drivers.

☐   2.   Clean the work area and return tools and equipment to storage.

☐   3.   Did you encounter any problems during this procedure? Yes ____ No ____

        If Yes, describe the problems: _____

        _____

        What did you do to correct the problems? _____

        _____

☐   4.   Have your instructor check your work and sign this job sheet.

Performance Evaluation—Instructor Use Only

Did the student complete the job in the time allotted? Yes ____ No ____

If No, which steps were not completed?_____

How would you rate this student's overall performance on this job?_____

5–Excellent, 4–Good, 3–Satisfactory, 2–Unsatisfactory, 1–Poor

Comments: _____

_____

_____

INSTRUCTOR'S SIGNATURE _____

Job 54—Reassemble an Automatic Transmission

After completing this job, you will be able to properly reassemble an automatic transmission.

Procedures

> **Note**
> This job begins with a partially disassembled transmission. Refer to Job 52 for procedures for disassembling an automatic transmission.

☐   1.   Obtain an automatic transmission to be used in this job. Your instructor may direct you to perform this job on a shop transmission.

☐   2.   Gather the tools needed to perform the following job. Refer to the tools and materials list at the beginning of the project.

☐   3.   Lightly lubricate all parts with the appropriate transmission fluid before proceeding.

Install Internal Components

☐   1.   Install the rear planetary gearsets, drums, holding members, and the output shaft, using snap rings as needed.

> **Note**
> Do not forget to install thrust washers at the proper places. After the snap rings (or bolts) are in place, make sure the output shaft will turn.

☐   2.   Install the center support (if used) and make sure the output shaft turns after the center support is installed.

☐   3.   Install the front bands, input shaft, forward clutch packs, front planetary gearset, and input shell, using snap rings as needed. Make sure all thrust washers are installed in their proper places.

☐   4.   Install the front pump. Use the proper gasket.

> **Note**
> As you slowly tighten the pump fasteners, check that the input and output shafts can be turned by hand.

☐   5.   Check the transmission endplay and compare it to specifications. Adjust endplay if necessary.

Measured input shaft endplay:_____

Specified input shaft endplay:_____

Measured output shaft endplay:_____

Specified output shaft endplay:_____

Is correction necessary? Yes ____ No ____

If Yes, what did you do to correct the endplay? _____

_____

Install Case Components

☐ 1. Install the band apply linkage.

☐ 2. Adjust all bands to specifications. See Job 45.

☐ 3. Install the governor in the case or on the output shaft, as applicable.

☐ 4. Install the extension housing, if necessary. Use the proper gasket.

☐ 5. Install a new rear seal on the extension housing.

☐ 6. Perform air-pressure tests to ensure the clutch apply pistons and band servos are working properly.

☐ 7. Install the valve body–to–case spacer plate, the spacer plate gaskets, and any check balls in the transmission case.

☐ 8. Install the valve body.

☐ 9. Install a new transmission oil filter.

☐ 10. Install any external case-mounted parts, such as electrical switches or the vacuum modulator.

☐ 11. Install the transmission oil pan.

☐ 12. Add 1–2 quarts (1–2 liters) of transmission fluid to the converter.

☐ 13. Install the converter in the converter housing, making sure the internal splines of the turbine hub and one-way clutch engage their shafts and the lugs on the pump drive hub fully engage the slots in the front pump.

Note

The transmission is ready to be installed in the vehicle. Refer to Job 58 for specific instructions on automatic transmission installation.

☐ 14. Clean the work area and return tools and equipment to storage.

☐ 15. Did you encounter any problems during this procedure? Yes ___ No ___

If Yes, describe the problems: _____

_____

What did you do to correct the problems? _____

_____

☐ 16. Have your instructor check your work and sign this job sheet.

Performance Evaluation—Instructor Use Only

Did the student complete the job in the time allotted? Yes ___ No ___

If No, which steps were not completed?_____

How would you rate this student's overall performance on this job?_____

5–Excellent, 4–Good, 3–Satisfactory, 2–Unsatisfactory, 1–Poor

Comments: _____

_____

_____

INSTRUCTOR'S SIGNATURE _____

Project

16

Rebuilding an Automatic Transaxle

Introduction

Automatic transaxles can be successfully rebuilt, but require careful disassembly and inspection of parts to determine what parts should be replaced. Thorough cleaning of all parts and careful reassembly are also vital steps in successfully rebuilding a transaxle. In Job 55, you will disassemble an automatic transaxle and identify worn or defective internal parts. You will rebuild the transaxle's subassemblies in Job 56 and reassemble the transaxle in Job 57.

Project 16 Jobs

- Job 55—Disassemble and Inspect an Automatic Transaxle
- Job 56—Service Automatic Transaxle Components and Rebuild Subassemblies
- Job 57—Reassemble an Automatic Transaxle

Tools and Materials

The following list contains the tools and materials that may be needed to complete the jobs in this project. The items used will depend on the make and model of vehicle being serviced.

- Automatic transaxle requiring overhaul.
- Applicable service information.
- Transaxle holding fixture (if available).
- Drain pan.
- Converter flusher.
- Oil cooler and line flusher.
- Slide hammer.
- Miscellaneous measuring instruments.
- Clutch pack spring compressor.
- Servo-cover depressor.
- Clutch/servo seal installers.

- Seal driver.
- Pump alignment tool.
- Band adjustment tools and gauges.
- Transmission pressure gauge(s).
- Transaxle replacement parts.
- Specified automatic transmission fluid.
- Hand tools as needed.
- Air powered tools as needed.
- Safety glasses and other protective equipment as needed.

Safety Notice

Before performing this job, review all pertinent safety information in the text, and review safety information with your instructor.

Notes

Job 55—Disassemble and Inspect an Automatic Transaxle

After completing this job, you will be able to disassemble an automatic transaxle and identify worn or defective parts.

Procedures

> **Note**
> There are many variations in automatic transaxle design. The following procedure is a general guide only. Always consult the manufacturer's service manual for exact disassembly procedures and specifications.

☐ 1. Obtain an automatic transaxle to be used in this job. Your instructor may direct you to perform this job on a shop transaxle.

☐ 2. Gather the tools needed to perform the following job. Refer to the tools and materials list at the beginning of the project.

Remove Case Components

☐ 1. Drain the transaxle lubricant.

☐ 2. Mount the transaxle in a special holding fixture, if available, or secure it on a clean workbench.

☐ 3. Remove the torque converter.

☐ 4. Check the transaxle endplay as explained in the manufacturer's instructions and compare it to specifications.

Measured endplay:_____

Specified endplay:_____

☐ 5. Remove any external transaxle components, such as vacuum modulators, case-mounted governors, and electrical connectors.

☐ 6. Remove the transaxle's oil pan and check it for metal particles, varnish buildup, and sludge.

☐ 7. Remove sheet metal or cast aluminum covers from the transaxle.

☐ 8. Remove the transaxle oil filter and valve body, carefully observing the position of all valve body parts.

> **Note**
> Some transaxles have more than one valve body.

☐ 9. Remove snap rings or threaded fasteners as necessary and remove the drive chain and sprockets, **Figure 55-1**.

or

Remove fasteners as necessary and remove the drive and transfer gears, **Figure 55-2**.

☐ 10. Remove the transmission oil pump, if it is separately mounted. If the side valve body is installed under the drive chain and sprockets, remove it at this time.

Figure 55-1. The sprockets are usually lifted from the case as an assembly. (DaimlerChrysler)

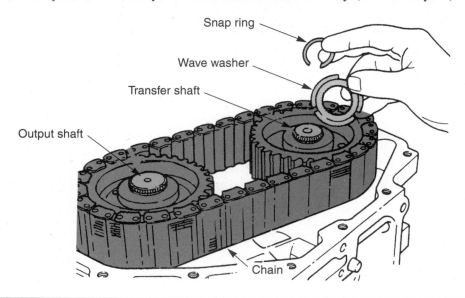

Figure 55-2. Many transfer gears are pressed in place and must be removed with a special puller. (DaimlerChrysler)

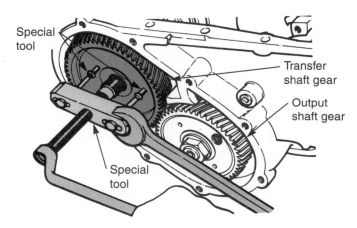

Remove Internal Components

 1. Remove the differential case assembly.

> **Note**
>
> In some designs, the differential case assembly is removed after the other internal transaxle components. Also, if the transaxle is used with a longitudinal engine, the differential assembly may be a separate unit that does not require disassembly unless defective.

 2. Remove the front bands, input shaft, forwardmost clutch packs, front planetary gearset, and input shell, removing snap rings as needed.

> **Note**
>
> Be sure to note the location of all thrust washers as they come out with the other internal parts.

Job 55—Disassemble and Inspect an Automatic Transaxle (continued)

☐ 3. If the transaxle has a center support, remove the attaching bolts or snap rings and remove the support from the case.

> **Caution**
> ⚠ Never attempt to drive the center support from the transaxle case without removing the retainers.

☐ 4. Remove any rear planetary gearsets, holding members, remaining clutch packs, and the output shaft, removing snap rings as needed.

☐ 5. Remove any internal case-mounted accessories, such as servos, accumulators, and governors.

Clean and Inspect Parts

☐ 1. Thoroughly drain the torque converter and inspect it for the following:
- Hub worn at the bushing area.
- Worn or damaged pump drive lugs (when used).
- Stripped threads.
- Improper stator one-way clutch operation.
- Cracked or deformed flex plate.
- Burrs or damages to converter pilot projection.
- Dents or obvious leakage (pressure test may be necessary).
- Stripped or damaged converter attaching bolts.
- Stripped converter attaching bolt holes.

Were any problems found? Yes _____ No _____

If Yes, describe the problem(s): _____

_____

_____

☐ 2. Check converter endplay as explained in the manufacturer's instructions and compare it against specifications.

Measured endplay:_____

Specified endplay:_____

Can the converter be reused? Yes ____ No ____

Explain your answer:_____

_____

> **Note**
> 📄 If the transmission is very dirty or coated with varnish, replace the converter or clean it with a converter flushing machine.

☐ 3. Inspect all parts for varnish and sludge deposits.

☐ 4. Carefully scrape old gasket material from the transaxle case, oil pan, and other parts.

☐ 5. Clean all metal parts in a safe solvent. Dry the parts with compressed air or allow them to air dry.

☐ 6. If there is evidence of fluid contamination, flush the transmission cooler lines using an oil cooler and line flusher. Follow the manufacturer's instructions for using the equipment. Blow the lines clear with compressed air.

> **Caution**
>
> Do not apply full air pressure to the cooler and lines, they could rupture.

Inspect Subassemblies

☐ 1. Check the drive chain and sprockets for wear.

or

Check the drive and transfer gears and related bearings for wear.

Did you find any excessive wear? Yes ___ No ___

If Yes, what components were excessively worn? _____

_____

☐ 2. Disassemble one of the clutch packs by removing the snap ring holding the steel plates and friction discs in the clutch drum.

☐ 3. Check the steel plates for leopard spots or discoloration caused by overheating, scoring, and other damage.

Describe the condition of the steel plates: _____

_____

_____

☐ 4. Check the friction discs for charred, glazed, or heavily pitted friction linings or linings separated from the steel backing.

Describe the condition of the friction discs: _____

_____

_____

☐ 5. Repeat steps 2 through 4 for all clutch packs.

☐ 6. Check the friction linings on transmission bands (when used) for burned, glazed, or worn friction material.

Describe the condition of the band friction material: _____

_____

☐ 7. Check drums where bands apply for scored, overheated, or worn mating band-contact surfaces.

Describe the condition of the drums: _____

_____

☐ 8. Check for worn or improperly functioning one-way clutches by visual inspection and by turning the clutches by hand.

Describe the condition of the clutches: _____

_____

☐ 9. Check the main valve body and any auxiliary valve bodies for the following:
   - Sticking valves.
   - Nicked, scored, or burred valves.
   - Cracked accumulator pistons.
   - Worn or broken accumulator seals or rings.
   - Worn, scored, or scratched accumulator bores.
   - Damaged pressure sensors or solenoids.
   - Plugged solenoid screens.
   - Porous or cracked valve body casting.
   - Damaged machined surfaces.
   - Damaged spacer plates or excessive wear where check balls operate.
   - Auxiliary pump wear or damage (when used).

Job 55—Disassemble and Inspect an Automatic Transaxle (continued)

Describe valve body condition: _____

_____

_____

_____

☐ 10. Check the sealing rings, ring grooves, and bores for wear and damage.

Describe sealing ring, groove, and bore condition: _____

_____

☐ 11. Check the oil pump for the following:
- Worn, scored, or pitted gears.
- Worn, scored, or pitted pump body, **Figure 55-3**.
- Excessive clearance between the pump gears.
- Excessive clearance between the gears and body.
- Worn pump bushing.

Describe oil pump condition: _____

_____

☐ 12. Inspect the planetary gearsets for the following:
- Worn gears.
- Damaged gear teeth.
- Gears loose on the carrier.
- Worn or damaged bushings and needle bearings.

Describe planetary gearset condition: _____

_____

☐ 13. Check all thrust washers for wear, scoring, or other damage.

Describe thrust washer condition: _____

_____

☐ 14. Check all snap rings for cracks and distortion.

Describe snap ring condition: _____

_____

Figure 55-3. Check the pump gears and body for excessive wear or scoring.

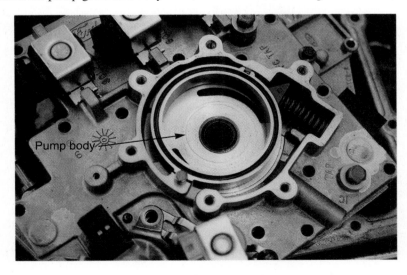

Figure 55-4. Check all transaxle bearings for obvious wear and roughness.

15. Check antifriction bearings for wear, scoring, and roughness. Do not forget to check the bearing installed in the case, **Figure 55-4**.

    Describe bearing condition: _____

    _____

16. Check the transmission case for cracks or distortion at sealing surfaces. Also check servo and accumulator bores for wear or other damage. See **Figure 55-5**. Badly damaged cases are usually replaced. However, some cases can be repaired.

    Describe case condition: _____

    _____

17. If a vacuum modulator is used, check for transmission fluid on the vacuum-hose side of the modulator.

    Is transmission fluid present? Yes ___ No ___

Figure 55-5. Carefully check the case for cracks, scoring in servo and accumulator bores, warped surfaces and other damage.

Job 55—Disassemble and Inspect an Automatic Transaxle (continued)

☐ 18. If a governor is used, check that the governor valve moves freely in the governor body.

Does the governor valve move freely? Yes ___ No ___

If No, consult your instructor.

☐ 19. Make ohmmeter checks to isolate any damaged or inoperative pressure or temperature sensors.

Did the sensor(s) pass? Yes ___ No ___

If No, list the sensors that must be replaced: _____

_____

☐ 20. Make ohmmeter checks to isolate any damaged or inoperative control solenoids.

Did the solenoid(s) pass? Yes ___ No ___

If No, list the solenoids that must be replaced: _____

_____

☐ 21. Check the final drive components for wear or damage. See the appropriate job in Project 12.

☐ 22. Report any transmission component defects that you found to your instructor and determine which parts must be replaced. If your instructor directs, list needed parts on a separate sheet of paper.

☐ 23. Clean the work area and return any equipment to storage.

☐ 24. Did you encounter any problems during this procedure? Yes ___ No ___

If Yes, describe the problems: _____

_____

What did you do to correct them?_____

_____

☐ 25. Have your instructor check your work and sign this job sheet.

Performance Evaluation—Instructor Use Only

Did the student complete the job in the time allotted? Yes ___ No ___

If No, which steps were not completed?_____

How would you rate this student's overall performance on this job?_____

5–Excellent, 4–Good, 3–Satisfactory, 2–Unsatisfactory, 1–Poor

Comments: _____

_____

_____

INSTRUCTOR'S SIGNATURE _____

Notes

Job 56—Service Automatic Transaxle Components and Rebuild Subassemblies

After completing this job, you will be able to service a transaxle's internal components and rebuild a transaxle's subassemblies.

Procedures

> **Note**
>
> This job begins with a partially disassembled transaxle. Refer to Job 55 for procedures for disassembling an automatic transaxle.

☐ 1. Obtain an automatic transaxle to be used in this job. Your instructor may direct you to perform this job on a shop transaxle.

☐ 2. Gather the tools needed to perform the following job. Refer to the tools and materials list at the beginning of the project.

☐ 3. Obtain the proper replacement parts and compare them with the old parts to ensure the replacements are correct.

☐ 4. Soak all replacement friction discs, bands, and seals in clean transmission fluid.

Service Clutch Packs

☐ 1. Use a clutch pack spring compressor to compress the clutch apply piston return spring(s), and remove the snap ring holding the return spring retainer.

☐ 2. Slowly release the spring compressor and remove the spring and retainer.

☐ 3. Remove the clutch apply piston.

☐ 4. Remove the internal piston seals.

☐ 5. Clean all clutch piston parts.

☐ 6. Compare the replacement piston seals to the old seals.

Are the new seals correct? Yes ___ No ___

☐ 7. Install the new seals on the clutch piston (and clutch drum hub, if applicable).

> **Note**
>
> Seals must be lubricated with transmission fluid, special assembly lube, or petroleum jelly.

☐ 8. Install the piston in the drum, being careful not to damage the seals.

☐ 9. Use the clutch spring compressor to install the return spring and snap ring.

☐ 10. Install the clutches in the clutch drum, alternating friction discs and steel plates.

> **Note**
>
> Install any reaction plates at the top and bottom as required.

☐ 11. Check clutch pack clearance. Refer to the manufacturer's instructions.

Is clearance correct? Yes ___ No ___

If clearance is too small or large, correct the clearance by changing to a different thickness spacer plate or snap ring.

☐ 12. Repeat steps 1 through 11 for all clutch packs.

Service Servos and Accumulators

☐ 1. Remove the snap ring or bolts securing the servo or accumulator cover.

> **Note**
>
> A special tool may be needed to depress the cover.

☐ 2. Remove the servo or accumulator cover.

☐ 3. Remove the servo or accumulator piston and return spring.

☐ 4. Remove the seals from the servo or accumulator piston.

☐ 5. Inspect and clean all servo or accumulator parts. Pay particular attention to the condition of the pistons and bores. Replace any cracked pistons, broken springs, or parts with scored bores.

☐ 6. Compare the old and new seals.

Are the new seals correct? Yes ___ No ___

☐ 7. Install the new piston seals on the servo or accumulator piston.

> **Note**
>
> Seals must be lubricated with transmission fluid, special assembly lube, or petroleum jelly.

☐ 8. Install the servo or accumulator piston.

☐ 9. Install the servo or accumulator cover, using the proper new O-ring or gasket.

☐ 10. Repeat steps 1 through 9 for all servos and accumulators.

Service Other Components

☐ 1. Clean, inspect, and rebuild other transaxle subassemblies according to manufacturer's instructions.
- **Valve body:** Ensure that all valves and springs are correctly installed and all fasteners are correctly tightened. Check that solenoids are installed correctly and clean or replace solenoid and other screens. Make a final check to ensure that the valves move freely.
- **Oil pump:** Replace worn gears and the front pump bushing. Lightly oil the pump before reassembling. Tighten the pump fasteners to the exact torque, following the proper sequence. Install a new pump seal.
- **Seal rings:** Always install new rings. Make sure that the rings are correct, and that the proper fit exists between the rings and ring grooves and the ring bore. Do not expand the rings any more than necessary to install them into the groove.
- **Planetary gears:** Replace worn gears, bushings, thrust washers, and needle bearings. Sometimes it is possible to replace one gear of a planetary gearset instead of replacing the entire gearset. Consult your instructor to determine whether individual planetary gears can be replaced.
- **Bushings:** Carefully press out worn bushings and press in new bushings using a hydraulic press and the proper bushing drivers.

Job 56—Service Automatic Transaxle Components and Rebuild Subassemblies (continued)

☐ 2. Clean the work area and return tools and equipment to storage.

☐ 3. Did you encounter any problems during this procedure? Yes ___ No ___

If Yes, describe the problems: _____

_____

What did you do to correct the problems? _____

_____

☐ 4. Have your instructor check your work and sign this job sheet.

Performance Evaluation—Instructor Use Only

Did the student complete the job in the time allotted? Yes ___ No ___

If No, which steps were not completed?_____

How would you rate this student's overall performance on this job?_____

5–Excellent, 4–Good, 3–Satisfactory, 2–Unsatisfactory, 1–Poor

Comments: _____

_____

_____

INSTRUCTOR'S SIGNATURE _____

Notes

Job 57—Reassemble an Automatic Transaxle

After completing this job, you will be able to assemble an automatic transaxle.

Procedures

> **Note**
> This job begins with a partially disassembled transaxle. Refer to Job 55 for procedures for disassembling an automatic transaxle.

☐ 1. Obtain an automatic transaxle to be used in this job. Your instructor may direct you to perform this job on a shop transaxle.

☐ 2. Gather the tools needed to perform the following job. Refer to the tools and materials list at the beginning of the project.

Install Internal Components

☐ 1. Lightly lubricate all parts before proceeding.

☐ 2. Install the differential case assembly. This might be done after assembling the transaxle, as required.

☐ 3. Install rear planetary gearsets, holding members, clutch packs, and the output shaft, using snap rings as needed.

> **Note**
> After the snap rings (or bolts) are in place, make sure the output shaft will turn.

☐ 4. If a center support is used, install it now. Make sure the output shaft still turns after the snap ring is installed.

☐ 5. Install the front bands, input shaft, forward clutch packs, front planetary gearset(s), and input shell, using snap rings as needed.

☐ 6. Install the side valve body if necessary, then install the drive chain and sprockets or the drive and transfer gears.

Install Case Components

☐ 1. After all the internal parts have been installed, install the transaxle oil pump, using the proper gasket.

> **Note**
> As you slowly tighten down the pump fasteners, check that the input and output shafts can be turned by hand.

☐ 2. Install the drive chain and sprockets or the drive gears as applicable.

> **Note**
> Ensure that all thrust washers are in place.

☐ 3. Check the transaxle endplay as explained in the manufacturer's instructions, and compare it against specifications. Adjust if necessary.

Measured endplay:_____

Specified endplay:_____

☐ 4. Install the band apply linkage.

☐ 5. Adjust all bands to specifications. A special tool may be needed.

☐ 6. Perform an air pressure test on the transaxle to ensure that the clutch apply pistons and band servos are working properly. Band movement or a sharp *thunk* from the clutch packs indicates that the air pressure is operating the holding members.

☐ 7. Install the valve body-to-case spacer plate, the spacer plate gaskets, and any check balls in the transaxle case.

☐ 8. Install the valve body or bodies.

☐ 9. Install a new transaxle oil filter.

☐ 10. Install any throttle valve and shift linkage components.

☐ 11. Install any external case-mounted parts, such as the vacuum modulator or electrical switches.

☐ 12. Install the transaxle oil pan.

☐ 13. Install any sheet metal or cast aluminum covers.

☐ 14. Position the converter over the turbine shaft and lower it until its internal splines mesh with splines on the turbine shaft, oil pump shaft, and stator shaft, as applicable. If a pump drive hub drives the oil pump, the drive lugs must fully engage the oil pump.

> **Note**
>
> The transaxle is ready to be installed in the vehicle. Refer to Job 59 for specific instructions on automatic transaxle installation.

☐ 15. Clean the work area and return tools and equipment to storage.

☐ 16. Did you encounter any problems during this procedure? Yes ___ No ___

If Yes, describe the problems: _____

_____

What did you do to correct the problems? _____

_____

☐ 17. Have your instructor check your work and sign this job sheet.

Performance Evaluation—Instructor Use Only

Did the student complete the job in the time allotted? Yes ___ No ___

If No, which steps were not completed?_____

How would you rate this student's overall performance on this job?_____

5–Excellent, 4–Good, 3–Satisfactory, 2–Unsatisfactory, 1–Poor

Comments: _____

_____

_____

INSTRUCTOR'S SIGNATURE _____

Project

17 Installing an Automatic Transmission or Transaxle

Introduction

Automatic transmission or transaxle installation is relatively simple. Automatic transmissions and transaxles are always heavy and precautions must be taken to avoid damage and injury. You will install an automatic transmission into a vehicle in Job 58. In Job 59, you will install an automatic transaxle into a vehicle.

Project 17 Jobs

- Job 58—Install an Automatic Transmission
- Job 59—Install an Automatic Transaxle

Tools and Materials

The following list contains the tools and materials that may be needed to complete the jobs in this project. The items used will depend on the make and model of vehicle being serviced.

- An automatic transmission.
- An automatic transaxle.
- Vehicle in need of an automatic transmission.
- Vehicle in need of an automatic transaxle.
- Applicable service information.
- Hydraulic lift (or jacks and jack stands).
- Flywheel turning tool.
- Transmission jack.
- Torque converter holding tool (or C-clamp).
- Linkage adjustment tools and gauges.
- Specified automatic transmission fluid.
- Hand tools as needed.
- Air-operated tools as needed.
- Safety glasses and other protective equipment as needed.

Safety Notice

Before performing this job, review all pertinent safety information in the text, and review safety information with your instructor.

Notes

Job 58—Install an Automatic Transmission

After completing this job, you will be able to install an automatic transmission.

Procedures

☐ 1. Obtain an automatic transmission and a vehicle to be used in this job. Your instructor may direct you to perform this job on a shop vehicle.

☐ 2. Gather the tools needed to perform the following job. Refer to the tools and materials list at the beginning of the project.

☐ 3. Raise the vehicle.

> **Warning**
> ⚠ Raise and support the vehicle in a safe manner. Always use approved lifts or jacks and jack stands.

☐ 4. Secure the torque converter with a torque converter holding tool or a C-clamp.

☐ 5. Lubricate the converter pilot hub and crankshaft bore with multipurpose grease.

> **Note**
> Before installing the transmission, perform a final inspection of the engine and transmission. Check engine core plugs and the rear crankshaft seal for leaks. Check the dowel pins and dowel pin holes to ensure they are in good condition. Also, ensure that the mating surfaces between the engine and transmission are properly cleaned. Correct any problems before installing the transmission.

☐ 6. Raise the transmission into position under the vehicle, using a transmission jack.

> **Warning**
> ⚠ Install the safety chain to ensure that the transmission does not fall from the transmission jack.

☐ 7. Slide the transmission converter housing into engagement with the rear of the engine.

☐ 8. Install at least two of the converter housing attaching bolts.

☐ 9. Ensure that the torque converter can turn freely. If it cannot, the converter is not properly installed.

☐ 10. If the converter turns freely, install the remainder of the converter housing–to–engine bolts and torque all bolts to specifications.

> **Note**
> If the upper bolts cannot be reached from under the vehicle, they may be installed once the vehicle is lowered.

☐ 11. Install the two lower converter-to-flywheel attaching bolts or nuts and tighten them to specifications. Then, rotate the flywheel and install the remaining bolts.

☐ 12. Install the starter if it was removed.

☐ 13. Install the flywheel dust cover.

☐ 14. Install the transmission cross member.

☐ 15. Install and tighten all cross member fasteners.

☐ 16. Connect the parking brake linkage.

☐ 17. Connect the throttle, kickdown, and shift linkages and make preliminary adjustments. See Job 45.

☐ 18. If the transmission has a modulator, install the vacuum line.

☐ 19. Install the exhaust pipes if they were removed.

☐ 20. Install the oil cooler lines and tighten the fittings.

☐ 21. Install the electrical connectors and the speedometer cable assembly.

☐ 22. Remove the cap at the rear of the transmission and install the drive shaft. Remember to line up the match marks made during removal.

☐ 23. Lower the vehicle, but leave the wheels off the ground so the drive wheels are free to turn.

☐ 24. Install the negative battery cable.

☐ 25. Check the throttle or kick down and shift linkages to ensure that they are properly adjusted. Readjust them if necessary.

☐ 26. Check that all vacuum lines and electrical connectors are in place on the engine.

☐ 27. Add the specified fluid to the transmission. Overfill the transmission by 2 quarts as measured on the dipstick.

☐ 28. Install the upper converter housing–to–engine attaching bolts if they have not yet been installed.

☐ 29. Start the engine and immediately recheck the transmission fluid. Quickly add transmission fluid to the transmission to bring it up to a safe level.

☐ 30. With the engine still running, move the shift selector lever through the forward and reverse quadrant positions and make the following checks.

Describe the action of the drive wheels with the selector lever in reverse position: _____
_____

Check upshifts and downshifts with the selector in the drive position at various throttle openings.

Do upshifts occur at the proper times? Yes ___ No ___

Do downshifts occur at the proper times? Yes ___ No ___

☐ 31. Raise the vehicle and check for leaks.

☐ 32. Lower the vehicle and recheck the fluid level.

☐ 33. Road test the vehicle, carefully observing transmission operation.

Are the shifts firm, without harshness or slippage? Yes ___ No ___

Do upshifts occur at proper times? Yes ___ No ___

Do downshifts occur at proper times? Yes ___ No ___

☐ 34. After the road test, readjust linkage if necessary to obtain the best shift speeds and feel.

☐ 35. Look for signs of leaks and recheck the fluid level.

☐ 36. Clean the work area and return tools and equipment to storage.

Job 58—Install an Automatic Transmission (continued)

☐ 37.   Did you encounter any problems during this procedure? Yes ___ No ___

If Yes, describe the problems: _____

_____

What did you do to correct the problems? _____

_____

☐ 38.   Have your instructor check your work and sign this job sheet.

Performance Evaluation—Instructor Use Only

Did the student complete the job in the time allotted? Yes ___ No ___

If No, which steps were not completed?_____

How would you rate this student's overall performance on this job?_____

5–Excellent, 4–Good, 3–Satisfactory, 2–Unsatisfactory, 1–Poor

Comments: _____

_____

_____

INSTRUCTOR'S SIGNATURE _____

Notes

Job 59—Install an Automatic Transaxle

After completing this job, you will be able to install an automatic transaxle.

Procedures

☐ 1. Obtain an automatic transaxle and a vehicle to be used in this job. Your instructor may direct you to perform this job on a shop vehicle.

☐ 2. Gather the tools needed to perform the following job. Refer to the tools and materials list at the beginning of the project.

☐ 3. Raise the vehicle.

> ### Warning
> Raise and support the vehicle in a safe manner. Always use approved lifts or jacks and jack stands.

> ### Note
> Before installing the transaxle, perform a final inspection of the engine and transaxle. Check engine core plugs and the rear crankshaft seal for leaks. Check the dowel pins and dowel pin holes to ensure they are in good condition. Also, ensure that the mating surfaces between the engine and transaxle are properly cleaned. Correct any problems before installing the transaxle.

☐ 4. Secure the torque converter in place on the transaxle with a torque converter holding tool or a C-clamp.

☐ 5. Install any brackets that were removed during transaxle service, **Figure 59-1**.

Figure 59-1. If any transaxle mounting brackets were removed during transaxle service, remember to reinstall them before installing the transaxle.

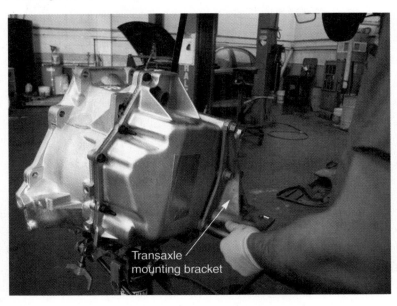

Transaxle mounting bracket

☐   6.   Use a transmission jack to raise the transaxle into position under the vehicle.

☐   7.   Slide the transaxle case into engagement with the rear of the engine, **Figure 59-2**.

☐   8.   Loosely install at least two of the converter housing–to-engine attaching bolts.

☐   9.   Ensure that the torque converter can turn freely. If it cannot, the converter is not properly installed.

☐ 10.   If the converter is correctly installed, install the remainder of the transaxle converter housing bolts and torque all bolts to specifications.

☐ 11.   Align the converter fastener holes with the holes in the flywheel, **Figure 59-3**, and then install and tighten the converter-to-flywheel bolts (or nuts) to specifications.

Figure 59-2. Slide the transaxle into position on the engine. If the engine and transaxle mounting surfaces do not mate easily, check for obstructions and retry.

Figure 59-3. Align the converter mounting holes with the holes in the flywheel.

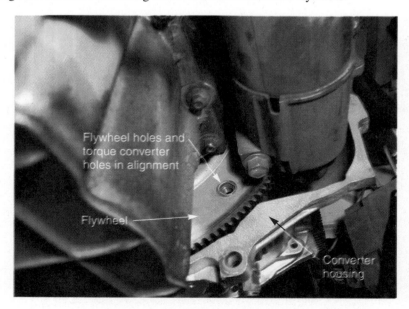

Job 59—Install an Automatic Transaxle (continued)

☐ 12.   Install the starter, if applicable, and the dust cover.

☐ 13.   Install the engine mounts and/or engine cradle.

☐ 14.   Install the front drive axles. Refer to Job 67 for CV axle installation procedures.

☐ 15.   Install the oil cooler lines and tighten the fittings.

☐ 16.   Install the exhaust pipes and the stabilizer bars and other suspension components previously removed.

☐ 17.   Reinstall any other frame and body parts that were removed for clearance.

☐ 18.   Remove the transmission jack from under the vehicle and lower the vehicle. Leave the wheels off the ground so the drive wheels are free to turn.

☐ 19.   Connect the TV and shift linkages and make preliminary adjustments.

☐ 20.   Reconnect the speedometer cable, electrical connectors, ground straps, modulator vacuum line, and other components that were reached from under the hood.

> **Caution**
>
> ⚠ Do not allow the wheels to turn without supporting the lower control arms. Raise the control arms to a position near their normal position when the vehicle is resting on the ground. If you do not raise the control arms to this position, the CV joints will operate at excessive angles, possibly causing joint damage.

☐ 21.   Remove the engine holding fixture.

☐ 22.   Install the negative battery cable.

☐ 23.   Check that all vacuum lines and electrical connectors are in place on the engine.

☐ 24.   Check all linkages to make sure they are properly adjusted. Adjust them if necessary.

☐ 25.   Add the proper transmission fluid until the transaxle dipstick reads two quarts over the full mark.

☐ 26.   Start the engine and immediately recheck the transaxle fluid level. Quickly add transmission fluid to the transaxle to bring it up to a safe level.

☐ 27.   With the engine still running, move the shift selector lever through the drive and reverse quadrant positions. Check out action of the drive wheels in reverse and allow the transaxle to upshift and downshift a few times by operating the throttle pedal in drive.

☐ 28.   Put the vehicle in park, raise the vehicle slightly, and remove the jack stands supporting the control arms. Lower the vehicle to the ground and recheck the fluid level. Add fluid as needed.

☐ 29.   Close the hood and road test the vehicle, carefully observing transaxle operation.

☐ 30.   After the road test, look for signs of leaks and recheck the fluid level.

☐ 31.   Clean the work area and return tools and equipment to storage.

☐ 32.   Did you encounter any problems during this procedure? Yes ___ No ___

If Yes, describe the problems: _____

_____

What did you do to correct the problems? _____

_____

☐ 33.   Have your instructor check your work and sign this job sheet.

Performance Evaluation—Instructor Use Only

Did the student complete the job in the time allotted? Yes ___ No ___

If No, which steps were not completed?_____

How would you rate this student's overall performance on this job?_____

5–Excellent, 4–Good, 3–Satisfactory, 2–Unsatisfactory, 1–Poor

Comments: _____

_____

_____

INSTRUCTOR'S SIGNATURE _____

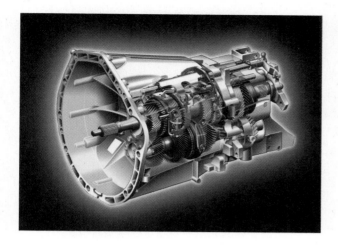

Project

Correcting Noise, Vibration, and Harshness Problems

18

Introduction

Many drive train noise and vibration problems have a simple cause; the trick is finding it. Diagnosis of noise, vibration, and harshness (NVH) problems can sometimes be time consuming. Using the proper diagnostic procedures can greatly reduce the amount of time you spend diagnosing drive train problems. In Job 60, you will isolate a noise and/or vibration problem in a drive train component. You will measure and correct drive shaft runout in Job 61, drive shaft angles in Job 62, and drive shaft balance in Job 63.

Project 18 Jobs

- Job 60—Diagnose a Drive Train Problem
- Job 61—Check and Correct Drive Shaft Runout
- Job 62—Check and Adjust Drive Shaft Angles
- Job 63—Check and Adjust Drive Shaft Balance

Tools and Equipment

The following list contains the tools and materials that may be needed to complete the jobs in this project. The items used will depend on the make and model of vehicle being serviced.

- Vehicle with noise, vibration, and/or leakage complaint.
- Applicable service information.
- Drive shaft runout measurement tools.
- Drive shaft angle measurement tools.
- Hand tools as needed.
- Air-powered tools as needed.
- Safety glasses and other protective equipment as needed.

Safety Notice

Before performing this job, review all pertinent safety information in the text, and review safety information with your instructor.

Notes

Job 60—Diagnose a Drive Train Problem

After completing this job, you will be able to diagnose a drive train problem.

Procedures

☐ 1. Obtain a vehicle to be used in this job. Your instructor may direct you to perform this job on a shop vehicle.

☐ 2. Gather the tools needed to perform the following job. Refer to the tools and materials list at the beginning of the project.

☐ 3. Determine the exact problem by performing the following actions:
- Question the vehicle driver.
- Consult previous service records.
- Road test the vehicle.

Briefly describe the problem: _____

_____

_____

_____

Does your description of the problem agree with the driver's original description of the problem? Yes ___ No ___

If No, explain: _____

_____

_____

☐ 4. Check for obvious problems by performing the following actions:
- Make a visual underhood inspection.
- Check the levels and conditions of fluids in all drive train components.
- Retrieve trouble codes.
- Make basic electrical and mechanical checks.

Describe any obvious problems found:_____

_____

_____

Do you think that this could be the cause of the problem determined in step 3?

Yes ___ No ___

Explain your answer:_____

_____

_____

_____

_____

☐ 5. Determine which component or system is causing the problem by performing the following actions:

 Note

In cases where an obvious problem is located before all steps are completed, the instructor can approve the elimination of some steps.

- Obtain the proper service literature and other service information.
- Use the service information to compare the problem with any trouble codes retrieved.

- Obtain needed test equipment.
- Check engine systems.
- Check the exhaust system.
- Check drive train components.
- Check the heater/air conditioner.

List any abnormal conditions or readings that were not within specifications: _____

_____

_____

_____

Could any of these be the cause of the problem? Yes ___ No ___

Explain your answer:_____

_____

_____

_____

6. Eliminate possible causes of the problem by checking the components that could create the abnormal conditions identified in step 4.
   - Make physical/visual checks.
   - Make electrical checks.

   List any defective components or systems that you identify:_____

   _____

   _____

   Do you think these defects could be the cause of the original problem? Yes ___ No ___

   Explain your answer:_____

   _____

   _____

7. Isolate and recheck the suspected causes of the problem:
   - Recheck damaged components.
   - Repeat electrical checks.

   Did the rechecking procedure reveal any new problems? Yes ___ No ___

   If Yes, describe the additional problems: _____

   _____

   Did the rechecking procedure establish that suspected components were in fact good?
   Yes ___ No ___

   If there were any differences with the conclusions in step 5, what was the reason? _____

   _____

   _____

   _____

8. Correct the defect by making necessary repairs or adjustments.

   Briefly describe the services performed: _____

   _____

   _____

   _____

9. Recheck system operation by performing either or both of the following actions:
   - Make checks using test equipment.
   - Road test vehicle.

Job 60—Diagnose a Drive Train Problem (continued)

Did the services in step 8 correct the problem? Yes ___ No ___

If No, what steps should you take now? _____

_____

_____

_____

☐ 10. Clean the work area and return tools and equipment to storage.

☐ 11. Did you encounter any problems during this procedure? Yes ___ No ___

If Yes, describe the problems: _____

_____

What did you do to correct the problems? _____

_____

☐ 12. Have your instructor check your work and sign this job sheet.

Performance Evaluation—Instructor Use Only

Did the student complete the job in the time allotted? Yes ___ No ___

If No, which steps were not completed?_____

How would you rate this student's overall performance on this job?_____

5–Excellent, 4–Good, 3–Satisfactory, 2–Unsatisfactory, 1–Poor

Comments: _____

_____

_____

INSTRUCTOR'S SIGNATURE _____

Notes

Job 61—Check and Correct Drive Shaft Runout

After completing this job, you will be able to measure and correct drive shaft runout.

Procedures

☐ 1. Obtain a vehicle to be used in this job. Your instructor may direct you to perform this job on a shop vehicle.

☐ 2. Gather the tools needed to perform the following job. Refer to the tools and materials list at the beginning of the project.

☐ 3. Place the vehicle's transmission in neutral.

☐ 4. Raise the vehicle on a hoist or with a jack so that the rear wheels are free to turn. Support the ends of the rear axle housing with jack stands, as necessary, to keep the transmission and rear axle assembly in normal alignment.

> **Warning**
> ⚠ Raise and support the vehicle in a safe manner. Always use approved lifts or jacks and jack stands.

> **Note**
> The drive shaft must *not* be at a sharp angle during runout measurement. If the vehicle is raised so the rear axle assembly is hanging unsupported, the test results will not be accurate. Also, if the wheels cannot rotate freely, the runout check cannot be made.

☐ 5. Select a spot on the front of the drive shaft to attach a dial indicator.

☐ 6. Manually rotate the drive shaft to make sure that the spot selected does not have any weights, flat spots, weld splatter, or any other variations that would cause an incorrect reading.

☐ 7. Mount the indicator so that its plunger contacts the drive shaft. The indicator can be mounted on a tall floor stand under the car or clamped to the underside of the vehicle.

☐ 8. Zero the indicator.

☐ 9. Rotate the drive shaft slowly by hand.

☐ 10. Observe and record the maximum variation of the dial indicator.

☐ 11. Reposition the dial indicator at the center and then the rear of the drive shaft, repeating steps 6 through 10 for each location. Maximum drive shaft runout, at any point, should typically be no more than 0.040″ (1 mm).
Reading at front:_____
Reading at center:_____
Reading at rear:_____
Is runout within specifications? Yes ___ No ___
If No, take the following steps:
a. Remove the drive shaft from the differential pinion yoke.
b. Rotate the drive shaft a half turn.
c. Reinstall the drive shaft.
d. Recheck runout.

Did this correct the problem? Yes ___ No ___

If No, replace the drive shaft or send it to a drive shaft specialty shop for repairs. If the specialty shop says that the drive shaft is okay, the pinion flange should be checked.

☐ 12. Clean the work area and return tools and equipment to storage.

☐ 13. Did you encounter any problems during this procedure? Yes ___ No ___

If Yes, describe the problems: _____

_____

What did you do to correct the problems? _____

_____

☐ 14. Have your instructor check your work and sign this job sheet.

Performance Evaluation—Instructor Use Only

Did the student complete the job in the time allotted? Yes ___ No ___

If No, which steps were not completed?_____

How would you rate this student's overall performance on this job?_____

5–Excellent, 4–Good, 3–Satisfactory, 2–Unsatisfactory, 1–Poor

Comments: _____

_____

_____

INSTRUCTOR'S SIGNATURE _____

Job 62—Check and Adjust Drive Shaft Angles

After completing this job, you will be able to measure and adjust drive shaft angles.

Procedures

☐ 1. Obtain a vehicle to be used in this job. Your instructor may direct you to perform this job on a shop vehicle.

☐ 2. Gather the tools needed to perform the following job. Refer to the tools and materials list at the beginning of the project.

☐ 3. Make sure that the rear axle housing is at its curb-height specification (normal position, or ride height) relative to the vehicle body. The wheels do not have to turn freely, but you must have some way to rotate the drive shaft by 90°. The gas tank should be full, and there should be nothing heavy in the car.

☐ 4. Bounce the vehicle up and down to take any windup out of the springs.

☐ 5. If your vehicle specifications call for checking the distance between the top of the rear axle housing axle tubes and the bottom of the chassis, do so now, before checking the U-joint angles. If the distance is too great, you may have to add some weight to the vehicle to obtain the proper readings. If the distance is too short, the springs are sagging and should be replaced.

☐ 6. Clean off any dirt, undercoating, and rust from all of the U-joint bearing cups. This will help ensure an accurate measurement.

☐ 7. Place the inclinometer on the rear U-joint bearing cup that passes through the drive shaft yoke. Make sure that the bearing cup is facing downward, or pointing straight up and down.

☐ 8. Center the bubble in the sight glass and record the measurement.

☐ 9. Rotate the drive shaft 90°. This can be done by raising one drive wheel and rotating the wheel to obtain 90° shaft movement. Place the gauge on the rear U-joint bearing cup that passes through the differential pinion yoke/flange, center the bubble in the sight glass, and record the measurement.

☐ 10. Compute the angle. If the slopes run in *opposite* directions, *add* the readings, **Figure 62-1**. If the slopes of the connected components run in the *same* direction, *subtract* the smaller reading from the larger reading. See **Figure 62-2**. The answer is the rear U-joint angle.

 Angle:_____

☐ 11. Repeat steps 6 through 10 at the front U-joint.

 Angle:_____

☐ 12. If applicable, repeat steps 6 through 10 at the center U-joint.

 Angle:_____

☐ 13. Compare both front and rear (and center if applicable) U-joint angles against specifications.

Note

If the U-joint angles are less than the maximum allowable angles, no adjustment is necessary. However, if the U-joint angles are greater than the maximum allowable angle, they must be adjusted.

Figure 62-1. In this example, the differential pinion yoke and the drive shaft yoke have opposite slope directions. The measured angles are added together to determine the drive shaft angle. A—When the drive shaft yoke lugs are aligned vertically, the bottom of the U-joint tilts toward the front of the vehicle. B—The drive shaft assembly is rotated 90°. With the differential yoke lugs aligned vertically, the bottom of the U-joint tilts toward the back of the vehicle.

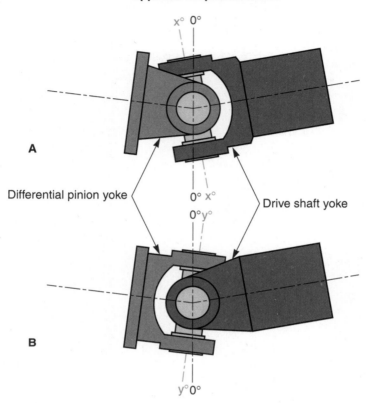

Opposite Slope Directions

Differential pinion yoke

Drive shaft yoke

x°+ y°= drive shaft angle

14. Use one of the following methods to adjust drive shaft angles as needed:
    a. Add or remove shims from the transmission mount.
    b. Loosen the control arms attaching the rear axle assembly to the vehicle chassis and move the rear axle assembly up or down.
    c. If a and b do not solve the problem, change the front engine mounts.

15. Recheck drive shaft angle to ensure that the adjustment was successful.

16. Clean the work area and return tools and equipment to storage.

17. Did you encounter any problems during this procedure? Yes ___ No ___
    If Yes, describe the problems: _____
    _____
    What did you do to correct the problems? _____
    _____

18. Have your instructor check your work and sign this job sheet.

Name _____ Date _____

Instructor _____ Period _____

Job 62—Check and Adjust Drive Shaft Angles (continued)

Figure 62-2. In this example, the differential pinion yoke and the drive shaft yoke have the same slope direction. The smallest measured angle is subtracted from the largest to determine the drive shaft angle. A—When the drive shaft yoke lugs are aligned vertically, the bottom of the U-joint tilts toward the front of the vehicle. B—After the drive shaft assembly is rotated 90°, the bottom of the U-joint still tilts toward the front of the vehicle.

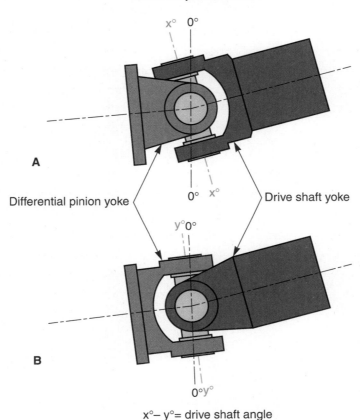

Same Slope Direction

A

Differential pinion yoke   Drive shaft yoke

B

x°− y°= drive shaft angle

Notes

Job 63—Check and Adjust Drive Shaft Balance

After completing this job, you will be able to measure and adjust drive shaft balance.

Procedures

☐ 1. Obtain a vehicle to be used in this job. Your instructor may direct you to perform this job on a shop vehicle.

☐ 2. Gather the tools needed to perform the following job. Refer to the tools and materials list at the beginning of the project.

☐ 3. Raise the vehicle on a hoist with the rear axle housing supported and the wheels free to rotate.

> **Warning**
>
> ⚠ The vehicle must be raised and supported in a safe manner. Always use approved lifts or jacks and jack stands.

☐ 4. Check the U-joints for wear and check the drive shaft for extreme bends or dents.
Are the U-joints in good condition? Yes ____ No ____
If No, replace the U-joints.
Is the drive shaft in good condition? Yes ____ No ____
If No, repair or replace the drive shaft as needed.

☐ 5. Remove both rear wheels. This way, the balance or imbalance of the wheels will not affect the rotation of the drive shaft.

☐ 6. Reinstall the wheel lug nuts with the flat sides inward. This will hold the brake drums or discs in place.

☐ 7. Mark and number the drive shaft at four spots, 90° apart. The marks should be made at the rear of the shaft, ahead of any balance weights.

☐ 8. Start the engine and shift the transmission into a direct (1:1) gear.

> **Warning**
>
> ⚠ Use extreme caution when working around a spinning drive shaft. Do not let any part of your body come in contact with the drive shaft. Make sure your hair does not get caught in a U-joint.

☐ 9. Run the vehicle at approximately 55 mph (89 km/h), noting the amount of vibration.

☐ 10. Turn off the engine.

☐ 11. Install two hose clamps near the rear of the drive shaft. Align the screws of both clamps at one of the four marks that you made on the shaft. Make sure that the clamps will not hit any stationary part of the vehicle when the drive shaft is rotating, then tighten the clamps securely.

☐ 12. Start the engine and shift the transmission into a direct (1:1) gear.

☐ 13. Run the vehicle at approximately 55 mph (89 km/h) and note the amount of vibration.

> **Caution**
>
> To avoid overheating the engine or transmission, do not operate the vehicle on the hoist for long periods of time.

☐ 14. Press on the brake pedal and push in the clutch pedal to stop drive shaft rotation.

☐ 15. Shift into neutral and stop the engine.

☐ 16. Rotate the clamps to the next mark on the drive shaft.

☐ 17. Repeat steps 12 through 16 until the vehicle has been tested with the clamps at all four spots on the shaft.

☐ 18. Decide which point produced the least vibration and reposition the clamps at that point.

☐ 19. Rotate the clamps away from each other about 45° by rotating each clamp approximately half the total distance between the point of minimum imbalance and the next mark on the drive shaft.

☐ 20. Run the vehicle at approximately 55 mph (89 km/h) and note whether the vibration has been reduced.

☐ 21. Repeat steps 18 through 20, moving the clamps away from each other in smaller amounts until the vibration is at its minimum.

☐ 22. Reinstall the wheels and road test the vehicle.

Is the problem solved? Yes ___ No ___

If No, refer to Job 60.

☐ 23. Clean the work area and return tools and equipment to storage.

☐ 24. Did you encounter any problems during this procedure? Yes ___ No ___

If Yes, describe the problems: _____

_____

What did you do to correct the problems? _____

_____

☐ 25. Have your instructor check your work and sign this job sheet.

Performance Evaluation—Instructor Use Only

Did the student complete the job in the time allotted? Yes ___ No ___
If No, which steps were not completed?_____
How would you rate this student's overall performance on this job?_____
5–Excellent, 4–Good, 3–Satisfactory, 2–Unsatisfactory, 1–Poor
Comments: _____

_____

_____

INSTRUCTOR'S SIGNATURE _____

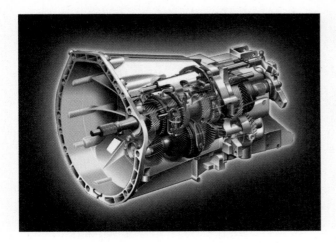

Project
Performing Manual Drive Train Service
19

Introduction

Often, what seems like a major drive train problem can have a simple cause. Sometimes, the problem is in the transmission, transaxle, or transfer case linkage or an electrical device. Adjusting the linkage or replacing a switch or sensor can eliminate the problem. In Job 64, you will diagnose a drive train problem.

Simple things such as loose fasteners or overfilling cause many leakage problems. Other leaks are caused by leaking gaskets or seals, or a damaged component. In many cases, finding the leak location is more difficult than repairing it. In Job 65, you will locate and correct a leak in a drive train component.

Project 19 Jobs

- Job 64—Diagnose and Service Linkage and Switches/Sensors
- Job 65—Diagnose Drive Train Component Leakage

Tools and Equipment

The following list contains the tools and materials that may be needed to complete the jobs in this project. The items used will depend on the make and model of vehicle being serviced.

- Vehicle with a linkage, switch, or sensor problem.
- Vehicle with a drive train leak.
- Applicable service information.
- Ohmmeter or multimeter.
- Hydraulic lift (or jacks and jack stands).
- Drain pan.
- Replacement electrical device.
- Replacement gaskets or seals.
- Specified lubricant.
- Hand tools as needed.
- Air-powered tools as needed.
- Safety glasses and other protective equipment as needed.

Safety Notice

Before performing this job, review all pertinent safety information in the text and discuss safety procedures with your instructor.

Notes

Job 64—Diagnose and Service Linkage and Switches/Sensors

Procedures

☐ 1. Obtain a vehicle to be used in this job. Your instructor may direct you to perform this job on a shop vehicle.

☐ 2. Gather the tools needed to perform the following job. Refer to the tools and materials list at the beginning of the project.

Check and Correct Linkage Complaint

☐ 1. Make sure that the problem is not caused by the clutch. Refer to Job 71.

☐ 2. Operate the shift linkage to check for binding or excessive looseness.
   • Binding. ___
   • Looseness. ___
   • Jumps out of gear. ___
   • Other. ___ Describe: _____
   List the gears in which problem occurs: _____

☐ 3. Raise the vehicle.

> **Warning**
> ⚠ The vehicle must be raised and supported in a safe manner. Always use approved lifts or jacks and jack stands.

☐ 4. Examine the linkage for the following problems:
   • Loose fasteners ___
   • Worn pivot points ___
   • Lack of lubrication ___
   • Bends or other damage ___
   • Misadjusted linkage ___
   Note the common wear and adjustment points shown in **Figure 64-1**.
   Were any problems found? Yes ___ No ___
   If Yes, could they be the cause of the shifting problem? Yes ___ No ___
   If problems were observed, go to step 6. If no problems were found, go to step 5.

☐ 5. If the above actions do not solve the problem, disconnect the linkage from the transmission, transaxle, or transfer case.
   Does the linkage now operate correctly? Yes ___ No ___
   If Yes, the unit has an internal problem. If No, repeat step 4 to isolate the problem.

☐ 6. Consult your instructor about what steps to take to correct the problem. Steps may include some of the following actions:
   • Tightening fasteners.
   • Lubricating moving parts.
   • Replacing worn or damaged parts.
   • Adjusting linkage.
   Steps to be taken: _____
   _____
   _____
   _____

Figure 64-1. Note the wear and adjustment points shown here. A—In this arrangement, control rods connect the shift mechanism to the shift levers on the transmission. B—This shift mechanism uses cables to transmit the driver's shifts to the transmission.

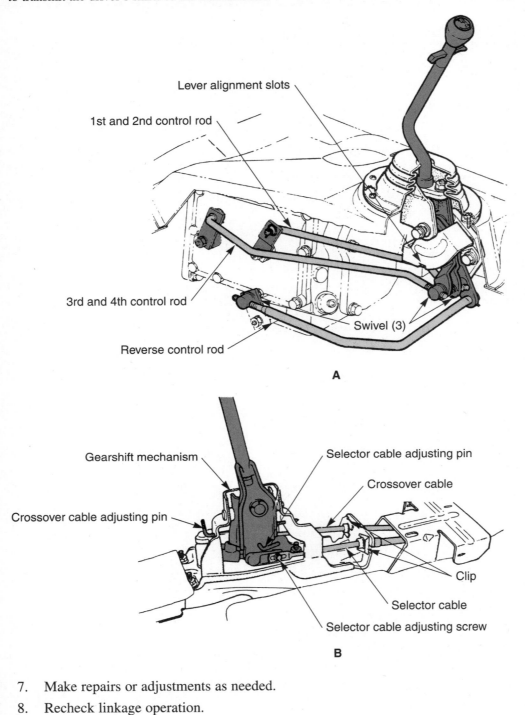

A

B

7. Make repairs or adjustments as needed.

8. Recheck linkage operation.
   Does the unit shift properly? Yes ___ No ___
   If No, consult with your instructor about further steps to be taken.

Job 64—Diagnose and Service Linkage and Switches/Sensors (continued)

Check and Correct Switch or Sensor Complaint

> **Note**
>
> Many drive train switch or sensor problems can be located with the proper scan tool. Refer to Job 11.

☐ 1. Locate the switch or sensor to be tested. Many electrical devices are threaded directly into the case, **Figure 64-2**.

☐ 2. Determine the following from the manufacturer's service literature:

Correct switch/sensor resistance reading:_____

Terminals at which to take readings:_____

_____

Position of manual linkage connected to the switch or sensor: _____

_____

☐ 3. Obtain an ohmmeter or multimeter.

☐ 4. Set the ohmmeter or multimeter to the appropriate range and calibrate (zero) the meter.

> **Note**
>
> If you are using a digital multimeter, it is usually not necessary to calibrate it.

☐ 5. Ensure that the ignition switch is in the *off* position.

☐ 6. Remove the electrical connector from the switch or sensor to be tested.

Figure 64-2. Many electrical devices used with manual transmissions or transaxles are threaded directly into the case.

> **Caution**
>
> Ensure that all electrical power is removed from the switch or sensor before beginning the ohmmeter test.

☐ 7. Place the ohmmeter test leads on the appropriate device terminals, **Figure 64-3**. Some devices can be tested by disconnecting a remote connector and accessing the correct connector pins, **Figure 64-4**.

☐ 8. Observe the meter reading and compare it against specifications.

Measured resistance:_____

Specified resistance:_____

Figure 64-3. Note the connections used to test this sensor. Some speed sensors, including the one shown here, can be turned to determine whether they are producing a signal. Other speed sensors are checked for proper resistance.

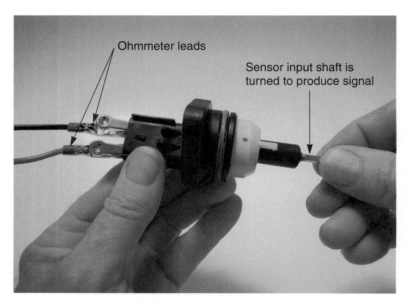

Figure 64-4. Sometimes an easily accessed electrical connector can be used to test a sensor or other electrical component. You must consult the proper service manual to be sure that you are accessing the proper pins.

Job 64—Diagnose and Service Linkage and Switches/Sensors (continued)

☐　9.　Determine from the ohmmeter readings whether the switch or sensor is good or bad.

☐　10.　Replace the switch or sensor, if needed. See the *Replace Sensor or Switch* section in this job.

☐　11.　Reconnect the electrical connector and reinstall any other components that were removed to gain access to the sensor or switch.

☐　12.　Ensure that the sensor or switch is operating properly.

Replace Sensor or Switch

☐　1.　Disconnect the battery negative cable.

☐　2.　Raise the vehicle if necessary.

> **Warning**
>
> ⚠ The vehicle must be raised and supported in a safe manner. Always use approved lifts or jacks and jack stands.

☐　3.　Remove any parts that prevent access to the sensor or switch.

☐　4.　If necessary, position a drain pan under the sensor or switch.

☐　5.　Remove the sensor/switch electrical connector.

☐　6.　Remove the sensor or switch by unthreading it from the drive train part or by removing attaching screws that hold the part in place.

☐　7.　Compare the old and new parts to ensure that the replacement is correct.

☐　8.　Install the replacement sensor or switch, using a new seal or gasket where required.

☐　9.　Reconnect the electrical connector.

☐　10.　Reinstall any other parts that were removed.

☐　11.　If necessary, refill the drive train component with the proper lubricant, being careful not to overfill.

☐　12.　Remove the drain pan if necessary and lower the vehicle.

☐　13.　Start the engine and check sensor or switch operation. Road test the vehicle as needed.

☐　14.　If applicable, recheck fluid level and check for fluid leaks at the new sensor or switch.

☐　15.　Clean the work area and return tools and equipment to storage.

☐　16.　Did you encounter any problems during this procedure? Yes ___ No ___

　　　　If Yes, describe the problems: _____

　　　　_____

　　　　What did you do to correct the problems? _____

　　　　_____

☐　17.　Have your instructor check your work and sign this job sheet.

Performance Evaluation—Instructor Use Only

Did the student complete the job in the time allotted? Yes ＿＿ No ＿＿

If No, which steps were not completed?＿＿＿＿＿＿＿＿＿＿＿＿

How would you rate this student's overall performance on this job?＿＿＿＿

5–Excellent, 4–Good, 3–Satisfactory, 2–Unsatisfactory, 1–Poor

Comments: ＿＿＿＿＿＿＿＿＿＿＿＿＿＿＿＿＿＿＿＿＿＿＿＿＿＿＿＿＿＿＿＿＿＿

＿＿＿＿＿＿＿＿＿＿＿＿＿＿＿＿＿＿＿＿＿＿＿＿＿＿＿＿＿＿＿＿＿＿＿＿＿＿＿

INSTRUCTOR'S SIGNATURE ＿＿＿＿＿＿＿＿＿＿＿＿＿＿＿＿＿＿＿＿＿＿＿＿＿

Job 65—Diagnose Drive Train Component Leakage

After completing this job, you will be able to locate and diagnose leaks in drive train components.

Procedures

☐ 1. Obtain a vehicle to be used in this job. Your instructor may direct you to perform this job on a shop vehicle.

☐ 2. Gather the tools needed to perform the following job. Refer to the tools and materials list at the beginning of the project.

Visually Check for Leakage

☐ 1. If possible, determine from the owner what unit is leaking.

> **Note**
> Usually the owner will not know what is leaking and will only be aware of oil spots on the pavement, oil on the underside of vehicle, or a performance problem related to a low oil level.

☐ 2. Raise the vehicle.

> **Warning**
> ⚠ The vehicle must be raised and supported in a safe manner. Always use approved lifts or jacks and jack stands.

☐ 3. Obtain a drop light or other source of illumination.

☐ 4. Observe the underside of the vehicle for evidence of oil or grease. Slight seepage is normal.
   Is excessive oil or grease observed? Yes ___ No ___
   If Yes, where does the oil/grease appear to be coming from? _____
   _____

> **Note**
> Airflow under the vehicle will blow leaking oil backwards. The leak may be some distance forward from where the oil appears.

Use the Powder Method to Check for Leakage

☐ 1. Refill the unit to be checked with the proper lubricant. This step applies to drive train parts with a reservoir for liquid lubricant, such as the transmission, transaxle, transfer case, and rear axle assembly.

☐ 2. Thoroughly clean the area around the suspected leak.

☐ 3. Apply talcum powder to the area around the suspected leak.

☐ 4. Lower the vehicle from the lift and operate the vehicle for several miles, or carefully run it on the lift for 10–15 minutes.

5. Raise the vehicle, if necessary, and check the area around the suspected leak.

   Does the powder show streaks of lubricant? Yes ___ No ___

   If Yes, where does the lubricant appear to be coming from? _____

   _____

   If No, repeat this procedure on another suspected component, or attempt to locate the leak using the black light method.

Use the Black Light Method to Check for Leakage

1. Refill the unit to be checked with the proper lubricant, and add fluorescent dye to the unit through the dipstick or filler plug, being careful not to spill dye on the outside of the case.

2. Thoroughly clean the area around the suspected leak.

3. Lower the vehicle from the lift and operate the vehicle for several miles, or carefully run it on the lift for 10–15 minutes.

4. Raise the vehicle, if necessary.

5. Turn on the black light and direct it toward the area around the suspected leak. See **Figure 65-1**.

   Does the black light show the presence of dye? Yes ___ No ___

   If Yes, where does the dye and lubricant appear to be coming from? _____

   _____

   If No, repeat this procedure on another suspected component, or attempt to locate the leak using another detection method.

Figure 65-1. Fluorescent dye and a black light can be used to locate leaks. (Tracer Products Division of Spectronics Corporation)

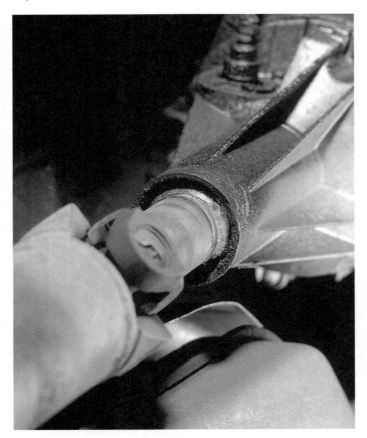

Job 65—Diagnose Drive Train Component Leakage (continued)

Repair the Leak

☐ 1. Consult your instructor about the steps to take to correct the leak. Steps may include some of the following actions:
- Tightening fasteners.
- Replacing gasket or seal. See Jobs 3 and 4.
- Replacing a cracked, broken, or punctured part.

What steps should be taken to correct the leak?_____

_____

_____

☐ 2. Make repairs as needed.

☐ 3. Recheck the unit for leaks.

Is the unit leaking? Yes ___ No ___

If No, go to step 4. If Yes, repeat one of the leak detection procedures until the leak has been located and corrected.

☐ 4. Clean the work area and return tools and equipment to storage.

☐ 5. Did you encounter any problems during this procedure? Yes ___ No ___

If Yes, describe the problems: _____

_____

What did you do to correct the problems? _____

_____

☐ 6. Have your instructor check your work and sign this job sheet.

Performance Evaluation—Instructor Use Only

Did the student complete the job in the time allotted? Yes ___ No ___
If No, which steps were not completed?_____
How would you rate this student's overall performance on this job?_____
5–Excellent, 4–Good, 3–Satisfactory, 2–Unsatisfactory, 1–Poor
Comments: _____

_____

_____

INSTRUCTOR'S SIGNATURE _____

Notes

_____
_____
_____
_____
_____
_____
_____
_____
_____
_____
_____
_____
_____
_____
_____
_____
_____
_____
_____
_____
_____
_____
_____
_____
_____
_____
_____
_____
_____
_____
_____
_____
_____
_____
_____
_____
_____

Project

Servicing Drive Shafts and CV Axles

20

Introduction

Rear-wheel drive vehicles always have a drive shaft. Drive shafts can be one-piece, two-piece, or three-piece designs. Almost all drive shafts use U-joints to provide a flexible coupling between the rotating parts. A slip yoke is installed somewhere on the drive shaft to allow for changes in drive shaft length.

Many modern vehicles, including all front-wheel drive vehicles, have CV axles equipped with CV joints. The CV axles must be removed in order to perform many service operations on front-wheel drive vehicles. Many technicians overlook the CV axle shaft bearings and seals when they service CV axles. Some technicians also forget that the front wheels have bearings that can cause problems. CV axle shaft bearings should always be suspected when the front end develops a whine or rumble. Seals can also fail, causing the bearings to quickly be ruined.

Drive shaft and CV axle service is relatively simple, but must be done carefully to avoid later vibration and noise problems or catastrophic part failures. In Job 66, you will remove and replace a one-piece or two-piece drive shaft and replace a U-joint. You will remove and replace a CV axle and replace a CV joint in Job 67. In Job 68, you will diagnose problems with, and replace, CV axle bearings and seals.

Project 20 Jobs

- Job 66—Remove, Inspect, and Reinstall U-Joints and a Drive Shaft
- Job 67—Remove, Inspect, and Reinstall a CV Axle
- Job 68—Diagnose and Service CV Axle Shaft Bearings and Seals

Tools and Materials

The following list contains the tools and materials that may be needed to complete the jobs in this project. The items used will depend on the make and model of vehicle being serviced.

- Vehicle with a suspected CV axle bearing problem.
- Vehicle with a U-joint requiring replacement.
- Hydraulic lift (or jacks and jack stands).
- Applicable service information.
- U-joint removal/installation tool.
- Hydraulic press, arbor press, or heavy-duty vise.
- Required press adapters.
- Specified lubricant.
- Bearing cup adapters.
- Front drive shaft nut socket.
- Bearing driver.
- Slide hammer with CV joint adapters.
- Wheel hub and bearing assembly puller.
- Ball joint separators.
- CV joint boot clamp installation tool.
- CV joint repair kit.
- Replacement CV joint boot.
- Front end alignment equipment.
- Hand tools as needed.
- Air-powered tools as needed.
- Safety glasses and other protective equipment as needed.

Safety Notice

Before performing this job, review all pertinent safety information in the text, and review safety information with your instructor.

Job 66—Remove, Inspect, and Reinstall U-Joints and a Drive Shaft

After completing this job, you will be able to remove and replace a drive shaft and U-joints.

Procedures

Note

The procedures presented in this job also apply to the front drive shaft of a four-wheel drive vehicle.

☐   1.   Obtain a vehicle to be used in this job. Your instructor may direct you to perform this job on a shop vehicle.

☐   2.   Gather the tools needed to perform the following job. Refer to the tools and materials list at the beginning of the project.

Remove the Drive Shaft

☐   1.   Place the vehicle's transmission in neutral.

☐   2.   Raise the vehicle.

Warning

⚠   The vehicle must be raised and supported in a safe manner. Always use approved lifts or jacks and jack stands.

☐   3.   Place match marks on the drive shaft and differential yokes to aid drive shaft reinstallation.

Caution

If the vehicle has a two piece drive shaft, mark the front and rear shafts so they can be reinstalled in the same position.

☐   4.   Remove the fasteners holding the rear of the drive shaft to the rear axle assembly. This is accomplished on most drive shafts by one of the following methods:

Remove the nuts and bolts holding the companion flange to the pinion flange. Then, separate the companion flange from the pinion flange.

or

Remove two U-bolts secured with nuts or two straps secured with cap screws.

Caution

Do not let loose bearing cups fall off their trunnions once the U-bolts or straps are removed.

☐   5.   If the vehicle has a two piece drive shaft, loosen and remove the center support bolts.

☐   6.   Remove the drive shaft assembly from the vehicle.

Note

On some vehicles with two piece drive shafts, the rear section of the drive shaft is removed first, and then the center support and front drive shaft section are removed.

7. Cap the rear of the transmission once the drive shaft is removed to prevent oil loss.

Repair a Cross-and-Roller U-Joint

Note

This procedure details the general removal and installation of a U-joint. If there is any variation between the procedures given here and the procedures in the manufacturer's service literature, use the manufacturer's procedure.

1. Attempt to rotate the two sides of the U-joint, **Figure 66-1**. Note any roughness, excessive play, noise, or binding.

   Describe the condition of the U-joints:_____

   _____

   _____

2. Place the drive shaft on a stand or workbench so that the shaft is horizontal.

3. Clamp the yoke of the U-joint to be replaced in a vise.

4. Mark the drive shaft yokes so they can be reassembled in the same position relative to each other. See **Figure 66-2**. This step is not necessary on a rear U-joint with only one yoke.

5. Remove the bearing cup retaining devices. Removal methods vary depending on the retaining device used:
   - Use a hammer and a small chisel to remove C-clips.
   - Use pliers to remove snap rings.
   - Use a heat gun or torch to soften injected plastic retainers.

6. Obtain a piece of pipe (or similar tool) having an inside diameter slightly larger in size than a bearing cup. Obtain a socket or other metal piece with an outside diameter slightly smaller in size than a bearing cup.

Figure 66-1. Attempt to move the drive shaft yokes by hand to detect problems.

Job 66—Remove, Inspect, and Reinstall U-Joints and a Drive Shaft (continued)

Figure 66-2. Mark the drive shaft yokes so that they can be reassembled in the same position as they were originally.

 7. Remove the bearing cup, using one of the following methods:

> **Caution**
>
> Be careful not to lose any needle bearings when you remove the bearing cups.

Press Method
a. Position the front yoke in a press to drive the bearing cups from the yoke. You can press the cups out with a vise, a C-clamp, or an arbor press.
b. Place the small diameter pressing tool against one of the front yoke bearing cups.
c. Place the larger-diameter tool over the opposite cup and against the yoke itself.
d. Slowly operate the press until the bearing cup is driven from the yoke as far as it will go. The small pressing tool will drive the bearing cup that is opposite it into the larger-diameter tool.

Hammer Method

> **Warning**
>
> Do not use a socket as a driver. Hammer blows can shatter a socket.

a. Place one of the front yoke bearing cups on top of the small pressing tool.
b. Place the larger-diameter tool over the top bearing cup and against the yoke.
c. Strike the larger-diameter tool with a hammer. This will cause the top cup to move up and out of the yoke and into the larger-diameter tool.
d. Rotate the drive shaft 180° and place the slip yoke in a vise.
e. With the slip yoke clamped across its lugs in a vise, place the larger-diameter tool over the top cup of the front yoke and against the yoke.

f. Strike the larger-diameter tool with a hammer. This will cause the front yoke lug to flex downward and the cup to move up out of the yoke and into the receiving piece (the larger-diameter tool).

C-Frame Driver Method

a. Place a C-frame driver (U-joint removal/installation tool) around the front drive shaft yoke, **Figure 66-3**.

b. Turn the driver handle to press the cup from the yoke.

8. If you are using the press or C-frame driver methods, repeat step 7 to remove the opposite bearing cup. To remove the remaining cup, place the small driver tool on the exposed trunnion and the large driver tool over the remaining cup.

9. Remove the slip yoke and U-joint from the drive shaft by slipping the front yoke lugs from the U-joint cross.

10. If necessary, complete the removal of the yoke bearing cups by tapping them out with a hammer.

11. Remove the bearing cups from the slip yoke by repeating steps 7 through 10.

12. Visually examine the U-joint and other drive shaft components, looking for the following:
   - Damaged or rusty needle bearings or bearing cups.
   - Worn, rough, rusty, or out-of-round U-joint trunnions.
   - Cracked, spalled, or pitted U-joint cross.
   - Burred yokes, caused by bearing cup removal, or damaged yokes.
   - Worn or damaged slip yoke splines.
   - Bent drive shaft.
   - Excessive pinion flange or companion flange runout.

   Inspect components using special measuring instruments, as required.

Figure 66-3. Install the C-frame driver around the drive shaft yoke. (OTC)

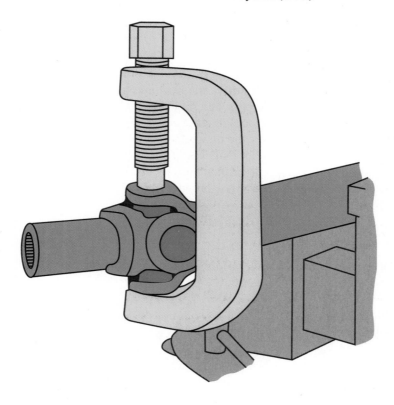

Job 66—Remove, Inspect, and Reinstall U-Joints and a Drive Shaft (continued)

Describe the condition of the components:_____

_____

_____

_____

_____

☐ 13. Report any drive shaft component defects to your instructor and determine which parts must be replaced.

> **Note**
> If either the cross or the bearings are worn, replace both. Never install new bearings on an old cross, or vice versa.

☐ 14. Obtain the proper replacement parts and compare them with the old parts to ensure that they are the correct replacements.

☐ 15. If bearing needles are loose, begin bearing cup installation by lining the cup with the needles, using grease to hold them in place. Then, install a new seal.

☐ 16. Push a bearing cup about one fourth of its length into either lug of the front yoke.

> **Note**
> Check for proper cup/yoke alignment and make sure the bearing needles are still in place.

☐ 17. Place the cross into the partially installed bearing cup.

☐ 18. Hammer or press the cup onto the cross trunnion, until the bearing cup is flush with the outside of the yoke and fully seated on the cross trunnion.

☐ 19. Repeat steps 15 through 18 to install the opposite bearing cup.

☐ 20. Install the retainer, snap rings, or clips, to ensure that the bearing cups are fully seated.

> **Note**
> If snap rings are used, install the ring gap so it points away from the yoke opening.

☐ 21. Check for free movement of the U-joint.

☐ 22. Repeat steps 15 through 21 to install the bearing cups in the slip yoke, making sure to align the match marks made before disassembly.

☐ 23. Ensure that the snap rings or clips are fully seated in their grooves.

Install the Drive Shaft

☐ 1. Lightly lubricate the outside of the slip yoke with automatic transmission fluid or gear oil, depending on the type of transmission the vehicle has.

☐ 2. Slide the slip yoke onto the transmission output shaft, being careful not to damage the rear seal.

> **Note**
>
> Ensure that the extension bushing and rear seal are in good condition before installing the slip yoke.

☐ 3. If the drive shaft has a center support bearing, align the center support with the cross-member and loosely install the center support bearing fasteners.

☐ 4. Install the rear drive shaft on the differential yoke. Make sure you align the match marks that were made during disassembly.

☐ 5. If the drive shaft is a two-piece type, be sure to align the front and rear drive shaft yokes, sometimes called phasing the drive shaft. A slip yoke installed between the two drive shafts may have a large spline that matches a missing spline in the shaft. This prevents improper installation.

> **Warning**
>
> ⚠ If the yokes are not aligned, the drive shaft will vibrate severely.

☐ 6. Tighten the drive shaft yoke and center support fasteners.

☐ 7. Once all drive shaft parts are installed, tighten the fasteners to the proper torque and check drive shaft alignment.

☐ 8. Lower the vehicle.

☐ 9. Road test the vehicle.

☐ 10. Clean the work area and return tools and equipment to storage.

☐ 11. Did you encounter any problems during this procedure? Yes ___ No ___

If Yes, describe the problems: _____

_____

What did you do to correct the problems? _____

_____

☐ 12. Have your instructor check your work and sign this job sheet.

Performance Evaluation—Instructor Use Only

Did the student complete the job in the time allotted? Yes ___ No ___

If No, which steps were not completed?_____

How would you rate this student's overall performance on this job?_____

5–Excellent, 4–Good, 3–Satisfactory, 2–Unsatisfactory, 1–Poor

Comments: _____

_____

_____

INSTRUCTOR'S SIGNATURE _____

Job 67—Remove, Inspect, and Reinstall a CV Axle

After completing this job, you will be able to remove, inspect, and reinstall a CV axle.

Procedures

☐ 1. Obtain a vehicle to be used in this job. Your instructor may direct you to perform this job on a shop vehicle.

☐ 2. Gather the tools needed to perform the following job. Refer to the tools and materials list at the beginning of the project.

> **Note**
>
> The following is a general procedure for CV axle service, including typical CV joint repair and CV joint boot replacement. Some Rzeppa and tripod joints are not repaired, and the entire assembly, including the boot, must be replaced. Also, some boots are made in two pieces and can be changed without removing the axle from the vehicle. Always consult the manufacturer's service manual for exact CV joint repair or boot replacement procedures.

Remove a CV Axle

☐ 1. Disconnect the battery negative cable.

☐ 2. Place the vehicle transmission in neutral.

☐ 3. Loosen the nut holding the CV axle stub shaft to the wheel hub and bearing assembly.

☐ 4. Raise the vehicle.

> **Warning**
>
> ⚠ The vehicle must be raised and supported in a safe manner. Always use approved lifts or jacks and jack stands.

☐ 5. Remove the wheel from the axle to be serviced.

☐ 6. Cover the outer CV boot with a shop cloth or boot protector.

☐ 7. If needed, remove the tie rod end nut and use a ball joint separator to free the tie rod end from the steering knuckle.

☐ 8. Remove the stub shaft nut. A special deep socket may be needed to remove the nut, **Figure 67-1**.

> **Note**
>
> On some vehicles it is necessary to remove the brake rotor, caliper, and wheel speed sensor to gain access to the CV axle shaft assembly.

☐ 9. Remove the CV axle shaft assembly from the steering knuckle. Some shafts will slide from the hub, while others must be pressed out.

Figure 67-1. A special deep socket is sometimes needed to remove the stub shaft nut. These sockets are usually sold in sets of three—30 mm, 32 mm, and 36 mm.

> **Note**
>
> Removal of the CV axle shaft assembly from the steering knuckle may require the removal of the tie rod end, strut assembly, or lower ball joint. Consult the proper manufacturer's service manual for exact procedures.

☐ 10. Remove the CV axle shaft assembly from the transaxle with a suitable puller or pry bar. If the transaxle uses a stub shaft that is attached to the CV axle by flanges, remove the bolts attaching the flanges.

> **Note**
>
> Cap the transaxle opening to prevent oil loss.

Repair a CV Joint and Replace the Boot

☐ 1. Lightly clamp the shaft in a vise with soft jaws.

☐ 2. Remove the clamps from the boot of the CV joint to be serviced (or from the boot to be replaced).

☐ 3. Pull the boot back from the CV joint and remove enough grease to further disassemble the CV joint.

☐ 4. If applicable, remove the clip holding the CV joint to the axle shaft.

☐ 5. Place alignment marks on the joint housing and the axle shaft to aid in reassembly.

☐ 6. Pull or tap the CV joint from the axle shaft.

☐ 7. Remove the CV joint boot from the axle shaft.

> **Note**
>
> Proceed to step 14 to continue with boot replacement only.

Job 67—Remove, Inspect, and Reinstall a CV Axle (continued)

☐　　8.　Disassemble the CV joint by one of the following procedures, depending on the CV joint type.

List the type of joint(s) to be serviced: _____

Rzeppa Joint
a. Tap the bearing cage downward with a brass drift, **Figure 67-2**, until the ball on the opposite side can be removed through the cage opening.
b. Repeat the previous step until all balls are removed.
c. Remove the cage and inner race from the housing.

Tripod Joint
a. Remove the snap ring or clip holding the joint together, **Figure 67-3**.
b. Pull the tripod assembly from the housing.
c. Remove the tripod assembly from the shaft. Tripod joints may be held in place by a clip or may be pressed on the shaft. Consult the service information for exact procedures to avoid damaging the joint or shaft.

> **Note**
> Some CV joints are not serviceable and must be replaced as a unit.

☐　　9.　Wash all parts in solvent and dry with compressed air.

☐　10.　Inspect all internal CV joint parts for excessive wear, cracked or chipped parts, or scoring.

Describe their condition: _____

_____

_____

Figure 67-2. Tap the bearing cage downward until the first ball can be removed. Repeat the process until all balls are out.

Figure 67-3. Many tripod joints are held in place by a clip, as shown here, or a retaining collar. (DaimlerChrysler)

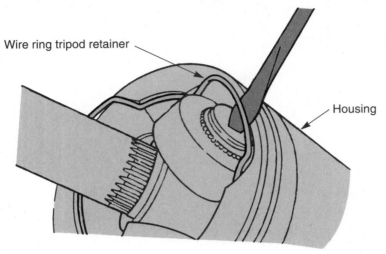

Wire ring tripod retainer

Housing

11. Check the joint splines and threads for damage.

Describe their condition: _____

_____

_____

12. Report any CV joint component defects that you found to your instructor. Then, determine which parts must be replaced.

13. Obtain the proper replacement parts and compare them with the old parts to ensure that they are the correct replacements.

14. Lightly lubricate all CV joint parts before proceeding.

15. Align and install all CV joint parts according to the manufacturer's instructions.

16. If the CV joint boot uses a one-piece clamp, slide the clamp onto the axle shaft. Often it is easier to place the clamps on the boot before installing it.

17. Install the CV joint boot on the axle shaft.

> **Note**
>
> It may be necessary to position the small end of the boot lip face in line with a mark or groove on the shaft.

18. Place the small clamp over the groove at the small end of the boot and tighten it by hand.

19. Place sufficient grease in the boot cavity to provide the CV joint with lubrication during vehicle operation.

> **Caution**
>
> CV joints must be lubricated with special grease, usually supplied with the replacement joint parts or boot. Do not use ordinary chassis or wheel bearing grease.

20. Position the CV joint over the splines of the axle shaft.

21. Engage the splines and tap the joint into position with a wooden or rubber hammer.

22. Install the joint-to-shaft clip, if used.

Job 67—Remove, Inspect, and Reinstall a CV Axle (continued)

☐ 23. Pull the CV joint boot over the CV joint housing. Ensure that the boot end fits snugly into the matching groove in the housing.

> **Caution**
>
> Ensure that the boot is not twisted or otherwise deformed during the installation process.

☐ 24. Place the large clamp over the groove in the boot. Tighten the clamp by hand.

☐ 25. Fully tighten both boot clamps using the correct special tool.

> **Caution**
>
> Do not cut through the boot by overtightening the clamp.

☐ 26. Check for free movement of the CV joint and otherwise ensure that it is properly assembled.

Install CV Axle

☐ 1. Install the inboard stub shaft of the CV axle assembly into the transaxle.

☐ 2. Install the outboard stub shaft of the CV axle assembly into the steering knuckle.

☐ 3. Install the stub shaft nut on the outboard shaft.

☐ 4. Adjust front wheel bearing preload, if necessary.

☐ 5. Install brake rotor, caliper, and wheel speed sensor as needed.

☐ 6. Reinstall the wheel.

☐ 7. Lower the vehicle.

☐ 8. Install the battery negative cable.

☐ 9. Start the engine.

☐ 10. Check brake operation.

☐ 11. Road test the vehicle.

☐ 12. Clean the work area and return tools and equipment to storage.

☐ 13. Did you encounter any problems during this procedure? Yes ___ No ___

If Yes, describe the problems: _____

_____

What did you do to correct the problems? _____

_____

☐ 14. Have your instructor check your work and sign this job sheet.

Performance Evaluation—Instructor Use Only

Did the student complete the job in the time allotted? Yes ___ No ___

If No, which steps were not completed?_____

How would you rate this student's overall performance on this job?_____

5–Excellent, 4–Good, 3–Satisfactory, 2–Unsatisfactory, 1–Poor

Comments: _____

_____

_____

INSTRUCTOR'S SIGNATURE _____

Job 68—Diagnose and Service CV Axle Shaft Bearings and Seals

After completing this job, you will be able to diagnose and service CV axle shaft bearings and seals.

Procedures

☐ 1. Obtain a vehicle to be used in this job. Your instructor may direct you to perform this job on a shop vehicle.

☐ 2. Gather the tools needed to perform the following job. Refer to the tools and materials list at the beginning of the project.

Diagnose Problems with CV Axle Shaft Bearings

☐ 1. Determine the exact nature of the complaint by discussing the problem with the driver.

☐ 2. Road test the vehicle listening for noises and vibration.

What condition did you observe? _____

_____

Could this be caused by the CV axle shaft bearings? Yes ___ No ___

If No, what other vehicle systems could be causing the problem? _____

_____

☐ 3. Place the vehicle's transaxle in neutral.

☐ 4. Raise the vehicle.

> **Warning**
>
> ⚠ The vehicle must be raised and supported in a safe manner. Always use approved lifts or jacks and jack stands.

☐ 5. Rotate the wheel on the side with the suspect bearings while listening for noise and feeling for roughness.

Is noise heard or roughness felt? Yes ___ No ___

☐ 6. Observe the wheel bearings seals for leakage.

Is any leakage present? Yes ___ No ___

☐ 7. Consult your instructor about the steps to take to correct bearing and seal problems.

Steps to be taken: _____

_____

_____

☐ 8. Make repairs as needed by following the steps in the second part of this job.

Remove and Replace CV Axle Shaft Bearings and Seals

☐ 1. Remove the CV axle assembly from the wheel hub. A special puller may be needed to remove the axle from the bearing, **Figure 68-1**. Removal of the CV axle assembly from the hub is covered in detail in Job 67. The CV axle can usually be placed to one side of the hub without removing it from the vehicle, **Figure 68-2**.

Figure 68-1. Many CV axle hubs are a press fit on the spindle and must be removed with a special puller. (OTC)

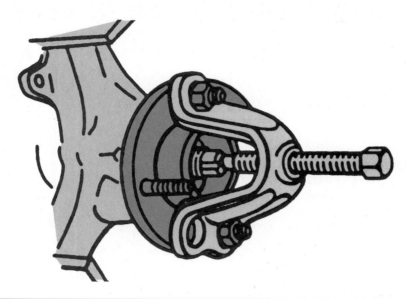

Figure 68-2. Once the CV axle is removed, place it to one side. Wire the CV axle out of the way if there is any danger that it could be damaged by other service operations.

> **Note**
>
> On some vehicles, it is not necessary to remove the hub from the vehicle to replace the bearings and seals. Consult the manufacturer's service literature before performing steps 2 and 3.

☐ 2. Remove any fasteners holding the spindle and hub to the suspension that were not removed when the CV axle was removed.

☐ 3. Remove the hub from the vehicle.

☐ 4. Place the hub on a clean bench.

☐ 5. Inspect the bearing seal and protector condition. Look for leakage and damage.

Job 68—Diagnose and Service CV Axle Shaft Bearings and Seals (continued)

> **Note**
> Seals are usually replaced when removed from the hub.

☐ 6. Remove the snap rings holding the bearings to the hub.

> **Note**
> Some hub and bearing assemblies are replaced as a unit and no disassembly is possible. If this type of assembly is used, proceed to step 15.

☐ 7. Remove the bearings from the hub. Most bearings must be pressed out using a hydraulic press and the proper adapters. A few bearings will slide out once the snap rings are removed.

☐ 8. Check the bearings for lack of lubricant, roughness, and visible damage.

☐ 9. Check the hub for wear and cracks.

☐ 10. Obtain new parts as necessary.

☐ 11. Compare the old and new parts.

   Are the new parts correct? Yes ___ No ___

☐ 12. Install the bearings in the hub.

> **Caution**
> Be careful when pressing in new bearings to avoid damage to the bearings and hub. Always use the proper adapters.

☐ 13. Install the snap rings that hold the bearing to the hub.

☐ 14. Install the seals and dust protector.

☐ 15. Reinstall the hub on the vehicle.

☐ 16. Reinstall the CV axle assembly into the hub. See Job 67.

☐ 17. Road test the vehicle.

   Has the problem been eliminated? Yes ___ No ___

   If No, consult with your instructor about further diagnostic measures to be taken.

☐ 18. Clean the work area and return tools and equipment to storage.

☐ 19. Did you encounter any problems during this procedure? Yes ___ No ___

   If Yes, describe the problems: _____

   _____

   What did you do to correct the problems? _____

   _____

☐ 20. Have your instructor check your work and sign this job sheet.

Performance Evaluation—Instructor Use Only

Did the student complete the job in the time allotted? Yes _____ No _____

If No, which steps were not completed?_____

How would you rate this student's overall performance on this job?_____

5–Excellent, 4–Good, 3–Satisfactory, 2–Unsatisfactory, 1–Poor

Comments: _____

_____

_____

INSTRUCTOR'S SIGNATURE _____

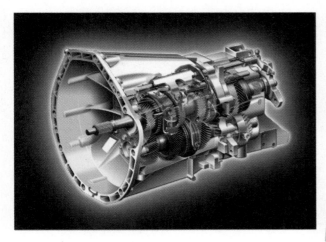

Project
Removing a Manual Transmission or Transaxle

21

Introduction

Once, manual transmissions could be found on almost any vehicle, from standard-sized pickups to luxury cars. Today, manual transmissions are usually found in a small percentage of vehicles, usually large trucks, small light-duty trucks, sports cars, and economy cars. The transmission often must be removed to change the clutch or perform internal repairs. In Job 69, you will remove and replace a manual transmission.

Manual transaxles are used in a few small cars and vans. Transaxles are also used on mid- and rear-engine cars. They are rarely found in trucks or large, front-engine cars. Transaxle removal is often more complex than rear-wheel drive transmission removal. Care must be taken not to damage the engine when the transaxle is removed. In Job 70, you will remove and replace a manual transaxle.

Project 21 Jobs

- Job 69—Remove a Manual Transmission
- Job 70—Remove a Manual Transaxle

Tools and Materials

The following list contains the tools and materials that may be needed to complete the jobs in this project. The items used will depend on the make and model of vehicle being serviced.

- Vehicle with a manual transmission.
- Hydraulic lift (or jacks and jack stands).
- Applicable service information.
- Engine holding fixture.
- Transmission jack(s) or floor jack(s).
- Drain pan.
- Specified manual transmission lubricant.
- CV axle removal tools.
- Hand tools as needed.
- Air-powered tools as needed.
- Safety glasses and other protective equipment as needed.

Safety Notice

Before performing this job, review all pertinent safety information in the text, and review safety information with your instructor.

Notes

Job 69—Remove a Manual Transmission

After completing this job, you will be able to remove a manual transmission.

Procedures

> **Note**
> The following procedure is a general guide only. This procedure assumes that the engine remains installed and only the transmission will be removed. Always refer to the manufacturer's service literature for specific procedures.

☐ 1. Obtain a vehicle to be used in this job. Your instructor may direct you to perform this job on a shop vehicle.

☐ 2. Gather the tools needed to perform the following job. Refer to the tools and materials list at the beginning of the project.

☐ 3. Identify the type of transmission being removed.

_____

Is the transmission electronically controlled? Yes _____ No _____
If Yes, briefly describe the function of the electronic controls.

_____

_____

_____

_____

> **Note**
> If the vehicle has a transfer case, removal and replacement procedures will differ from those given below. Refer to the manufacturer's service literature.

☐ 4. Disconnect the battery negative cable.

☐ 5. If the bell housing is integral with the transmission case, remove the upper bell housing–to-engine bolts.

> **Note**
> **Figure 69-1** shows the fasteners that must be removed to remove a typical manual transmission. Refer to the proper service information for exact procedures for the transmission that you are removing.

☐ 6. Remove the gearshift lever knob and remove the screws holding the rubber boot to the floor. If the vehicle has an internal floor-shift linkage (the shift lever enters directly into the transmission), remove the gearshift lever if called for by the manufacturer's instructions.

> **Note**
> If directed by the manufacturer, install an engine holding fixture before proceeding to step 6.

Figure 69-1. This exploded view shows the fasteners, connectors, and components that must be removed on most manual transmissions. Other transmissions will require that other parts be removed. (TTC/Tremec)

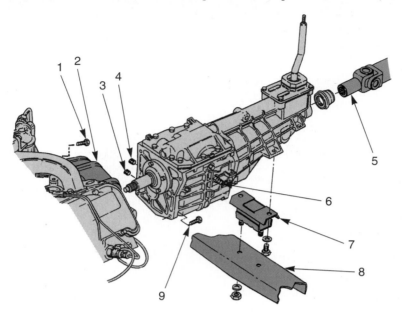

1. Bell housing–to-engine fasteners. These are removed if the bell housing is being removed with the transmission.
2. Bell housing. The bell housing is removed if the clutch requires service or if it is an integral part of the transmission housing.
3. Drain plug.
4. Fill plug.
5. Propeller shaft.
6. Back-up switch electrical connector. Make sure all electrical connectors are disconnected before removing the transmission.
7. Transmission mount. Note the various fasteners securing the mount between the transmission and crossmember.
8. Crossmember. The crossmember may be removed by unbolting it from the frame.
9. Transmission-to-bell housing fasteners.

Not Shown: Speedometer cable.

    ☐   7.   Raise the vehicle.

> **Warning**
>
> ⚠ The vehicle must be raised and supported in a safe manner. Always use approved lifts or jacks and jack stands.

☐   8.   Inspect the engine mounts to ensure that they are not broken. A broken engine mount can allow the engine to fall backward when the rear crossmember is removed.

☐   9.   Place a drip pan under the transmission and locate the oil drain plug.

☐ 10. Remove the drain plug and allow the transmission oil to drain into the pan.

☐ 11. Remove the drive shaft. Refer to Job 66 for specific instructions on drive shaft removal.

☐ 12. Remove the speedometer cable and any speed sensor wiring.

☐ 13. If necessary, remove the exhaust pipes.

Job 69—Remove a Manual Transmission (continued)

☐ 14. Disconnect the shift linkage at the transmission, if necessary.

☐ 15. If the bell housing is integral with the transmission case, remove any parts connected to the housing, such as the starter, clutch linkage, or hydraulic lines.

☐ 16. Remove any electrical connectors and mark them for reinstallation.

☐ 17. Disconnect or loosen the parking brake linkage.

☐ 18. If an engine holding fixture was not installed, support the engine with a jack placed under the rear of the engine or bell housing (if the bell housing is separate from transmission).

> **Warning**
>
> ⚠ Do not let the engine hang suspended by the front engine mounts. Failure to support the engine before removing the transmission can result in serious damage or injury.

☐ 19. Place a transmission jack (or floor jack) under the transmission. Raise the transmission slightly to remove weight from the crossmember.

☐ 20. Remove the fasteners holding the crossmember to the vehicle frame and to the transmission mount.

☐ 21. Remove the crossmember.

☐ 22. Remove the attaching bolts holding the transmission to the bell housing or engine, as applicable.

> **Caution**
>
> ◇ The transmission jack will support the transmission and will prevent damage to the clutch disc hub or transmission input shaft once attaching bolts are removed. Never let a transmission hang unsupported once these bolts are removed.

☐ 23. Slide the transmission straight back until the input shaft clears the clutch disc and bell housing.

☐ 24. Lower the transmission from the vehicle.

☐ 25. Clean the work area and return tools and equipment to storage.

☐ 26. Did you encounter any problems during this procedure? Yes ___ No ___

If Yes, describe the problems: _____

_____

What did you do to correct the problems? _____

_____

☐ 27. Have your instructor check your work and sign this job sheet.

Performance Evaluation—Instructor Use Only

Did the student complete the job in the time allotted? Yes ____ No ____

If No, which steps were not completed?_____

How would you rate this student's overall performance on this job?_____

5–Excellent, 4–Good, 3–Satisfactory, 2–Unsatisfactory, 1–Poor

Comments: _____

_____

_____

INSTRUCTOR'S SIGNATURE _____

Job 70—Remove a Manual Transaxle

After completing this job, you will be able to remove a manual transaxle used in a front-wheel drive or a rear-/mid-engine vehicle.

Procedures

> **Note**
>
> The following procedure is a general guide only. This procedure assumes that the engine remains installed and only the transaxle will be removed. Always refer to the manufacturer's service literature for specific procedures.

- [ ] 1. Obtain a vehicle to be used in this job. Your instructor may direct you to perform this job on a shop vehicle.
- [ ] 2. Gather the tools needed to perform the following job. Refer to the tools and materials list at the beginning of the project.
- [ ] 3. Disconnect the battery negative cable.
- [ ] 4. Support the engine from the top to take the weight off of the bottom engine mounts.

> **Caution**
>
> Special fixtures are used to support the engine. Do not attempt to remove the transaxle unless you have the tools to perform this step. **Figure 70-1** shows a typical holding fixture.

- [ ] 5. While under the hood, remove the upper transaxle case–to-engine bolts, clutch and shift linkages, electrical connectors, and ground straps. If necessary, mark electrical connectors for reinstallation.

Figure 70-1. Use the proper holding fixture to ensure that the engine does not fall out of place when the transaxle is removed.

☐ 6. Inspect the engine mounts to ensure that they are not broken. If one or more of the engine mounts are accessed from under the hood, inspect those mounts before raising the vehicle.

☐ 7. Raise the vehicle.

> **Warning**
> ⚠ The vehicle must be raised and supported in a safe manner. Always use approved lifts or jacks and jack stands.

☐ 8. Remove the splash shields and any other hardware that will be in the way during CV axle or transaxle removal.

☐ 9. Place a drip pan under the transaxle drain plug.

☐ 10. Remove the drain plug and allow the transaxle fluid to drain from the unit.

☐ 11. Remove the CV axles.

> **Note**
> ▤ Refer to Job 67 for specific instructions on CV axle removal.

☐ 12. Remove the speedometer cable and any electrical connectors.

☐ 13. Remove the dust cover in front of the clutch and flywheel.

☐ 14. Remove the starter, if it is bolted to the transaxle clutch housing.

☐ 15. Remove the stabilizer bars or other suspension components that block transaxle removal.

☐ 16. Remove the exhaust pipes, if necessary for clearance.

☐ 17. Loosen the remaining transaxle case–to-engine bolts.

> **Caution**
> ◇ Leave at least two bolts holding the transaxle to the engine until you have a transmission jack under the transaxle.

☐ 18. Place the transmission jack under the transaxle and raise it slightly to remove transaxle weight from the lower transaxle mounts.

☐ 19. Remove the lower engine mount fasteners.

☐ 20. Remove the engine cradle, if the vehicle design makes cradle removal necessary.

☐ 21. Remove the last bolts holding the transaxle to the engine.

☐ 22. Raise or lower the transmission jack until there is no binding between the splines on the transaxle input shaft and the clutch disc.

☐ 23. Slide the transaxle away from the engine until the input shaft clears the clutch disc.

☐ 24. Lower the transaxle from the vehicle.

☐ 25. Clean the work area and return tools and equipment to storage.

☐ 26. Did you encounter any problems during this procedure? Yes ___ No ___

If Yes, describe the problems: _____

_____

What did you do to correct the problems? _____

_____

☐ 27. Have your instructor check your work and sign this job sheet.

Name _____   Date_____

Instructor _____   Period _____

Job 70—Remove a Manual Transaxle (continued)

Performance Evaluation—Instructor Use Only

Did the student complete the job in the time allotted? Yes ___ No ___

If No, which steps were not completed?_____

How would you rate this student's overall performance on this job?_____

5–Excellent, 4–Good, 3–Satisfactory, 2–Unsatisfactory, 1–Poor

Comments: _____

_____

_____

INSTRUCTOR'S SIGNATURE _____

Notes

Project
Servicing a Clutch

22

Introduction

Manual transmission clutch diagnosis is relatively simple. The difficult part of clutch diagnosis is often ensuring that the problem is not caused by another vehicle system. Many clutches will require some service during the life of the vehicle. Clutch life is often increased by adjustment and service operations, such as those described in this project. In Job 71, you will diagnose clutch problems. You will adjust clutch linkage in Job 72 and bleed a hydraulic clutch in Job 73.

Manual transmissions and transaxles are less common than their automatic counterparts. There are enough around, however, that the technician will eventually be called on to change a clutch. Clutch replacement is relatively simple and modern practice is to replace the pressure plate and throwout bearing while the clutch is disassembled. In Job 74, you will remove a clutch. You will inspect and repair the clutch in Job 75 and reinstall it in Job 76.

Project 22 Jobs

- Job 71—Diagnose Clutch Problems
- Job 72—Adjust Clutch Pedal Free Play
- Job 73—Bleed a Hydraulic Clutch System
- Job 74—Remove a Clutch
- Job 75—Inspect and Repair a Clutch
- Job 76—Install a Clutch

Tools and Materials

The following list contains the tools and materials that may be needed to complete the jobs in this project. The items used will depend on the make and model of vehicle being serviced.

- Vehicle with manual clutch in need of service.
- Applicable service information.
- Approved vacuum-collection or liquid-cleaning system.
- Ruler or clutch adjustment gauge.
- Clutch alignment tool.
- Pilot bearing puller tool (if needed).
- Appropriate brake fluid.
- Drip pan.
- Hand tools as needed.
- Air-powered tools as needed.
- Safety glasses and other protective equipment as needed.

Safety Notice

Before performing this job, review all pertinent safety information in the text, and review safety information with your instructor.

Job 71—Diagnose Clutch Problems

After completing this job, you will be able to diagnose clutch problems.

Procedures

☐ 1. Obtain a vehicle to be used in this job. Your instructor may direct you to perform this job on a shop vehicle.

☐ 2. Gather the tools needed to perform the following job. Refer to the tools and materials list at the beginning of the project.

☐ 3. Check for a slipping clutch using the following procedure:
   a. Set the vehicle's emergency brake.
   b. Start the engine.
   c. Put the transmission or transaxle in high gear.
   d. Slowly release the clutch.

   If the engine stalls, the clutch is experiencing little or no slippage. If the engine slows down but does not stall, the clutch is experiencing moderate to severe slippage. If the engine is unaffected, the clutch is experiencing severe slippage or a failure to engage.

   What degree of slippage is the clutch experiencing? _____

☐ 4. Use the following procedure to check for clutch noises:
   a. Set the vehicle's emergency brake.
   b. Ensure that the transmission or transaxle is in neutral.
   c. Start the engine.
   d. Listen for noise with the clutch engaged.

   Did you hear any noise? Yes ___ No ___

   If Yes, describe the noise: _____

   _____

   e. Slowly push the clutch pedal down while listening for noise.

   Did you hear any noise? Yes ___ No ___

   If Yes, describe the noise: _____

   _____

   f. Hold the clutch in while listening for noise.

   Did you hear any noise? Yes ___ No ___

   If Yes, describe the noise: _____

   _____

   Could these noises be caused by another vehicle component?

   Yes ___ No ___

   Explain your answer:_____

   _____

☐ 5. Check for other defects:

   Is there any clutch chatter (jerking or bumping sensation when the clutch is applied)?

   Yes ___ No ___

   Is there incomplete pedal movement or clutch release? Yes ___ No ___

   Is the clutch pedal difficult to operate? Yes ___ No ___

   Could any of these problems be caused by another vehicle component? Yes ___ No ___

   Explain your answer:_____

   _____

☐ 6. Visually inspect the clutch linkage, including the following components:
  - Clutch pedal and pivot bushings.
  - Clutch linkage attaching brackets and fasteners.
  - On a hydraulic clutch, the master cylinder, slave cylinder, and any related hoses or lines.
  - Automatic adjuster device, if used.
  - Clutch return and anti-rattle springs.

Were any defects found? Yes ___ No ___

If Yes, describe the defects:_____

☐ 7. Consult with your instructor about the defects you identified and determine which service and repair steps are necessary.

What, if any, service and repairs are needed?_____

_____

☐ 8. If instructed to do so, make the needed adjustments or repairs. Refer to the other jobs in this project as needed.

Describe the repairs made: _____

_____

☐ 9. Recheck clutch operation.

☐ 10. Clean the work area and return any equipment to storage.

☐ 11. Did you encounter any problems during this procedure? Yes ___ No ___

If Yes, describe the problems: _____

_____

What did you do to correct the problems? _____

_____

☐ 12. Have your instructor check your work and sign this job sheet.

Performance Evaluation—Instructor Use Only

Did the student complete the job in the time allotted? Yes ___ No ___

If No, which steps were not completed?_____

How would you rate this student's overall performance on this job?_____

5–Excellent, 4–Good, 3–Satisfactory, 2–Unsatisfactory, 1–Poor

Comments: _____

_____

_____

INSTRUCTOR'S SIGNATURE _____

Job 72—Adjust Clutch Pedal Free Play

After completing this job, you will be able to adjust clutch pedal free play.

Procedures

☐ 1. Obtain a vehicle to be used in this job. Your instructor may direct you to perform this job on a shop vehicle.

☐ 2. Gather the tools needed to perform the following job. Refer to the tools and materials list at the beginning of the project.

> **Note**
>
> Many clutches have self-adjusting linkage and no manual adjustment is possible. A badly worn or glazed clutch assembly cannot be restored to normal operation by adjustment.

☐ 3. Measure the amount of free play at the clutch pedal and compare it against specifications. See **Figure 72-1**.
Measured free play:_____
Specified free play:_____

> **Note**
>
> If free play specifications are not available, clutch adjustment should be about 1/2″ to 1″ (13 mm to 25 mm) of free play. If free play is within specifications, consult your instructor.

☐ 4. If the vehicle has hydraulic linkage, check the master cylinder reservoir and ensure that fluid level is sufficient.

Figure 72-1. Free play can be measured with a ruler.

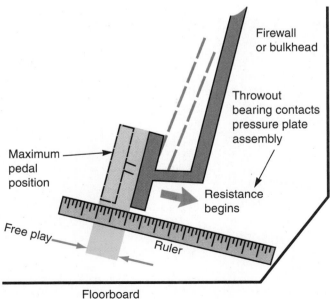

> **Note**
>
> If the fluid level is excessively low, locate and correct any leaks before proceeding.

☐ 5. To adjust free play, loosen the locking device that holds the clutch adjustment mechanism.
Describe the locking device: _____

_____

☐ 6. Turn the adjustment mechanism to obtain the proper free play at the clutch pedal.

☐ 7. Tighten the locking device.

☐ 8. Recheck the free play at the clutch pedal.
Measured free play:_____
Is free play within specifications? Yes ___ No ___
If No, repeat steps 5 through 7.

☐ 9. Road test the vehicle and check clutch operation.

☐ 10. If free play is not right, repeat steps 4 through 9.

> **Note**
>
> If adjustment cannot correct the free play, or if the clutch is slipping, the clutch must be disassembled and any worn parts replaced.

☐ 11. Clean the work area and return any equipment to storage.

☐ 12. Did you encounter any problems during this procedure? Yes ___ No ___
If Yes, describe the problems: _____

_____

What did you do to correct the problems? _____

_____

☐ 13. Have your instructor check your work and sign this job sheet.

Performance Evaluation—Instructor Use Only

Did the student complete the job in the time allotted? Yes ___ No ___
If No, which steps were not completed?_____
How would you rate this student's overall performance on this job?_____
5–Excellent, 4–Good, 3–Satisfactory, 2–Unsatisfactory, 1–Poor
Comments: _____

_____

_____

INSTRUCTOR'S SIGNATURE _____

Job 73—Bleed a Hydraulic Clutch System

After completing this job, you will be able to bleed a hydraulic clutch system.

Procedures

☐ 1. Obtain a vehicle to be used in this job. Your instructor may direct you to perform this job on a shop vehicle.

☐ 2. Gather the tools needed to perform the following job. Refer to the tools and materials list at the beginning of the project.

> **Caution**
>
> Brake fluid can damage paint. Cover painted surfaces near the master cylinder before beginning this procedure.

> **Warning**
>
> Use the proper type of fluid in the clutch hydraulic system. Most systems use brake fluid. Fluid type is usually stamped on the reservoir cover. Always use the correct type of fluid.

☐ 3. Locate the clutch master cylinder and check the fluid level in the reservoir. See **Figure 73-1**.

Is the level near the full mark? Yes ___ No ___

If Yes, go to step 5. If No, add fluid and go to step 4.

☐ 4. If the fluid level was low, inspect the hydraulic clutch's master cylinder, slave cylinder, and all hoses and lines for leaks.

Did you find any leaks? Yes ___ No ___

If Yes, consult with your instructor about what steps should be taken to correct the leak(s). If No, go to step 5.

☐ 5. Place a drip pan under the master and slave cylinders.

Figure 73-1. Check the fluid in the clutch master cylinder reservoir before proceeding. In this photograph the smaller reservoir next to the brake fluid reservoir is the clutch master cylinder reservoir.

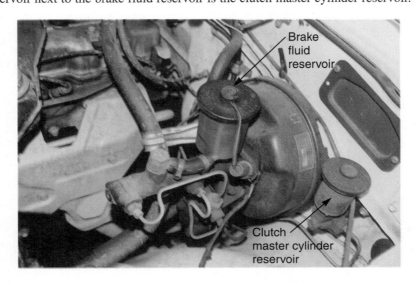

Brake fluid reservoir

Clutch master cylinder reservoir

☐ 6. Have an assistant depress and hold the clutch pedal.

☐ 7. Open the bleeder on the master cylinder.

Does only fluid exit, or does air exit also? Exiting air will make a spitting sound and will show up as bubbles in the fluid. Fluid only ___ Air and fluid ___

☐ 8. Close the bleeder screw before allowing your assistant to release the pedal. Once the bleeder screw is seated, have your assistant pump and then depress and hold the clutch pedal.

☐ 9. Repeat steps 7 and 8 until only fluid exits from the bleeder.

> **Note**
>
> Keep the clutch master cylinder reservoir filled with the proper fluid during this procedure.

☐ 10. Road test the vehicle and check clutch operation.

☐ 11. If the clutch pedal operation continues to feel soft or spongy, repeat steps 6 through 10 until the pedal becomes firm.

☐ 12. Clean the work area and return any equipment to storage.

☐ 13. Did you encounter any problems during this procedure? Yes ___ No ___

If Yes, describe the problems: _____

_____

What did you do to correct the problems? _____

_____

☐ 14. Have your instructor check your work and sign this job sheet.

Performance Evaluation—Instructor Use Only

Did the student complete the job in the time allotted? Yes ___ No ___

If No, which steps were not completed?_____

How would you rate this student's overall performance on this job?_____

5–Excellent, 4–Good, 3–Satisfactory, 2–Unsatisfactory, 1–Poor

Comments: _____

_____

_____

INSTRUCTOR'S SIGNATURE _____

Job 74—Remove a Clutch

After completing this job, you will be able to remove a manual transmission or transaxle clutch.

Procedures

☐ 1. Obtain a vehicle to be used in this job. Your instructor may direct you to perform this job on a shop vehicle.

☐ 2. Gather the tools needed to perform the following job. Refer to the tools and materials list at the beginning of the project.

☐ 3. Disconnect the battery negative cable.

☐ 4. Raise the vehicle.

> **Warning**
> ⚠ The vehicle must be raised and supported in a safe manner. Always use approved lifts or jacks and jack stands.

☐ 5. Remove the drive shaft or front drive axles, as required. Be sure to cap the extension housing openings to reduce oil leaks.

> **Note**
> Refer to Job 66 for specific instructions on drive shaft removal. Refer to Job 67 for specific instructions on CV axle removal.

☐ 6. Disconnect the clutch linkage:
> - Pushrod or cable and return spring connected to the clutch fork or clutch shaft.
> - Slave cylinder or hydraulic lines to internal slave cylinder.

☐ 7. Remove the transmission or transaxle, as applicable.

> **Note**
> Refer to Job 69 for specific instructions on manual transmission removal. Refer to Job 70 for specific instructions on manual transaxle removal.

☐ 8. Remove the throwout bearing and operating mechanism using one of the following methods:
> - **If the clutch fork is held to a ball stud by spring clips:** Remove the rubber dust boot and pull the fork straight out from the bell housing, **Figure 74-1**.
> - **If the clutch fork is mounted on a clutch shaft:** Remove the clips holding the shaft to the bell housing and slide the shaft from the housing as shown in **Figure 74-2**. Then, remove the bearing and release fork.
> - **If the clutch uses an internal (also called concentric or ring) hydraulic slave cylinder installed on the transmission:** Remove fasteners holding the bearing and slave cylinder assembly to the transmission and remove the assembly. See **Figure 74-3**.

☐ 9. Remove the bell housing from the back of the engine, if not removed as part of step 7.

> **Note**
> Bell housing removal is not necessary for all clutch replacement jobs. Consult the service literature before removing the housing.

Figure 74-1. Pull the clutch fork out to remove it from the ball stud. (DaimlerChrysler)

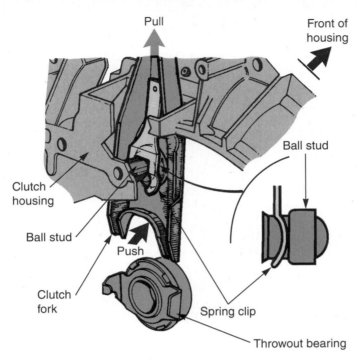

10. Remove any dust inside the bell housing and on the pressure plate assembly.

> **Warning**
>
> ⚠ Clutch discs contain asbestos, a cancer-causing substance. Do not breathe the dust inside the bell housing or clutch assembly. Do not blow dust off with compressed air. Use a vacuum-collection system or liquid-cleaning system for dust removal.

11. If the pressure plate assembly and flywheel do not have dowels or offset bolts holes for alignment purposes, use a punch to mark the original position of the pressure plate relative to the flywheel.

12. Insert a clutch alignment tool through the clutch disc hub and into the pilot bearing, to keep the clutch disc from falling out of the pressure plate assembly as the pressure plate attaching bolts are removed.

13. Loosen the pressure plate attaching bolts one turn at a time until all apply spring pressure is released.

14. Carefully remove the pressure plate assembly and the clutch disc as a unit with the alignment tool.

Inspect Engine Core Plugs, Rear Crankshaft Seal, Dowel Pins, Dowel Pin Holes, and Mating Surfaces

1. Once the clutch housing, clutch plate, and flywheel have been removed, inspect the following parts at the rear of the engine block.

   Are the core plugs corroded? Yes _____ No _____

   Do the core plugs have leaks? Yes _____ No _____

   Are the core plugs installed improperly? Yes _____ No _____

   Does the rear seal have leaks? Yes _____ No _____

Job 74—Remove a Clutch (continued)

Figure 74-2. A—Remove the clip holding the throwout bearing shaft to the bell housing. B—Slide the shaft from the bell housing. Actual attachment methods may vary. (DaimlerChrysler)

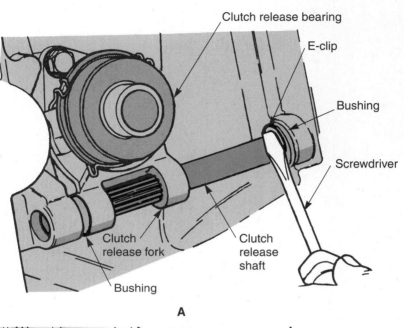

A

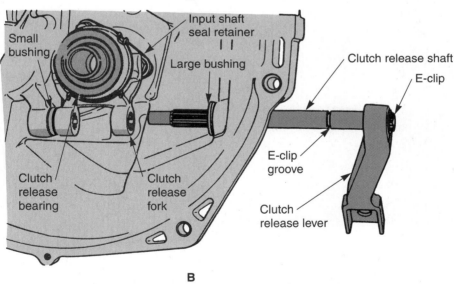

B

Is the rear seal damaged? Yes _____ No _____

Are there missing dowels? Yes _____ No _____

Do the dowel pins have burrs? Yes _____ No _____

Are there bent dowels? Yes _____ No _____

Are the dowel pin holes elongated? Yes _____ No _____

Do the dowel pin holes have burrs? Yes _____ No _____

☐ 2. Clean the work area and return any equipment to storage.

Figure 74-3. The internal slave cylinder and throwout bearing assembly is attached to the transmission.

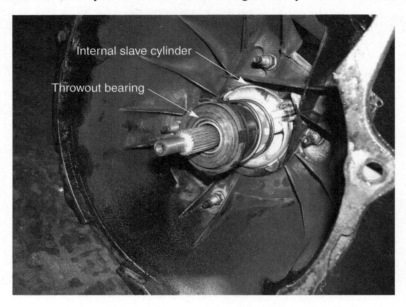

3   Did you encounter any problems during this procedure? Yes ___ No ___
     If Yes, describe the problems: _____

     _____

     What did you do to correct the problems? _____
     _____

4.   Have your instructor check your work and sign this job sheet.

Performance Evaluation—Instructor Use Only

Did the student complete the job in the time allotted? Yes ___ No ___
If No, which steps were not completed?_____
How would you rate this student's overall performance on this job?_____
5–Excellent, 4–Good, 3–Satisfactory, 2–Unsatisfactory, 1–Poor
Comments: _____

_____

INSTRUCTOR'S SIGNATURE _____

Job 75—Inspect and Repair a Clutch

After completing this job, you will be able to inspect and repair a manual transmission or trans-axle clutch.

Procedures

☐ 1. Obtain a clutch assembly to be used in this job. Your instructor may direct you to perform this job on a shop clutch assembly.

☐ 2. Gather the tools needed to perform the following job. Refer to the tools and materials list at the beginning of the project.

> **Note**
>
> This job begins with the clutch assembly removed from the engine and disassembled. Refer to Job 74 for procedures for removing and disassembling a clutch.

☐ 3. Wash the pressure plate assembly and all other metal parts in cleaning solvent.

> **Caution**
>
> Do not wash the throwout bearing or clutch disc in cleaning solvent. Doing so will ruin these components.

☐ 4. Ensure that the clutch disc can slide freely on the external splines of the transmission input shaft and inspect the clutch disc for the following defects. Refer to **Figure 75-1**.
- Worn, glazed, or oil-soaked friction facings.
- Loose rivets.
- Broken torsion or cushion springs.
- Worn hub flange splines.

Figure 75-1. Carefully inspect the surfaces of the clutch disc and pressure plate.

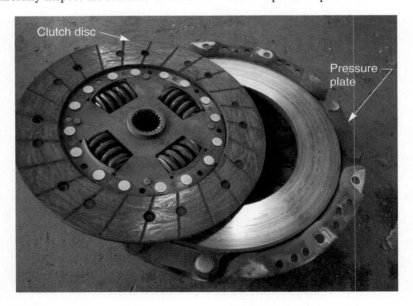

Describe the condition of the clutch disc:_____

_____

_____

☐ 5. Inspect the pressure plate for the following defects. Refer to **Figure 75-1**.
 • Discolored, scored, or extremely shiny surfaces.
 • Cracks.
 • Weak or overheated springs.
 • Worn or misadjusted release fingers or levers.
 • Loose rivets or bolts.
 • Elongated fastener openings.
 • Any other sign of damage.
 Describe the condition of the pressure plate. _____

 _____

 _____

> **Note**
>
> If any part of the pressure plate assembly is damaged, it is usually replaced rather than rebuilt.

☐ 6. Inspect the flywheel for the following defects. See **Figure 75-2**.
 • Discolored, scored, or extremely shiny surfaces.
 • Cracks.
 • Stripped threads.
 • Broken gear teeth.
 • Any other sign of damage.
 Describe the condition of the flywheel: _____

 _____

 _____

☐ 7. Check for warping of the flywheel contact surfaces, using a straightedge and feeler gauge. Compare the measurement against specifications.
 Specified maximum allowable warpage:_____
 Measured warpage (feeler gauge size):_____

Figure 75-2. Check the flywheel surface for scoring, overheating, and cracks.

Job 75—Inspect and Repair a Clutch (continued)

☐ 8. Check the flywheel runout using a dial indicator and compare it against specifications.

Specified maximum runout:_____

Measured runout:_____

> **Note**
> If the flywheel is damaged, it must be removed from the engine and resurfaced by a machine shop. Some shops prefer to install a new flywheel.

☐ 9. If necessary, check crankshaft endplay using the dial indicator.

☐ 10. Attempt to rotate the throwout bearing by hand to check for roughness.

Describe the condition of the throw-out bearing: _____

_____

_____

> **Note**
> A rough bearing cannot be restored by greasing. If the throwout bearing rotation is rough, obtain a new bearing. Most technicians prefer to install a new bearing while the transmission is out.

☐ 11. Check the bearing collar for a free, but not excessively loose fit on the transmission's front bearing retainer hub. Inspect the hub for wear and check the collar for wear where it contacts the clutch fork.

Describe the fit and the condition of the bearing collar and hub: _____

_____

_____

☐ 12. Inspect the clutch fork and related parts for damage and check the pivot points and other contact areas for wear.

Describe the condition of the parts: _____

_____

_____

☐ 13. Clean the clutch pilot bearing and perform the following inspections:

- **Check roller or ball bearings for lack of lubrication and roughness.** Turn the bearing and feel for roughness. If it does not turn smoothly, it should be replaced.
- **Check bushing-type pilot bearings for scoring and looseness between the bushing and the input shaft pilot.** Insert a good used input shaft into the bearing and try to wiggle it. If it has too much play, the bushing is worn out, and it should be replaced. Some pilot bushings can be checked with an inside micrometer.

Describe pilot bearing condition: _____

_____

_____

☐ 14. Check the bell housing for cracks and damage to the mounting surfaces.

Describe the bell housing's condition: _____

_____

15. Report any clutch component defects that you found to your instructor. Determine which parts must be replaced or, in the case of the flywheel, resurfaced.

16. Obtain the proper replacement parts and compare them with the old parts to ensure that they are the correct replacements.

17. Clean any parts that were not cleaned previously.

18. Use the following procedure to replace the pilot bearing, skipping steps as applicable:
    a. If the bearing is held by a snap ring (usually roller or ball bearings are held this way), remove the snap ring and pull the bearing from the rear of the crankshaft.
    b. If the bearing or bushing is pressed in, use a pilot bearing puller tool to remove it, or pack the recess behind the bearing with heavy grease and use a driver to push the bearing from the crankshaft.
    c. Slip the new pilot bearing over the input shaft to ensure that it is the proper bearing.
    d. If the bearing is held by a snap ring, insert it in the rear of the crankshaft and install the snap ring.
    e. If the bearing is pressed in, drive it into the crankshaft bore with the proper size driver. Make sure that the new bearing is installed to the proper depth.
    f. Lubricate the new pilot bearing with high-temperature grease. Do not add too much lubricant or it will be thrown onto the clutch disc.

19. Install the flywheel if it was removed.

20. Clean the work area and return any equipment to storage.

21. Did you encounter any problems during this procedure? Yes ___ No ___
    If Yes, describe the problems: _____
    _____

    What did you do to correct the problems? _____
    _____

22. Have your instructor check your work and sign this job sheet.

Performance Evaluation—Instructor Use Only

Did the student complete the job in the time allotted? Yes ___ No ___
If No, which steps were not completed?_____
How would you rate this student's overall performance on this job?_____
5–Excellent, 4–Good, 3–Satisfactory, 2–Unsatisfactory, 1–Poor
Comments: _____
_____
_____

INSTRUCTOR'S SIGNATURE _____

Job 76—Install a Clutch

After completing this job, you will be able to install a manual transmission or transaxle clutch.

Procedures

☐ 1. Obtain a vehicle to be used in this job. Your instructor may direct you to perform this job on a shop vehicle.

☐ 2. Gather the tools needed to perform the following job. Refer to the tools and materials list at the beginning of the project.

> **Note**
>
> This job begins with a disassembled clutch assembly. Refer to Job 74 for procedures for removing and disassembling a clutch.

☐ 3. Place the clutch disc and pressure plate assembly on the flywheel. Use a clutch alignment tool to align the clutch disc hub with the pilot bearing. The alignment tool may be a special tool or an old input shaft, **Figure 76-1**.

Figure 76-1. A special tool is best for aligning the clutch plate, but an old input shaft can also be used. Note the dowels used to align the pressure plate with the flywheel. (DaimlerChrysler)

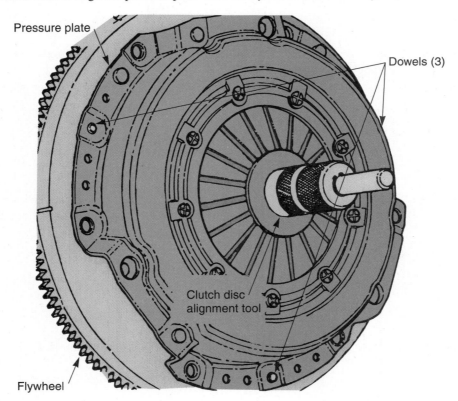

> **Note**
>
> If an alignment tool is not used when tightening the pressure plate attaching bolts, the transmission input shaft may bind in the clutch assembly when installing the transmission. Many replacement clutch and pressure plate assemblies are sold with an alignment tool.

☐ 4. Align the pressure plate assembly on the flywheel by matching punch marks, offset bolt holes, or holes for flywheel dowels, as applicable.

☐ 5. Begin threading in the pressure plate attaching bolts by installing two directly opposite each other and finger-tightening them. Thread in and finger-tighten the remaining bolts, making sure that all bolts are started properly.

☐ 6. Tighten the pressure plate attaching bolts gradually in a crisscross pattern.

☐ 7. To ensure that the clutch hub and pilot bearing are aligned properly, slide the clutch alignment tool in and out of its installation and verify that it moves freely. Then, remove the clutch alignment tool.

☐ 8. Lightly lubricate the moving parts of the pressure plate fork and ball stud or shaft with high-temperature grease.

☐ 9. Install the clutch fork by placing the fork on the ball stud in the bell housing, using the spring clip to hold it in place.

or

Assemble the clutch fork and shaft by passing the release shaft through the bell housing and clutch fork and locking it in place with the clip or lock bolts, as required.

or

Install the throwout bearing and internal slave cylinder assembly on the transmission and install the assembly fasteners.

☐ 10. Coat the inside of the throwout bearing collar and the outer groove of the throwout bearing with a small amount of high-temperature grease.

☐ 11. Ensure that the throwout bearing fork is installed properly on the bearing outer groove.

☐ 12. Install the bell housing, if it is separate from the transmission case.

☐ 13. Install the transmission or transaxle, as applicable.

> **Note**
>
> Refer to Job 78 for specific instructions on manual transmission installation. Refer to Job 80 for specific instructions on manual transaxle installation.

☐ 14. Connect the pushrod or cable and return spring connected to the clutch fork.

☐ 15. Install the drive shaft or front drive axles, as required.

> **Note**
>
> Refer to Job 66 for specific instructions on drive shaft installation. Refer to Job 67 for specific instructions on CV axle installation.

☐ 16. Lower the vehicle.

Job 76—Install a Clutch (continued)

☐ 17. Check and adjust the clutch pedal free play.

   Measured free play:_____

   Specified free play:_____

☐ 18. Install the battery negative cable.

☐ 19. Road test the vehicle. During the road test, make sure that the clutch engages and disengages properly. Check that there is no slippage and notice the clutch pedal free play. Apply the clutch at least 25 times to seat the linings. Be careful not to overheat the clutch.

☐ 20. Readjust the clutch pedal free play if necessary.

☐ 21. Clean the work area and return any equipment to storage.

☐ 22. Did you encounter any problems during this procedure? Yes ___ No ___

   If Yes, describe the problems: _____

   _____

   What did you do to correct the problems? _____

   _____

☐ 23. Have your instructor check your work and sign this job sheet.

Performance Evaluation—Instructor Use Only

Did the student complete the job in the time allotted? Yes ___ No ___
If No, which steps were not completed?_____
How would you rate this student's overall performance on this job?_____
5–Excellent, 4–Good, 3–Satisfactory, 2–Unsatisfactory, 1–Poor
Comments: _____

_____

_____

INSTRUCTOR'S SIGNATURE _____

Notes

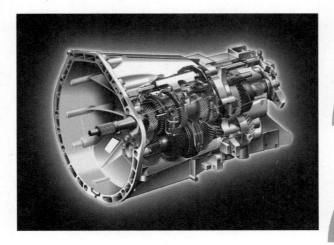

Project 23

Rebuilding and Installing a Manual Transmission or Transaxle

Introduction

While it is not an everyday occurrence, most technicians will eventually have to overhaul a manual transmission. The process of overhauling a manual transmission requires thorough cleaning, inspection, and adjustment, as well as careful attention to detail. In Job 77, you will disassemble and inspect a manual transmission. You will then reassemble and install the manual transmission in Job 78.

Although transaxles are widely used, they are usually automatic versions. Manual transaxles, however, are sometimes encountered and sometimes require overhaul. The internal parts of a manual transaxle resemble their counterparts in a rear-wheel drive transmission. However, many other parts are different. Unlike the manual transmission, the transaxle incorporates the differential and final drive unit, which also must be serviced. In Job 79, you will disassemble and inspect a manual transaxle. You will reassemble and install the manual transaxle in Job 80.

Many times, a technician will carefully repair a manual transmission or transaxle, only to spoil the job by careless reinstallation procedures. It is easy to become rushed and damage the transmission or transaxle or a related part. Common mistakes include failing to line up the clutch disc and input shaft, pinching wires or hoses between the transmission/transaxle housing and engine block, failing to tighten drive shaft or CV axle fasteners, and incorrectly attaching the shift linkage.

Use the same caution when reinstalling a transmission or transaxle as you did when repairing it. Ensure that the clutch disc is properly positioned, then carefully line up the input shaft as you move the transmission or transaxle housing into position. Make sure that there are no gaps between the housing and the block. As you install the fasteners, carefully observe all related parts, ensuring that the housing fasteners tighten easily with no sign of binding. Once the transmission or transaxle is installed on the engine, carefully install all other parts and make sure that all fasteners are correctly tightened. Keep these points in mind as you install a manual transmission in Job 78 and a manual transaxle in Job 80.

Project 23 Jobs

- Job 77—Disassemble and Inspect a Manual Transmission
- Job 78—Reassemble and Install a Manual Transmission
- Job 79—Disassemble and Inspect a Manual Transaxle
- Job 80—Reassemble and Install a Manual Transaxle

Tools and Materials

The following list contains the tools and materials that may be needed to complete the jobs in this project. The items used will depend on the make and model of vehicle being serviced.

- Vehicle with a manual transmission in need of overhaul.
- Vehicle with a manual transaxle in need of overhaul.
- Applicable service information.
- Hydraulic lift (or jacks and jack stands).
- Transmission jack(s) or floor jack(s).
- Transmission holding fixture.
- Engine holding fixture.
- Bearing driver.
- Seal driver.
- Measuring instruments.
- Hydraulic press and proper adapters.
- Gear and bearing puller.
- Dummy shaft.
- Drain pan.
- Specified manual transmission lubricant.
- Hand tools as needed.
- Air-powered tools as needed.
- Safety glasses and other protective equipment as needed.

Safety Notice

Before performing this job, review all pertinent safety information in the text, and review safety information with your instructor.

Job 77—Disassemble and Inspect a Manual Transmission

After completing this job, you will be able to disassemble a manual transmission, identify worn or defective manual transmission parts, and replace those parts.

Procedures

> **Note**
>
> There are many variations in manual transmission design. The following procedure is a general guide only and some steps will not apply to every transmission. Always consult the manufacturer's service literature for exact disassembly procedures and specifications.

- [ ] 1. Obtain a manual transmission to be used in this job. Your instructor may direct you to perform this job on a shop transmission.
- [ ] 2. Gather the tools needed to perform the following job. Refer to the tools and materials list at the beginning of the project.

Remove the External Components

- [ ] 1. Mount the transmission in a special holding fixture, if available, or secure it on a clean workbench.
- [ ] 2. Drain the transmission lubricant, if not already drained.
- [ ] 3. Remove the transmission top access cover, if there is one.
- [ ] 4. Check the operation of the gears, synchronizers, and shift linkage by moving the shift forks while rotating the input shaft.

     Were shifting problems evident? Yes ___ No ___

     If Yes, describe the problems: _____

     _____

     _____

- [ ] 5. Remove any access covers holding the shift forks.
- [ ] 6. If the front bearing retainer could be installed in the wrong position, place match marks on the case and bearing retainer for reinstallation.
- [ ] 7. Remove the throwout bearing from the input shaft, if it is still installed. If the transmission uses an internal-type (concentric or ring) hydraulic slave cylinder, remove it with the throwout bearing.
- [ ] 8. Unbolt the front bearing retainer and slide it off of the input shaft. Remove the bell housing if it is bolted to the transmission. **Figure 77-1** shows removal of the bearing retainer and bell housing bolts.
- [ ] 9. Remove the front bearing retainer seal from its seat within the retainer.
- [ ] 10. Remove any external components that would prevent, or would otherwise be damaged by, extension housing removal, such as the speedometer driven gear and housing and rear-mounted shift rails or detent assemblies.
- [ ] 11. Remove the extension housing. If necessary, rotate the housing so that it clears the reverse gear, while pulling the housing from the transmission case.
- [ ] 12. Remove external electrical devices such as speed sensors and backup light switches.

Figure 77-1. After the fasteners have been removed, remove the bearing retainer and bell housing from the transmission.

Remove the Internal Components

> **Note**
>
> Steps 1 and 2 may not apply to all transmissions.

1. Locate endplay specifications for the mainshaft and mainshaft gears, and the counter gear and case.

> **Note**
>
> The mainshaft is also referred to as the output shaft.

2. Measure gear endplay between each mainshaft gear by attempting to insert various size feeler gauges into the gap between gears. Then, note the size of the thickest feeler gauge that can be inserted. This is the measured endplay.

   Measured endplay:_____ Specified endplay:_____

   Measured endplay:_____ Specified endplay:_____

   Measured endplay:_____ Specified endplay:_____

3. Check endplay in the remaining transmission gears by installing a dial indicator on the case so that the pointer touches the gear to be measured. Next, pull the gear forward and zero the dial indicator. Then, push the gear rearward and read the endplay on the dial indicator.

   Measured endplay:_____ Specified endplay:_____

   Measured endplay:_____ Specified endplay:_____

   Measured endplay:_____ Specified endplay:_____

Job 77—Disassemble and Inspect a Manual Transmission (continued)

> **Note**
>
> If endplay exceeds specifications, look for wear on the sides of the gears or thrust washers after the shaft is disassembled. Replace these components as necessary.

- [ ] 4. If necessary, use a puller to remove any gears that prevent further transmission disassembly.
- [ ] 5. If the internal parts of the shift linkage and the shift forks were not installed in an access cover, remove them from the case now.
- [ ] 6. If necessary, remove the reverse idler gear from the case.
- [ ] 7. If necessary, remove the speedometer drive gear from the mainshaft.
- [ ] 8. Remove the mainshaft assembly from the case.

   or

   If the transmission uses a vertical split case design, separate the front and rear housings, or remove the front (and rear if applicable) housings from the center support after removing any fasteners.

   or

   If the transmission uses a horizontal split case design, separate the upper and lower case halves after removing any fasteners.

   Which type of case does this transmission use? _____

- [ ] 9. If necessary, remove the bolts or snap ring holding the front bearing to the transmission housing.
- [ ] 10. Remove the input shaft from the case.
- [ ] 11. Remove the key or pin holding the countershaft in place and remove the shaft.
- [ ] 12. Remove the thrust washers (one on each end, if used) and the countershaft gear.

Disassemble Transmission Components

- [ ] 1. Remove the snap ring from the front end of the output shaft.
- [ ] 2. Remove the synchronizers and mainshaft gears. Remove other snap rings along the mainshaft, as required, so that all gears and synchronizers may be removed.

> **Note**
>
> Some gears may need to be pressed from the shaft.

- [ ] 3. Remove the countershaft gear from the case if necessary.
- [ ] 4. Disassemble the shift linkage, if not done earlier.
- [ ] 5. Remove any seals and bearings that are still in place in the case.

Inspect Transmission Parts

- [ ] 1. Clean all metal parts in a safe solvent and dry the parts with compressed air or let them air dry.

> **Warning**
>
> Do *not* spin bearings with compressed air.

☐ 2. Recover all loose parts, such as needle bearings, from the bottom of the case.

> **Note**
>
> Almost all manual transmissions use needle bearings between the input shaft and mainshaft and between the countershaft gear and countershaft. These parts must be located and saved.

☐ 3. Visually examine the disassembled transmission components for the following list of defects. Some defects, such as the chipped teeth in **Figure 77-2** will be obvious. Others, such as the worn bearing race inside of the gear in **Figure 77-3**, will be harder to spot. Look carefully!

- Worn or broken teeth on synchronizer hubs and sleeves.
- Wear on tapered surfaces or worn teeth on blocking rings.
- Worn shift fork grooves on synchronizer outer sleeves.
- Worn or distorted shift forks.
- Bent shift rails.
- Worn, cracked, or chipped teeth on transmission gears.
- Worn gear bushings.
- Worn races on the transmission shafts.
- Worn or damaged thrust washers and snap rings.
- Worn or rough bearings.
- Flat spots on needle bearings.
- A cracked or distorted transmission case.
- Worn front bearing retainer hub.
- Bent or missing oil slingers.
- Blocked oil passages.

Inspect components using special measuring instruments as required.

Figure 77-2. Most chipped gear teeth will be easy to spot. This gear must be replaced.

Job 77—Disassemble and Inspect a Manual Transmission (continued)

Figure 77-3. Wear on this bearing race would have been missed if the part had not been thoroughly cleaned and carefully inspected.

Describe the condition of the transmission components: _____
_____
_____
_____
_____
_____
_____
_____

4. Report any transmission component defects that you found to your instructor. Then, determine which parts must be replaced.

> **Note**
>
> A synchronizer showing *any* wear should be replaced.

List needed parts: _____
_____
_____
_____
_____

5. Obtain the proper replacement parts. Compare replacement parts with the old parts to ensure that they are the correct replacements.

6. Clean the work area and return tools and equipment to storage.

☐ 7. Did you encounter any problems during this procedure? Yes ___ No ___

If Yes, describe the problems: _____

_____

What did you do to correct the problems? _____

_____

☐ 8. Have your instructor check your work and sign this job sheet

Performance Evaluation—Instructor Use Only

Did the student complete the job in the time allotted? Yes ___ No ___

If No, which steps were not completed?_____

How would you rate this student's overall performance on this job?_____

5–Excellent, 4–Good, 3–Satisfactory, 2–Unsatisfactory, 1–Poor

Comments: _____

_____

_____

INSTRUCTOR'S SIGNATURE _____

Job 78—Reassemble and Install a Manual Transmission

After completing this job, you will be able to properly assemble and install a manual transmission.

Procedures

> **Note**
> There are many variations in manual transmission design. The following procedure is a general guide only and some steps will not apply to every transmission. Always consult the manufacturer's service literature for exact disassembly procedures and specifications.

☐ 1. Obtain a disassembled manual transmission to be used in this job. Your instructor may direct you to perform this job on a shop transmission.

☐ 2. Gather the tools needed to perform the following job. Refer to the tools and materials list at the beginning of the project.

Install the Transmission's Internal Components

☐ 1. Lightly lubricate all parts before proceeding.

☐ 2. Assemble all synchronizers and hubs.

☐ 3. Install the synchronizers, hubs, and gears on the mainshaft. Carefully observe the direction of the blocking rings and all other parts during reinstallation.

> **Note**
> The mainshaft is also referred to as the output shaft.

☐ 4. Install snap rings along the mainshaft, as required. Snap rings should fit snugly in their grooves.

☐ 5. Install the snap ring at the front end of the mainshaft.

☐ 6. Ensure that all parts on the mainshaft turn without binding.

☐ 7. Check mainshaft gear endplay as was done in Job 77.

Is endplay within specifications? Yes ____ No ____

If No, repeat earlier steps to change to thicker or thinner shims to produce correct endplay.

> **Note**
> On some transmissions, steps 8 through 13 do not apply.

☐ 8. If necessary, install the needle bearings in the bearing bores of the countershaft gear, using a dummy shaft to hold the bearings in place.

☐ 9. Grease the countershaft gear thrust washers (if used) to hold them in place.

☐ 10. Insert the countershaft gear thrust washers (if used) into each end of the transmission housing.

☐ 11. Insert the countershaft gear through the rear opening of the transmission case and set the gear into position.

☐ 12. Insert the countershaft through the case and countershaft gear. As the countershaft is pushed in, the dummy shaft will be pushed out.

☐ 13. If the shaft uses a retainer pin or Woodruff key, install it. Then, push the shaft into the fully installed position.

☐ 14. Check countershaft gear endplay, as was done in Job 77, and compare it against manufacturer's specifications.

Measured endplay:_____

Specified endplay:_____

Is the endplay within specifications? Yes ___ No ___

If No, adjust the endplay by using thicker or thinner thrust washers or snap rings, as called for by the transmission manufacturer.

☐ 15. If necessary install the reverse idler gear, shaft, and thrust washer(s).

☐ 16. Install the needle bearings in the inner bore of the input shaft, using grease to hold the bearings in place, as was done with the countershaft gear.

☐ 17. Place the input shaft in the case. Then, install the front bearing, being careful not to dislodge the uncaged needle bearings.

☐ 18. Install the mainshaft gear assembly through the rear of the transmission case and slide the nose of the mainshaft into the pilot bore behind the main drive gear.

or

If the case is a vertical split case type, install the assembled shafts as applicable. Next, install the front and rear housings over the center support using new gaskets. Then install the fasteners and tighten them to the proper torque.

or

If the case is a horizontal split case type, place the assembled shafts in the lower case half. Then place new case gaskets in position and install the upper case half over the lower case half. Install the fasteners and tighten them to the proper torque.

☐ 19. Install the rear bearing and rear bearing snap ring (if used).

☐ 20. Install case-mounted shift forks or rails.

Install the Transmission's External Components

☐ 1. Install new extension housing gaskets.

☐ 2. If the reverse gear is contained in the extension housing, install it now.

☐ 3. Place the extension housing over the mainshaft and reverse gear, engaging the reverse shift fork with the reverse gear, if necessary.

☐ 4. Install and tighten the extension housing bolts.

☐ 5. Install the front bearing retainer, matching the alignment marks made earlier.

☐ 6. Tighten the fasteners to the proper torque.

☐ 7. Check that the transmission input shaft and mainshaft turn with no binding.

Caution

⚠ If the shafts do not turn freely, find out why and correct the problem.

Job 78—Reassemble and Install a Manual Transmission (continued)

☐ 8. Check the transmission shaft endplay as explained in the manufacturer's instructions, and compare it against specifications. See **Figure 78-1**.

Measured endplay:_____

Specified endplay:_____

If the endplay is incorrect, adjust it by using thicker or thinner thrust washers or snap rings, as called for by the transmission manufacturer.

☐ 9. Replace the access covers, being sure to align the shift forks with the synchronizer sleeves where necessary. Use new gaskets at all locations.

☐ 10. Install speed sensors, electrical switches, and any other external case components removed as part of transmission disassembly.

☐ 11. Install and tighten the transmission drain plug.

☐ 12. Replace the side or top access covers. If necessary, align the shift forks as the cover is being moved into position.

☐ 13. Install the internal gearshift lever assembly, if applicable.

☐ 14. Check internal shift linkage operation and verify that the input shaft, mainshaft, and the countershaft gear all turn without binding.

Install a Manual Transmission

> **Note**
>
> If the vehicle has a transfer case, transmission replacement procedures will differ from those given in this job. Refer to the manufacturer's service literature for more information.

☐ 1. Raise the vehicle.

> **Note**
>
> If the vehicle has a transfer case, transmission replacement procedures will differ from those given in this job. Refer to the manufacturer's service literature for more information.

Figure 78-1. One setup for checking endplay. Note that in this setup, the extension housing has been removed. Exact setup and procedures will vary with each transmission. (TTC/Tremec)

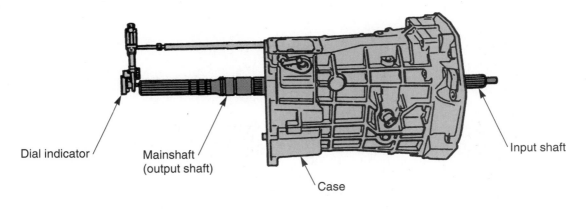

> **Warning**
>
> ⚠ The vehicle must be raised and supported in a safe manner. Always use approved lifts or jacks and jack stands.

- [ ] 2. Make sure the throw-out bearing and clutch disc are properly positioned, **Figure 78-2**.
- [ ] 3. Make sure the transmission input shaft is clean and undamaged, **Figure 78-3**.
- [ ] 4. Make sure the clutch ball (when used) is undamaged. See **Figure 78-4**.
- [ ] 5. Make sure the throw-out bearing's moving parts are lightly lubricated.
- [ ] 6. Lightly grease the pilot bearing if not done as part of clutch service.

Figure 78-2. The clutch disc should be checked for proper alignment. Alignment should be correct if the clutch has not been disturbed.

Figure 78-3. Carefully inspect the input shaft for damage. Remove any burrs with a small file and carefully remove any metal chips.

Job 78—Reassemble and Install a Manual Transmission (continued)

Figure 78-4. Check the condition of the clutch fork ball. Some manufacturers recommend lightly lubricating the ball and clutch fork.

Clutch fork ball

☐ 7. Raise the transmission into position under the vehicle.

☐ 8. Place the transmission in a gear and turn the output shaft to align the input shaft and clutch disc splines. Slide the transmission input shaft into the clutch disc and ensure that the shaft is fully engaged. If necessary, slightly shift or wiggle the transmission to install the shaft through the clutch disc and into the pilot bearing.

☐ 9. Install the attaching bolts that hold the transmission to the bell housing or engine.

> **Caution**
>
> Do not attempt to draw a binding transmission into the clutch assembly by tightening the attaching bolts. This can bend the clutch disc or break the transmission case.

☐ 10. Install the transmission crossmember.

☐ 11. Install and tighten all crossmember fasteners.

☐ 12. Connect or tighten the parking brake linkage.

☐ 13. Install all electrical connectors.

☐ 14. If the bell housing is integral with the transmission case, install the starter, clutch linkage or hydraulic lines, and any other parts connected to the bell housing.

☐ 15. Install the shift linkage.

☐ 16. Install the exhaust pipes, if they were removed.

☐ 17. Install the speedometer cable assembly.

☐ 18. Remove the cap at the rear of the transmission. Then, install the drive shaft, making sure to line up the match marks made during removal. Refer to Job 66 for specific instructions for drive shaft installation.

☐ 19. Fill the transmission to the proper level with the specified lubricant. Install the fill plug and tighten.

☐ 20. If the transmission has external shift linkage, install the linkage to the transmission.

☐ 21. Lower the vehicle.

☐ 22. Install the upper bell housing bolts if they were not installed from under the vehicle.

☐ 23. Install the battery's negative cable.

☐ 24. Install the passenger compartment dust boot to the floor or console and install the gearshift lever knob.

☐ 25. Move the clutch pedal and gearshift lever to check operation of the clutch and shift linkage. Adjust the shift and clutch linkage as necessary.

☐ 26. Road test the vehicle, carefully observing transmission operation. After the road test, look for signs of leaks and recheck the fluid level.

☐ 27. Readjust the shift linkage as needed.

☐ 28. Clean the work area and return tools and equipment to storage.

☐ 29. Did you encounter any problems during this procedure? Yes ____ No ____

   If Yes, describe the problems: _____

   _____

   What did you do to correct the problems? _____

   _____

☐ 30. Have your instructor check your work and sign this job sheet.

Performance Evaluation—Instructor Use Only

Did the student complete the job in the time allotted? Yes ____ No ____

If No, which steps were not completed?_____

How would you rate this student's overall performance on this job?_____

5–Excellent, 4–Good, 3–Satisfactory, 2–Unsatisfactory, 1–Poor

Comments: _____

_____

_____

INSTRUCTOR'S SIGNATURE _____

Job 79—Disassemble and Inspect a Manual Transaxle

After completing this job, you will be able to disassemble a manual transaxle. You will also be able to identify and replace worn or defective manual transaxle parts.

Procedures

> **Note**
>
> There are many variations in manual transaxle design. The following procedure is a general guide only. Always consult the manufacturer's service literature for exact disassembly procedures and specifications.

- [ ] 1. Obtain a manual transaxle to be used in this job. Your instructor may direct you to perform this job on a shop transaxle.
- [ ] 2. Gather the tools needed to perform the following job. Refer to the tools and materials list at the beginning of the project.

Remove the Transaxle's External Components

- [ ] 1. Drain the transaxle lubricant, if not already drained.
- [ ] 2. Mount the transaxle in a special holding fixture, if available, or secure it on a clean workbench.
- [ ] 3. Remove the throwout bearing from the hub and clutch fork.
- [ ] 4. Remove the clutch fork:
  - If the fork rides on a pivot ball, pull the fork outward to remove it from the ball.
  - If the fork is operated by a clutch shaft, remove the clip or lock bolt holding the shaft and slide the shaft from the bell housing. Then, remove the fork.
  - If the transaxle uses an internal-type (concentric or ring) hydraulic slave cylinder, remove the fasteners and remove the slave cylinder and throwout bearing as a unit.
- [ ] 5. If possible, check the operation of the gears, synchronizers, and shift linkage by moving the shift forks while rotating the input shaft.
- [ ] 6. Remove any electrical switches or speed sensors from the case.
- [ ] 7. Gain access to the transaxle's internal parts by removing the through bolts and splitting the case halves apart, or by removing sheet metal or cast aluminum covers. Once this is done, the gears, shafts, and other internal parts will be exposed, **Figure 79-1**.

Remove the Transaxle's Internal Components

- [ ] 1. Remove the differential assembly.

> **Note**
>
> On some designs, the differential case assembly is removed after the transaxle's internal parts. If the transaxle is used with a longitudinal engine, the differential assembly is a separate unit and does not require disassembly unless defective.

- [ ] 2. Remove the snap rings or other fasteners holding the shafts and gears to the case.
- [ ] 3. Remove the main drive gears or drive chain and sprockets, if used.

Figure 79-1. Once the case on this particular transaxle is split, all internal parts are in view.

- ☐ 4. Remove the shaft assemblies and internal shift linkage assemblies from the case.
- ☐ 5. Remove the reverse idler gear and shaft, if separately mounted.

Disassemble the Transaxle's Internal Components

- ☐ 1. If tapered roller bearings are to be replaced, press them from their shafts.
- ☐ 2. Remove third and fourth (and fifth and sixth, if equipped) gear(s) and synchronizer assemblies from their shafts by removing snap rings, **Figure 79-2**, and pressing them off as required.
- ☐ 3. Remove any countershaft gears and synchronizer assemblies from their shafts.
- ☐ 4. Disassemble the internal shift linkage, if not done earlier.
- ☐ 5. Disassemble the differential and final drive assembly for inspection, removing the side bearings only if they are going to be replaced.
- ☐ 6. Remove the transaxle seals from their housings once all other parts are removed.

Figure 79-2. Check for snap rings before attempting to pull or press any gear from a transaxle shaft. (DaimlerChrysler)

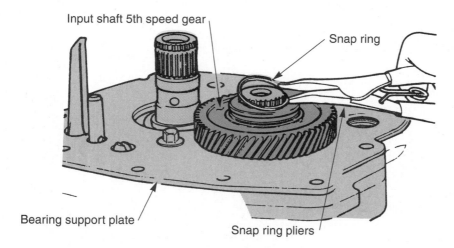

Input shaft 5th speed gear

Snap ring

Bearing support plate

Snap ring pliers

Job 79—Disassemble and Inspect a Manual Transaxle (continued)

Inspect the Transaxle Parts

☐ 1. Thoroughly clean the transaxle case and all parts.

> **Warning**
> ⚠ Do *not* spin bearings with air pressure.

☐ 2. Recover all loose parts, such as needle bearings, from the bottom of the case.

> **Note**
> 📄 All manual transaxles use needle bearings. These parts *must* be located and saved.

☐ 3. Visually examine the disassembled transaxle components, looking for the following:
- Worn or broken teeth on synchronizer hubs and sleeves.
- Wear on tapered surfaces or worn teeth on blocking rings.
- Worn shift fork grooves on synchronizer outer sleeves.
- Worn or distorted shift forks.
- Bent shift rails.
- Worn, cracked, or chipped teeth on transaxle gears.
- Worn gear bushings.
- Worn races on transaxle shafts.
- Worn or damaged thrust washers and snap rings.
- Worn or rough bearings.
- Flat spots on needle bearings.
- Cracked or distorted transaxle case.
- Worn front bearing retainer hub.
- Bent or missing oil slingers.
- Blocked oil passages.
- Worn differential and final drive components.

Inspect the components with special measuring instruments as required.

Describe the condition of the transaxle components: _____

_____

_____

_____

_____

☐ 4. Report any transaxle component defects that you found to your instructor. Then, determine which parts must be replaced.

> **Note**
> 📄 A synchronizer showing *any* wear should be replaced.

List the parts needed: _____

_____

_____

_____

_____

☐ 5. Obtain the proper replacement parts and compare them with the old parts to ensure that they are the correct replacements.

☐ 6. Clean the work area and return tools and equipment to storage.

☐ 7. Did you encounter any problems during this procedure? Yes ___ No ___

If Yes, describe the problems: _____

_____

What did you do to correct the problems? _____

_____

☐ 8. Have your instructor check your work and sign this job sheet.

Performance Evaluation—Instructor Use Only

Did the student complete the job in the time allotted? Yes ___ No ___

If No, which steps were not completed?_____

How would you rate this student's overall performance on this job?_____

5–Excellent, 4–Good, 3–Satisfactory, 2–Unsatisfactory, 1–Poor

Comments: _____

_____

_____

INSTRUCTOR'S SIGNATURE _____

Job 80—Reassemble and Install a Manual Transaxle

After completing this job, you will be able to properly assemble and install a manual transaxle.

Procedures

> **Note**
>
> There are many variations in manual transaxle design. The following procedure is a general guide only. Always consult the manufacturer's service literature for exact disassembly procedures and specifications.

☐ 1. Obtain a disassembled manual transaxle to be used in this job. Your instructor may direct you to perform this job on a shop transaxle.

☐ 2. Gather the tools needed to perform the following job. Refer to the tools and materials list at the beginning of the project.

Install Internal Components

☐ 1. Lightly lubricate all parts.

☐ 2. Assemble the synchronizers.

☐ 3. Install the synchronizers and gears on the transaxle shafts, using snap rings as required.

> **Note**
>
> Carefully observe the direction of the synchronizers during reinstallation. Snap rings should fit snugly in their grooves. Some parts may need to be tapped into place, as shown in **Figure 80-1**.

☐ 4. Check that plastic inserts are in place on the shift forks, **Figure 80-2**.

☐ 5. Reassemble the shift forks and shift rails and install them on the shaft gear assemblies.

Figure 80-1. It may be necessary to lightly tap some gears into position on the shaft.

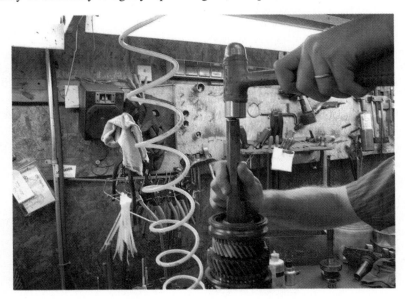

Figure 80-2. Plastic inserts should be in good condition and in place on the shift forks.

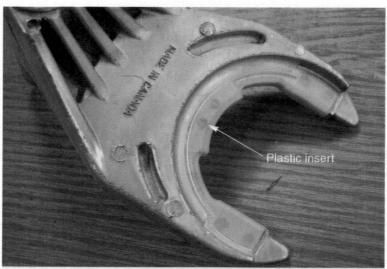

6. Place the shaft gear assemblies in position in the case.

> ### Note
> Some gear assembly parts may need to be installed individually into the case. Consult the appropriate service literature.

7. Install the reverse idler gear and shaft, if separately mounted.

8. Reinstall all washers in the position and direction shown in the factory service literature.

9. Check endplay of the various shafts and compare against specifications. Adjust if necessary.

Shaft	Measured Endplay	Specified Endplay
_____	_____	_____
_____	_____	_____
_____	_____	_____

10. Install the differential case assembly, before or after assembling the transaxle, as required.

Install the Transaxle's External Components

1. Reassemble transaxle case halves and/or install sheet metal or cast aluminum covers, as applicable. Use the proper kind of gasket or seal.

2. After tightening the case halves, make sure that the transaxle shafts turn without binding. If the shafts do not turn freely, find out why and correct the problem.

3. Check shaft endplay as described in the manufacturer's instructions, and compare it against specifications. Adjust if necessary.

   Transaxle input shaft:
   Measured endplay:_____
   Specified endplay:_____

   Transaxle output shaft (when applicable):
   Measured endplay:_____
   Specified endplay:_____

   Final drive shafts or gears (when applicable):
   Measured endplay:_____
   Specified endplay:_____

Job 80—Reassemble and Install a Manual Transaxle (continued)

> **Note**
>
> If the endplay is incorrect, adjust it by using thicker or thinner thrust washers or snap rings, as called for by the transaxle manufacturer.

☐ 4. If the transaxle uses a separate housing for the internal shift linkage, install it now. Check that the linkage operates properly when installed.

☐ 5. Reinstall any external case components, including electrical devices and linkage parts.

☐ 6. Install the transaxle drain plug.

☐ 7. Install the clutch throwout bearing and clutch fork or clutch shaft, if they were removed.

☐ 8. Ensure that all synchronizers and shift forks operate without binding.

Install a Manual Transaxle

☐ 1. Raise the vehicle.

> **Warning**
>
> ⚠ The vehicle must be raised and supported in a safe manner. Always use approved lifts or jacks and jack stands.

☐ 2. Inspect the throw-out bearing and clutch disc to ensure that they are properly positioned. See **Figure 80-3**.

☐ 3. Lightly grease the pilot bearing if it was not done as part of clutch service.

☐ 4. Ensure that the transmission input shaft is clean and undamaged, and that the clutch fork pivot ball (if used) is undamaged. See **Figure 80-4**.

Figure 80-3. Make sure that the clutch disc has not slipped out of position. The disc will remain aligned if the pressure plate has not been released.

Figure 80-4. Check the condition of the input shaft and the clutch fork pivot ball before continuing.

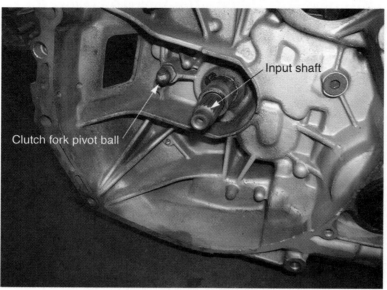

5. Raise the transaxle into position under the vehicle.

6. Place the transaxle in a gear and turn the output shaft to align the input shaft and clutch disc splines. Slide the transaxles input shaft into the clutch disc and ensure that the shaft is fully engaged. If necessary, slightly shift or wiggle the transaxle to install the shaft through the clutch disc and into the pilot bearing.

7. Install the attaching bolts that hold the transaxle to the engine.

8. Install the engine mounts and/or engine cradle.

9. Install the starter, if it was removed, and the dust cover.

10. Install the speedometer cable and any electrical connectors.

11. Install the exhaust pipes, if they were removed.

12. Install the front drive axles.

> **Note**
>
> Refer to Job 67 for specific instructions on CV axle installation.

13. Install stabilizer bars or other suspension components previously removed.

14. Install any other frame and body parts that were removed for clearance.

15. Lower the vehicle.

16. Remove the engine holding fixture.

17. Reconnect the transaxle shift and clutch linkages, electrical connectors, ground wires, and other components that were reached from under the hood.

18. Make preliminary clutch and shift linkage adjustments.

19. Fill the transaxle with the proper type of lubricant. If the unit has a separate reservoir for differential oil, refill it also.

20. Install the battery negative cable.

Job 80—Reassemble and Install a Manual Transaxle (continued)

☐ 21. Move the clutch pedal and gearshift lever to check the operation of the clutch and shift linkage.

☐ 22. Road test the vehicle and check transaxle operation.

☐ 23. Adjust the shift linkage, if necessary, and recheck transaxle's oil level.

☐ 24. Clean the work area and return tools and equipment to storage.

☐ 25. Did you encounter any problems during this procedure? Yes ___ No ___

If Yes, describe the problems: _____

_____

What did you do to correct the problems? _____

_____

☐ 26. Have your instructor check your work and sign this job sheet.

Performance Evaluation—Instructor Use Only

Did the student complete the job in the time allotted? Yes ___ No ___

If No, which steps were not completed?_____

How would you rate this student's overall performance on this job?_____

5–Excellent, 4–Good, 3–Satisfactory, 2–Unsatisfactory, 1–Poor

Comments: _____

_____

_____

INSTRUCTOR'S SIGNATURE _____

Notes

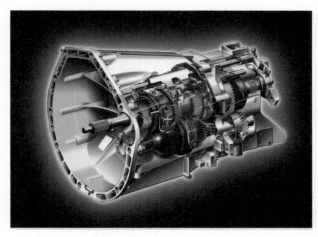

Project

Servicing a Transfer Case

24

Introduction

Transfer cases are commonly used on SUVs and pickup trucks. Removing and replacing a transfer case is similar to removing a manual transmission. One major difference is that most transfer cases are awkwardly shaped and prone to fall from the transmission jack. Extreme caution must be exercised. In Job 81, you will remove, inspect, and replace a transfer case.

Because four-wheel drive vehicles are becoming increasingly popular, it is inevitable that a technician will eventually have to overhaul a transfer case. Overhauling a transfer case requires careful attention to detail. Parts must be thoroughly cleaned and inspected. Repairs and adjustments must be made carefully since transfer cases are often operated under rough conditions. In Job 82, you will overhaul a transfer case. Four-wheel drive vehicles that are used off-road or in severe weather will require more attention to linkage, vacuum diaphragms, locking hubs, and other front axle components. In Job 83, you will test and adjust transfer case linkage and service diaphragms, locking hubs, and wheel bearings.

Project 24 Jobs

- Job 81—Inspect, Remove, and Replace a Transfer Case
- Job 82—Overhaul a Transfer Case
- Job 83—Service Transfer Case Shift Controls, Locking Hubs, and Wheel Bearings

Tools and Materials

The following list contains the tools and materials that may be needed to complete the jobs in this project. The items used will depend on the make and model of the vehicle being serviced.

- Vehicle in need of transfer case service.
- Hydraulic lift (or jack and jack stands).
- Applicable service information.
- Bearing driver.
- Seal driver.
- Gear and bearing puller.
- Measuring instruments.
- Dummy shaft.
- Drain pan.
- Hydraulic press and proper adapters.
- Hand tools.
- Air-powered tools.
- Safety glasses and other protective equipment.

Safety Notice

Before performing these jobs, review all pertinent safety information in the text and review safety information with your instructor.

Job 81—Inspect, Remove, and Replace a Transfer Case

After completing this job, you will be able to inspect, remove, and replace a transfer case.

Procedures

> **Note**
>
> . The following procedure is a general guide only. Always refer to the manufacturer's service literature for specific procedures.

☐ 1. Obtain a vehicle to be used in this job. Your instructor may direct you to perform this job on a shop vehicle.

☐ 2. Gather the tools needed to perform the following job. Refer to the tools and materials list at the beginning of the project.

Inspect a Transfer Case

☐ 1. Check the general condition of the transfer case by performing the following inspections:
   * Check the case for cracks, extreme corrosion, and loose fasteners. ___
   * Check for worn or damaged mounts. ___
   * Check the input and output shafts and flanges for damage or loose fasteners. ___
   * Check the linkage for binding, damage, or missing parts. ___
   * Check for plugged or missing vent tube(s) and related hoses. ___

   Is any damage found? Yes ___ No ___

   If Yes, describe the problem:_____

   _____

   _____

   _____

☐ 2. Check for leaks at the following points:
   * Front and rear pinion seals. ___
   * Case gaskets. ___
   * Linkage and sensor gaskets and seals. ___

   Are any leaks found? Yes ___ No ___

   If Yes, describe the problem:_____

   _____

   _____

☐ 3. Check the fluid level and the general condition of the fluid.

   Is the fluid level OK? Yes ___ No ___

   Describe the fluid condition:_____

   _____

   _____

☐ 4. Are any transfer case problems found while performing the above steps? Yes ___ No ___

   If Yes, what should be done to correct them?_____

   _____

   _____

Remove a Transfer Case

Note

The following procedure is a general guide only. This procedure assumes that the transmission remains installed and only the transfer case will be removed. Always refer to the manufacturer's service literature for specific procedures.

☐ 1. Disconnect the negative battery cable.

☐ 2. Raise the vehicle.

Warning

⚠ The vehicle must be raised and supported in a safe manner. Always use an approved lift or jack and jack stands.

☐ 3. If the vehicle has an off-road skid plate, remove it. **Figure 81-1** shows a skid plate. With the skid plate removed, the transfer case should be clearly visible and accessible, **Figure 81-2**.

☐ 4. Drain the lubricant from the transfer case.

Note

It is not necessary to drain the lubricant from a transfer case that does not have a slip yoke entering the case.

☐ 5. Place a transmission jack or stand under the transmission to help support the transmission and transfer case during removal.

☐ 6. Remove any ground straps and electrical connectors from the transfer case.

☐ 7. Remove the transfer case linkage, if used.

☐ 8. Remove the speedometer cable assembly, if necessary.

Figure 81-1. This is a typical skid plate, installed under the transfer case.

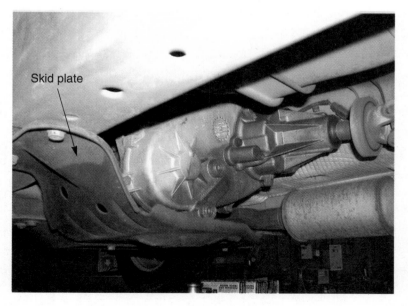

Job 81—Inspect, Remove, and Replace a Transfer Case (continued)

Figure 81-2. This transfer case can be removed after the drive shafts and the fasteners holding the case to the rear of the transmission are removed. Any electrical connectors and linkages must also be removed.

☐ 9. Remove the front and rear drive shafts.

Note

Refer to Job 66 for specific drive shaft removal instructions.

☐ 10. If necessary, remove the transmission mount and rear cross member.

☐ 11. Remove any brackets or fasteners holding the transfer case to the vehicle frame.

☐ 12. Place a transmission jack (or floor jack) under the transfer case.

Warning

⚠ If the transfer case is not properly supported and secured to the jack, it will drop when the bolts holding it to the transmission are removed, causing injury or damage.

☐ 13. Remove the bolts holding the transfer case to the transmission.

☐ 14. Slide the transfer case straight back from the transmission until it clears the output shaft.

☐ 15. Lower the transfer case from the vehicle. Since transfer cases have a shape that can be awkward to handle, it is usually best to place the transfer case on the ground where it cannot fall, **Figure 81-3**.

Install a Transfer Case

☐ 1. Using a transmission jack or floor jack, raise the transfer case into position under the vehicle.

☐ 2. Slide the transfer case into engagement with the rear of the transmission, **Figure 81-4**.

Figure 81-3. Place the transfer case on the ground to prevent damage or injury.

Figure 81-4. Slide the transfer case onto the back of the transmission case, making sure that the gasket (if used) is in position. (TTC/Tremic)

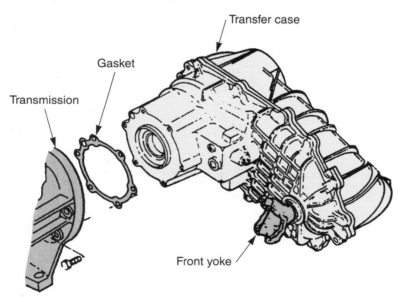

Transfer case

Gasket

Transmission

Front yoke

Caution

 If a gasket is used between the transmission and transfer case, be sure a new gasket is in place before installing the transfer case.

☐ 3. Install the attaching bolts that hold the transfer case to the transmission.

Note

 Make sure that the drive shaft yokes can turn after the transfer case fasteners are tightened.

Job 81—Inspect, Remove, and Replace a Transfer Case (continued)

☐ 4. Install the brackets or fasteners holding the transfer case to the vehicle frame.

☐ 5. If applicable, install the crossmember and transmission mount.

☐ 6. Install the drive shafts, making sure to line up the match marks made during removal.

> **Note**
> Refer to Job 66 for specific instructions on drive shaft installation.

☐ 7. Install the transfer case linkage, if used.

☐ 8. Install the speedometer cable assembly, if necessary.

☐ 9. Reconnect any ground straps and electrical connectors removed from the transfer case.

☐ 10. Fill the transfer case to the proper level with the specified lubricant.

☐ 11. Install and tighten the filler plug.

> **Note**
> If the transmission and transfer case use the same oil, check the oil level at the transmission.

☐ 12. Install the skid plate, if used.

☐ 13. Lower the vehicle.

☐ 14. Install the negative battery cable.

☐ 15. Road test the vehicle, carefully observing transfer case operation.

☐ 16. After the road test, look for signs of leaks and recheck the fluid level.

☐ 17. Adjust the shift linkage, if necessary.

Check for Matching Axle Sizes

☐ 1. Raise the vehicle.

☐ 2. Check the axle ratios to ensure that they are right for the vehicle by one of the following methods.

Use Serial Numbers
- Check the serial numbers on the front and rear axles and compare them with vehicle specifications.
  Do the serial numbers indicate that the ratios of the front and rear axles are the same?
  Yes ___ No ___

Or Check Manually
- Place the transmission in neutral.
  Turn the rear drive shaft while holding one of the rear drive wheels stationary.
  Record the number of turns needed to make the free drive wheel complete one revolution. ____
  Turn the front drive shaft while holding one of the front drive wheels stationary.
  Record the number of turns needed to make the free drive wheel complete one revolution. ____

> **Note**
>
> For an accurate reading, repeat the previous step at least once.

Compare the number of turns. Is the number of front and rear turns the same?
Yes ___ No ___
If No, explain: _____
_____
_____

Check for Matching Tire Sizes

1. Inspect all four tires and record the sizes and ratings.

	Size (For example: 225 × 15)	Load Index	Speed Rating (may not be present on some tires)
RF			
LF			
RR			
LR			

Do the tire sizes and ratings match on all four tires? Yes ___ No ___
If No, explain: _____
_____
_____
_____

2. Clean the work area and return any equipment to storage.

3. Did you encounter any problems during this procedure? Yes ___ No ___
If Yes, describe the problems: _____
_____
What did you do to correct the problems? _____
_____

4. Have your instructor check your work and sign this job sheet.

Performance Evaluation—Instructor Use Only

Did the student complete the job in the time allotted? Yes ___ No ___
If No, which steps were not completed?_____
How would you rate this student's overall performance on this job?_____
5–Excellent, 4–Good, 3–Satisfactory, 2–Unsatisfactory, 1–Poor
Comments: _____
_____

INSTRUCTOR'S SIGNATURE _____

Job 82—Overhaul a Transfer Case

After completing this job, you will be able to disassemble and reassemble a transfer case. You will also be able to identify, repair, and replace worn or defective transfer case parts.

Procedures

☐ 1. Obtain a vehicle to be used in this job. Your instructor may direct you to perform this job on a shop vehicle.

☐ 2. Gather the tools needed to perform the following job. Refer to the tools and materials list at the beginning of the project.

> **Note**
> There are many variations in transfer case designs. The following procedure is only a general guide. Always consult the manufacturer's service literature for exact disassembly procedures and specifications.

Disassemble a Transfer Case

☐ 1. Secure the transfer case on a clean workbench.

☐ 2. Remove the drive yoke nuts and then remove the yokes.

☐ 3. Remove the speedometer gear, electrical switches, sensors, motors, solenoids, and other external transfer case parts, as outlined in the service literature.

☐ 4. Remove bearing retainers or split the case halves as necessary to gain access to the internal components of the transfer case.

> **Note**
> If the transfer case has a chain drive, check the chain for looseness before continuing disassembly, **Figure 82-1**.

Figure 82-1. Check for excessive transfer case chain wear before removing the chain and sprockets.

5. Remove snap rings, retainer plates, or other fasteners holding the gears to their shaft, the shafts to the housing, or the shafts to the drive chain, **Figure 82-2**. Consult the service literature for exact procedures.

6. Remove the drive chain and sprockets, and any thrust washers located under the sprockets.

   or

   Remove the drive, idler, and driven transfer gears from the case, as shown in **Figure 82-3**.

   Does the transfer case use a chain or transfer gears? _____

7. Remove the remaining internal components from the transfer case. Some internal parts are held with large snap rings, as in **Figure 82-4**.

   Which of the following parts does the transfer case contain?

   Clutch pack. ____
   Differential unit. ____
   Input and output shafts. ____
   Oil passages or slingers. ____
   Oil pump. ____
   Planetary gears. ____
   Shift forks. ____
   Shift rails. ____
   Sliding gears. ____
   Synchronizers. ____
   Viscous coupling. ____

Inspect Transfer Case Parts

1. Carefully scrape all old gasket material from the transfer case housings.

2. Clean all metal parts in a safe solvent and dry the parts with compressed air or let them air dry.

> **Warning**
>
> ⚠ Do *not* spin anti-friction bearings with compressed air.

3. Recover all loose parts, such as needle bearings, from the bottom of the housing.

Figure 82-2. A—Remove any retainer plates that prevent removal of the transfer case internal parts. B—Remove snap rings that hold the shafts to the transfer case.

A   B

Job 82—Overhaul a Transfer Case (continued)

Figure 82-3. Once the case halves are split, the transfer gears can be lifted from the case. (Arvin/Meritor)

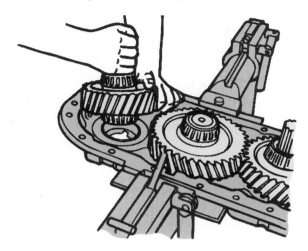

Figure 82-4. Many internal parts, such as this planetary gear assembly, are held in place with large snap rings. These rings can usually be removed with a large screwdriver.

4. Visually examine the disassembled transfer case components for the following:
   - Worn gear bushings.
   - Worn races on transfer case shafts.
   - Worn or rough bearings.
   - Damaged bearing retainers.
   - Worn, cracked, or chipped gear teeth.
   - Worn shaft splines.
   - Worn tapered surfaces or worn teeth on blocking rings.
   - Worn shift fork grooves on sliding clutch sleeves.
   - A worn or damaged differential assembly.
   - A worn or leaking viscous coupling.
   - Plugged or leaking vents.
   - Leaking seals.
   - Worn clutch plates.
   - A cracked or distorted transfer case housing or extension housing.

Inspect components using special measuring instruments, as required.

Describe the condition of the transfer case components: _____

_____

_____

_____

_____

☐ 5. Report any transfer case component defects to your instructor. Then, determine which parts must be replaced.

Which parts will be replaced?_____

_____

_____

_____

_____

☐ 6. Obtain the proper replacement parts and compare them with the old parts to ensure that they are the correct replacements.

Assemble a Transfer Case

☐ 1. Lightly lubricate all parts before proceeding. Use the same fluid as will be used in the transfer case.

☐ 2. Assemble the input and output shaft gears and hubs onto the shaft.

☐ 3. Install the internal transfer case parts into the transfer case housing.

> **Note**
>
> The shift forks are usually installed at the same time as the gears that they operate. Make sure that the shafts are properly seated in the housing to avoid binding.

☐ 4. Install the chain drive. Make sure that the sprockets and shafts are properly aligned with the chain and that all thrust washers are installed. **Figure 82-5** shows an assembled chain drive.

or

Install the drive, idler, and driven transfer gears.

Figure 82-5. An assembled transfer case chain.

Job 82—Overhaul a Transfer Case (continued)

☐   5.   Install the snap rings or retainers that hold the chain and shafts in place.

☐   6.   Place a gasket or appropriate sealer on the front housing or bearing retainer mating surface, as applicable. Then, reassemble the transfer case housing.

☐   7.   Check internal shift linkage operation and verify that the shafts turn without binding.

> **Caution**
>
> ⚠️   If the shafts do not turn freely, find out why and correct the problem.

☐   8.   Adjust bearing preload, if necessary.

☐   9.   If required, check shaft endplay as explained in the manufacturer's instructions. **Figure 82-6** shows a typical procedure for checking endplay on a transfer case with a gear drive. Compare the measured endplay to the specification.

     Measured endplay:_____

     Specified endplay:_____

☐ 10.   Adjust helical drive and driven transfer gears, if applicable.

☐ 11.   Install the speedometer gear, electrical switches, and other external transfer case parts. Consult the service literature for exact procedures and information on adjusting sensors.

☐ 12.   Install the front and rear drive yokes.

☐ 13.   Clean the work area and return any equipment to storage.

☐ 14.   Did you encounter any problems during this procedure? Yes ___ No ___

     If Yes, describe the problems: _____

     _____

     What did you do to correct the problems? _____

     _____

☐ 15.   Have your instructor check your work and sign this job sheet.

Figure 82-6. If specified, check the endplay before finishing reassembly of the transfer case. (Arvin/Meritor)

Performance Evaluation—Instructor Use Only

Did the student complete the job in the time allotted? Yes ____ No ____

If No, which steps were not completed?_____

How would you rate this student's overall performance on this job?_____

5–Excellent, 4–Good, 3–Satisfactory, 2–Unsatisfactory, 1–Poor

Comments: _____

_____

_____

INSTRUCTOR'S SIGNATURE _____

Job 83—Service Transfer Case Shift Controls, Locking Hubs, and Wheel Bearings

After completing this job, you will be able to service transfer case shift controls, locking hubs, and wheel bearings.

Procedures

> **Note**
>
> The following procedure is a general guide only. Always refer to the manufacturer's service literature for specific procedures.

☐ 1. Obtain a vehicle to be used in this job. Your instructor may direct you to perform this job on a shop vehicle.

☐ 2. Gather the tools needed to perform the following job. Refer to the tools and materials list at the beginning of the project.

Check Manual Linkage

☐ 1. Operate the shift linkage to check for binding, excessive looseness, or other problem.
- Binding. ___
- Looseness. ___
- Jumps out of gear. ___
- Other. ___ Describe:
Positions in which the problem occurs: _____
_____
_____
_____

☐ 2. Raise the vehicle.

> **Warning**
>
> ⚠ The vehicle must be raised and supported in a safe manner. Always use an approved lift or jack and jack stands.

☐ 3. Locate the transfer case or front axle manual linkage (similar to **Figure 83-1**) and examine it for the following:
- Lack of lubrication. ___
- Bent rods or levers. ___
- Missing parts. ___
- Worn or binding bushings or other moving parts. ___
- Binding shift cables or damage to cable sheaths. ___
- Loose or missing adjusters. ___

Figure 83-1. Most transfer case linkages will resemble the one shown here. Older transfer cases may have two or more rod-and-lever linkages. Many linkages do not have adjustment provisions.

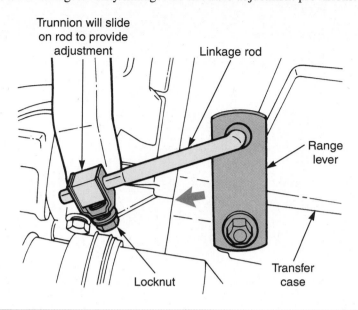

Trunnion will slide on rod to provide adjustment

Linkage rod

Range lever

Transfer case

Locknut

Note

If you find an obvious problem, consult your instructor. Your instructor may direct you to correct it before proceeding.

Are any problems found? Yes ___ No ___

Could they be the cause of the complaint? Yes ___ No ___

If problems are found, go to step 5.

If no problems are found, go to step 4.

4. If the previous steps do not reveal the problem, disconnect the linkage from the transfer case or transaxle.

Does the linkage now correctly operate? Yes ___ No ___

If Yes, the transfer case has an internal problem. Refer to Jobs 81 and 82.

If No, repeat step 3 to isolate the problem.

5. Consult your instructor as to what steps to take to correct the problem. Steps may include the following:
   - Tightening fasteners. ___
   - Lubricating moving parts. ___
   - Replacing worn or damaged parts. ___
   - Adjusting the linkage. ___

   Steps to be taken: _____

   _____

   _____

6. Make repairs or adjustments as needed.

7. Recheck linkage operation.

Does the transfer case operate properly? Yes ___ No ___

If No, repeat the previous steps until the problem has been located and corrected.

Job 83—Service Transfer Case Shift Controls, Locking Hubs, and Wheel Bearings (continued)

Check Vacuum Shift Diaphragms

☐ 1. Check that the vacuum diaphragm operates during the proper selector positions with the engine running.

Does the diaphragm operate? Yes ___ No ___

If Yes, the transfer case or front axle may have an internal problem.

If No, go to step 2.

☐ 2. Check that vacuum is available to the diaphragm from the control system at the proper times.

Is vacuum available? Yes ___ No ___

If No, go to step 3.

If Yes, go to step 4.

☐ 3. Verify that manifold vacuum is available to the control system by checking the hose connections from the engine intake manifold through the control valves and switches.

Is vacuum available? Yes ___ No ___

If No, repair the vacuum system as needed.

If Yes, go to step 4.

☐ 4. Use a vacuum pump to test the diaphragm. See **Figure 83-2**.

Does the diaphragm hold vacuum? Yes ___ No ___

> **Note**
>
> Some systems have two diaphragms. Both should hold vacuum.

Does the diaphragm linkage move when vacuum is applied? Yes ___ No ___

> **Note**
>
> On some designs, the linkage is not visible.

If the answer to either of the above questions is No, replace the diaphragm and recheck operation.

Figure 83-2. Test the diaphragm at both vacuum ports.

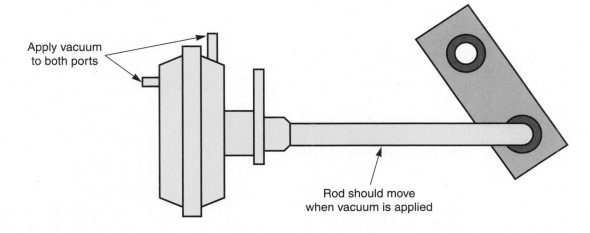

Apply vacuum to both ports

Rod should move when vacuum is applied

Check Electrical Controls Using a Scan Tool

☐   1.   Obtain a scan tool and related service literature for the vehicle.

Type of scan tool: _____

☐   2.   Attach the proper test connector cable and power lead to the scan tool.

☐   3.   Ensure that the ignition switch is in the *off* position.

☐   4.   Locate the correct diagnostic connector.

Diagnostic connector location: _____

☐   5.   Attach the scan tool test connector cable to the diagnostic connector. If necessary, attach the scan tool power lead to the cigarette lighter or battery terminals.

☐   6.   Observe the scan tool screen to ensure that it is working properly.

☐   7.   Enter vehicle information as needed to program the scan tool. If the scan tool can read the VIN from the vehicle computer, skip this step.

☐   8.   Turn the ignition switch to the *on* position.

☐   9.   Determine whether the scan tool is communicating with the computer module.

Are the module and scan tool communicating? Yes ___ No ___

If Yes, go to step 10.

If No, go to step 12.

☐   10.   Retrieve trouble codes and list them in the spaces provided.

Trouble Codes

_____ _____ _____

_____ _____ _____

_____ _____ _____

☐   11.   Turn the ignition switch to the *off* position.

Do the codes apply to the transfer case? Yes ___ No ___

If Yes, proceed to step 12.

If No, or if no codes are present, consult your instructor.

☐   12.   Make further electrical checks:
- Check system fuses. ___
- Check for battery voltage at system input connections. ___
- Check system and vehicle grounds. ___
- Check for disconnected or wiring with high resistance. ___
- Check sensor resistance readings, as explained in the next section of this job. ___

What problems are found? _____

_____

What should you do to correct them? _____

_____

Check Electrical Controls Using an Ohmmeter

☐   1.   Locate the switch or sensor to be tested. Many electrical devices are threaded directly into the case, as shown in **Figure 83-3**.

☐   2.   Determine the following from the manufacturer's service literature:
- Correct switch/sensor resistance reading:_____
- Terminals at which to take readings: _____
- Position of manual linkage connected to the switch or sensor: _____

Job 83—Service Transfer Case Shift Controls, Locking Hubs, and Wheel Bearings (continued)

Figure 83-3. This linkage position switch is threaded into the case next to the front output shaft.

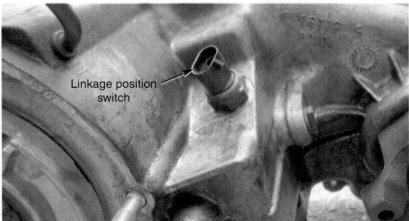

Linkage position switch

☐ 3. Obtain an ohmmeter or multimeter. Set the ohmmeter to the appropriate range and calibrate (zero) the meter.

> **Note**
>
> Calibration is usually not necessary when using digital meters.

☐ 4. Ensure that the ignition switch is in the *off* position.

☐ 5. Remove the electrical connector from the switch or sensor to be tested.

> **Caution**
>
> Ensure that all electrical power is removed from the switch or sensor before beginning the ohmmeter test.

☐ 6. Place the ohmmeter test leads on the appropriate device terminals. Some devices can be tested by disconnecting a remote connector and accessing the correct connector pins.

☐ 7. Observe the meter reading and compare it to specifications.

Measured resistance:_____

Specified resistance:_____

☐ 8. Determine from the ohmmeter readings whether switch or sensor is good or bad.

☐ 9. Replace the switch or sensor if needed. See the *Replace Sensors and Switches* section later in this job.

☐ 10. Reconnect the electrical connector and reinstall any other components that were removed to gain access to the sensor or switch.

☐ 11. Ensure that the sensor or switch is operating properly.

Service Locking Hubs and Front Wheel Bearings

> **Warning**
>
> ⚠ The vehicle must be raised and supported in a safe manner. Always use an approved lift or jack and jack stands.

☐ 1. Raise the wheel corresponding to the locking hub that is to be replaced.

☐ 2. Place transfer case in two-wheel drive.

☐ 3. Lock the hub (manual).

☐ 4. Remove the outer cover.

☐ 5. Remove the axle cotter pin/locknut and the axle nut, if applicable.

☐ 6. Remove snap rings and other retainer rings as applicable.

☐ 7. Slide the internal parts from the axle shaft.

☐ 8. Clean the internal parts and inspect them for the following:
  - Wear. ___
  - Damage. ___
  - Rust. ___
  - Lack of lubrication. ___
  - Dirt contamination. ___

  What problems are found? _____

  _____

  What should you do to correct them? _____

  _____

☐ 9. Obtain new parts as needed.

> **Note**
>
> ▤ Always replace seals and gaskets when the locking hub is disassembled. Many technicians prefer to replace tension springs as well since the springs can lose tension without showing obvious defects.

☐ 10. Thoroughly clean the axle housing.

☐ 11. Heavily lubricate all internal parts with the proper grease.

☐ 12. Reassemble the internal parts.

> **Note**
>
> ▤ Adjust the wheel bearings according to manufacturer's procedures and specifications.

☐ 13. Install a new axle cotter pin/locknut and the axle nut, if applicable.

☐ 14. Reinstall the cover using a new gasket.

☐ 15. Lower the wheel.

☐ 16. Check locking hub operation.

Job 83—Service Transfer Case Shift Controls, Locking Hubs, and Wheel Bearings (continued)

Linkage Service

> **Warning**
>
> The vehicle must be raised and supported in a safe manner. Always use an approved lift or jack and jack stands.

☐ 1. Raise the vehicle.

☐ 2. Remove the defective linkage or cable.

☐ 3. Compare the old and replacement parts.

☐ 4. Install the new part.

☐ 5. Adjust the linkage as necessary.

☐ 6. Lower the vehicle.

☐ 7. Check the linkage and overall four-wheel drive operation.

Replace a Vacuum Shift Diaphragm

> **Warning**
>
> The vehicle must be raised and supported in a safe manner. Always use an approved lift or jack and jack stands.

☐ 1. Raise the vehicle.

☐ 2. Remove the vacuum line from the vacuum diaphragm.

☐ 3. Remove the linkage and the fasteners holding the vacuum diaphragm to the housing.

> **Note**
>
> Some front axle vacuum diaphragms can only be removed by disconnecting the linkage inside of the axle housing. Drain the lubricant and remove the differential cover to access the linkage. Some vacuum diaphragms are part of an assembly that contains the shift fork and linkage, **Figure 83-4**. The assembly must be removed from the axle, after which the clips can be detached and the diaphragm assembly slid from the housing, **Figure 83-5**.

Figure 83-4. The vacuum diaphragm may be part of an assembly containing the shift fork. (Chrysler)

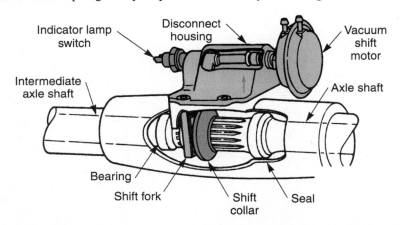

Figure 83-5. Once the retainer clips are removed, the diaphragm will slip from the housing. (Chrysler)

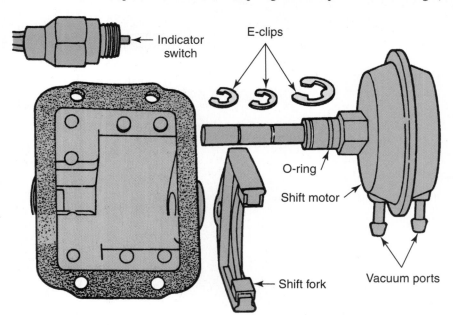

4. Compare the old and new diaphragms to ensure that the replacement is correct.

5. Place the vacuum diaphragm in position on the housing.

6. Attach the diaphragm linkage as necessary.

7. Install and tighten the vacuum diaphragm fasteners.

8. Reattach the vacuum line.

9. Lower the vehicle.

10. Check the operation of the diaphragm and overall four-wheel drive operation.

Replace Sensors and Switches

1. Disconnect the negative battery cable.

2. Raise the vehicle, if necessary.

> **Warning**
> The vehicle must be raised and supported in a safe manner. Always use an approved lift or jack and jack stands.

3. Remove any parts that prevent access to the sensor or switch.

4. If necessary, position a drain pan under the sensor or switch.

5. Remove the sensor/switch electrical connector.

6. Remove the sensor or switch by unthreading it from the drive train part or by removing the attaching screws that hold the part in place.

7. Compare the old and new parts to ensure that the replacement is correct.

8. Install the replacement sensor or switch, using a new seal or gasket where required.

9. Reconnect the electrical connector.

10. Reinstall any other parts that were removed.

Job 83—Service Transfer Case Shift Controls, Locking Hubs, and Wheel Bearings (continued)

☐ 11. If necessary, refill the drive train component with the proper lubricant, being careful not to overfill the reservoir.

☐ 12. Remove the drain pan and lower the vehicle.

☐ 13. Start the engine and check sensor or switch operation. Road test the vehicle if necessary.

☐ 14. Recheck fluid level and check for fluid leaks at the new sensor or switch.

☐ 15. Clean the work area and return any equipment to storage.

☐ 16. Did you encounter any problems during this procedure? Yes ___ No ___

If Yes, describe the problems: _____

_____

What did you do to correct the problems? _____

_____

☐ 17. Have your instructor check your work and sign this job sheet.

Performance Evaluation—Instructor Use Only

Did the student complete the job in the time allotted? Yes ___ No ___

If No, which steps were not completed?_____

How would you rate this student's overall performance on this job?_____

5–Excellent, 4–Good, 3–Satisfactory, 2–Unsatisfactory, 1–Poor

Comments: _____

_____

_____

INSTRUCTOR'S SIGNATURE _____

Notes

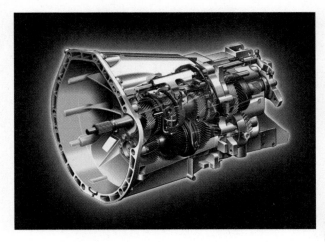

Project

25

Servicing Rear-Wheel Drive Axles and Differentials

Introduction

Modern rear-wheel drive vehicles use one of three types of axles: the retainer-type axle, the C-lock axle, and the independent axle. Service procedures for all three vary. Although rear axles are seldom overhauled, every drive train technician will eventually be confronted with the task of servicing a rear drive axle. This job should not be tackled without the special measuring tools needed. The tooth contact pattern between the differential pinion gear and ring gear is vital to quiet operation and long life.

You will remove, service, and install a retainer-type and C-lock axles in Job 84, and an independent rear axle shaft in Job 85. In Job 86, you will remove an entire rear axle assembly from a vehicle. You will disassemble and inspect the differential in Job 87, and reassemble and adjust the differential in Job 88. In Job 89, you will reinstall the rear axle assembly.

Project 25 Jobs

- Job 84—Remove, Service, and Install Retainer-Type and C-Lock Axles
- Job 85—Remove and Replace an Independent Rear Axle Shaft
- Job 86—Remove a Rear Axle Assembly
- Job 87—Disassemble and Inspect a Differential
- Job 88—Reassemble and Adjust a Differential
- Job 89—Install a Rear Axle Assembly

Tools and Materials

The following list contains the tools and materials that may be needed to complete the jobs in this project. The items used will depend on the make and model of vehicle being serviced.

- Vehicle in need of service.
- Hydraulic lift (or jacks and jack stands).
- Applicable service information.
- Appropriate brake tools.
- Slide hammer.
- Bearing driver.
- Seal driver.
- Companion flange spanner wrench.
- Companion flange holding tool.
- Gear and bearing puller.
- Miscellaneous measuring instruments.
- Drain pan.
- Hydraulic press and proper adapters.
- Replacement gaskets and seals.
- Appropriate sealers and adhesive.
- Specified rear end lubricant.
- Hand tools as needed.
- Air-powered tools as needed.
- Safety glasses and other protective equipment as needed.

Safety Notice

Before performing this job, review all pertinent safety information in the text, and review safety information with your instructor.

Job 84—Remove, Service, and Install Retainer-Type and C-Lock Axles

After completing this job, you will be able to remove and install retainer-type and C-lock axle shafts used in rear-wheel drive vehicles. You will also be able to replace axle bearings and axle seals.

Procedures

☐ 1. Obtain the vehicles to be used in this job. Your instructor may direct you to perform this job on shop vehicles.

☐ 2. Gather the tools needed to perform the following job. Refer to the tools and materials list at the beginning of the project.

Service Retainer-Type Axle

In the following section of the job, you will service a retainer-type axle.

Remove the Axle

☐ 1. Disconnect the battery negative cable.

☐ 2. Raise the vehicle.

> **Warning**
> ⚠ The vehicle must be raised and supported in a safe manner. Always use approved lifts or jacks and jack stands.

☐ 3. Remove the rear wheel on the side of the rear axle being serviced.

☐ 4. Remove the brake drum, marking it so it can be reinstalled in the same position.

or

Remove the brake caliper and rotor. Use wire to support the caliper so it will not hang and damage the brake line. Mark the rotor and hub so the rotor can be installed in its original position in the axle flange.

> **Note**
> 📄 Marking the original position of each drum or rotor and reinstalling it in the same position ensures that slight tolerance variations will not cause a vibration that was not present before repairs.

☐ 5. Remove the drive axle retainer plate fasteners.

☐ 6. Remove the drive axle and place it on a clean workbench. It may be necessary to use a slide hammer puller to loosen and remove the axle, **Figure 84-1**.

> **Caution**
> ◇ When pulling the axle out, make sure that the retainer plate does not hang up on the brake backing plate, causing the backing plate to pull away from the rear axle housing.

Figure 84-1. Axle bearings often freeze in the axle housing and it may be necessary to use a slide hammer puller to loosen and remove the axle. (OTC)

Inspect and Repair Axle Parts

 1. Grind or chisel the axle collar from the axle. **Figure 84-2** shows how the chisel should be placed to crack the collar.

> ## Warning
> ⚠ Always use eye protection when removing the collar.

☐ 2. Press the axle bearing from the axle.

☐ 3. Slide the retainer plate from the axle.

☐ 4. Inspect the axle components for damage.

Describe their condition: _____

_____

_____

Figure 84-2. Use a sharp chisel and hammer to crack the collar, after which it can be removed from the shaft. Always wear eye protection when using a chisel.

Job 84—Remove, Service, and Install Retainer-Type and C-Lock Axles (continued)

☐ 5. Report any axle component defects that you found to your instructor. Then, determine which parts must be replaced.

List the parts that must be replaced: _____

_____

_____

☐ 6. Obtain the proper replacement parts and compare them with the old parts to ensure that they are the correct replacements.

☐ 7. Replace the axle seal in the axle tube bore.

☐ 8. Clean the retainer plate.

☐ 9. Install a new seal (and gasket if called for) in the retainer plate.

☐ 10. Install the retainer plate on the axle.

☐ 11. Press the new axle bearing onto the axle.

☐ 12. Press the new axle collar onto the axle.

Install the Axle

☐ 1. Remove the brake backing plate and clean the axle tube flange and backing plate of dirt and old gasket material.

☐ 2. Place a new gasket over the axle tube flange and reinstall the brake backing plate.

☐ 3. Clean any old gasket material from the outside surface of the brake backing plate, at the site of the retainer plate.

☐ 4. Place a new gasket on the retainer plate or over the brake backing plate on the axle tube flange studs.

☐ 5. Lubricate the splines on the drive axle with the proper lubricant.

☐ 6. Install the drive axle into the axle tube, being careful not to damage the axle seal. Engage the axle splines with the side gear splines, pushing in the axle as far as it will go.

☐ 7. Install and tighten the retainer plate fasteners.

☐ 8. Check the axle endplay with a dial indicator. Follow the general procedure given here:
a. Install the dial indicator on the axle flange.
b. Push the axle inward.
c. Zero the dial indicator.
d. Pull the axle outward.
e. Note the dial indicator reading.
Measured endplay:_____
Specified endplay:_____
Is the reading within specifications? Yes ___ No ___
If No, what should you do to correct it? _____

_____

f. Reposition the dial indicator to measure axle flange runout.
g. Zero the dial indicator.
h. Rotate the axle and note the dial indicator reading.
Is the reading within specifications? Yes ___ No ___
If No, what should you do to correct it? _____

_____

☐ 9. Install the brake drum or rotor and caliper.

10. Install the rear wheel.

11. Lower the vehicle.

12. Install the battery negative cable.

13. Start the engine.

14. Check that the brakes operate properly.

15. Road test the vehicle, carefully observing rear axle operation.

Service a C-Lock Axle

In this section of the job, you will service a C-lock axle.

Remove the Axle

1. Disconnect the battery negative cable.

2. Raise the vehicle.

> **Warning**
>
> ⚠ The vehicle must be raised and supported in a safe manner. Always use approved lifts or jacks and jack stands.

3. Remove the rear wheel on the side of the rear axle being serviced.

4. Remove the brake drum, marking it so that it can be reinstalled in the same position.

   or

   Remove the brake caliper and rotor. Use wire to support the caliper so it will not hang and damage the brake line. Mark the rotor and hub so the rotor can be installed in its original position in the axle flange.

> **Note**
>
> 📄 Marking the original position of each drum or rotor and reinstalling it in the same position ensures that slight tolerance variations will not cause a vibration that was not present before repairs.

5. Place a drain pan beneath the differential carrier. Loosen cover bolts or drain plug and allow the lubricant to drain.

6. Remove the differential cover.

7. Begin removal of the pinion shaft, which holds the differential spider gears, by removing the lock bolt or pin holding it to the differential case, **Figure 84-3**.

8. Slide the shaft from the case, leaving the spider gears in place.

9. Push the axle flange inward.

10. Remove the C-lock, **Figure 84-4**.

11. Remove the drive axle.

Inspect and Repair Axle Parts

1. Remove the bearing seal or bearing seal plate from the axle housing, as applicable.

2. Pry the axle bearing from the axle tube or use a slide hammer if the bearing is stuck.

3. Clean the axle, seal plate (if used), and axle bearing.

Job 84—Remove, Service, and Install Retainer-Type and C-Lock Axles (continued)

Figure 84-3. Remove the lock that holds the pinion shaft in place. (DaimlerChrysler)

Figure 84-4. Remove the C-lock from the shaft. (DaimlerChrysler)

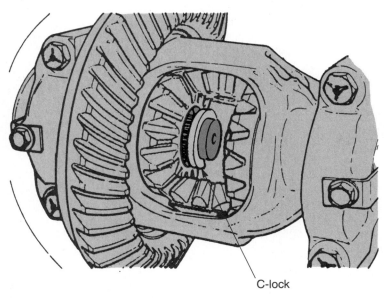

C-lock

☐ 4. Inspect the axle components for damage.

Describe their condition: _____

_____

_____

_____

☐ 5. Report any axle component defects that you found to your instructor. Then, determine which parts must be replaced.

List the parts that must be replaced: _____

_____

_____

_____

☐ 6. Obtain the proper replacement parts and compare them with the old parts to ensure that they are the correct replacements.

☐ 7. Install a new seal in the bearing seal plate, if applicable.

☐ 8. Lubricate the axle bearing with the proper lubricant.

☐ 9. Use the correct bearing driver to install the bearing in the axle tube.

☐ 10. Install the bearing seal or bearing seal plate, as applicable.

Install the Axle

☐ 1. Lubricate the splines on the drive axle with the proper lubricant.

☐ 2. Install the drive axle into the axle tube, being careful not to damage the axle seal. Engage axle splines with side gear splines and push the axle in until the end extends past the side gear.

☐ 3. Insert the C-lock in the groove in the axle.

☐ 4. Pull the axle flange outward to seat the C-lock.

☐ 5. Install the pinion shaft.

☐ 6. Install the pinion shaft lock bolt or pin. Torque the bolt to specifications.

☐ 7. Check axle endplay and axle flange runout with a dial indicator. Follow the general procedures given here:
   a. Install the dial indicator on the axle flange.
   b. Push the axle inward.
   c. Zero the dial indicator.
   d. Pull the axle outward.
   e. Note the dial indicator reading.
   Measured endplay:_____
   Specified endplay:_____
   Is the reading within specifications? Yes ___ No ___
   If No, what should you do to correct it?_____
   _____
   f. Reposition the dial indicator to measure axle flange runout.
   g. Zero the dial indicator.
   h. Rotate the axle and note the dial indicator reading.
   Is the reading within specifications? Yes ___ No ___
   If No, what should you do to correct it?_____
   _____

☐ 8. Scrape off any old gasket material from the differential cover and its mating surface on the rear axle housing.

☐ 9. Install a new gasket on the differential cover using nonhardening sealer, or use form-in-place gaskets, as required.

☐ 10. Tighten the differential cover bolts in a crisscross pattern, torquing the bolts to manufacturer's specification.

☐ 11. Fill the differential with the proper type of lubricant. Replace the fill plug and torque it to specifications.

☐ 12. Install the brake drum or rotor and caliper.

☐ 13. Adjust the brakes if necessary.

☐ 14. Install the rear wheel.

☐ 15. Lower the vehicle.

Job 84—Remove, Service, and Install Retainer-Type and C-Lock Axles (continued)

☐ 16. Install the battery negative cable.

☐ 17. Start the engine.

☐ 18. Check the brakes.

☐ 19. Road test the vehicle, carefully observing rear axle operation.

☐ 20. Clean the work area and return tools and equipment to storage.

☐ 21. Did you encounter any problems during this procedure? Yes ___ No ___

If Yes, describe the problems: _____

_____

What did you do to correct the problems? _____

_____

☐ 22. Have your instructor check your work and sign this job sheet.

Performance Evaluation—Instructor Use Only

Did the student complete the job in the time allotted? Yes ___ No ___

If No, which steps were not completed? _____

How would you rate this student's overall performance on this job? _____

5–Excellent, 4–Good, 3–Satisfactory, 2–Unsatisfactory, 1–Poor

Comments: _____

_____

_____

INSTRUCTOR'S SIGNATURE _____

Notes

Job 85—Remove and Replace an Independent Rear Axle Shaft

After completing this job, you will be able to remove and install independently suspended rear axle shafts used in rear-wheel drive vehicles. You will also be able to replace axle bearings and axle seals.

Procedures

☐ 1. Obtain a vehicle to be used in this job. Your instructor may direct you to perform this job on a shop vehicle.

☐ 2. Gather the tools needed to perform the following job. Refer to the tools and materials list at the beginning of the project.

Note

The following paragraphs give a general outline of the independently suspended drive axle removal process. Consult the proper service manual for exact removal procedures of independently suspended drive axles.

Remove the Axle

☐ 1. Disconnect the battery negative cable.

☐ 2. Raise the vehicle.

Warning

⚠ The vehicle must be raised and supported in a safe manner. Always use approved lifts or jacks and jack stands.

☐ 3. Remove the rear wheel that is driven by the rear axle being serviced.

☐ 4. Remove the brake drum, marking it so that it can be reinstalled in the same position.

or

Remove the brake caliper and rotor. Use wire to support the caliper so it will not hang and damage the brake line. Mark the rotor and hub so the rotor can be installed in its original position in the axle flange.

Note

Marking the original position of each drum or rotor and reinstalling it in the same position ensures that slight tolerance variations will not cause a vibration that was not present before repairs.

☐ 5. If the axle is attached with flanges, mark the mating flanges so that they can be reinstalled in their original positions.

☐ 6. Remove the fasteners that hold the drive axle flange to the differential carrier.

or

Remove the bolt or roll pin that holds the axle to the differential stub shaft.

☐ 7. Remove the fasteners that hold the drive axle flange to the wheel hubs.

or

Remove the bolt or roll pin that holds the axle to the wheel hub.

☐ 8. Have an assistant hold the drive axle to keep it from falling.

☐ 9. Separate the flanges by one of the following methods:

If the drive axle uses U-joints, move the wheel assembly outward slightly to separate the flanges or splined connections and gain clearance for removal.

or

If the drive axle uses CV joints, compress the axle slightly to reduce its length and separate the flanges or splined connections.

> **Note**
>
> Separation can usually be performed without removing suspension fasteners.

☐ 10. Once one side of the axle is clear of the mating flange, lower the axle to gain clearance and remove it from the vehicle.

> **Note**
>
> Most independent rear axle shafts use CV joints. A few have U-joints. For U-joint service procedures, see Job 66. For CV joint service procedures, see Job 67.

Install the Axle

☐ 1. Ensure that the mounting flanges or mating splines are clean and have no burrs, dents, or other damage that would prevent the proper mating of parts. If necessary, clean the parts and use a hand file to remove any burrs. If the parts cannot be repaired, replace them.

☐ 2. Place the drive axle inner flange or spline in position on the differential carrier and align the match marks made before removal. If a roll pin or through bolt is used, make sure the matching holes are aligned.

☐ 3. On flanged axles, install the fasteners but do not tighten them at this time.

> **Note**
>
> Some manufacturers specify the use of new fasteners when reinstalling axles. Consult the service literature before reusing the old fasteners.

☐ 4. Install the outer drive axle flange or splined part to the wheel hub by one of the following procedures.

If the drive axle uses U-joints, move the wheel assembly outward slightly to gain enough clearance and place the axle in position.

or

If the drive axle uses CV joints, compress the axle slightly to reduce its length and place the axle and wheel parts in position.

☐ 5. If flanges are used, align the match marks and loosely install the fasteners.

☐ 6. Once all fasteners are started, begin tightening the fasteners on one flange in a cross or star sequence.

or

Install the bolt or roll pin that holds the axle to the wheel hub.

Job 85—Remove and Replace an Independent Rear Axle Shaft (continued)

> **Note**
>
> Make sure that the flanges pull together evenly and are flush with each other when all fasteners are tightened. The gap between the flanges should be the same on all sides of the flanges.

☐ 7. Repeat step 6 on the other side of the axle as necessary.

☐ 8. Tighten all fasteners to the proper torque using a torque wrench. It is critical that the fasteners be properly torqued to prevent damage or injury.

☐ 9. Clean the work area and return tools and equipment to storage.

☐ 10. Did you encounter any problems during this procedure? Yes ___ No ___

If Yes, describe the problems: _____

_____

What did you do to correct the problems? _____

_____

☐ 11. Have your instructor check your work and sign this job sheet.

Performance Evaluation—Instructor Use Only

Did the student complete the job in the time allotted? Yes ___ No ___

If No, which steps were not completed?_____

How would you rate this student's overall performance on this job?_____

5–Excellent, 4–Good, 3–Satisfactory, 2–Unsatisfactory, 1–Poor

Comments: _____

_____

_____

INSTRUCTOR'S SIGNATURE _____

Notes

Job 86—Remove a Rear Axle Assembly

After completing this job, you will be able to remove a rear axle assembly.

Procedures

1. Obtain a vehicle to be used in this job. Your instructor may direct you to perform this job on a shop vehicle.

2. Gather the tools needed to perform the following job. Refer to the tools and materials list at the beginning of the project.

> **Note**
>
> Some rear axle assemblies are constructed so that the differential carrier (also referred to as the chunk or pumpkin) can be removed without removing the entire rear axle assembly from the vehicles. There are many other variations in rear end design. Always consult the manufacturer's service literature for exact disassembly procedures and specifications.

3. Disconnect the battery negative cable.

4. Raise the vehicle.

> **Warning**
>
> ⚠ The vehicle must be raised and supported in a safe manner. Always use approved lifts or jacks and jack stands.

5. Place a drain pan beneath the differential carrier. Loosen cover bolts or drain plug and allow the lubricant to drain.

6. Remove the rear wheels. Remove the brake drums if there is danger of them falling off the axle flanges.

7. Remove the drive shaft and cap the rear of the transmission to reduce leakage.

> **Note**
>
> Refer to Job 66 for specific instructions on drive shaft removal.

8. Support the rear axle housing with a transmission jack or other support device.

9. Remove the brake lines and any electrical connectors.

10. Remove the bottom shock absorber mountings.

11. Disconnect any suspension mountings at the rear axle assembly.

12. Remove any additional brackets or fasteners holding the rear axle assembly to the vehicle.

13. Lower the rear axle assembly from the vehicle and place it on a clean workbench.

14. Clean the work area and return tools and equipment to storage.

☐ 15. Did you encounter any problems during this procedure? Yes ___ No ___

If Yes, describe the problems: _____

_____

What did you do to correct the problems? _____

_____

☐ 16. Have your instructor check your work and sign this job sheet.

Performance Evaluation—Instructor Use Only

Did the student complete the job in the time allotted? Yes ___ No ___

If No, which steps were not completed?_____

How would you rate this student's overall performance on this job?_____

5–Excellent, 4–Good, 3–Satisfactory, 2–Unsatisfactory, 1–Poor

Comments: _____

_____

_____

INSTRUCTOR'S SIGNATURE _____

Job 87—Disassemble and Inspect a Differential

After completing this job, you will be able to disassemble a differential and identify worn or defective parts.

Procedures

- [ ] 1. Obtain a rear axle assembly to be used in this job.
- [ ] 2. Gather the tools needed to perform the following job. Refer to the tools and materials list at the beginning of the project.
- [ ] 3. If the rear axle assembly has retainer-type axles, remove the drive axles.

Note

Refer to Job 84 for specific instructions on retainer-type drive axle removal.

Disassemble the Differential

- [ ] 1. Remove the differential cover or differential carrier, as necessary.
- [ ] 2. Remove the differential pinion shaft.
- [ ] 3. Rotate the differential spider gears out of the case.
- [ ] 4. If the axles are the C-lock type, remove the C-locks and remove the drive axles.

Note

Refer to Job 84 for specific instructions on C-lock drive axle removal.

- [ ] 5. Remove the differential side gears and thrust washers from the case.

Note

If the differential is a limited-slip type, remove the clutches or cones at this time.

- [ ] 6. Unbolt the differential carrier bearing caps. Keep bearing cups with their respective bearings.
- [ ] 7. Remove any side bearing shims or threaded adjusters from the differential carrier.
- [ ] 8. Remove the differential carrier from the differential case.
- [ ] 9. Remove the pinion flange or yoke nut.
- [ ] 10. Remove the pinion flange or yoke from the drive pinion gear shaft.
- [ ] 11. Remove the drive pinion gear from the rear axle housing.
- [ ] 12. Pry the pinion seal from the differential carrier housing with a suitable tool.
- [ ] 13. Remove the front pinion bearing and any adjusting shims from the front of the carrier.

Inspect the Rear Axle and Differential Parts

☐ 1. Thoroughly clean the differential carrier and all parts.

☐ 2. Visually examine the disassembled rear axle and differential components, looking for the following:
  • Worn, pitted, or rough bearings.
  • Worn, pitted, or otherwise damaged bearing cups.
  • Worn, scored, or pitted gears.
  • Worn side gear or axle splines.

  Inspect the components with special measuring instruments, as required.

  Describe the condition of the rear axle components: _____

  _____

  _____

  _____

  _____

☐ 3. If the differential contains a limited slip assembly, inspect the clutches or cones for:
  • Worn or damaged clutch or cone faces.
  • Insufficient clutch thickness.
  • Damaged clutch splines.
  • Damaged hub splines.
  • Damaged cone mating faces.

> **Note**
> Most technicians prefer to replace the limited-slip clutches or cones whenever the differential is disassembled.

☐ 4. Report any differential component defects that you found to your instructor. Then, determine which parts must be replaced.

  List the parts that must be replaced: _____

  _____

  _____

  _____

  _____

> **Note**
> The ring and pinion are a matched set. They should always be replaced as a set. See **Figure 87-1**.

☐ 5. Obtain the proper replacement parts and compare them with the old parts to ensure that they are the correct replacements.

☐ 6. Clean the work area and return tools and equipment to storage.

☐ 7. Did you encounter any problems during this procedure? Yes ___ No ___

  If Yes, describe the problems: _____

  _____

  What did you do to correct the problems? _____

  _____

☐ 8. Have your instructor check your work and sign this job sheet.

Job 87—Disassemble and Inspect a Differential (continued)

Figure 87-1. The ring and pinion should always be replaced as a set. This illustration shows a used set that was replaced.

Performance Evaluation—Instructor Use Only

Did the student complete the job in the time allotted? Yes ___ No ___

If No, which steps were not completed?_____

How would you rate this student's overall performance on this job?_____

5–Excellent, 4–Good, 3–Satisfactory, 2–Unsatisfactory, 1–Poor

Comments: _____

_____

_____

INSTRUCTOR'S SIGNATURE _____

Notes

Job 88—Reassemble and Adjust a Differential

After completing this job, you will be able to assemble and adjust a differential.

Procedures

☐   1.   Obtain a disassembled rear axle assembly to be used in this job.

☐   2.   Gather the tools needed to perform the following job. Refer to the tools and materials list at the beginning of the project.

Reassemble the Differential and Rear Axle

☐   1.   Lightly lubricate all parts before proceeding.

☐   2.   Replace the pinion bearing cups, if needed. Be careful not to mar the surfaces of new cups as they are driven into place.

☐   3.   Install the front pinion bearing and pinion seal in the carrier, installing any bearing spacers that may have been present between the bearing and the carrier.

☐   4.   If the ring and pinion are being replaced, set the pinion depth and select the proper shim for the replacement pinion.

> **Note**
>
> Special tools are needed to set pinion depth. Consult the manufacturer's service literature for the tools and procedures needed for this task.

☐   5.   Press the rear pinion bearing onto the drive pinion gear shaft, installing any required shims between the bearing and pinion gear.

☐   6.   Place a new collapsible or solid spacer on the drive pinion gear shaft.

☐   7.   Slide the pinion shaft through the front pinion bearing.

☐   8.   Install the pinion yoke or flange and a new nut.

☐   9.   Check and adjust the pinion bearing preload according to manufacturer's specifications.

      Measured pinion bearing preload:_____

      Specified pinion bearing preload:_____

☐ 10.   If a new ring gear is being used, install it now on the differential case by lightly tapping it into place and installing the attaching bolts in a crisscross pattern.

☐ 11.   Install new side bearings by pressing them into place.

> **Caution**
>
> ⚠ Install new bearings with a suitable press. Do not hammer the bearings into place.

☐ 12.   Install the differential case into the differential carrier, installing adjusting shims, if used, and bearing cups with the case.

☐ 13. With the side gears, spider gears, and thrust washers positioned in the differential case assembly, check internal part clearances with a dial indicator or feeler gauges as needed:
- Attempt to insert various size feeler gauges into the gap between gears. The thickness of the thickest feeler gauge that can be inserted is the endplay.
- Install a dial indicator on the case so that the pointer touches the part to be checked. Then, pull the part to the limit of its travel in one direction and zero the dial indicator. Push the part to the limit of its travel in the opposite direction and read the endplay.

> **Note**
>
> If the differential is a limited-slip type, install the clutches or cones. If clutches are used, be sure to alternate steel and friction clutches. Follow manufacturer's directions exactly.

☐ 14. Permanently install the differential case assembly internals by inserting the pinion shaft through the spider gears and case and installing the lock bolt or pin.

☐ 15. If the axle assembly is an independent rear axle, install the stub axles at this time.

Make Differential and Axle Adjustments

☐ 1. Check and adjust the side bearing preload according to manufacturer's specifications.

Measured side bearing preload:_____

Specified side bearing preload:_____

☐ 2. Check and adjust the ring gear backlash according to manufacturer's specifications. See **Figure 88-1**.

Measured ring gear backlash:_____

Specified ring gear backlash:_____

☐ 3. Check the ring gear runout according to manufacturer's specifications and correct if necessary.

Measured ring gear runout:_____

Specified ring gear runout:_____

☐ 4. Check the ring gear contact pattern and compare it to manufacturer's specifications. Correct it if necessary.

Figure 88-1. Backlash must be as specified before the contact pattern can be checked. (DaimlerChrysler)

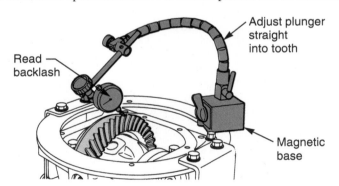

Adjust plunger straight into tooth

Read backlash

Magnetic base

Job 88—Reassemble and Adjust a Differential (continued)

Reassemble the Differential

> **Note**
>
> If the rear axle assembly is used with an independent suspension, skip this section.

☐ 1. If the differential carrier is a removable type, install it using a new gasket at the mating surfaces of the carrier and rear axle housing.

☐ 2. Install the drive axles.

> **Note**
>
> For specific installation instructions, refer to Job 84 for retainer-type and C-lock axles, and Job 85 for independent rear axles.

☐ 3. Scrape off any old gasket material from the differential cover and its mating surface on the rear axle housing.

☐ 4. Install a new gasket on the differential cover, using nonhardening gasket sealer, or use form-in-place gaskets, as required.

☐ 5. Tighten the differential cover bolts in a crisscross pattern, torquing the bolts to manufacturer's specifications.

☐ 6. Clean the work area and return tools and equipment to storage.

☐ 7. Did you encounter any problems during this procedure? Yes ___ No ___

If Yes, describe the problems: _____

_____

What did you do to correct the problems? _____

_____

☐ 8. Have your instructor check your work and sign this job sheet.

Performance Evaluation—Instructor Use Only

Did the student complete the job in the time allotted? Yes ___ No ___

If No, which steps were not completed?_____

How would you rate this student's overall performance on this job?_____

5–Excellent, 4–Good, 3–Satisfactory, 2–Unsatisfactory, 1–Poor

Comments: _____

_____

_____

INSTRUCTOR'S SIGNATURE _____

Notes

_____

_____

_____

_____

_____

_____

_____

_____

_____

_____

_____

_____

_____

_____

_____

_____

_____

_____

_____

_____

_____

_____

_____

_____

_____

_____

_____

_____

_____

_____

_____

_____

_____

_____

_____

_____

_____

_____

_____

Job 89—Install a Rear Axle Assembly

After completing this job, you will be able to install a rear axle assembly.

Procedures

☐ 1. Obtain a vehicle to be used in this job. Your instructor may direct you to perform this job on a shop vehicle.

☐ 2. Gather the tools needed to perform the following job. Refer to the tools and materials list at the beginning of the project.

Install the Differential and Axle Assembly

☐ 1. Raise the vehicle.

> **Warning**
> ⚠ The vehicle must be raised and supported in a safe manner. Always use approved lifts or jacks and jack stands.

☐ 2. Raise the rear axle assembly into position under the vehicle.

> **Caution**
> ◇ Use a floor jack or transmission jack to lift the assembly. Have a helper assist you.

☐ 3. Support the rear axle housing with jack stands.

☐ 4. Install the suspension members holding the rear axle housing to the vehicle frame.

☐ 5. Install all rear axle assembly attaching hardware and reconnect the brake lines and any electrical connectors.

☐ 6. Remove the cap at the rear of the transmission and install the drive shaft. Be sure to line up the match marks made during removal.

> **Note**
> 📄 Refer to Job 66 for specific instructions on drive shaft installation.

☐ 7. Install the shock absorber bottom mounts.

☐ 8. If the differential is a limited-slip type, use a torque wrench to measure the turning torque at each wheel. If the torque is not correct, disassemble the unit and correct any problems.

☐ 9. Reinstall the brake drums or brake calipers and rotors. Ensure that the drums or rotors are installed in the same position as they were originally.

☐ 10. Add the proper type of lubricant to the rear end.

> **Caution**
> If the differential is a limited-slip type, follow the manufacturer's recommendation and use limited-slip oil, or add limited-slip additives to conventional gear oil.

☐ 11. Replace the fill plug and torque to specifications.

☐ 12. Lower the vehicle.

☐ 13. Install the battery negative cable.

☐ 14. Bleed the rear brakes. See Job 115.

☐ 15. Start the engine.

☐ 16. Check that the brakes operate properly.

☐ 17. Road test the vehicle, carefully observing rear end operation.

☐ 18. Clean the work area and return tools and equipment to storage.

☐ 19. Did you encounter any problems during this procedure? Yes ____ No ____

If Yes, describe the problems: _____

_____

What did you do to correct the problems? _____

_____

☐ 20. Have your instructor check your work and sign this job sheet.

Performance Evaluation—Instructor Use Only

Did the student complete the job in the time allotted? Yes ____ No ____

If No, which steps were not completed?_____

How would you rate this student's overall performance on this job?_____

5–Excellent, 4–Good, 3–Satisfactory, 2–Unsatisfactory, 1–Poor

Comments: _____

_____

_____

INSTRUCTOR'S SIGNATURE _____

Project

26

Diagnosing and Servicing General Steering and Suspension Concerns

Introduction

Steering and suspension systems can develop many problems, including loose steering, vibration, noises, and pulling to one side. These problems can be caused by several suspension and steering parts, as well as other vehicle systems. Diagnosis involves determining the exact problem and isolating possible defective parts.

Steering and suspension maintenance primarily consists of lubricating ball-and-socket points through grease fittings, as well as checking CV boot condition and power steering fluid level. Another maintenance job that you may be called on to perform is cleaning and greasing tapered roller wheel bearings used on nondriving front axles. Refer to Job 68 for service of flat and ball bearings used on driving front axles.

In Jobs 90 through 92, you will diagnose steering and suspension problems, and perform steering and suspension maintenance.

Project 26 Jobs

- Job 90—Diagnose Steering and Suspension Problems
- Job 91—Lubricate Steering and Suspension
- Job 92—Service Tapered Roller Wheel Bearings

Tools and Materials

The following list contains the tools and materials that may be needed to complete the jobs in this project. The items used will depend on the make and model of vehicle being serviced.

- Vehicle in need of steering and suspension diagnosis.
- Lift or floor jack and jack stands.
- Service information.
- Replacement parts.
- Power steering fluid.
- Scan tool.
- Tape measure.
- Pressure gauges.
- Dial indicators and/or micrometers.
- Hand tools.
- Air-powered tools.
- Safety glasses and other necessary protective equipment.

Safety Notice

Before performing this job, review all pertinent safety information in the text and discuss safety procedures with your instructor.

Notes

Job 90—Diagnose Steering and Suspension Problems

After completing this job, you will be able to diagnose steering and suspension problems.

Procedures

☐ 1. Obtain a vehicle to be used in this job. Your instructor may direct you to perform this job on a shop vehicle.

☐ 2. Gather the tools needed to perform this job. Refer to the project's tools and materials list.

> **Note**
>
> This job contains most, but not all, steering and suspension diagnosis procedures.

- If the complaint involves the power steering system, see Job 97.
- If the complaint involves ride height or alignment, see Job 103.
- For wheel and tire related problems, see Job 104.

Initial Diagnosis

☐ 1. Obtain a description of the problem from the vehicle operator.

Problem as described by the operator: _____

_____

☐ 2. Obtain the proper service literature for the vehicle.

☐ 3. Check ride height according to manufacturer's procedures and record the results below. Indicate the units of measurement.

Left Front	**Right Front**
Specification:_____ inches/mm	Specification:_____ inches/mm
Actual:_____ inches/mm	Actual:_____ inches/mm
Left Rear	**Right Rear**
Specification:_____ inches/mm	Specification:_____ inches/mm
Actual:_____ inches/mm	Actual:_____ inches/mm

Does any ride height measurement exceed the manufacturer's specifications?
Yes ___ No ___

☐ 4. Road test the vehicle to duplicate the original complaint.

Describe abnormal sounds and indicate the conditions under which they occur: _____

_____

_____

Describe any vibrations, shocks or bumping/looseness sensations, body sway on turns or at cruising speeds, wander or loose sensation at cruising speeds, or sensations of harshness and the conditions under which the problems occur: _____

_____

_____

_____

Did the road test confirm that the problem described by the operator exists?

Yes ___ No ___

If the answer to the question in steps 3 or 4 is Yes, go to step 5.

If the answer is No, discuss your findings with your instructor before proceeding.

5. Raise the front of the vehicle so the front suspension is correctly suspended for checking the ball joints. Make sure the steering linkage is free to turn.

Warning

⚠ Raise and support the vehicle in a safe manner. Always use approved lifts or jacks and jack stands.

---

6. Shake each front wheel and note any looseness.

a. Grasp the wheel at the top and bottom and shake it in and out to check the suspension parts. See **Figure 90-1**.

Does the right wheel feel loose? Yes ___ No ___

Does the left wheel feel loose? Yes ___ No ___

b. Grasp the wheel at the front and back and shake it back and forth to check the steering linkage parts, **Figure 90-2**.

Does the right wheel feel loose? Yes ___ No ___

Does the left wheel feel loose? Yes ___ No ___

c. On some MacPherson strut vehicles, grasp the wheel at approximately 45° from the top and bottom and shake it in and out to check the suspension.

Does the right wheel feel loose? Yes ___ No ___

Does the left wheel feel loose? Yes ___ No ___

7. Have a helper shake each front wheel while you observe the front suspension.

Were any parts loose? Yes ___ No ___

If Yes, list them in the following blanks.

Front left side: _____

Front right side: _____

Figure 90-1. When checking for loose suspension parts, grasp the wheel at the top and bottom, and try to shake it. Slight movement is normal.

Job 90—Diagnose Steering and Suspension Problems (continued)

Figure 90-2. When checking the steering linkage for looseness, try to move the wheel back and forth.

Check Suspension Parts

☐ 1. Check all front suspension parts and their mounts for damage and complete the following chart. Write "NA" in the first blank when the part is not used on the vehicle you are checking.

	Passed	Worn/Loose	Bent/Broken	Leaking
Upper ball joints	___	___	___	___
Upper control arms	___	___	___	___
Lower ball joints	___	___	___	___
Lower control arms	___	___	___	___
Control arm bushings	___	___	___	___
Rebound bumpers	___	___	___	___
Steering knuckle	___	___	___	___
Strut rod	___	___	___	___
Strut rod bushings	___	___	___	___
Stabilizer bar	___	___	___	___
Stabilizer bar bushings	___	___	___	___
Shock absorbers	___	___	___	___
Strut cartridges	___	___	___	___
Springs/torsion bars	___	___	___	___
Insulators	___	___	___	___
Fasteners	___	___	___	___
Frame/subframe	___	___	___	___

☐ 2. Obtain a dial indicator and use it to check ball joint play.

Note

Checking ball joint play may not be necessary on the vehicle you are servicing. Check with your instructor.

Specified maximum ball joint play:_____

Actual ball joint play:_____

Is ball joint play within specifications? Yes ___ No ___

3. If not already done, raise the rear of the vehicle so the rear suspension parts can be checked.

> **Warning**
>
> ⚠ Raise and support the vehicle in a safe manner. Always use approved lifts or jacks and jack stands.

4. Check all rear suspension parts and their mounts for damage and complete the following chart. Write "NA" in the first blank when the part is not used on the vehicle you are checking.

	Passed	Worn/Loose	Bent/Broken	Leaking
Upper ball joints	___	___	___	___
Upper control arms	___	___	___	___
Lower ball joints	___	___	___	___
Lower control arms	___	___	___	___
Rebound bumpers	___	___	___	___
Strut rod	___	___	___	___
Strut rod bushings	___	___	___	___
Stabilizer bar	___	___	___	___
Stabilizer bar bushings	___	___	___	___
Shock absorbers	___	___	___	___
Strut cartridges	___	___	___	___
Springs/torsion bars	___	___	___	___
Insulators	___	___	___	___
Fasteners	___	___	___	___
Axle	___	___	___	___

5. If the vehicle is equipped with rear lower ball joints, check them for looseness as specified by the service manual.

Specified maximum ball joint play:_____

Actual ball joint play:_____

Is ball joint play within specifications? Yes ___ No ___

Check Steering Parts

1. Check the steering linkage, steering gear, and steering shaft for damage. Complete the following chart. Write "NA" in the first column when the part is not used on the vehicle you are checking.

	Passed	Worn/Loose	Bent/Broken	Leaking
Steering arms	___	___	___	___
Outer tie rod ends	___	___	___	___
Inner tie rod ends	___	___	___	___
Pitman arm	___	___	___	___
Idler arm	___	___	___	___
Relay rod/drag link	___	___	___	___
Steering gear	___	___	___	___
Steering coupler	___	___	___	___
Steering shaft	___	___	___	___
Power steering pump	___	___	___	___
Power steering belt	___	___	___	___
Power steering hoses	___	___	___	___
Fasteners	___	___	___	___

Name _____   Date_____

Instructor _____   Period _____

> **Note**
>
> In many vehicles equipped with power steering, a power steering pressure switch sends a signal to the ECM when the system pressure exceeds a preset limit. In response, the ECM will raise the idle speed or turn off the air conditioning compressor to prevent the engine from stalling. If the engine is experiencing performance problems that seem to coincide with steering actions, suspect the power steering pressure switch.

☐ 2. If necessary, check the rear steering linkage for damage. Complete the following chart. Write "NA" in the first column when the part is not used on the vehicle you are checking.

	Passed	Worn/Loose	Bent/Broken	Leaking
Steering arms	___	___	___	___
Outer tie rod ends	___	___	___	___
Inner tie rod ends	___	___	___	___
Steering gear	___	___	___	___
Power steering hoses	___	___	___	___
Hydraulic valves	___	___	___	___
Electrical controls	___	___	___	___
Fasteners	___	___	___	___

☐ 3. Discuss with your instructor the results obtained in the preceding charts. Determine which suspension and steering parts are defective.

List the defective parts: _____

_____

_____

☐ 4. After obtaining your instructor's approval, make needed repairs based on your front and rear suspension inspections.

What repairs were made? _____

_____

_____

☐ 5. Recheck the condition of the suspension and steering system. If the problem has been corrected, go to step 6. If the problem still exists, repeat the procedures in this job until the problem is located and corrected.

☐ 6. Return all tools and equipment to storage.

☐ 7. Clean the work area.

☐ 8. Did you encounter any problems during this procedure? Yes ___ No ___

If Yes, describe the problems: _____

_____

What did you do to correct the problems? _____

_____

☐ 9. Have your instructor check your work and sign this job sheet.

Performance Evaluation—Instructor Use Only

Did the student complete the job in the time allotted? Yes ___ No ___

If No, which steps were not completed?_____

How would you rate this student's overall performance on this job?_____

5–Excellent, 4–Good, 3–Satisfactory, 2–Unsatisfactory, 1–Poor

Comments: _____

_____

_____

INSTRUCTOR'S SIGNATURE _____

Job 91—Lubricate Steering and Suspension

After completing this job, you will be able to perform steering and suspension maintenance.

Procedures

☐ 1. Obtain a vehicle to be used in this job. Your instructor may direct you to perform this job on a shop vehicle.

☐ 2. Gather the tools needed to perform this job. Refer to the project's tools and materials list.

Lubricate Steering and Suspension

☐ 1. Raise the vehicle.

> ### Warning
> ⚠ Raise and support the vehicle in a safe manner. Always use approved lifts or jacks and jack stands.

☐ 2. Locate the steering and suspension lubrication fittings. If servicing a rear-wheel drive vehicle, be sure to check for grease fittings on the drive shaft universal joints.

How many fittings are used?_____

> ### Note
> 📄 Some vehicles are not equipped at the factory with grease fittings. On these vehicles, the fitting plugs can be removed and replaced with grease fittings.

☐ 3. If necessary, remove the fitting plugs and install grease fittings.

☐ 4. Obtain an air-operated or manual grease gun and make sure it contains a sufficient amount of appropriate grease for this job.

☐ 5. Install the grease gun nozzle on the first fitting, **Figure 91-1**.

Figure 91-1. Install the grease gun nozzle securely on the fitting.

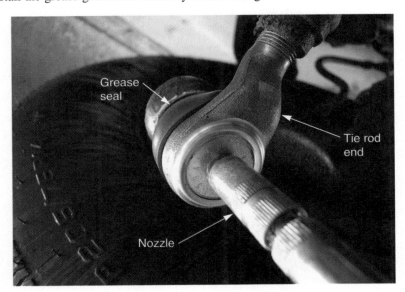

☐ 6. Inject grease until the grease seal begins to bulge.

☐ 7. Repeat steps 5 and 6 on all grease fittings.

☐ 8. If the vehicle is equipped with CV axles, check the boots for tears and signs of lost lubricant.

☐ 9. Check the fluid level in the power steering reservoir.

☐ 10. Add fluid to the reservoir if necessary. Be sure to add the proper type of fluid. The type of fluid required is usually stamped on the reservoir cover.

Was fluid added? Yes _____ No _____

If Yes, what type? _____

☐ 11. Clean the work area and return any equipment to storage.

☐ 12. Did you encounter any problems during this procedure? Yes _____ No _____

If Yes, describe the problems: _____

_____

What did you do to correct the problems? _____

_____

☐ 13. Have your instructor check your work and sign this job sheet.

Performance Evaluation—Instructor Use Only

Did the student complete the job in the time allotted? Yes _____ No _____

If No, which steps were not completed?_____

How would you rate this student's overall performance on this job?_____

5–Excellent, 4–Good, 3–Satisfactory, 2–Unsatisfactory, 1–Poor

Comments: _____

_____

_____

INSTRUCTOR'S SIGNATURE _____

Job 92—Service Tapered Roller Wheel Bearings

After completing this job, you will be able to service tapered roller wheel bearings.

Procedures

☐ 1. Obtain a vehicle to be used in this job. Your instructor may direct you to perform this job on a shop vehicle.

☐ 2. Gather the tools needed to perform this job. Refer to the project's tools and materials list.

☐ 3. Safely raise the vehicle.

> **Warning**
> ⚠ Raise and support the vehicle in a safe manner. Always use approved lifts or jacks and jack stands.

☐ 4. Remove the wheel from the vehicle.

☐ 5. On a disc brake system, remove the brake caliper.

☐ 6. Remove the rotor if it is not integral with the hub.

☐ 7. Remove the cotter key from the spindle nut, if used.

☐ 8. Remove the spindle nut and washer.

☐ 9. Remove the hub assembly.

> **Caution**
> ⚠ If the hub is a one-piece assembly with the drum or rotor, be careful not to get grease on the friction surfaces.

☐ 10. Remove the outer bearing from the hub.

☐ 11. Remove the grease seal from the inner hub and discard it.

☐ 12. Remove the inner bearing from the hub.

☐ 13. Clean the bearings thoroughly and blow dry.

☐ 14. Inspect the bearings for wear or damage. **Figure 92-1** shows the most common type of bearing damage.

List any signs of bearing damage:_____

Can this bearing be reused? Yes ___ No ___

If No, consult your instructor before proceeding.

☐ 15. Clean the bearing races and the hub interior.

☐ 16. Check the races for wear and damage.

List any signs of race damage. _____

Can this race be reused? Yes ___ No ___

If No, consult your instructor before proceeding.

Figure 92-1. Check the wheel bearing and race for damage. A—Typical bearing damage. B—Typical outer race damage.

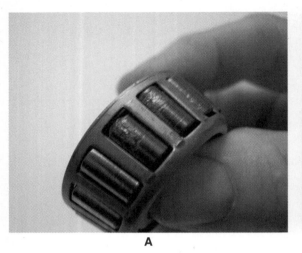

A   B

☐ 17. If a bearing race must be replaced, drive out the old race and install the new race.

☐ 18. Place a small amount of grease in the hub cavity.

☐ 19. Grease the bearings by hand or with a packing tool.

☐ 20. Place the inner bearing in the hub and install the new grease seal.

☐ 21. Clean the spindle.

☐ 22. Reinstall the hub on the spindle.

☐ 23. Install the outer bearing.

☐ 24. Install the washer and spindle nut.

☐ 25. Adjust bearing preload:
a. Tighten the nut to about 100 foot pounds to seat all of the components.
b. Back off the nut completely.
c. Tighten the nut until proper preload is obtained. Depending on the manufacturer, this is measured by determining the endplay of the hub or by measuring the amount of force needed to turn the hub.

☐ 26. Install a new cotter pin and bend it to ensure that the nut does not loosen.

☐ 27. Reinstall the rotor or drum if it was not part of the hub assembly.

☐ 28. Reinstall the caliper if necessary.

☐ 29. Reinstall the wheel.

☐ 30. Clean the work area and return any equipment to storage.

☐ 31. Did you encounter any problems during this procedure? Yes ___ No ___

If Yes, describe the problems: _____

_____

What did you do to correct the problems? _____

_____

☐ 32. Have your instructor check your work and sign this job sheet.

Job 92—Service Tapered Roller Wheel Bearings (continued)

Performance Evaluation—Instructor Use Only

Did the student complete the job in the time allotted? Yes ___ No ___

If No, which steps were not completed?_____

How would you rate this student's overall performance on this job?_____

5–Excellent, 4–Good, 3–Satisfactory, 2–Unsatisfactory, 1–Poor

Comments: _____

_____

_____

INSTRUCTOR'S SIGNATURE _____

Notes

Project
Servicing the Steering System

27

Introduction

The steering column connects the steering wheel to the steering gear. The steering gear converts the rotation of the steering wheel into a linear motion to move the wheels. Steering gears can be conventional worm-and-sector types or rack-and-pinion designs. Movement of the steering gear is transferred through steering linkage to the wheels. The term steering linkage covers many parts including tie rod ends, idler arms, pitman arms, drag links/relay rods, and toe adjusting sleeves. For safety, steering linkage service must be done accurately and all fasteners must be tightened properly. Power steering systems can be checked for proper operation with a few simple test procedures. Following these procedures will enable you to pinpoint the problem or eliminate the power steering system as a source of complaints.

In Jobs 93 through 97, you will inspect and service a steering wheel and column, as well as rack-and-pinion and conventional steering gears. You will also replace steering linkage and ensure that the vehicle is properly aligned. Finally, you will check the power steering system and related components for leaks and pressure problems.

Project 27 Jobs

- Job 93—Inspect a Steering Column
- Job 94—Service a Steering Column
- Job 95—Remove and Replace a Rack-and-Pinion Steering Gear and Inner Tie Rod
- Job 96—Remove and Replace Steering Linkage Components
- Job 97—Test a Power Steering System

Tools and Materials

The following list contains the tools and materials that may be needed to complete the jobs in this project. The items used will depend on the make and model of vehicle being serviced.

- Vehicle(s) in need of service.
- Applicable service information.
- Steering wheel removal tool.
- Outer tie rod removal tool.
- Rack-and-pinion inner tie rod removal tool.
- Miscellaneous hand tools.
- Air-powered tools.
- Safety glasses and other protective equipment.

Safety Notice

Before performing this job, review all pertinent safety information in the text and discuss safety procedures with your instructor.

Job 93—Inspect a Steering Column

After completing this job, you will be able to inspect a steering wheel and steering column, and diagnose noises, looseness, and binding concerns.

Procedures

☐ 1. Obtain a vehicle to be used in this job. Your instructor may direct you to perform this job on a shop vehicle.

☐ 2. Gather the tools needed to perform this job. Refer to the project's tools and materials list.

☐ 3. Locate a paved area, such as an empty parking lot, where Step 4 can be performed safely.

☐ 4. Drive the vehicle slowly while turning the steering wheel from side to side.

Are there any spots that are harder to turn through (binding) as the steering wheel is turned? Yes ___ No ___

Are any noises apparent as the steering wheel is turned? Yes ___ No ___

If Yes, describe the noise (such as knocks, rattles, scraping noises) _____

_____

_____

☐ 5. Park the vehicle and turn off the engine.

☐ 6. Check for looseness by turning the steering wheel while observing the shaft at the point where it enters the steering gear.

> **Note**
>
> It may be necessary to raise the vehicle and observe the shaft as an assistant turns the wheel.

There should be virtually no delay between moving the steering wheel and movement of the shaft at the steering gear.

Does the column appear to have excessive looseness? Yes ___ No ___

☐ 7. Inspect the steering shaft parts for the following problems. **Figure 93-1** shows typical steering system parts.
- Damaged steering shaft universal joint yokes.
- Loose steering shaft universal joint fasteners.
- Damaged flexible coupling yokes.
- Loose flexible coupling fasteners.
- Bent steering shaft(s).
- Misaligned steering shaft(s).
- Loose instrument panel to steering column fasteners.

Describe any problems found:_____

_____

☐ 8. Remove under-dashboard covers as necessary to gain access to the collapsible steering column parts.

☐ 9. Visually inspect the collapsible steering assembly for the following conditions.
- Incorrect steering column jacket length or other signs of collapse.
- Missing or sheared off plastic pins on steering shaft.

Figure 93-1. The parts of a typical steering column are shown here. (Federal-Mogul)

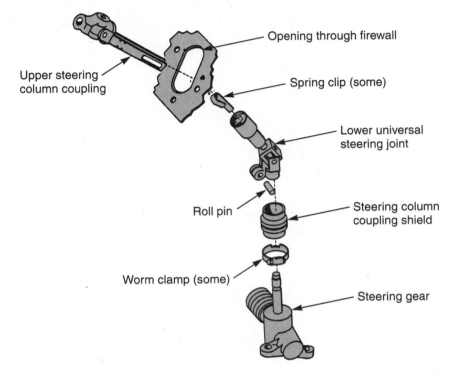

Opening through firewall

Upper steering column coupling

Spring clip (some)

Lower universal steering joint

Roll pin

Steering column coupling shield

Worm clamp (some)

Steering gear

☐ 10. Inspect the tilt wheel mechanism (when equipped) by operating it through all positions. There should be no noise or looseness.

Were any defects found? Yes ___ No ___

☐ 11. Inspect the lock cylinder mechanism:
   a. Make sure the key is not in the lock cylinder.
   b. Apply the parking brake.
   c. Attempt to move the shift lever out of Park.

Does the lever move out of Park? Yes ___ No ___
   d. Install the ignition key and turn the switch to the *on* position without starting the engine.
   e. Attempt to move the shift lever out of Park.

Does the lever move out of Park? Yes ___ No ___
   f. Check that the key cannot be removed from the lock cylinder when the shift lever is in any position other than Park.

Can the key be removed? Yes ___ No ___

☐ 12. Replace the under-dashboard covers if necessary.

☐ 13. Based on your findings in steps 1 through 8, determine whether the steering column is in need of service.

Is service needed? Yes ___ No ___

If Yes, list the defect(s) and probable repair: _____

_____

_____

Job 93—Inspect a Steering Column (continued)

Identify Hybrid Vehicle Steering System Electrical Circuits

> **Warning**
>
> ⚠ Do not assume that any high-voltage wiring is safe. Always wear protective gloves. This is true even when the battery is discharged. A high-voltage battery that will no longer operate the vehicle may still produce several hundred volts.

☐ 1. Consult the proper service information and determine whether the high-voltage battery drives the steering motor.

Does the high-voltage battery drive the steering motor? Yes ___ No ___

> **Caution**
>
> ◇ Check the service information for exact information. Put on insulated rubber gloves when near any of the high-voltage electrical components.

☐ 2. Using the manufacturer's service information, identify the vehicle's high- and low-voltage circuits.

☐ 3. Locate the high-voltage components and cables on the vehicle.

> **Note**
>
> 📄 High-voltage cables on most hybrid vehicles are covered with orange insulation.

☐ 4. Check the vehicle service information for the exact location of the disconnect device. Disconnect device shape and location varies between vehicles.

Describe the disconnect device: _____

_____

☐ 5. While wearing insulated rubber gloves, remove the disconnect device.

☐ 6. Wait at least five minutes to allow the full discharge of the high-voltage condensers.

☐ 7. Use a voltmeter or multimeter to test the high-voltage cables for low or zero voltage. Less than 12 volts is considered safe.

Voltage _____

☐ 8. Reinstall the disconnect device as directed by your instructor.

☐ 9. Clean the work area and return any equipment to storage.

☐ 10. Did you encounter any problems during this procedure? Yes ___ No ___

If Yes, describe the problems: _____

_____

What did you do to correct the problems? _____

_____

☐ 11. Have your instructor check your work and sign this job sheet.

Performance Evaluation—Instructor Use Only

Did the student complete the job in the time allotted? Yes ___ No ___

If No, which steps were not completed?_____

How would you rate this student's overall performance on this job?_____

5–Excellent, 4–Good, 3–Satisfactory, 2–Unsatisfactory, 1–Poor

Comments: _____

_____

_____

INSTRUCTOR'S SIGNATURE _____

Job 94—Service a Steering Column

After completing this job, you will be able to service a steering wheel and steering column.

Procedures

☐ 1. Obtain a vehicle to be used in this job. Your instructor may direct you to perform this job on a shop vehicle.

☐ 2. Gather the tools needed to perform this job. Refer to the project's tools and materials list.

> **Note**
> The procedures in the *Remove the Steering Wheel* and *Replace the Steering Wheel* sections are a necessary part of a variety of steering column service tasks. Review these sections before performing the other service tasks covered in this job.

Remove the Steering Wheel

☐ 1. Disable the air bag system. Refer to Job 147.

> **Note**
> Some manufacturers call for disconnecting the battery negative cable as part of air bag service.

☐ 2. Remove the air bag from the steering wheel.

☐ 3. Disconnect any electrical connectors attached to the steering wheel, such as the horn and air bag clockspring connectors.

☐ 4. Remove the large nut holding the steering wheel to the steering column. Save any washers for reinstallation.

☐ 5. If necessary, mark the steering wheel and column to ensure that the wheel will be installed in the same position.

☐ 6. Use a puller and the proper adapters to remove the wheel from the column, **Figure 94-1**.

Replace the Steering Wheel

☐ 1. Place the steering wheel in position on the steering column.

☐ 2. Install and tighten the nut holding the steering wheel to the steering column. Be sure to reinstall any washers that were removed.

☐ 3. Reinstall any steering wheel electrical connectors.

☐ 4. Reinstall the air bag on the steering wheel.

☐ 5. Enable the air bag system and reconnect the vehicle battery negative cable if necessary. Refer to Job 147.

Remove and Replace the Tilt Wheel Mechanism

☐ 1. Remove the steering wheel.

☐ 2. If necessary, use a special tool to depress the lock plate and remove the lock plate retaining ring, **Figure 94-2**.

Figure 94-1. Use a puller to remove the steering wheel. Do not attempt to remove the steering wheel by hammer or prying.

Figure 94-2. Many steering columns have a lock plate secured by a retaining ring. A special tool like the one shown makes retaining ring removal and installation much easier. (DaimlerChrysler)

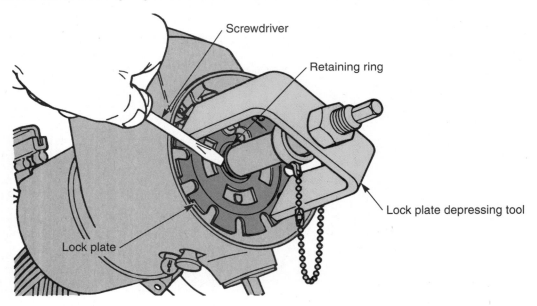

3. Remove the turn signal lever and clockspring if necessary.

4. Remove any electrical connectors attached to the tilt mechanism.

5. Remove the fasteners holding the tilt mechanism to the steering column jacket.

6. Slide the tilt mechanism from the steering column.

> **Note**
> Some tilt mechanisms are attached to the upper part of the steering column. The column must be disconnected at the steering shaft flexible coupling or universal joint as explained earlier in this job.

Job 94—Service a Steering Column (continued)

☐ 7. Compare the old and new tilt mechanisms.

☐ 8. Place the new tilt mechanism in position on the steering column.

☐ 9. Reinstall the electrical connectors as needed.

☐ 10. Reinstall the tilt mechanism fasteners.

☐ 11. Reinstall the turn signal lever and clockspring if necessary.

☐ 12. If necessary, depress the lock plate and install the lock plate retaining ring.

☐ 13. Reinstall the steering wheel.

Replace and Time Clockspring

☐ 1. Determine the number of steering wheel rotations from the full-left to full-right positions.

Number of rotations:_____

☐ 2. Remove the steering wheel.

☐ 3. Remove the lock plate, if necessary.

☐ 4. Remove the clockspring electrical connectors.

☐ 5. Remove the fasteners holding the clockspring to the steering column and remove the clockspring, **Figure 94-3**.

☐ 6. Compare the old and new clocksprings.

> **Note**
>
> Most replacement clocksprings will be pre-centered. However, the procedure for centering is outlined in steps 7 through 11. If the replacement clockspring is pre-centered, skip these steps.

☐ 7. Make sure the coil ribbon is wound snugly against the center hub (part that contacts the steering wheel).

☐ 8. Depress or unlock the locking mechanism as necessary.

☐ 9. Divide the number of steering wheel rotations recorded in step 1 by two.

Record your answer:_____

☐ 10. Unwind the coil ribbon approximately the number of rotations recorded in the previous step.

☐ 11. Lock the locking mechanism as necessary.

Figure 94-3. This clockspring is being removed from the steering column. (Federal-Mogul)

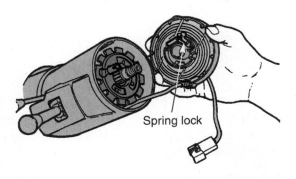

Spring lock

☐ 12. Install the clockspring on the steering column.

☐ 13. Install the clockspring fasteners and electrical connectors.

☐ 14. Reinstall the steering wheel.

Remove and Replace the Steering Shaft Flexible Coupling

☐ 1. Mark the relative position of the upper and lower steering columns and the coupling.

☐ 2. Loosen the steering column as necessary to allow the coupling to be removed.

☐ 3. Remove both coupling-to-shaft fasteners.

☐ 4. Slide the coupling in one direction to clear one of the shafts, then move it to one side of the shaft and pull it in the other direction to remove it.

☐ 5. Slide the new coupling on one shaft, and then slide it into position on the other shaft.

> **Caution**
>
> ⬦ Be sure that the original shaft and coupling alignment has been maintained.

☐ 6. Install and tighten the coupling-to-shaft fasteners.

☐ 7. Tighten the steering column fasteners.

Remove and Replace the Steering Shaft Universal Joint

☐ 1. Mark the relative position of the upper and lower steering columns and the universal joint.

☐ 2. Loosen the steering column as necessary to allow the universal joint to be removed.

☐ 3. Remove both universal joint-to-shaft fasteners.

☐ 4. Slide the universal joint in one direction to clear one of the shafts, then move it to one side to clear the shaft and pull it in the other direction to remove it.

☐ 5. Slide the new universal joint on one shaft, and then slide it into position on the other shaft.

> **Caution**
>
> ⬦ Be sure that the original shaft and universal joint alignment has been maintained.

☐ 6. Install and tighten the universal joint-to-shaft fasteners.

☐ 7. Tighten the steering column fasteners.

Remove and Replace Collapsible Steering Shaft

> **Note**
>
> 📄 Some steering shafts can be removed from the steering column jacket without removing the steering column from the vehicle. Consult the appropriate service literature before proceeding.

☐ 1. Disable the air bag and remove the steering wheel.

☐ 2. Remove any under-dashboard covers that limit access to the steering column.

☐ 3. Remove the steering shaft from the coupler or universal joint as explained in the *Remove and Replace Steering Shaft Flexible Coupling* and the *Remove and Replace Steering Shaft Universal Joint* sections of this job.

Job 94—Service a Steering Column (continued)

- [ ] 4. Remove electrical connectors as necessary to allow removal of the steering column.
- [ ] 5. Remove the ignition and brake interlock linkage or cables as applicable.
- [ ] 6. Remove the fasteners holding the steering column to the underside of the dashboard.
- [ ] 7. Remove the steering column from the vehicle and place it on a clean workbench.
- [ ] 8. Disassemble the steering column as necessary to gain access to the steering shaft.
- [ ] 9. Remove the old shaft and compare it to the replacement.

> **Note**
> If the steering shaft is being replaced, inspect the collapsible part of the steering jacket and replace it if necessary.

- [ ] 10. Install the new shaft and reassemble the steering column.
- [ ] 11. Place the steering column in position in the vehicle.
- [ ] 12. Install the steering column fasteners but do not tighten them at this time.
- [ ] 13. Reinstall the electrical connectors and ignition and brake interlock linkage or cables as necessary.
- [ ] 14. Reinstall the coupler or universal joint on the steering shaft as explained in the *Remove and Replace the Steering Shaft Flexible Coupling* and the *Remove and Replace the Steering Shaft Universal Joint* sections of this job.
- [ ] 15. Install the steering wheel and enable the air bag.
- [ ] 16. Reinstall any under-dashboard covers.

Remove and Replace a Lock Cylinder Mechanism

- [ ] 1. Disconnect the battery negative cable.
- [ ] 2. Disarm the air bag and remove the steering wheel.
- [ ] 3. Disassemble the steering column as necessary to access the lock cylinder mechanism.
- [ ] 4. Remove the cylinder-to-ignition switch link.
- [ ] 5. Remove the lock cylinder mechanism fastener. The fastener will be a long machine screw or a splined pin. See **Figure 94-4**.

> **Note**
> Some locking pins must be drilled out. Consult the appropriate service literature.

- [ ] 6. Slide the lock cylinder from the steering column.
- [ ] 7. Slide the new cylinder into place.
- [ ] 8. Install the lock cylinder fastener.
- [ ] 9. Install the cylinder-to-ignition switch link.
- [ ] 10. Reassemble the steering column.
- [ ] 11. Install the steering wheel and arm the air bag.
- [ ] 12. Reconnect the battery negative cable.

Figure 94-4. After accessing the lock cylinder assembly, remove the locking pin and pull the lock cylinder from the housing. (DaimlerChrysler)

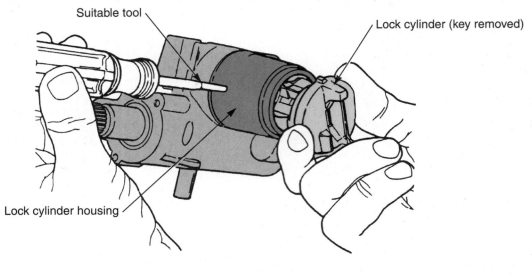

Suitable tool

Lock cylinder (key removed)

Lock cylinder housing

☐ 13. Clean the work area and return any equipment to storage.

☐ 14. Did you encounter any problems during this procedure? Yes ___ No ___

If Yes, describe the problems: _____

_____

What did you do to correct the problems? _____

_____

☐ 15. Have your instructor check your work and sign this job sheet.

Performance Evaluation—Instructor Use Only

Did the student complete the job in the time allotted? Yes ___ No ___
If No, which steps were not completed?_____
How would you rate this student's overall performance on this job?_____
5–Excellent, 4–Good, 3–Satisfactory, 2–Unsatisfactory, 1–Poor
Comments: _____
_____
_____

INSTRUCTOR'S SIGNATURE _____

Job 95—Remove and Replace a Rack-and-Pinion Steering Gear and Inner Tie Rods

After completing this job, you will be able to replace a rack-and-pinion gear and rack-and-pinion inner tie rod ends.

Procedures

☐ 1.   Obtain a vehicle to be used in this job. Your instructor may direct you to perform this job on a shop vehicle.

☐ 2.   Gather the tools needed to perform this job. Refer to the project's tools and materials list.

Remove and Replace a Rack-and-Pinion Steering Gear

> **Note**
>
> The vehicle must be aligned after the steering gear is replaced.

☐ 1.   Raise the vehicle.

> **Warning**
>
> ⚠ Raise and support the vehicle in a safe manner. Always use approved lifts or jacks and jack stands.

> **Note**
>
> Refer to **Figure 95-1** as you perform the following steps.

☐ 2.   Disconnect the steering shaft or universal as necessary. If the steering gear shaft or universal can be reinstalled in more than one position, mark the mating parts to ensure that they are reinstalled in their original positions.

Figure 95-1. This illustration shows the parts that must be removed when changing a rack-and-pinion steering gear.

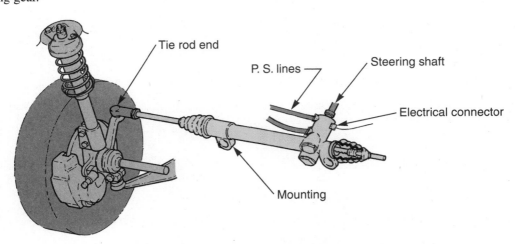

Tie rod end

P. S. lines

Steering shaft

Electrical connector

Mounting

☐ 3. If removing a power rack-and-pinion assembly, place a drip pan under the hose connections and remove the hoses.

> **Caution**
>
> Use line wrenches to loosen the hose fittings. This will prevent damage to the fittings or lines.

> **Note**
>
> If the vehicle has electronic-assist steering, remove any electrical connectors according to manufacturer's procedures.

☐ 4. Remove the outer tie rods.

☐ 5. Remove the mounting fasteners.

☐ 6. Remove the steering gear from the vehicle.

☐ 7. Check the steering gear mounts, brackets, and insulators for the following:
- Damaged rubber mounts or insulators.
- Bent or rusted brackets.
- Stripped or damaged fasteners.

Did you find damaged parts? Yes ___No ___

If Yes, list the damaged parts: _____

_____

Consult your instructor and arrange to obtain replacements for damaged parts.

☐ 8. Place the steering gear in position on the vehicle and loosely attach the mounting fasteners.

☐ 9. Install the power steering hoses and electrical connectors as necessary.

☐ 10. Tighten the mounting fasteners.

☐ 11. Install the outer tie rods.

☐ 12. Install the steering shaft or universal as required.

☐ 13. Lower the vehicle.

☐ 14. Start the engine.

☐ 15. Refill and bleed the power system as necessary.

☐ 16. Road test the vehicle and test steering system operation.

☐ 17. Check and adjust vehicle alignment as necessary. See Job 103.

Remove and Replace Rack-and-Pinion Inner Tie Rods

☐ 1. Raise the vehicle on a lift or place it on jack stands.

> **Warning**
>
> Raise and support the vehicle in a safe manner. Always use approved lifts or jacks and jack stands.

☐ 2. Measure the length of the tie rod assembly at the outer tie rod end. See **Figure 95-2**.

☐ 3. Loosen the outer tie rod locknut and unthread the outer tie rod end from the inner tie rod.

☐ 4. Remove the clamps holding the bellows to the rack-and-pinion assembly.

☐ 5. Fold the bellows away from the rack-and-pinion assembly to gain access to the inner tie rod.

Job 95—Remove and Replace a Rack-and-Pinion Steering Gear and Inner Tie Rods (continued)

Figure 95-2. A tape measure can be used to check the length of the tie rod assembly at the outer tie rod end.

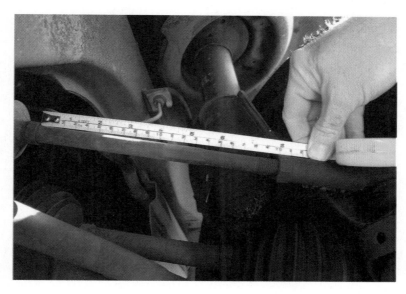

☐ 6. If necessary, turn the steering wheel to move the tie rod away from the rack-and-pinion body. The inner tie rod should be fully exposed before continuing.

☐ 7. Remove the pin, washer, or crimp holding the inner tie rod to the rack assembly.

☐ 8. Install a wrench or a special tool over the inner tie rod. **Figure 95-3** shows a tool needed to remove some inner tie rod.

☐ 9. Place a large adjustable wrench on the rack to keep it from turning.

☐ 10. Unscrew the inner tie rod from the rack.

☐ 11. Compare the old and new tie rods.

Do the tie rods match? Yes ___No ___

If No, how do you explain the difference? _____

_____

_____

☐ 12. Inspect the bellows boot for tears, cracking, or other signs of damage.

Can the bellows boot be reused? Yes ___ No ___

If No, obtain a new boot before proceeding.

Figure 95-3. Some inner tie rods can only be removed by using a special tool. (OTC)

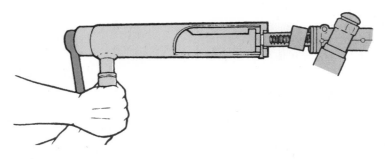

☐ 13. Using a wrench or the special tool, install the inner new tie rod.

☐ 14. Crimp the new inner tie rod to the rack or install the locking pin or washer.

☐ 15. Install the bellows boot over the inner tie rod and install the boot fasteners.

☐ 16. Install the outer tie rod end on the inner tie rod, ensuring that the linkage is adjusted to the original length.

☐ 17. Repeat steps 2 through 16 to change the inner tie rod on the opposite side of the rack-and-pinion assembly (if necessary).

☐ 18. Lower the vehicle and check steering system operation.

☐ 19. Check and adjust vehicle alignment as necessary. See Job 103

☐ 20. Clean the work area and return any equipment to storage.

Did you encounter any problems during this procedure? Yes ___ No ___

If Yes, describe the problems: _____

_____

What did you do to correct the problems? _____

_____

☐ 21. Have your instructor check your work and sign this job sheet.

Performance Evaluation—Instructor Use Only

Did the student complete the job in the time allotted? Yes ___ No ___

If No, which steps were not completed?_____

How would you rate this student's overall performance on this job?_____

5–Excellent, 4–Good, 3–Satisfactory, 2–Unsatisfactory, 1–Poor

Comments: _____

_____

_____

INSTRUCTOR'S SIGNATURE _____

Job 96—Remove and Replace Steering Linkage Components

After completing this job, you will be able to remove and replace steering linkage components.

Procedures

☐ 1. Obtain a vehicle to be used in this job. Your instructor may direct you to perform this job on a shop vehicle.

☐ 2. Gather the tools needed to perform this job. Refer to the project's tools and materials list.

☐ 3. Raise the vehicle on a lift or by using a floor jack and jack stands.

> **Warning**
> Raise and support the vehicle in a safe manner. Always use approved lifts or jacks and jack stands.

☐ 4. List the steering part(s) that will be replaced.

_____

_____

> **Note**
> If a tie rod is being replaced, measure the overall length of the tie rod assembly. Installing the new tie rod so the length is the same as that of the old tie rod will make alignment easier.

☐ 5. Remove the cotter pin(s) from the nut(s) holding ball socket(s) of the part to be replaced. See **Figure 96-1**.

Figure 96-1. Diagonal pliers are handy for removing cotter pins.

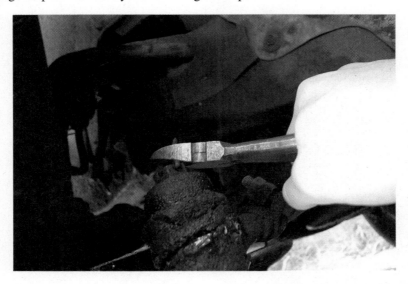

6. Remove the nut(s) from the ball socket(s).

7. If necessary break the stud tapers using a puller, **Figure 96-2**, or a hammer and ball joint separator (pickle fork).

   How did you break the tapers? _____

   _____

   _____

8. Loosen and remove the fasteners holding the part to the frame or steering gear, as applicable. If the pitman arm is being removed, a pitman arm puller will be needed.

   How is the part fastened to the frame or gear? _____

   _____

   _____

9. Remove the part from the vehicle.

10. Compare the old and new parts as shown in **Figure 96-3**.

11. Install the new part on the vehicle.

12. Install and tighten all fasteners and install new cotter pins as necessary. Refer to Figure **96-4** for cotter pin installation.

13. Lubricate the new part if it is equipped with a grease fitting.

14. Repeat steps 5 through 13 for other linkage parts to be changed.

15. Lower the vehicle and check steering system operation.

> **Note**
>
> After any steering linkage part is replaced, the vehicle must be aligned. Check with your instructor to determine whether you should perform Job 103 after completing this job.

16. Return all tools and equipment to storage.

17. Clean the work area and dispose of old parts.

Figure 96-2. This special puller can be used to break the taper holding steering parts together.

Job 96—Remove and Replace Steering Linkage Components (continued)

Figure 96-3. Compare the old and new parts to make sure that they match, as with the old and new tie rod ends shown here. Check for overall size and length and make sure that the thread size and direction are the same.

Figure 96-4. Install the cotter pin correctly so that it will hold the part securely. A—Incorrect. Although this installation method is sometimes used on nuts installed on rotating parts, it is almost always incorrect for steering and suspension parts. B—Correct.

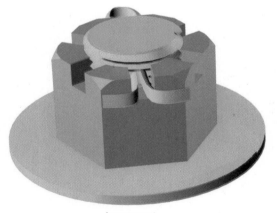

Incorrect

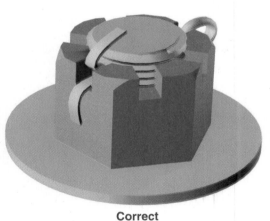

Correct

☐ 18. Did you encounter any problems during this procedure? Yes ___ No ___
If Yes, describe the problems: _____
_____
What did you do to correct the problems? _____
_____

☐ 19. Ask your instructor to check your work and sign this job sheet.

Performance Evaluation—Instructor Use Only

Did the student complete the job in the time allotted? Yes ___ No ___

If No, which steps were not completed?_____

How would you rate this student's overall performance on this job?_____

5–Excellent, 4–Good, 3–Satisfactory, 2–Unsatisfactory, 1–Poor

Comments: _____

_____

_____

INSTRUCTOR'S SIGNATURE _____

Job 97—Test a Power Steering System

After completing this job, you will be able to check power steering components.

Procedures

1. Obtain a vehicle to be used in this job. Your instructor may direct you to perform this job on a shop vehicle.

2. Gather the tools needed to perform this job. Refer to the project's tools and materials list.

> **Note**
>
> For information on removing, inspecting, replacing, and adjusting a power steering pump belt, refer to Jobs 31 and 33.

Check Power Steering Fluid Level and Condition

1. Locate the power steering reservoir.

2. Remove the cap and observe the fluid level. Most power steering reservoirs have the dipstick built into the cap. A few reservoirs are made of clear plastic and the cap does not have to be removed to observe fluid level.

   Is the fluid level low? Yes ___ No ___

3. Place some of the fluid on a clean white rag and check for the following conditions.
   - Discoloration.
   - Burned odor.

   Describe the condition of the power steering fluid: _____

   _____

Diagnose Power Steering System Leaks

Visual Observation Method

1. Visually observe the power steering system components for evidence of fluid leakage. It may be necessary to raise the vehicle to observe all power steering components.

   Did you notice any fluid leaks? Yes ___ No ___

   If Yes, where does the fluid appear to be coming from? _____

   _____

> **Note**
>
> Slight seepage from power steering components is normal.

Powder Method

1. Make sure the power steering system has enough fluid. Add fluid as necessary.

2. Thoroughly clean the area around the suspected leak.

3. Apply talcum powder to the area around the suspected leak.

4. Start and run the vehicle in the shop 10–15 minutes, turning the steering wheel occasionally. You can also start the vehicle and drive it for several miles before moving on to the next step.

> **Warning**
>
> ⚠ Never operate a vehicle in an enclosed area without proper ventilation. A running engine emits carbon monoxide, a deadly gas that builds up quickly.

☐ 5. Raise the vehicle (if necessary) and check the area around the suspected leak.

> **Warning**
>
> ⚠ Raise and support the vehicle in a safe manner. Always use approved lifts or jacks and jack stands.

Does the powder show fluid streaks? Yes ___ No ___

If Yes, where does the fluid appear to be coming from? _____

_____

Black Light Method

☐ 1. Make sure the power steering system has enough fluid. Add fluid if necessary.

☐ 2. Add fluorescent dye to the power steering reservoir, being careful not to spill dye on the outside of the reservoir.

☐ 3. Start the vehicle and run it in the shop 10–15 minutes, turning the steering wheel occasionally.

> **Warning**
>
> ⚠ Never operate a vehicle in an enclosed area without proper ventilation.

☐ 4. Turn on the black light and direct it toward the area around the suspected leak. See **Figure 97-1**.

Does the black light show the presence of dye? Yes ___ No ___

If Yes, where does the dye appear to be coming from? _____

_____

☐ 5. Consult your instructor for the steps to take to correct the leak. Steps may include some of the following:
   - Tightening fasteners.
   - Replacing a gasket or seal. See Jobs 3 and 4.
   - Replacing a cracked, broken, or punctured hose.

Steps to be taken: _____

_____

_____

☐ 6. Make repairs as needed.

☐ 7. Recheck the system for leaks.

Flush, Fill, and Bleed Power Steering System

☐ 1. Place a pan under the power steering reservoir.

☐ 2. Remove the return hose at the reservoir and allow fluid to drain from the reservoir and hose into the pan.

Job 97—Test a Power Steering System (continued)

Figure 97-1. Hard-to-find power steering fluid leaks can often be spotted with fluorescent dye and a black light. (Tracer Products Division of Spectronics Corporation)

☐ 3. Plug the return port and add fluid to the reservoir. Be sure to use the proper type of fluid. What type of fluid was added? _____

☐ 4. Start the engine and add clean fluid to the reservoir as the old fluid flows from the return hose.

☐ 5. When only clean fluid flows from the return hose, turn off the engine.

☐ 6. Remove the plug from the return port and reinstall the return hose.

☐ 7. Refill the reservoir with the proper type of fluid.

☐ 8. Start the engine and add fluid to the reservoir until it stabilizes at the full mark.

☐ 9. Stop the engine and wait approximately two minutes.

☐ 10. Add fluid to the reservoir, if necessary.

☐ 11. Start the engine and turn the steering wheel from side to side for about one minute.

> **Caution**
> Raise the wheels off the ground or move the vehicle a few feet after making each complete turn to avoid rubbing flat spots on the tires.

☐ 12. Repeat Steps 9, 10, and 11 until the fluid has no visible bubbles or foam.

Replace Power Steering System Filter

☐ 1. Drain fluid from the power steering system as necessary.

☐ 2. Place a drip pan under the vehicle.

☐   3.   Remove the clamps holding the filter to the return hose and remove the filter.

         or

         Remove the housing containing the filter and remove the filter.

☐   4.   Install the new filter.

Note

If the filter is an in-line type, make sure that it is installed so that fluid will flow through the filter in the proper direction.

☐   5.   Refill and bleed the power steering system as necessary.

☐   6.   Recheck power steering system operation.

Inspect and Test Electric/Electronic Power Steering

☐   1.   Road test the vehicle to verify problems and determine steering effort from side to side.

☐   2.   Ensure that binding or looseness problems involve the electric power steering system and are not mechanical.

Caution

Some manufacturers recommend removing the power steering control fuse and road testing the vehicle to isolate problems. Do not, however, remove any fuses before retrieving trouble codes.

☐   3.   Use the scan tool to retrieve trouble codes.

         List the retrieved codes: _____

         _____

☐   4.   Check service information to determine whether any of the codes are electric power steering–related codes. Do the steering trouble codes involve any of the following?

         Steering angle sensors? Yes ____ No ____ If yes, list: _____

         Torque sensors? Yes ____ No ____ If yes, list: _____

         Motor amperage? Yes ____ No ____ If yes, list: _____

☐   5.   Use the scan tool to check sensor inputs, motor amperage draw, and other parameters as necessary.

         Results of tests: _____

         _____

         _____

☐   6.   Consult with your instructor about needed repairs.

Pressure Test a Power Steering System

☐   1.   Check the power steering fluid and check for leaks as explained in the above steps.

☐   2.   Check the power steering pump drive belt following the belt inspection procedure described in Job 31. Replace or adjust belt as necessary.

☐   3.   Start the engine and listen for noises from the power steering system.

Job 97—Test a Power Steering System (continued)

> **Warning**
>
> ⚠ Never operate a vehicle in an enclosed area without proper ventilation.

Are any noises present? Yes ___ No ___

If Yes, what could be the source? _____

> **Note**
>
> 📄 If you located a power steering system problem while performing steps 1 through 3, consult your instructor before proceeding with the rest of this job.

☐ 4. Attach a power steering pressure gauge to the pump according to the gauge manufacturer's instructions.

☐ 5. Fill and bleed the power steering system as necessary.

☐ 6. Measure and record the fluid temperature, if called for by the manufacturer.

   Temperature reading:_____

   Specified temperature:_____

> **Note**
>
> 📄 If the fluid temperature is too low, turn the steering wheel to the right or left full lock position for no more than 5 seconds to heat the fluid.

☐ 7. Start the engine and observe pressure readings with the steering wheel in the straight-ahead position.

   Pressure reading:_____ psi/kPa

   Factory specification:_____ psi/kPa

> **Caution**
>
> ◇ The following readings should be taken as quickly as possible to prevent overheating the fluid.

☐ 8. With the engine running, turn the steering wheel all the way to the right and observe pressure readings.

   Pressure reading:_____ psi/kPa

   Factory specification:_____ psi/kPa

☐ 9. With the engine running, turn the steering wheel all the way to the left and observe pressure readings.

   Pressure reading:_____ psi/kPa

   Factory specification:_____ psi/kPa

☐ 10. With the engine running, close the pressure gauge valve and observe pressure readings.

Pressure reading:_____ psi/kPa

Factory specification:_____ psi/kPa

Warning

⚠ Take the closed valve readings quickly to avoid system damage or fluid overheating.

☐ 11. Compare readings with factory specifications and determine whether there is a problem in the power steering hydraulic system.

☐ 12. With your instructor's approval, make needed repairs.

Repairs made: _____

☐ 13. Recheck power steering system operation.

Has the problem been corrected? Yes ___ No ___

If Yes, go to step 14.

If No, what should you do next?_____

Replace a Power Steering Pump

☐ 1. Place a drain pan under the pump.

☐ 2. Remove the drive belt if necessary, referring to Job 33. If the pump is electric motor driven, separate the pump from the motor.

☐ 3. Remove the high- and low-pressure hoses at the pump.

☐ 4. Allow the fluid to drain from the pump.

☐ 5. Remove the fasteners holding the pump to the engine.

☐ 6. Remove the pump from the vehicle.

☐ 7. Compare the old and new pumps.

☐ 8. If necessary, remove the reservoir from the old pump and install it on the new pump using new O-rings and gaskets as necessary.

☐ 9. Place the new pump in position on the engine.

☐ 10. Install the pump fasteners.

☐ 11. Install the high- and low-pressure hoses, using new seals and clamps when necessary.

☐ 12. Reinstall the drive belt or install the pump on the drive motor.

Note

📄 If the pump is electric motor driven, skip steps 13 and 14.

☐ 13. Adjust the belt. Skip this step if the belt is a serpentine type.

☐ 14. Tighten the pump and belt adjustment fasteners.

☐ 15. Add fluid and bleed the system as explained earlier in this Job.

☐ 16. Road test the vehicle.

Job 97—Test a Power Steering System (continued)

Replace a Power Steering Pump Pulley

☐ 1. Remove the power steering belt as explained earlier in this Job.

> **Note**
>
> It is often possible to loosen the belt and position it out of the way without removing other belts. Remember not to crank the engine with the belt removed.

☐ 2. Attach a removal tool to the pulley.

☐ 3. Pull the pulley from the shaft.

☐ 4. Compare the old and new pulleys.

☐ 5. Use a special tool to install the new pulley.

> **Warning**
>
> Do not hammer the pulley into place.

☐ 6. Reinstall the drive belt.

☐ 7. Start the engine and check pulley alignment.

☐ 8. Return all tools and equipment to storage.

☐ 9. Clean the work area and dispose of old parts.

☐ 10. Did you encounter any problems during this procedure? Yes ___ No ___

If Yes, describe the problems: _____

_____

What did you do to correct the problems? _____

_____

☐ 11. Have your instructor check your work and sign this job sheet.

Performance Evaluation—Instructor Use Only

Did the student complete the job in the time allotted? Yes ___ No ___

If No, which steps were not completed? _____

How would you rate this student's overall performance on this job? _____

5–Excellent, 4–Good, 3–Satisfactory, 2–Unsatisfactory, 1–Poor

Comments: _____

_____

_____

INSTRUCTOR'S SIGNATURE _____

Notes

Project
Servicing a Suspension System

28

Introduction

The front suspension is the source of most suspension-related complaints. While modern suspensions are generally simpler than those used in the past, many suspension components have been made smaller and lighter. The use of front-wheel drive has added another layer of complexity to the front suspension, while placing more stresses on the components. Nevertheless, rear suspension components should not be overlooked. The increasing use of independent rear suspension has placed more importance on thoroughly checking the components of the rear suspension. In Jobs 98 through 102, you will replace a strut rod, a stabilizer bar, ball joints, control arms, bushings, and coil springs. You will also inspect and replace MacPherson struts and shock absorbers.

Project 28 Jobs

- Job 98—Replace Strut Rods, Stabilizer Bars, and Bushings
- Job 99—Replace Ball Joints, Control Arms, and Bushings
- Job 100—Replace Coil Springs
- Job 101—Inspect and Replace MacPherson Strut Components
- Job 102—Inspect and Replace Shock Absorbers

Tools and Materials

The following list contains the tools and materials that may be needed to complete the jobs in this project. The items used will depend on the make and model of vehicle being serviced.

- Vehicle in need of suspension service.
- Applicable service information.
- Lift or floor jack and jack stands.
- Replacement parts.
- Motor oil or transmission fluid.
- Coil spring compressor.
- MacPherson strut compressor.
- Bushing drivers.
- Ball joint removal and installation tools.
- Air-operated tools as needed.
- Hand tools as needed.
- Safety glasses and other protective equipment as needed.

Safety Notice

Before performing this job, review all pertinent safety information in the text, and review safety information with your instructor.

Notes

Job 98—Replace Strut Rods, Stabilizer Bars, and Bushings

After completing this job, you will be able to replace strut rods, stabilizer bars, and bushings.

Procedures

☐ 1. Obtain a vehicle to be used in this job. Your instructor may direct you to perform this job on a shop vehicle.

☐ 2. Gather the tools needed to perform this job. Refer to the tools and materials list at the beginning of the project.

Identify Track Bar, Strut Rod, Radius Rod, and Related Mounts and Bushings

☐ 1. Raise the vehicle.

☐ 2. Locate the following suspension parts.

> **Note**
> Not every vehicle will have all of the components in the following list.

- Track bar.

Location _____ Describe: _____

_____

Identify the type of mounting. Check both types if applicable.

Through bushings ____ Direct to chassis ____

- Strut rod.

Location _____ Describe: _____

_____

Identify the type of mounting. Check both types if applicable.

Through bushings ____ Direct to chassis ____

- Radius rod.

Location _____ Describe: _____

_____

Identify the type of mounting. Check both types if applicable.

Through bushings ____ Direct to chassis ____

- Stabilizer bar.

Location _____ Describe: _____

_____

Identify the type of mounting. Check both types if applicable.

Through bushings ____ Direct to chassis ____

Replace a Strut Rod and/or Bushing

> **Note**
>
> Manufacturing tolerances result in slight differences in strut rod length, mounting hole placement, and bushing size. Therefore, the vehicle should be aligned after strut rods or bushings are replaced. Check with your instructor to determine if you should perform Job 103 after completing this job.

☐ 1. Safely raise the vehicle and determine the strut rod mounting type. Many strut rods are mounted as shown in **Figure 98-1**.

> **Warning**
>
> ⚠ Raise and support the vehicle in a safe manner. Always use approved lifts or jacks and jack stands.

☐ 2. Remove the strut rod mounting at the control arm.

Describe the mounting method: _____

Is this where the bushing is located? Yes ___ No ___

Are there any provisions for adjustment? Yes ___ No ___

☐ 3. Remove the strut rod mounting at the frame.

> **Note**
>
> If the strut rod is threaded for adjustment at the frame, mark the relative position of the parts for reassembly.

☐ 4. Remove the strut rod from the vehicle.

☐ 5. Remove the bushings from the frame or control arm as necessary.

☐ 6. Compare the old and new bushings.

☐ 7. Install the new bushings in the frame or control arm.

Figure 98-1. This strut rod is bolted to the control arm and passes through a hole in the frame. Many vehicles use this mounting.

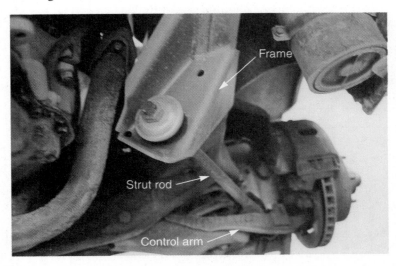

Job 98—Replace Strut Rods, Stabilizer Bars, and Bushings (continued)

☐ 8. Reinstall the strut rod on the vehicle.

☐ 9. Reinstall and tighten all fasteners.

☐ 10. Repeat steps 2 through 9 to replace other strut rods as needed.

☐ 11. Lower the vehicle.

Replace a Stabilizer Bar, Track Bar, Links, and/or Bushings

☐ 1. Raise the vehicle on a lift or place it on jack stands.

> **Warning**
>
> ⚠ Raise and support the vehicle in a safe manner. Always use approved lifts or jacks and jack stands.

☐ 2. Remove the stabilizer bar mounting and the links, if applicable, at each control arm.

Describe the stabilizer bar mounting method: _____

_____

_____

_____

☐ 3. Remove the stabilizer bar mountings at the frame.

☐ 4. Remove the stabilizer bar from the vehicle.

☐ 5. Remove the bushings from the stabilizer bar mountings.

☐ 6. Compare the old and new bushings.

☐ 7. Install the new bushings on the stabilizer bar mountings.

☐ 8. Reinstall the stabilizer bar on the vehicle.

☐ 9. Reinstall and tighten all fasteners.

☐ 10. Lower the vehicle.

☐ 11. Clean the work area and return any equipment to storage.

☐ 12. Did you encounter any problems during this procedure? Yes ___ No ___

If Yes, describe the problems: _____

_____

What did you do to correct the problems? _____

_____

☐ 13. Have your instructor check your work and sign this job sheet.

Performance Evaluation—Instructor Use Only

Did the student complete the job in the time allotted? Yes ____ No ____

If No, which steps were not completed?_____

How would you rate this student's overall performance on this job?_____

5–Excellent, 4–Good, 3–Satisfactory, 2–Unsatisfactory, 1–Poor

Comments: _____

_____

_____

INSTRUCTOR'S SIGNATURE _____

Job 99—Replace Ball Joints, Control Arms, and Bushings

After completing this job, you will be able to remove and replace upper and lower ball joints, control arm bushings, and related parts on a front or rear suspension system.

> ### Note
> The vehicle must be aligned after any ball joint, control arm, or bushing is replaced. After you have completed this job, your instructor may direct you or another student to perform Job 103.

Procedures

☐ 1. Obtain a vehicle to be used in this job. Your instructor may direct you to perform this job on a shop vehicle.

☐ 2. Gather the tools needed to perform this job. Refer to the tools and materials list at the beginning of the project.

Replace Ball Joints

☐ 1. Raise the vehicle on a lift or with a floor jack and jack stands.

> ### Warning
> Raise and support the vehicle in a safe manner. Always use approved lifts or jacks and jack stands.

☐ 2. Place a check mark next to the ball joints that will be replaced.

Front

Left upper ___   Right upper ___

Left lower ___   Right lower ___

Rear

Left upper ___   Right upper ___

Left lower ___   Right lower ___

☐ 3. Remove the wheel at the first ball joint to be replaced.

☐ 4. Place a jack stand under the lower control arm.

☐ 5. Determine whether the ball joint is loaded (weight of the vehicle passes through the ball joint) or unloaded. Consult the service literature or your instructor as necessary.

Loaded ball joint ___Unloaded ball joint ___

If the ball joint is loaded, is it tension loaded or compression loaded? _____

_____

☐ 6. Remove the cotter pin from the nut of the ball joint to be replaced.

☐ 7. Loosen the ball joint nut but do not remove it.

☐ 8. Break the ball joint-to-steering knuckle taper by one of the following methods:
  • Use a special tool.
  • Strike the spindle at the taper area using a large hammer.

Method used: _____

> **Warning**
>
> ⚠ If the spring is installed on the lower control arm, make sure the control arm is properly supported before removing the nut. Do this even when the upper ball joint is being replaced.

☐ 9. Remove the nut from the ball joint stud.

☐ 10. Remove the steering knuckle from the ball joint stud.

☐ 11. Remove the ball joint from the control arm by one of the following methods:
- Press the ball joint from the control arm using a special tool.
- Unthread the ball joint from the control arm using a special socket.
- Chisel the ball joint rivets from the control arm and remove the joint.
- Remove the fasteners holding the ball joint to the control arm and remove the joint.

> **Note**
>
> 📄 Refer to **Figure 99-1**, which shows the different types of ball joint fasteners. Consult the service manual if you have any doubts as to the correct removal method.

Method used: _____

☐ 12. Compare the old and new ball joints.

Is the new ball joint the same design as the old ball joint? Yes ___No ___

If No, consult with your instructor before proceeding.

> **Caution**
>
> ◇ Do not attempt to install an incorrect ball joint.

Figure 99-1. The four ways in which ball joints are installed in the control arm are shown here. A—Pressed-in ball joint. B—Threaded ball joint. C—Riveted ball joint. D—Ball joint attached with nuts and bolts.

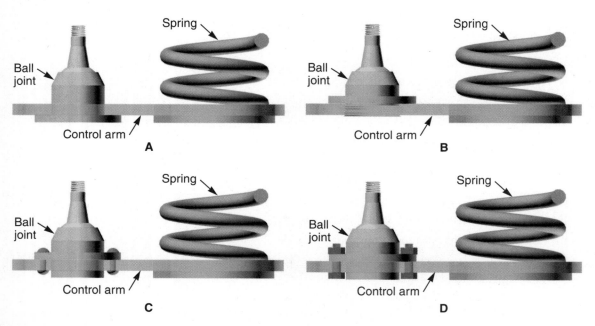

Job 99—Replace Ball Joints, Control Arms, and Bushings (continued)

☐ 13. Install the new ball joint in the control arm.

What is the attachment method?_____

Does this differ from the original attachment method? Yes ___No ___

If Yes, how? _____

☐ 14. Install the ball joint stud in the steering knuckle.

☐ 15. Install the nut on the ball joint stud and tighten it to specifications.

☐ 16. Install a new cotter pin and bend it as shown in **Figure 99-2** to keep it from falling out.

> **Caution**
>
> If the cotter pin holes in the nut and stud do not line up, slightly tighten the nut to align them. Do not loosen the nut.

☐ 17. If the new ball joint has a grease fitting, lubricate it at this time.

Does the ball joint have a grease fitting? Yes ___No ___

☐ 18. Reinstall the wheel.

☐ 19. Repeat steps 3 through 18 to replace other ball joints as needed.

☐ 20. Lower the vehicle and check steering and suspension system operation.

> **Note**
>
> Check with your instructor to determine whether you should perform Job 103 after completing this job.

Replace Control Arms and Bushings

☐ 1. Raise the vehicle so the control arm is free.

> **Warning**
>
> Raise and support the vehicle in a safe manner. Always use approved lifts or jacks and jack stands.

Figure 99-2. Be sure to bend the new cotter pin as shown to keep it from falling out.

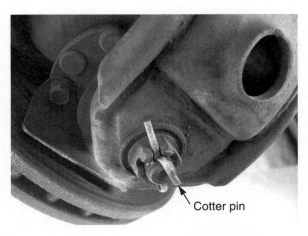

Cotter pin

☐ 2. Remove the wheel.

☐ 3. Remove the ball joint stud from the steering knuckle using the procedures in the *Replace Ball Joints* section of this job.

> **Warning**
>
> ⚠ If you are changing a control arm that is under spring tension, this tension must be removed before loosening any fasteners. Refer to Job 100 for instruction on using a spring compressor to remove tension.

☐ 4. Remove the strut rod, stabilizer bar, alignment shims, and shock absorber mounting as necessary.

List the parts that require removal: _____

_____

_____

☐ 5. Remove the control arm or cross bar attaching bolts.

Are there any eccentric cams on the control arm bolts? Yes ____ No ____

Is the cross bar attached by bolts passing through slotted holes? Yes ____ No ____

Did you mark the relative position of the parts for reassembly? Yes ____ No ____

☐ 6. Remove the control arm from the vehicle.

☐ 7. Place the control arm in a vise.

☐ 8. Remove the bushings using a special tool or an air chisel.

☐ 9. Compare the old and new bushings.

☐ 10. Install the new bushings using the proper driver.

☐ 11. Replace the control arm on the vehicle.

☐ 12. Install the fasteners at the control arm bushings or cross bar.

☐ 13. Install all related parts.

☐ 14. Install the ball joint on the steering knuckle.

☐ 15. Install the ball joint nut and a new cotter pin.

☐ 16. Install the wheel.

☐ 17. Repeat steps 2 through 16 to replace other control arms or control arm bushings as needed.

☐ 18. Lower the vehicle.

☐ 19. Return all tools and equipment to storage.

☐ 20. Clean the work area and dispose of old parts.

☐ 21. Did you encounter any problems during this procedure? Yes ____ No ____

If Yes, describe the problems: _____

_____

What did you do to correct the problems? _____

_____

☐ 22. Have your instructor check your work and sign this job sheet.

Job 99—Replace Ball Joints, Control Arms, and Bushings (continued)

Performance Evaluation—Instructor Use Only

Did the student complete the job in the time allotted? Yes ____ No ____

If No, which steps were not completed?_____

How would you rate this student's overall performance on this job?_____

5–Excellent, 4–Good, 3–Satisfactory, 2–Unsatisfactory, 1–Poor

Comments: _____

_____

_____

INSTRUCTOR'S SIGNATURE _____

Notes

Job 100—Replace Coil Springs

After completing this job, you will be able to remove and replace coil springs on a front or rear suspension system. The procedure presented in this job can also be used to remove the coil spring on a MacPherson strut suspension with a separate coil spring.

> **Note**
> The vehicle must be aligned after a coil spring is replaced. After you have completed this job, your instructor may direct you to perform Job 103.

Procedures

> **Note**
> The procedure for checking ride height is covered in Job 90.

☐ 1. Obtain a vehicle to be used in this job. Your instructor may direct you to perform this job on a shop vehicle.

☐ 2. Gather the tools needed to perform this job. Refer to the tools and materials list at the beginning of the project.

☐ 3. Raise the vehicle on a lift or with a floor jack and jack stands.

> **Warning**
> ⚠ Raise and support the vehicle in a safe manner. Always use approved lifts or jacks and jack stands.

Replacing a Spring on a Lower Control Arm

☐ 1. Remove the wheel and tire at the first spring to be replaced.

☐ 2. Remove the shock absorber if it is installed inside the coil spring.

☐ 3. Install the spring compressor and tighten it to remove spring pressure from the control arm.

☐ 4. Place a jack or jack stand under the lower control arm.

☐ 5. Remove the ball joint from the steering knuckle by following the procedures in Job 99.

☐ 6. Slowly lower the floor jack and control arm to gain clearance for the coil spring.

☐ 7. Remove the coil spring and compressor from the vehicle as a unit. See **Figure 100-1**.

☐ 8. Remove the spring compressor from the coil spring.

☐ 9. Compare the old and new springs.

Is the new spring the same design as the old spring? Yes ___ No ___

If No, how do you account for the differences? _____

☐ 10. Install the spring compressor on the new spring and tighten it to compress the spring.

☐ 11. Place the new coil spring in position on the vehicle.

☐ 12. Raise the control arm with the floor jack, ensuring that the coil spring seats correctly in the vehicle frame and control arm pockets.

☐ 13. Install the ball joint stud in the steering knuckle and install the nut.

Figure 100-1. A coil spring compressor should be installed and tightened to compress a coil spring.

14. Tighten the ball joint stud nut to specifications and install a new cotter pin.
15. Remove the floor jack from under the control arm.
16. Slowly decompress the spring. Then remove the coil spring compressor from the spring.
17. Reinstall the shock absorber if necessary.
18. Reinstall the wheel.
19. Repeat steps 1 through 18 to replace other coil springs as needed.
20. Lower the vehicle and check steering and suspension system operation.

> **Note**
>
> Check with your instructor to determine whether you should perform Job 103 after completing this job.

Replacing a Spring on an Upper Control Arm

> **Note**
> A few vehicles have the spring installed on the upper control arm. Use the following procedure to replace a spring mounted on the upper control arm.

Job 100—Replace Coil Springs (continued)

☐ 1. Remove the wheel at the spring to be replaced.

☐ 2. Remove the shock absorber if it is installed inside the coil spring.

☐ 3. Install the spring compressor and tighten it to remove pressure from the control arm.

☐ 4. Remove the coil spring from the vehicle.

☐ 5. Compare the old and new springs.

Is the new spring the same design as the spring? Yes ___ No ___

If No, how do you account for any differences?_____

_____

☐ 6. Install the spring compressor on the new spring and tighten it to compress the spring.

☐ 7. Install the coil spring in the vehicle, being sure that the spring is properly seated in the spring pockets and that any rubber insulators are in place.

☐ 8. Reinstall the shock absorber if it is installed inside the coil spring.

☐ 9. Reinstall the wheel at the spring to be replaced.

Replacing Springs on a Solid Rear Axle

☐ 1. Raise the vehicle on a lift or support the rear axle on jack stands.

or

Raise the vehicle by lifting the rear axle assembly and place jack stands under the rear of the vehicle frame.

> **Warning**
> ⚠ Raise and support the vehicle in a safe manner. Always use approved lifts or jacks and jack stands.

☐ 2. If necessary, remove any brake lines or electrical connectors that could be damaged when the axle assembly is lowered.

☐ 3. Remove the lower shock absorber mounting fasteners.

☐ 4. Rise the vehicle lift until the axle drops just enough to provide clearance to remove the springs.

or

Lower the axle just enough to provide clearance to remove the springs.

☐ 5. Remove the coil spring from the vehicle.

☐ 6. Compare the old and new springs.

Is the new spring the same design as the old spring? Yes ___ No ___

How do you account for any differences?_____

_____

☐ 7. Install the coil spring in the vehicle, making sure the spring is properly seated in the spring pockets and that any rubber insulators are in place.

☐ 8. Lower the vehicle lift until the shock absorber mounts can be reconnected.

or

Raise the axle until the shock absorber mounts can be reconnected.

☐ 9. Reinstall the shock absorber mounting fasteners.

☐ 10. Reconnect any brake lines or electrical connectors that were removed earlier.

☐ 11. Lower the vehicle and test spring operation.

☐ 12. Return all tools and equipment to storage.

☐ 13. Clean the work area and dispose of old parts.

☐ 14. Did you encounter any problems during this procedure? Yes ___ No ___

If Yes, describe the problems: _____

_____

What did you do to correct the problems? _____

_____

☐ 15. Have your instructor check your work and sign this job sheet.

Performance Evaluation—Instructor Use Only

Did the student complete the job in the time allotted? Yes ___ No ___

If No, which steps were not completed?_____

How would you rate this student's overall performance on this job?_____

5–Excellent, 4–Good, 3–Satisfactory, 2–Unsatisfactory, 1–Poor

Comments: _____

_____

_____

INSTRUCTOR'S SIGNATURE _____

Name _____   Date _____

Instructor _____   Period _____

Job 101—Inspect and Replace MacPherson Strut Components

After completing this job, you will be able to replace front or rear MacPherson strut components.

Procedures

☐ 1. Obtain a vehicle to be used in this job. Your instructor may direct you to perform this job on a shop vehicle.

☐ 2. Gather the tools needed to perform this job. Refer to the tools and materials list at the beginning of the project.

Check MacPherson Struts

☐ 1. Vigorously bounce one corner of the vehicle up and down by hand; then remove your hand at the bottom of a down stroke.

How many times does the vehicle bounce?_____

If the vehicle bounces more than once, the MacPherson strut cartridges are worn out.

☐ 2. Repeat step 1 at each corner of the vehicle.

Does the vehicle bounce excessively at any corner? Yes ___ No ___

If Yes, which corners? _____

☐ 3. Visually inspect the MacPherson strut assemblies for the following conditions. Check all that apply.
   • Dents or other damage to the strut. ___
   • Cartridge leaks. ___
   • Damage to the spring, strut tops, or other related parts. ___
   • Damaged, worn, or loose bushings or spring cushions. ___
   • Damaged, worn, or loose mounting brackets. ___

Describe your findings: _____

_____

Do the strut cartridges require replacement? Yes ___ No ___

Replace MacPherson Struts

> **Note**
>
> This is a general procedure. On some vehicles, the spring is separate from the strut assembly. On other vehicles, the strut shock cartridge can be removed from under the hood. Always consult the proper service manual before proceeding with strut replacement. After strut replacement, the vehicle must be aligned.

☐ 1. Safely raise the vehicle.

> **Warning**
>
> Raise and support the vehicle in a safe manner. Always use approved lifts or jacks and jack stands.

☐ 2. Remove the wheel at the strut to be replaced.

3. If the strut mountings contain alignment adjusters, mark the adjusters so they can be reinstalled in the same position.

Describe the adjuster type, if applicable: _____

4. Remove the fasteners holding the strut assembly to the vehicle.

Describe the fastening method: _____

5. Remove the strut assembly from the vehicle.

6. Place the strut assembly in the strut compressor. See **Figure 101-1**.

7. Mark the spring, strut, and upper bearing mount so they can be reinstalled in the same position.

8. Compress the spring.

9. Remove the upper cartridge shaft nut. This is usually best done with an impact wrench, **Figure 101-2**.

Figure 101-1. The strut must be mounted in an appropriate compressor before disassembly.

Figure 101-2. An impact wrench is generally used to remove the strut shaft nut.

Job 101—Inspect and Replace MacPherson Strut Components (continued)

☐ 10. Remove the strut from the spring.

> **Note**
>
> Steps 11 through 17 are performed only if the strut can be disassembled to remove and replace the internal shock cartridge in the strut body. If the strut is not serviceable, go to step 18.

☐ 11. Remove the large gland nut holding the shock cartridge in the strut body, **Figure 101-3**.

☐ 12. Remove the old shock cartridge.

☐ 13. Compare the old and new shock cartridges.

Do the old and new cartridge diameters and lengths match? Yes ___ No ___

Are the mounting shafts the same? Yes ___ No ___

If No, why not? _____

> **Note**
>
> If the shock cartridges do not match, obtain the correct cartridges before proceeding.

☐ 14. Add oil or transmission fluid to the strut body if necessary. Use the amount and type of lubricant specified.

☐ 15. Place the new shock cartridge in position in the strut body.

☐ 16. Install and tighten the gland nut.

☐ 17. Go to step 19.

Figure 101-3. This gland nut is being removed with a special tool. On some vehicles, the nut can be removed with a pipe wrench.

☐ 18. Obtain a new strut and compare the new strut to the old one.

Do the old and new strut diameters and lengths match? Yes ___ No ___

Are the mounting brackets and fasteners the same? Yes ___ No ___

If No, why not? _____

Note

If the struts do not match, obtain the correct replacement strut before proceeding.

☐ 19. Check the spring cushions and other related strut parts.

Are any parts damaged or worn out? Yes ___ No ___

If any parts are defective, consult your instructor before proceeding.

☐ 20. Place the new or rebuilt strut in position inside the spring.

☐ 21. Install and tighten the upper shaft nut.

☐ 22. Refer to the marks made during disassembly and ensure that all parts are properly aligned.

☐ 23. Decompress the spring slowly, ensuring that all parts are in their proper position.

☐ 24. Remove the strut assembly from the compressor and place it in position on the vehicle, aligning any marks made during disassembly. See **Figure 101-4**.

☐ 25. Install and tighten the strut fasteners and alignment devices.

☐ 26. Reinstall the wheel and tire.

☐ 27. Repeat steps 2 through 26 to replace other vehicle strut assemblies.

☐ 28. Lower the vehicle and check strut operation.

☐ 29. Clean the work area and return any equipment to storage.

☐ 30. Did you encounter any problems during this procedure? Yes ___ No ___

If Yes, describe the problems: _____

What did you do to correct the problems? _____

☐ 31. Have your instructor check your work and sign this job sheet.

Figure 101-4. It is usually easier to align the upper strut mount and start the fasteners before attaching the bottom of the strut.

Job 101—Inspect and Replace MacPherson Strut Components (continued)

Performance Evaluation—Instructor Use Only

Did the student complete the job in the time allotted? Yes ___ No ___

If No, which steps were not completed?_____

How would you rate this student's overall performance on this job?_____

5–Excellent, 4–Good, 3–Satisfactory, 2–Unsatisfactory, 1–Poor

Comments: _____

_____

_____

INSTRUCTOR'S SIGNATURE _____

Notes

Job 102—Inspect and Replace Shock Absorbers

After completing this job, you will be able to replace front or rear shock absorbers.

Procedures

☐ 1. Obtain a vehicle to be used in this job. Your instructor may direct you to perform this job on a shop vehicle.

☐ 2. Gather the tools needed to perform this job. Refer to the tools and materials list at the beginning of the project.

Check Shock Absorbers

☐ 1. Vigorously bounce one corner of the vehicle up and down by hand; then remove your hand at the bottom of a down stroke. If the vehicle bounces more than once after your hand is removed, the shock absorber is worn out. Perform the bounce test on all four corners of the vehicle.

☐ 2. Visually inspect the shock absorbers for the following problems:
   - Dents or other damage to the shock absorber body.
   - Bent shaft.
   - Leaks. See **Figure 102-1**.
   - Damaged, worn, or loose bushings.
   - Damaged, worn, or loose mountings and fasteners.

> **Note**
>
> On some vehicles, the shock absorber must be removed to make these inspections.

Describe your findings: _____

Do the shock absorbers require replacement? Yes ___ No ___

Figure 102-1. This shock absorber is obviously leaking and should be replaced.

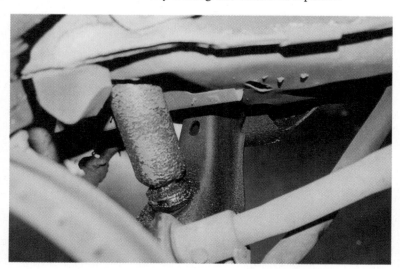

Replace Shock Absorbers

☐ 1. Raise the vehicle on a lift or with a floor jack and jack stands.

> **Warning**
>
> ⚠ Raise and support the vehicle in a safe manner. Always use approved lifts or jacks and jack stands.

☐ 2. If replacing shock absorbers on a rear axle, make sure the axle will not drop when the shock absorbers are removed. If necessary, support the rear axle with jack stands.

> **Note**
>
> If the rear shock absorbers are air operated, depressurize the system and remove the air lines at this time.

☐ 3. Remove the fasteners holding the shock absorber to the vehicle. See **Figure 102-2**.
   Describe how the shock absorbers are attached to the vehicle:
   Upper: _____
   Lower: _____

☐ 4. Remove the shock absorber from the vehicle.

☐ 5. Compare the old and new shock absorbers.

> **Note**
>
> If the shock absorbers do not match, obtain the correct shocks before proceeding. Shock absorbers that are the incorrect length will not operate properly.

☐ 6. Place the new shock absorber in position.

> **Note**
>
> It may be necessary to compress the new shock absorber slightly to place it in position.

☐ 7. Install the shock absorber fasteners.

☐ 8. Repeat steps 2 through 7 for other shock absorbers to be changed.

Figure 102-2. Shock absorber mountings vary from one type of vehicle to another. This illustration shows the attachment methods used on most modern shock absorbers.

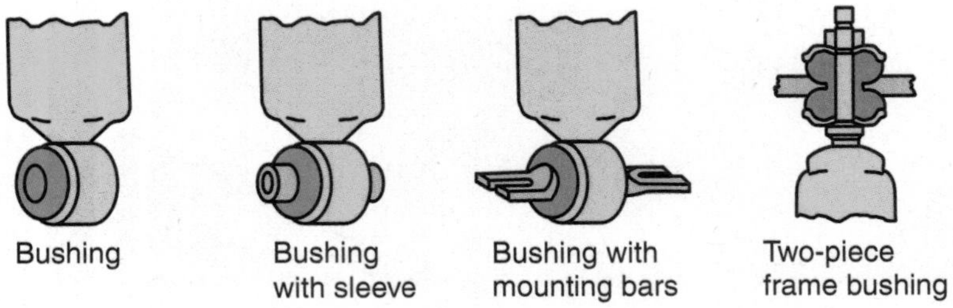

Bushing　　Bushing with sleeve　　Bushing with mounting bars　　Two-piece frame bushing

Job 102—Inspect and Replace Shock Absorbers (continued)

☐ 9. Lower the vehicle and check shock absorber operation.

☐ 10. Clean the work area and return any equipment to storage.

☐ 11. Did you encounter any problems during this procedure? Yes ___ No ___

If Yes, describe the problems: _____

_____

What did you do to correct the problems? _____

_____

☐ 12. Have your instructor check your work and sign this job sheet.

Performance Evaluation—Instructor Use Only

Did the student complete the job in the time allotted? Yes ___ No ___

If No, which steps were not completed?_____

How would you rate this student's overall performance on this job?_____

5–Excellent, 4–Good, 3–Satisfactory, 2–Unsatisfactory, 1–Poor

Comments: _____

_____

_____

INSTRUCTOR'S SIGNATURE _____

Notes

Project
Performing Alignment and Tire Service
29

Introduction

After any suspension or steering part is replaced, the vehicle must be aligned. Vehicles should also be aligned periodically to prevent excessive tire wear and handling difficulties. Modern alignment is usually done on all four wheels. In Job 103, you will check and, if necessary, adjust vehicle alignment.

Tires and wheels are a common and often overlooked source of problems. Common tire problems are under inflation, punctures, abnormal tread wear patterns and internal problems such as belt separation. In Job 104, you will check tires and rims. You will remove tire from a rim and repair a punctured tire in Job 105. In Job 106, you will rotate and balance tires.

Project 29 Jobs

- Job 103—Perform an Alignment
- Job 104—Inspect Tires and Rims
- Job 105—Remove and Install a Tire on a Rim; Repair a Punctured Tire
- Job 106—Rotate and Balance Tires

Tools and Materials

The following list contains the tools and materials that may be needed to complete the jobs in this project. The items used will depend on the make and model of vehicle being serviced.

- Vehicle in need of service.
- Applicable service literature.
- Alignment machine and alignment rack.
- Tire changer.
- Valve core tool.
- Brake pedal lock.
- Tire repair materials.
- Steering wheel lock.
- Alignment adjustment tools.
- Wheel-weight pliers.
- Tire balance equipment.
- Tread depth gauge.
- Tire pressure gauge.
- Tire lubricant.
- Hand tools as needed.
- Air-operated tools as needed.
- Safety glasses and other protective equipment as needed.

Safety Notice

Before performing this job, review all pertinent safety information in the text, and review safety information with your instructor.

Job 103—Perform an Alignment

After completing this job, you will be able to perform a two- or four-wheel alignment on a vehicle.

Procedures

☐   1.   Obtain a vehicle to be used in this job. Your instructor may direct you to perform this job on a shop vehicle.

☐   2.   Gather the tools needed to perform this job. Refer to the tools and materials list at the beginning of the project.

Preliminaries

☐   1.   Consult the driver to determine the exact nature of the complaint or, if there is no complaint, ask why the driver feels that an alignment is necessary.

> **Note**
>
> If necessary, road test the vehicle with the driver to determine the exact nature of the complaint.

☐   2.   Identify the owner's complaint. Check all that apply.

Pulling or drifting ___ Hard steering ___ Bump steer ___ Wander ___

Memory steer ___ Torque steer ___ Poor return ___ Uneven tire wear ___

Other (describe): _____

Briefly describe the results of the road test (if performed): _____

_____

☐   3.   Determine whether the complaint could be caused by alignment concerns. List possible causes of the problem.

Defective parts (list): _____

_____

Incorrect steering geometry (list): _____

_____

☐   4.   Obtain the necessary specifications for alignment. Also determine whether the vehicle requires a 2- or 4-wheel alignment.

> **Note**
>
> Many modern alignment machines contain the vehicle specifications in the machine software.

Prepare the Vehicle and Rack for Alignment

☐   1.   Make sure the alignment rack is in safe working order and that all turntables are locked in position.

☐   2.   Drive the vehicle onto the alignment rack and raise the rack to a comfortable working height.

☐   3.   Measure vehicle ride height at all four wheels. Record the readings in the following chart:

Left Front:_____   Right Front:_____

Left Rear:_____   Right Rear:_____

Compare your measurements to specifications. Were any ride height measurements out of specification? Yes ___ No ___

Is there more than 1/4″ (0.6 mm) variation from side to side or from front and rear?
Yes ___ No ___

If the answer to either of the above questions is Yes, consult your instructor before proceeding.

☐ 4. Raise the front of the vehicle so the front suspension is correctly suspended for checking the ball joints.

☐ 5. Raise the rear of the vehicle.

☐ 6. Perform the shake test on the front wheels:
   a. Grasp the wheel at the top and bottom and shake it back and forth. This checks the suspension parts.
      Do the wheels feel loose? Yes ___ No ___

> **Note**
> To check the suspension on some MacPherson strut vehicles, grasp the wheel at approximately 45° from the top and bottom and shake it back and forth.

   b. Grasp the wheel at the front and back and shake it back and forth. This checks steering linkage parts.
      Do the wheels feel loose? Yes ___ No ___

☐ 7. If the front wheels feel loose during the shake test, check the individual front suspension and steering parts, wheels, and tires for wear or damage.

   Problems found: _____

   What should you do to correct them? _____

☐ 8. Make a visual inspection of the front suspension and steering parts. Look carefully for loose or bent parts, loose fasteners, torn boots or seals, and a bent or misaligned frame or subframe/engine cradle.

   Problems found: _____

   What should you do to correct the problems? _____

> **Note**
> Be sure to check the front and rear frame components and cradles for damage.

☐ 9. Make a visual inspection of the rear suspension and steering parts, as well as the rear wheels and tires.

> **Note**
> If necessary, raise the rear of the vehicle to check for worn suspension and steering parts. Perform the shake test as described in step 6.

☐ 10. Check tire condition and pressure, and complete the following tire condition chart:

	Tire Pressure	Condition
Left front:	_____	_____
Right front:	_____	_____
Right rear:	_____	_____
Left rear:	_____	_____

   Problems found: _____

   What should you do to correct them? _____

   _____

Job 103—Perform an Alignment (continued)

☐ 11. After any problems found in steps 6 through 10 have been corrected, enter the vehicle information into the alignment machine.

> **Note**
>
> This step is only possible in newer electronic alignment machines.

☐ 12. Remove the wheel covers if necessary.

☐ 13. Attach the alignment heads or target boards to the wheel rims.

☐ 14. If necessary, compensate the heads.

Briefly describe how you compensated the heads: _____

_____

☐ 15. Remove the lock pins from the turntables.

☐ 16. If necessary, place the rack legs in the down position and lower the rack until it is resting on the legs.

☐ 17. Lock the brake pedal.

Check Vehicle Alignment

☐ 1. Observe the camber and toe as displayed on the alignment machine. Record the readings in the following chart:

Left front camber:_____ Right front camber:_____

Left front toe:_____ Right front toe:_____

Total front toe:_____

Left rear camber:_____ Right rear camber:_____

Left rear toe:_____ Right rear toe:_____

Total rear toe:_____ Thrust angle (tracking):_____

☐ 2. Compare the camber and toe readings with specifications.

Are readings correct? Yes ____ No ____

List any incorrect readings: _____

☐ 3. Operate the alignment machine and heads and turn the wheels as necessary to measure the following:

- Front caster.
- Steering axis inclination (SAI).
- Toe out on turns.

Record the readings:

Left front caster:_____ Right front caster:_____

Left front SAI:_____ Right front SAI:_____

Left toe out:_____ Right toe out:_____

Compare the readings with specifications.

Are readings correct? Yes ____ No ____

If No, list any incorrect readings: _____

☐ 4. Check wheel setback using the alignment machine or by making manual measurements.

Is setback acceptable? Yes ____ No ____

If No, what can be done to correct it?_____

Adjust Vehicle Alignment

> **Note**
>
>
>
> The type of adjustment device used to set various alignment angles may vary from one vehicle to another. See **Figure 103-1**. Refer to an appropriate service manual for specific adjustment instructions.

> **Caution**
>
>
>
> Some vehicle alignment angles may not be adjustable, including front and rear camber and toe, front camber and SAI, toe out on turns, and wheel setback. Some of these values can only be corrected by replacing parts or by frame straightening. Consult your instructor if any nonadjustable alignment value is incorrect.

Figure 103-1. These illustrations show the most common alignment adjusting devices. Other devices are also used. A—Eccentric cam. B—Slotted hole. C—Shims. D—Threaded rod. (Specialty Equipment Corp., Hunter)

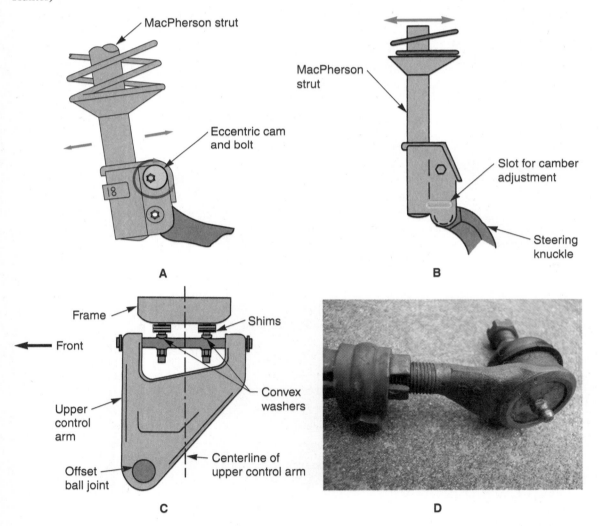

Job 103—Perform an Alignment (continued)

☐ 1. Adjust the rear wheel camber as necessary.

Type of adjustment device: _____

Briefly describe the adjustment process: _____

_____

_____

Were any special tools needed? Yes ___ No ___

If Yes, describe the tools: _____

_____

_____

☐ 2. Adjust the rear toe as necessary.

> **Note**
>
> To obtain the best thrust angle, or tracking, rear toe should be equal on both sides. If toe is equal on both sides, the thrust angle is as close as possible to the centerline of the vehicle. If toe cannot be equalized between sides, it may be possible to compensate by adjusting the front toe. Check the vehicle and alignment equipment manufacturers' service information.

Type of adjustment device: _____

Briefly describe the adjustment process: _____

_____

_____

Were any special tools needed? Yes ___ No ___

If Yes, describe the tools: _____

_____

> **Note**
>
> On some vehicles, steps 1 and 2 can be performed in one operation.

☐ 3. Adjust front camber as necessary.

Type of adjustment device: _____

Briefly describe the adjustment process: _____

_____

_____

Were any special tools needed? Yes ___ No ___

If Yes, describe the tools: _____

☐ 4. Adjust front caster as necessary.

Type of adjustment device: _____

Briefly describe the adjustment process: _____

_____

_____

Were any special tools needed? Yes ___ No ___

If Yes, describe the tools: _____

_____

_____

☐ 5. Center the steering wheel and install the steering wheel lock.

☐ 6. Adjust the front toe as necessary.

How are the tie rods held in position? Locknuts ___ Clamps ___

Briefly describe the adjustment process: _____

_____

_____

Were any special tools needed? Yes ___ No ___

If Yes, describe the tools: _____

_____

_____

☐ 7. Recheck all alignment readings.

Are the readings within specifications? Yes ___ No ___

If No, what should you do next?_____

_____

☐ 8. Remove the alignment heads or target boards from the vehicle and reinstall the wheel covers.

Note

Store the alignment heads properly.

☐ 9. Lower the alignment rack and drive the vehicle from the rack.

☐ 10. Road test the vehicle and observe the following:

Is the steering wheel centered? Yes ___ No ___

Does the vehicle track correctly (no pulling)? Yes ___ No ___

Is steering wheel return acceptable? Yes ___ No ___

Is there a minimal amount of road wander? Yes ___ No ___

If the answer to any of the above questions is No, what should you do next?_____

_____

☐ 11. If necessary, recenter the steering wheel using the following procedure:

Note

It is often necessary to recenter the steering wheel to compensate for excessive rear wheel thrust angle.

a. Determine in which direction the steering wheel is off center.

b. Loosen the tie rod adjuster lock nuts without moving the adjusters.

c. Turn the adjusters the same amount and in the same direction to center the steering wheel.

Job 103—Perform an Alignment (continued)

> **Caution**
>
> This procedure must be done very carefully to ensure that toe is not changed. Some technicians prefer to reinstall the alignment equipment to make this adjustment.

    d. Recheck steering wheel position and repeat step c as necessary.

    e. Retighten the tie rod adjuster lock nuts without moving the adjusters.

    f. Road test the vehicle to ensure that the steering wheel is straight.

☐ 12. If the vehicle is equipped with electronic stability control, reset the steering angle sensor in accordance with the manufacturer's instructions.

> **Note**
>
> Steering angle sensors must be reset after alignment work or any time a part in the steering system is replaced. Some steering angle sensors can be manually reset. Others must be reset using a scan tool.

☐ 13. Return all tools and equipment to storage.

☐ 14. Clean the work area.

☐ 15. Did you encounter any problems during this procedure? Yes ___ No ___

If Yes, describe the problems: _____

_____

What did you do to correct the problems? _____

_____

☐ 16. Have your instructor check your work and sign this job sheet.

Performance Evaluation—Instructor Use Only

Did the student complete the job in the time allotted? Yes ___ No ___

If No, which steps were not completed?_____

How would you rate this student's overall performance on this job?_____

5–Excellent, 4–Good, 3–Satisfactory, 2–Unsatisfactory, 1–Poor

Comments: _____

_____

_____

INSTRUCTOR'S SIGNATURE _____

Notes

Job 104—Inspect Tires and Rims

After completing this job, you will be able to inspect tires and rims for damage.

Procedures

☐ 1. Obtain a vehicle to be used in this job. Your instructor may direct you to perform this job on a shop vehicle.

☐ 2. Gather the tools needed to perform this job. Refer to the tools and materials list at the beginning of the project.

Inspect Tires on a Vehicle

☐ 1. Raise the vehicle in a safe manner.

> **Warning**
>
> ⚠ The vehicle must be raised and supported in a safe manner. Always use approved lifts or jacks and jack stands.

☐ 2. From the information printed on the sidewall of the tire, fill in the following tire data for each time.

	Left Front	Right Front	Left Rear	Right Rear
Manufacturer	_____	_____	_____	_____
Section width	_____	_____	_____	_____
Aspect ratio	_____	_____	_____	_____
Speed rating	_____	_____	_____	_____
Construction type	_____	_____	_____	_____
Rim diameter	_____	_____	_____	_____

Do all tires match? Yes ___ No ___

Maximum tire pressure as listed on the side wall of the tire:_____

☐ 3. Compare the tire pressure specification on the sidewall with the vehicle manufacturer's tire pressure recommendation. Vehicle manufacturer's tire pressure recommendations are listed on the tire loading information sticker. Common sticker locations are the driver's side door, driver's side door jamb, or inside of the trunk lid.

Vehicle manufacturer's tire pressure recommendation for this vehicle:_____

☐ 4. Check tire pressure and fill in the following chart. Add or remove pressure as necessary.

	Left Front	Right Front	Left Rear	Right Rear
Recommended pressure	_____	_____	_____	_____
Measured pressure	_____	_____	_____	_____

☐ 5. Measure tread depth of all the tires with a tire depth gauge. See **Figure 104-1**. Measure the tread at its deepest and shallowest points. If the tread depth is less than 1/16″ (1.59 mm) deep, the tire should be replaced.

	Left Front	Right Front	Left Rear	Right Rear
Maximum tread depth	_____	_____	_____	_____
Minimum tread depth	_____	_____	_____	_____

Figure 104-1. Use a depth gauge to check tread wear.

> **Note**
> A quick and relatively accurate method of checking tire wear is to insert a Lincoln penny into the tire tread at various locations with the top of the head pointing downward. If the top of Lincoln's head shows in any of the test locations, the tire is worn out.

> **Note**
> Many modern tires have tread ribs at the bottom of the tire tread, at a right angle (90°) to the tread. When the tread wears down so that these ribs can be seen to extend across the tire tread, the tire is worn excessively and should be replaced.

☐ 6. Check all the vehicle's tires for abnormal tire wear patterns and compare the vehicle's tire wear patterns to the patterns shown in **Figure 104-2**. Describe the tire wear patterns.

Left front: _____

Right front: _____

Left rear: _____

Right rear: _____

What could cause these wear patterns?_____

_____

☐ 7. Check all tires for damage, such as cuts, cracks, and tread separation.

Is any damage visible? Yes ___ No ___

If Yes, describe the damage: _____

_____

Job 104—Inspect Tires and Rims (continued)

Figure 104-2: This figure shows the cause, effect, and correction of several abnormal tire wear patterns. (DaimlerChrysler)

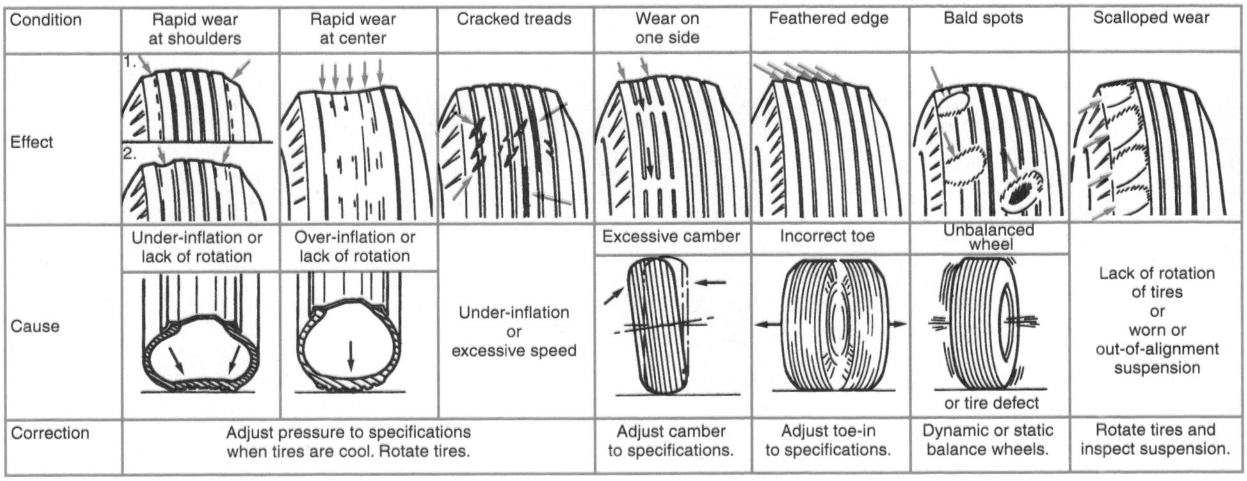

Condition	Rapid wear at shoulders	Rapid wear at center	Cracked treads	Wear on one side	Feathered edge	Bald spots	Scalloped wear
Effect							
Cause	Under-inflation or lack of rotation	Over-inflation or lack of rotation	Under-inflation or excessive speed	Excessive camber	Incorrect toe	Unbalanced wheel — or tire defect	Lack of rotation of tires or worn or out-of-alignment suspension
Correction	Adjust pressure to specifications when tires are cool. Rotate tires.			Adjust camber to specifications.	Adjust toe-in to specifications.	Dynamic or static balance wheels.	Rotate tires and inspect suspension.

☐ 8. Based on the information gathered in steps 5 through 7, decide whether any tire needs replacement.

Do any tires require replacement? Yes ____ No ____

If Yes, fill out the form below. Place a check mark next to the tire(s) that need to be replaced and explain why replacement is necessary.

Left front ____ Reason: _____

Right front ____ Reason: _____

Left rear ____ Reason: _____

Right rear ____ Reason: _____

Check Tire for Air Loss

☐ 1. Remove the wheel and tire assembly from the vehicle.

☐ 2. Inflate the tire to its maximum allowable air pressure.

☐ 3. Place the wheel and tire in a water tank. Most shops will have a special tank for this purpose.

☐ 4. Slowly rotate the tire through the water while watching for air bubbles. Bubbles indicate a leak. Carefully check all areas of the tire and wheel, including:

- Tread area. ____
- Sidewalls. ____
- Bead area. ____
- Rim. ____
- Valve and valve core. ____

Were any leaks found? Yes ____ No ____

If Yes, describe the location: _____

_____

☐ 5. If air leaks were found, consult your instructor. Your instructor may direct you to perform Job 105 at this time.

☐ 6. If no air leaks were found, reduce tire air pressure to the recommended amount and reinstall the wheel and tire on the vehicle. See Job 105 for lug nut tightening procedures.

Measure Tire Runout

☐ 1. Select a spot on the vehicle to attach a dial indicator.

☐ 2. Mount the dial indicator so its plunger contacts the tire tread.

☐ 3. Zero the dial indicator.

☐ 4. Rotate the wheel slowly by hand.

☐ 5. Observe and record the maximum variation of the dial indicator.

Maximum variation:_____

Is this within specifications? Yes ___ No ___

☐ 6. Reposition the dial indicator so its plunger contacts the tire sidewall.

☐ 7. Zero the dial indicator.

☐ 8. Observe and record the maximum variation of the dial indicator.

Maximum variation:_____

Is this within specifications? Yes ___ No ___

Measure Rim Runout

☐ 1. Remove the tire and rim assembly from the vehicle.

☐ 2. Remove the tire from the rim.

☐ 3. Install the rim assembly on a tire machine so that it can be turned by hand.

☐ 4. Install a dial indicator so the indicator plunger contacts the inner surface of the rim.

☐ 5. Zero the indicator.

☐ 6. Rotate the rim slowly by hand.

☐ 7. Observe and record the maximum variation of the dial indicator.

Maximum variation:_____

Is this within specifications? Yes ___ No ___

☐ 8. Reposition the dial indicator so that its plunger contacts the outer side surface of the rim.

☐ 9. Zero the indicator.

☐ 10. Rotate the rim slowly by hand.

☐ 11. Observe and record the maximum variation of the dial indicator.

Maximum variation:_____

Is this within specifications? Yes ___ No ___

☐ 12. If the tire and/or wheel runout is out of specification, replace the tire and/or wheel as necessary.

Measure Wheel Hub Runout

☐ 1. Remove the tire and rim assembly from the wheel to be checked.

☐ 2. Place the transmission in neutral if a rear hub is to be checked.

☐ 3. Install a dial indicator so the indicator plunger contacts the outer surface of the hub flange.

☐ 4. Zero the indicator.

☐ 5. Rotate the hub slowly by hand.

☐ 6. Observe and record the maximum variation of the dial indicator.

Maximum variation: _____

Is this within specifications? Yes _____ No _____

Job 104—Inspect Tires and Rims (continued)

☐ 7. If the hub runout is out of specification, replace the axle as necessary.

☐ 8. Did you encounter any problems during this procedure? Yes ___ No ___

If Yes, describe the problems: _____

_____

What did you do to correct the problems? _____

_____

☐ 9. Have your instructor check your work and sign this job sheet.

Performance Evaluation—Instructor Use Only

Did the student complete the job in the time allotted? Yes ___ No ___

If No, which steps were not completed?_____

How would you rate this student's overall performance on this job?_____

5–Excellent, 4–Good, 3–Satisfactory, 2–Unsatisfactory, 1–Poor

Comments: _____

_____

_____

INSTRUCTOR'S SIGNATURE _____

Notes

Job 105—Remove and Install a Tire on a Rim; Repair a Punctured Tire

After completing this job, you will able to mount and dismount tires on rims and repair a tire puncture.

Procedures

☐ 1. Obtain a vehicle to be used in this job. Your instructor may direct you to perform this job on a shop vehicle.

☐ 2. Gather the tools needed to perform this job. Refer to the tools and materials list at the beginning of the project.

Remove a Tire from a Rim

> **Warning**
> ⚠ If the vehicle is equipped with a tire pressure monitoring system, remove the tire carefully to ensure that the rim mounted pressure sensors are not damaged.

☐ 1. Remove the wheel cover of the tire and wheel to be removed.

☐ 2. Remove the valve stem core (or valve stem if stem is to be replaced).

> **Note**
> 📄 It is easier to remove the valve core or stem when the tire is solidly mounted on the vehicle.

☐ 3. Loosen the lug nuts.

☐ 4. Raise the vehicle.

> **Warning**
> ⚠ The vehicle must be raised and supported in a safe manner. Always use approved lifts or jacks and jack stands.

☐ 5. Remove the lug nuts.

☐ 6. Remove the wheel and tire from the vehicle.

☐ 7. Remove any wheel weights with wheel weight pliers.

> **Note**
> 📄 The wheel weights can be left on the rim if great care is taken not to move them during tire removal and the tire is reinstalled in exactly the same position on the rim.

☐ 8. Position the tire at the tire changing machine bead breaker and use the bead breaker to loosen both tire beads at the wheel flanges, **Figure 105-1**.

> **Caution**
> ◇ Avoid catching the bead breaker on the edge of the wheel. The bead breaker can bend a steel wheel or break an alloy wheel.

Figure 105-1. Many modern wheels are made of aluminum or composite materials and can be damaged if the bead breaker is not properly positioned.

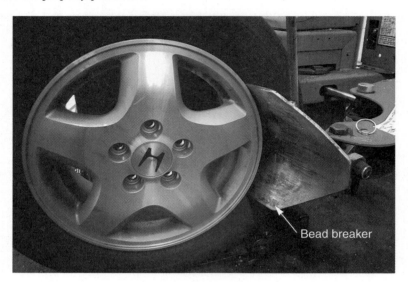

Bead breaker

9. Place the tire and wheel on the tire-changing machine and tightly install the wheel hold-down cone.

10. Place a generous quantity of rubber lubricant or soapy water on the tire bead and the wheel flange to ease tire removal.

11. Use the tire-changing machine inner spindle and the tire changer bar to remove the upper tire bead from the wheel.

>
> **Note**
> Use one hand to hold the tire bead in the center of the wheel while removing the opposite side of the tire.

12. Pull the bottom bead up to the top of the rim and repeat step 11 to remove the bottom tire bead from the wheel. See **Figure 105-2**.

Figure 105-2. Removing the upper tire bead. Use caution and wear safety goggles.

Job 105—Remove and Install a Tire on a Rim; Repair a Punctured Tire (continued)

Install a Tire on a Rim

> **Warning**
> ⚠ If the vehicle is equipped with a tire pressure monitoring system, install the tire carefully to ensure that the rim mounted pressure sensors are not damaged.

☐ 1. Clean the sealing edge of the wheel rim as needed. If the rim is rusted or dirty, clean it with steel wool.

☐ 2. Check the rim for cracks or dents at the sealing surface.

Was any damage found? Yes ___ No ___

If Yes, describe: _____

☐ 3. Check the condition of the valve stem. Bend it sideways and look for weather cracks or splits.

Can the tire valve be reused? Yes ___ No ___

If No, go to step 4.

If Yes, go to step 6.

☐ 4. Check the tire pressure sensor assembly for damage and make sure that it is solidly mounted on the rim. If the sensor and tire valve are a single assembly, visually inspect any mounting hardware and sealing rings.

☐ 5. Install a new tire valve if necessary. Remove the valve core after installing the valve.

☐ 6. Place tire lubricant or soapy water on the tire bead and the flange of the wheel.

☐ 7. Place the tire on top of the wheel.

> **Caution**
> ◇ Make sure that the tire is facing in the proper direction.

☐ 8. Use the tire-changing machine inner spindle and the tire changer bar to install the lower tire bead on the wheel.

☐ 9. Repeat step 8 to install the top tire bead on the wheel.

☐ 10. Pull up on the installed tire while twisting to partially seal the tire on the upper flange.

☐ 11. Install the air nozzle on the valve stem.

☐ 12. Fill the tire with air.

> **Note**
> If air rushes out of the tire as you try to fill it, push in lightly on the leaking part of the tire. If the tire will not hold enough air to expand and seal on the wheel, clamp a bead expander around the outside of the tire to push the tire bead against the wheel flange; then fill the tire with air.

Warning

⚠️ Release the bead expander as soon as the tire begins to inflate. If this is not done, the expander can break and fly off the tire with dangerous force.

☐ 13. Loosen the wheel hold-down cone as soon as the tire begins to inflate.

Warning

⚠️ Do not inflate the tire to over 40 psi (276 kPa). Lean away from the tire as it inflates to prevent possible injury.

☐ 14. Make sure the tire has popped into place over the safety ridges on the rim.

☐ 15. Remove the air nozzle from the valve stem.

☐ 16. Install the valve core.

☐ 17. Inflate the tire to the correct air pressure.

☐ 18. Rebalance the tire if necessary.

☐ 19. Place the tire in position on the vehicle.

☐ 20. Install the lug nuts using the following method:
   - Tighten the lug nuts in an alternating, or star, pattern. Do not tighten the lug nuts in a circle around the rim.
   - Tighten the lug nuts in steps. Do not tighten any lug nut down completely before the others have been partially tightened.

☐ 21. Lower the vehicle.

☐ 22. Use a torque wrench to perform the final tightening sequence in an alternating pattern. Check the service information for the proper lug nut torque for the vehicle.

Note

📄 Recheck the lug nut torque of aluminum and magnesium wheels after the vehicle has been driven 25 to 50 miles.

☐ 23. Reinstall the wheel cover if necessary.

Repair a Tire Puncture

Caution

◇ The following procedure is for repairing a puncture with a combination patch and plug called a mushroom patch. Consult your instructor if another type of patch is to be used. Tires can be repaired with rubber plugs inserted into the puncture, but tire manufacturers do not recommend this type of repair.

☐ 1. Remove the tire and rim assembly from the vehicle.

☐ 2. Remove the tire from the rim.

☐ 3. Inspect the inside of the tire for splits, cracks, punctures, or previous patches.

☐ 4. Locate and remove the puncturing object, then mark the damaged area with a tire crayon or other easily visible marker.

Job 105—Remove and Install a Tire on a Rim; Repair a Punctured Tire (continued)

> **Warning**
> ⚠ Do not attempt to repair a puncture on the tire shoulder or sidewall. Flexing will cause the patch to leak immediately. In addition, a puncture on the shoulder or sidewall may have structurally damaged the tire. Replace tires punctured in these areas.

5. Mount the tire on a tire repair fixture so the puncture area can be easily reached.

6. Use a scuffing tool to roughen the inside surface of the tire at the puncture site. Roughen an area larger than the size of the patch and remove all rubber particles.

7. Apply patch cement to the scuffed area.

8. Place the patch over the puncture area.

9. Pull the plug portion of the patch through the puncture.

10. Apply pressure to the patch to seal it to the inner surface of the tire. A special tool called a stitching tool is available for this operation.

11. Reinstall the tire on the rim and inflate to the proper pressure.

12. Check for leaks at the puncture to ensure that the repair is successful.

13. Trim off the portion of the mushroom patch so that it is even with the tread surface.

14. Reinstall the tire assembly on the vehicle.

Replace Tire Pressure Monitors

1. Remove the tire and rim assembly from vehicle.

2. Depressurize tire and remove the tire from the rim using the proper tire machine.

3. Remove the pressure sensor assembly from tire valve opening. Install a new pressure sensor assembly in the tire valve opening and tighten the fastener to the correct value.

   or

   Remove the strap holding the pressure sensor assembly on the inner part of the tire rim. Install the replacement pressure sensor assembly. Install and tighten the strap.

4. Reinstall the tire on the rim and inflate the tire to the proper pressure.

5. Reinstall the tire and rim assembly on the vehicle and reprogram the sensing system as needed.

Calibrate a Tire Pressure Monitoring System

> **Note**
> 📄 This procedure can be performed with the vehicle on or off the ground.

1. On the scan tool or instrument panel display, locate the menu for calibrating the tire pressure monitoring system.

> **Note**
> 📄 Most tire pressure monitoring systems are accessed through the scan tool. A few systems, however, use the instrument panel display to calibrate the system. Consult the appropriate service literature to determine which is used.

☐ 2. If necessary, attach a scan tool to the vehicle diagnostic connector.

☐ 3. Enter the tire pressure monitoring system calibration mode.

> **Note**
>
> Most calibration sequences must be completed in a set time period, such as one minute to calibrate the first tire and 30 seconds each to calibrate the remaining tires. If the calibration does not begin within this time frame, the system will abort the calibration procedure and all information will be lost.

☐ 4. Begin the calibration procedure at the first tire indicated by the calibration mode instructions.

> **Note**
>
> Some systems require that a special relearn magnet be placed over the tire valve to recalibrate the system. See the manufacturer's instructions for exact procedures.

☐ 5. Allow the sensor on the first wheel to transmit its code. The horn will sound very briefly to indicate completion of the transmission.

☐ 6. Repeat steps 3 and 4 for all four wheels, and the spare tire if a 5-tire system is used. When one tire calibration sequence is finished, the system software will move to the next tire to be calibrated. Tires are calibrated in a rotating pattern, such as LF, LR, RR, RF. Be sure to move the calibration magnet, if used, to the next tire to be calibrated.

☐ 7. Exit calibration mode.

☐ 8. Drive the vehicle to confirm that the calibration process was successful.

☐ 9. Clean the work area and return any equipment to storage.

☐ 10. Did you encounter any problems during this procedure? Yes ___ No ___

If Yes, describe the problems: _____

_____

What did you do to correct the problem? _____

_____

☐ 11. Have your instructor check your work and sign this job sheet.

Performance Evaluation—Instructor Use Only

Did the student complete the job in the time allotted? Yes ___ No ___

If No, which steps were not completed?_____

How would you rate this student's overall performance on this job?_____

5–Excellent, 4–Good, 3–Satisfactory, 2–Unsatisfactory, 1–Poor

Comments: _____

_____

_____

INSTRUCTOR'S SIGNATURE _____

Job 106—Rotate and Balance Tires

After completing this job, you will able to rotate and balance tires.

Procedures

☐ 1. Obtain a vehicle to be used in this job. Your instructor may direct you to perform this job on a shop vehicle.

☐ 2. Gather the tools needed to perform this job. Refer to the tools and materials list at the beginning of the project.

Rotate Tires

☐ 1. Remove all the wheel covers if necessary.

☐ 2. Loosen the lug nuts on all wheels.

☐ 3. Raise the vehicle in a safe manner.

> **Warning**
>
> ⚠ The vehicle must be raised and supported in a safe manner. Always use approved lifts or jacks and jack stands.

☐ 4. Remove all lug nuts.

☐ 5. Rotate the tires. Rotation patterns vary from vehicle to vehicle. Be sure to check the service information for the proper rotation pattern.

☐ 6. Reinstall the lug nuts.

☐ 7. Lower the vehicle.

☐ 8. Tighten the lug nuts to the proper torque.

☐ 9. Reinstall the wheel covers if necessary.

Balance Tires

☐ 1. Remove the tire and wheel from the vehicle.

☐ 2. Remove all wheel weights from the rim.

Static Balance

☐ 1. Install the tire and wheel assembly over the center cone of the bubble balancer.

☐ 2. Observe the position of the bubble.

☐ 3. If the bubble is not centered, place weights on the rim until the bubble is centered.

☐ 4. Once the bubble is centered, affix the weights to the rim.

☐ 5. Remove the tire and wheel assembly from the bubble balancer.

☐ 6. Install the assembly on the vehicle and tighten the lug nuts to the proper torque.

Dynamic Balance

☐ 1. Install the tire on the balance machine, **Figure 106-1**.

☐ 2. Set the tire balance machine to the proper wheel rim diameter and width.

☐ 3. Start the balance machine and determine whether the rim and tire are turning without excessive wobbling. If excessive wobble is noted, the rim may be improperly installed or it may be bent.

Figure 106-1. Tire mounted on a balance machine.

> **Note**
>
> Correct wobble before proceeding. A wobbling tire cannot be balanced. If the rim is bent, it must be replaced.

☐ 4. Note the amount and location of needed weights by reading the information on the tire balancer console.

☐ 5. Install the weights on the rim.

☐ 6. Start the balance machine and observe whether installing the weights corrected the balance.

Is the tire balanced? Yes ___ No ___

If No, remove the weights and repeat steps 3 through 6.

☐ 7. Reinstall the tire and wheel on the vehicle.

☐ 8. Clean the work area and return any equipment to storage.

Did you encounter any problems during this procedure? Yes ___ No ___

If Yes, describe the problems: _____

_____

What did you do to correct the problems? _____

_____

☐ 9. Have your instructor check your work and sign this job sheet.

Performance Evaluation—Instructor Use Only

Did the student complete the job in the time allotted? Yes ___ No ___

If No, which steps were not completed?_____

How would you rate this student's overall performance on this job?_____

5–Excellent, 4–Good, 3–Satisfactory, 2–Unsatisfactory, 1–Poor

Comments: _____

_____

_____

INSTRUCTOR'S SIGNATURE _____

Project

Performing General Brake System Diagnosis and Service

30

Introduction

Almost all technicians perform occasional brake system service. Some technicians become brake specialists and make a career of repairing brake systems. As part of brake service, the technician must diagnose the brake system to isolate braking problems. This process involves visual inspection and the use of various types of test equipment. In Job 107, you will diagnose problems in a vehicle brake system.

Wheel studs do not wear out in service, but they are often damaged when the tires and wheels are removed. Almost all studs are pressed in and can easily be replaced. In Job 108, you will inspect and replace wheel studs.

Wheel bearing service is part of overall brake repair. Tapered roller bearings require periodic cleaning, lubrication, and adjustment. Bearings often wear out and require replacement. Tapered roller bearings are replaced by driving the damaged races from the drum or rotor and installing new bearings and races. Sealed bearing assemblies can also be replaced. In Job 109, you will service wheel bearings.

Parking brakes are installed on the rear wheels and provide a way to hold the vehicle stationary when it is not being driven. They also provide an extra stopping device in the event of an emergency. Most parking brakes use linkage to mechanically apply the existing drum or disc brakes. A few systems are completely separate from the base brakes. They consist of small brake shoes located in a drum that is part of the rear brake rotor. Parking brakes are relatively trouble free, but require occasional inspection, lubrication, and adjustment. You will service parking brake mechanisms and linkage in Job 110.

Project 30 Jobs

- Job 107—Diagnose a Brake System
- Job 108—Inspect and Replace Wheel Studs
- Job 109—Service Wheel Bearings
- Job 110—Service a Parking Brake

Tools and Materials

The following list contains the tools and materials that may be needed to complete the jobs in this project. The items used will depend on the make and model of vehicle being serviced.

- Vehicle in need of service.
- Applicable service information.
- Pressure gauges.
- Brake micrometers.
- Dial indicators.
- Wheel bearing grease.

- Replacement stud.
- Closed cleaning equipment.
- Hand tools as needed.
- Air-operated tools as needed.
- Safety glasses and other protective equipment as needed.

Safety Notice

Before performing this job, review all pertinent safety information in the text and review safety information with your instructor.

Job 107—Diagnose a Brake System

After completing this job, you will be able to diagnose problems in a vehicle's brake system.

Procedures

> **Note**
>
> This procedure addresses problems in the base brake system. For problems in the power brake system, see Job 114. For problems in the anti-lock brake system, see Jobs 121 through 124.

☐ 1. Obtain a vehicle to be used in this job. Your instructor may direct you to perform this job on a shop vehicle.

☐ 2. Gather the tools needed to perform this job. Refer to the tools and materials list at the beginning of the project.

☐ 3. Obtain a detailed description of the brake problem from the vehicle's operator.

Explain the problem as described by the operator:_____

_____

_____

_____

☐ 4. Obtain the proper service literature for the vehicle being diagnosed and determine the type of brake system:

Four-wheel disc brakes: ___

Front disc, rear drum brakes: ___

Four-wheel drum brakes: ___

Is the vehicle equipped with ABS? Yes ___ No ___

☐ 5. Check the fluid level in the master cylinder. See **Figure 107-1**.

Is the level at or near the top? Yes ___ No ___

Figure 107-1. If the master cylinder reservoir is made of translucent plastic, compare the fluid level to markings on the reservoir. If the reservoir is not translucent or there are no markings, the fluid level should be approximately 1/4″ (6 mm) from the top of the reservoir opening.

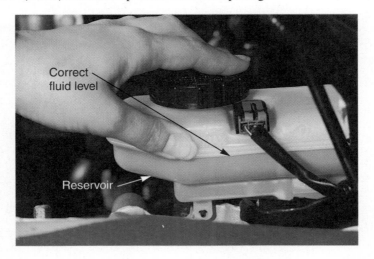

Correct fluid level

Reservoir

☐ 6. If the fluid level in the master cylinder is low, visually inspect the entire system for leaks. Were any leaks found? Yes ___ No ___

Warning

⚠ Do not perform a road test if the vehicle has no brakes or seriously malfunctioning brakes. Always check the brake fluid level before beginning any road test.

☐ 7. Drive the vehicle on a lightly traveled, level road until a speed of 30 mph (48 kph) is reached.

☐ 8. Apply the brakes lightly.

☐ 9. Release the brakes and accelerate to 30 mph (48 kph) again.

☐ 10. Apply the brakes firmly.

☐ 11. Observe whether any of the following occur during braking:
- Pulling to one side: Yes ___ No ___ To which side? _____
- Vibration throughout the vehicle body/seats: Yes ___ No ___
- Brake pedal pulsation: Yes ___ No ___
- Low brake pedal/excessive pedal travel: Yes ___ No ___
- Hard pedal: Yes ___ No ___
- Warning light illumination: Yes ___ No ___
- Noises, such as squealing, grinding, or knocking: Yes ___ No ___
  If yes, describe: _____
  _____
  _____

- Other apparent brake problems: Yes ___ No ___
  If yes, describe: _____
  _____
  _____

☐ 12. Release the brakes and accelerate to 60 mph (97 kph).

☐ 13. Ensure that no vehicles are behind you and apply the brakes hard.

☐ 14. Observe the following during hard braking:
- Slight pedal pulsation: Yes ___ No ___
- Straight stop, with no skidding or noise: (slight chirping is normal): Yes ___ No ___
- Wheel lockup on any wheel: Yes ___ No ___
- Illumination of the ABS light (some vehicles): Yes ___ No ___

☐ 15. Stop the vehicle on a slight incline.

☐ 16. Apply the parking brake.

Does the vehicle remain stationary? Yes ___ No ___

☐ 17. Release the parking brake.

Does the parking brake mechanism release easily, allowing the vehicle to move?
Yes ___ No ___

Job 107—Diagnose a Brake System (continued)

☐ 18. Remove the wheels, as well as the drums and/or calipers. Visually inspect the brake system friction materials.

Pads and/or Shoes

	Worn	Glazed	Cracked	Overheated	Oil or Brake Fluid Soaked
Left front:	___	___	___	___	___
Right front:	___	___	___	___	___
Left rear:	___	___	___	___	___
Right rear:	___	___	___	___	___
Parking brake (if separate):	___	___	___	___	___

Drums and/or Rotors

	Worn	Glazed	Scored	Overheated
Left front:	___	___	___	___
Right front:	___	___	___	___
Left rear:	___	___	___	___
Right rear:	___	___	___	___

☐ 19. Measure the rotors or drums using appropriate micrometers. See **Figure 107-2**.

	Specifications	Actual Reading	Out-of-Round	Taper
Left front:	_____	_____	_____	_____
Action needed:				
Right front:	_____	_____	_____	_____
Action needed:				
Left rear:	_____	_____	_____	_____
Action needed:				
Right rear:	_____	_____	_____	_____
Action needed:				
Parking brake (if separate):	_____	_____	_____	_____
Action needed:				

☐ 20. Use a dial indicator to check the rotors and hubs for warping.

	Variation	Maximum Variation Allowed
Left front:	_____	_____
Action needed:		
Right front:	_____	_____
Action needed:		
Left rear:	_____	_____
Action needed:		
Right rear:	_____	_____
Action needed:		
Parking brake (if separate):	_____	_____
Action needed:		

Figure 107-2. Brake drum diameter can be measured with a drum micrometer. Measure the drum diameter in several locations. Variations in diameter indicate that the drum is out-of-round.

☐ 21. Visually inspect other brake system components for problems.

> **Note**
>
> Not every part listed below will be found on every vehicle. Write NA in the space provided if a part is not used.

Front Brakes—Left

Component	Bent/ Broken	Leaking	Other Defect(s) (Describe)
Caliper:	___	___	_____
Hardware:	___	___	_____
Brake lines/hoses:	___	___	_____
Wheel lugs/studs:	___	___	_____
Bearings/races:	___	___	_____
Seals:	___	___	_____

Front Brakes—Right

Component	Bent/ Broken	Leaking	Other Defect(s) (Describe)
Caliper:	___	___	_____
Hardware:	___	___	_____
Brake lines/hoses:	___	___	_____
Wheel lugs/studs:	___	___	_____
Bearings/races:	___	___	_____
Seals:	___	___	_____

Job 107—Diagnose a Brake System (continued)

Rear Brakes—Left

Component	Bent/ Broken	Leaking	Other Defect(s) (Describe)
Calipers:	___	___	_____
Wheel cylinders:	___	___	_____
Hardware:	___	___	_____
Brake lines/hoses:	___	___	_____
Backing plate:	___	___	_____
Self-adjuster hardware:	___	___	_____
Parking brake cable:	___	___	_____
Parking brake lever:	___	___	_____
Wheel lugs/studs:	___	___	_____
Bearings/races:	___	___	_____
Seals:	___	___	_____

Rear Brakes—Right

Component	Bent/ Broken	Leaking	Other Defect(s) (Describe)
Calipers:	___	___	_____
Wheel cylinders:	___	___	_____
Hardware:	___	___	_____
Brake lines/hoses:	___	___	_____
Backing plate:	___	___	_____
Self-adjuster hardware:	___	___	_____
Park brake cable:	___	___	_____
Park brake lever:	___	___	_____
Wheel lugs/studs:	___	___	_____
Bearings/races:	___	___	_____
Seals:	___	___	_____

Hydraulic System

Component	Stuck	Leaking	Other (Describe)
Master cylinder:	___	___	_____
Metering valve:	___	___	_____
Proportioning valve:	___	___	_____
Combination valve:	___	___	_____
Booster and hoses:	___	___	_____

Brake Pedal

Is the pedal height correct? Yes ___ No ___

Is the pedal travel/free play about 1/8″ (3.1 mm) when the pedal is pressed down by hand until resistance is felt? Yes _____ No _____

Is the linkage adjustment correct? Yes ___ No ___

Are the linkage and hardware in good condition? Yes ___ No ___

Is the brake light switch adjustment correct? Yes ___ No ___

☐ 22. Attach pressure gauges to the vehicle following instructions in the appropriate service literature.

List the gauge connection point(s): _____

_____

☐ 23. Apply the brakes and record pressure readings.

_____ psi (kPa) with gauge connected to _____ .

_____ psi (kPa) with gauge connected to _____ .

_____ psi (kPa) with gauge connected to _____ .

_____ psi (kPa) with gauge connected to _____ .

Do these pressure readings indicate a hydraulic system problem? Yes ___ No ___

☐ 24. Based on the finding in steps 3 through 23, determine needed repairs.

Parts needed:_____

_____

Services needed (such as machining rotors/drums):_____

_____

Note

If your instructor approves, proceed to the other jobs in this manual and make the needed repairs to the brake system.

☐ 25. Recheck system operation.

Do any brake problems exist? Yes ___ No ___

If Yes, return to the beginning of this job and recheck the brake system.

☐ 26. Return the tools and test equipment to storage.

☐ 27. Did you encounter any problems during this procedure? Yes ___ No ___

If Yes, describe the problems: _____

_____

What did you do to correct the problems? _____

_____

☐ 28. Have your instructor check your work and sign this job sheet.

Performance Evaluation—Instructor Use Only

Did the student complete the job in the time allotted? Yes ___ No ___

If No, which steps were not completed?_____

How would you rate this student's overall performance on this job?_____

5–Excellent, 4–Good, 3–Satisfactory, 2–Unsatisfactory, 1–Poor

Comments: _____

_____

_____

INSTRUCTOR'S SIGNATURE _____

Job 108—Inspect and Replace Wheel Studs

After completing this job, you will be able to inspect and replace wheel studs.

Procedures

☐ 1. Obtain a vehicle to be used in this job. Your instructor may direct you to perform this job on a shop vehicle.

☐ 2. Gather the tools needed to perform this job. Refer to the project's tools and materials list.

☐ 3. Raise the vehicle.

> **Warning**
>
> ⚠ The vehicle must be raised and supported in a safe manner. Always use approved lifts or jacks and jack stands.

Inspect Wheel Studs

☐ 1. Remove the wheel on the axle containing the stud(s) to be inspected or replaced.

☐ 2. Check the wheel studs for the following conditions. Check all that apply.
  • Stripped threads: ___
  • Corrosion: ___
  • Broken stud: ___

Replace a Wheel Stud

☐ 1. If necessary, remove the rotor and splash shield to allow the stud to be removed from the rear of the wheel assembly.

☐ 2. Knock out the damaged stud with a large hammer.

  or

  Use a special stud removal tool to press out the damaged stud.

☐ 3. Compare the old and new studs.

> **Note**
>
> 📄 If the studs do not match, obtain the correct stud before proceeding. Stud length, diameter, and thread size must all match. Never attempt to install an incorrect stud.

☐ 4. Place the new stud in position on the hub.

☐ 5. Lightly lubricate the stud splines.

☐ 6. Install washers and a lug nut on the stud as shown in **Figure 108-1**.

☐ 7. Slowly tighten the nut to draw the stud into the hub.

☐ 8. Check that the stud is fully seated in the hub. The back of the wheel stud should be even with the other studs installed in the hub.

☐ 9. Remove the nut and washers from the stud.

☐ 10. Install the wheel.

☐ 11. Clean the work area and return equipment to storage.

Figure 108-1. Arrange washers and a lug nut on the new wheel stud as shown. Tighten the lug nut to draw the new stud into position.

☐ 12. Did you encounter any problems during this procedure? Yes ___ No ___

If Yes, describe the problems: _____

_____

What did you do to correct the problems? _____

_____

☐ 13. Have your instructor check your work and sign this job sheet.

Performance Evaluation—Instructor Use Only

Did the student complete the job in the time allotted? Yes ___ No ___

If No, which steps were not completed?_____

How would you rate this student's overall performance on this job?_____

5–Excellent, 4–Good, 3–Satisfactory, 2–Unsatisfactory, 1–Poor

Comments: _____

_____

_____

INSTRUCTOR'S SIGNATURE _____

Job 109—Service Wheel Bearings

After completing this job, you will be able to service tapered roller and sealed wheel bearings.

Procedures

☐ 1. Obtain a vehicle to be used in this job. Your instructor may direct you to perform this job on a shop vehicle.

☐ 2. Gather the tools needed to perform this job. Refer to the project's tools and materials list.

Diagnose Wheel Bearing Problems

☐ 1. Raise the vehicle.

> **Warning**
> ⚠ The vehicle must be raised and supported in a safe manner. Always use approved lifts or jacks and jack stands.

☐ 2. Shake the suspect wheel from side to side and from top to bottom and note any excessive movement. If the wheel moves excessively in one direction only, the problem is in the suspension or steering systems. If the wheel moves excessively in all directions, the wheel bearing is loose or worn.

Was any looseness detected? Yes ____ No ____

If Yes, describe: _____

☐ 3. Rotate each wheel while listening for noises and checking for roughness. Grinding or scraping noises indicate a dry or damaged bearing. Roughness or binding as the wheel is turned indicates that at least one roller or ball bearing has been damaged.

Was roughness or noise detected? Yes ____ No ____

If Yes, describe: _____

_____

☐ 4. Inform your instructor of your findings. Your instructor may direct you to perform the steps below to further inspect, adjust, or replace the bearings.

Service Tapered Roller Wheel Bearings

☐ 1. Safely raise the vehicle.

☐ 2. Remove the wheel from the vehicle.

☐ 3. On a disc brake system, remove the brake caliper.

☐ 4. Remove the dust cap and then remove the cotter key from the spindle nut.

☐ 5. Remove the spindle nut and washer.

☐ 6. Remove the brake drum or rotor and hub.

☐ 7. If necessary, separate the hub from the drum or rotor.

☐ 8. Remove the outer bearing from the hub.

☐ 9. Remove the grease seal from the inner hub and discard it.

☐ 10. Remove the inner bearing from the hub.

☐ 11. Clean the bearings thoroughly and blow dry.

> **Warning**
>
> ⚠ Do not spin the bearings with air pressure.

☐ 12. Inspect the bearings for wear or damage.

List any signs of bearing damage:_____

_____

Can this bearing be reused? Yes ___ No ___
If No, consult your instructor before proceeding.

☐ 13. Clean the bearing races and the hub interior.

☐ 14. Check the races for wear and damage.

List any signs of race damage: _____

_____

Can this race be reused? Yes ___ No ___
If No, consult your instructor before proceeding.

☐ 15. If a bearing race must be replaced, drive out the old race and install the new race. See **Figure 109-1**.

☐ 16. Place a small amount of grease in the hub cavity.

☐ 17. Grease the bearings by hand or with a packing tool.

☐ 18. Place the inner bearing in the hub and install the new seal.

☐ 19. Clean the spindle.

☐ 20. Reinstall the hub on the spindle.

☐ 21. Install the outer bearing.

☐ 22. Install the washer and spindle nut.

Figure 109-1. Note that the rear of the bearing race is visible through the notches in the rear of the hub. Place the punch or other removal tool on the bearing at the notch and drive it out. Alternate between sides to avoid cocking the race on the hub.

Job 109—Service Wheel Bearings (continued)

☐ 23. Adjust bearing preload:
   a. Tighten the nut to about 100 ft lb to seat all the components.
   b. Back off the nut until it is only finger tight.
   c. Tighten the nut until the proper preload is obtained. Depending on the manufacturer, preload is measured by determining the endplay of the hub or by measuring the amount of force needed to turn the hub.

☐ 24. Install a new cotter pin.

☐ 25. Reinstall the caliper, if necessary.

☐ 26. Reinstall the wheel.

Remove and Replace a Sealed Wheel Bearing Assembly

☐ 1. Raise the vehicle and support it in a safe manner.

> **Warning**
> ⚠ The vehicle must be raised and supported in a safe manner. Always use approved lifts or jacks and jack stands.

☐ 2. Remove the wheel at the bearing assembly to be replaced.

☐ 3. If the vehicle has disc brakes, remove the caliper and rotor.

   or

   If the vehicle has drum brakes, remove the wheel bearing nut and remove the drum.

> **Note**
> If the vehicle has a wheel speed sensor, remove it now to prevent damage. Some speed sensors are attached to the bearing assembly and cannot be removed.

☐ 4. If the speed sensor is attached to the bearing assembly remove the electrical connector.

☐ 5. Remove the fasteners holding the bearing assembly to the axle and remove the bearing.

☐ 6. Compare the old and new bearing assemblies to ensure that the new bearing is the correct replacement.

☐ 7. Place the new bearing in position on the axle and install the fasteners. See **Figure 109-2**.

☐ 8. If necessary, reinstall the speed sensor and/or attach the speed sensor electrical connector.

☐ 9. Reinstall the drum or rotor and caliper as applicable.

> **Caution**
> ◇ Use a new retaining nut if required by the manufacturer.

☐ 10. Reinstall the wheel.

☐ 11. Road test the vehicle to ensure that the bearing assembly replacement has been successful.

☐ 12. Clean the work area and return any equipment to storage.

Figure 109-2. The new sealed wheel bearing is positioned on the vehicle before the fasteners are installed.

□ 13. Did you encounter any problems during this procedure? Yes ___ No ___

If Yes, describe the problems: _____

_____

What did you do to correct the problems? _____

_____

□ 14. Have your instructor check your work and sign this job sheet.

Performance Evaluation—Instructor Use Only

Did the student complete the job in the time allotted? Yes ___ No ___

If No, which steps were not completed?_____

How would you rate this student's overall performance on this job?_____

5–Excellent, 4–Good, 3–Satisfactory, 2–Unsatisfactory, 1–Poor

Comments: _____

_____

_____

INSTRUCTOR'S SIGNATURE _____

Job 110—Service a Parking Brake

After completing this job, you will be able to service parking brakes.

Procedures

☐ 1. Obtain a vehicle to be used in this job. Your instructor may direct you to perform this job on a shop vehicle.

☐ 2. Gather the tools needed to perform this job. Refer to the project's tools and materials list.

Test Parking Brake Operation

☐ 1. Start the vehicle and move it to a flat area with enough room to allow it to move forward.

☐ 2. If the vehicle has an automatic transmission, place it in Park. If the vehicle has a manual transmission, place it in Neutral.

☐ 3. Apply the parking brake.

☐ 4. Remove your hand or foot from the parking brake lever and observe whether the lever remains in the applied position.

Does the lever remain applied? Yes ___ No ___

If Yes, proceed to the next step.

If No, inspect and repair the lever before moving to the next step.

☐ 5. Observe the parking brake warning light.

Is the warning light on? Yes ___ No ___

☐ 6. Place the transmission in drive (automatic) or place the shift lever in first gear and slowly release the clutch (manual). Do not apply the service brake pedal.

> **Note**
> If the vehicle has an automatic parking brake release, hold the parking brake in the applied position with your foot.

☐ 7. Observe vehicle movement.

Automatic transmission: Does the vehicle move? Yes ___ No ___

Manual transmission: Does the engine die when the clutch is released ? Yes ___ No ___

☐ 8. Release the parking brake and observe vehicle movement.

Automatic transmission: Does the vehicle move? Yes ___ No ___

Manual transmission: Does the engine die when the clutch is released ? Yes ___ No ___

☐ 9. Observe the warning light. Is the warning light on? Yes ___ No ___

☐ 10. Based on the above tests, does the parking brake require service? Yes ___ No ___

Inspect Parking Brake Components

> **Note**
> This procedure assumes the service brakes have been checked or are known to be in good condition and properly adjusted. See Jobs 116 and 118 for service brake repair and adjustment procedures.

☐   1.   In the passenger compartment, observe the condition of the parking brake lever components:

	Worn	Broken	Sticking	Other (Describe)
Lever assembly:	___	___	___	_____
Release mechanism:	___	___	___	_____
Cable end and fastener:	___	___	___	_____

☐   2.   Raise the vehicle.

Warning

⚠   The vehicle must be raised and supported in a safe manner. Always use approved lifts or jacks and jack stands.

☐   3.   Check the condition of the parking brake linkage:

	Worn	Broken	Sticking	Other (Describe)
Equalizer/multiplier levers:	___	___	___	_____
Cables/sheaths:	___	___	___	_____
Adjuster:	___	___	___	_____

Adjust Parking Brake Linkage

Manually Adjusted Linkage

☐   1.   Engage the hand or foot lever two or three notches.

☐   2.   Safely raise the vehicle.

Warning

⚠   The vehicle must be raised and supported in a safe manner. Always use approved lifts or jacks and jack stands.

☐   3.   If a locknut is used to hold the equalizer adjuster, loosen it now.

☐   4.   Turn the adjusting nut until there is a slight drag at the rear wheels.

       Number of turns needed: _____

☐   5.   Loosen the nut until the wheels turn freely. See **Figure 110-1**.

☐   6.   Tighten the locknut if applicable.

☐   7.   Firmly apply the parking brake and release.

☐   8.   Repeat steps 3 through 9 in the *Test Parking Brake Operation* section to ensure that the parking brake operates properly.

Self-Adjusting Linkage

☐   1.   Make sure the parking brake is fully released.

☐   2.   Release the parking brake adjuster lock clip. Most lock clips are pulled away from the cable to release them. The tensioner spring will take up any slack in the cable.

☐   3.   Reinstall the brake adjuster lock clip. If the clip does not line up with the nearest groove in the cable, move the cable slightly so the clip aligns with the nearest groove.

☐   4.   Apply and release the parking brake.

☐   5.   Test the parking brakes to ensure they operate properly.

Job 110—Service a Parking Brake (continued)

Figure 110-1. To adjust the parking brake, turn the adjusting nut until there is a slight drag at the rear wheels and then loosen the nut until the wheels turn freely.

Adjust Separate Drum-Type Parking Brakes

> **Note**
>
> This procedure assumes the brakes are in good condition, with sufficient lining.

☐ 1. Remove the wheels from the rear axle.

☐ 2. Remove the calipers and rotors from the rear axle.

☐ 3. Using a shoe-to-drum gauge, measure the drum diameter of one rotor.

Drum diameter: _____

☐ 4. Measure the shoe diameter and adjust the shoes to the proper clearance.

☐ 5. Reinstall the drum and recheck the adjustment. You should feel a slight drag when the drum is rotated through one revolution.

☐ 6. Repeat steps 3 through 5 on the other drum and shoes.

☐ 7. Apply and release the parking brake.

☐ 8. Test the parking brakes to ensure they operate properly.

Adjust a Parking Brake in a Rear Disc Brake Assembly

> **Note**
>
> This procedure assumes the brakes are in good condition, with sufficient lining and no caliper leaks.

☐ 1. Safely raise the vehicle.

> **Warning**
>
> ⚠ The vehicle must be raised and supported in a safe manner. Always use approved lifts or jacks and jack stands.

☐ 2. Note the amount of drag present on the brakes.

☐ 3. Remove the rear wheels.

☐ 4. Check the clearance between the inner brake pad and the rotor on one brake assembly.

   Is there a noticeable space between the inner pad and the rotor? Yes ___ No ___

☐ 5. If needed, adjust the inner pad-to-rotor clearance. Consult the proper service literature for exact adjustment procedures.

☐ 6. Repeat steps 4 and 5 for the other brake assembly.

☐ 7. Apply and release the parking brake.

☐ 8. Test the parking brakes to ensure they operate properly.

Retract an Integrated Parking Brake Caliper Piston

> **Caution**
>
> ◇ Special tools may be needed to perform this job.

> **Note**
>
> 📄 Turning the disc brake apply piston retracts most disc-type parking brakes. Consult the service information to determine the exact retraction procedure.

☐ 1. If necessary, remove the caliper.

☐ 2. Remove the disc brake pads from the caliper and remove the brake pads.

☐ 3. Put firm pressure on the piston face. This is accomplished with a C-clamp or by prying on the piston face with a pry bar.

☐ 4. Rotate the piston into the caliper body. This can be done by using a special tool, operating the use apply lever at the back of the caliper assembly, or turning the piston with a large pair of pliers. Maintain pressure on the piston face as the piston is rotated.

> **Note**
>
> 📄 Most pistons are rotated clockwise to move them back into the caliper body. A few pistons rotate counterclockwise. Always consult the proper service literature for the proper turning direction. Tighten the C-clamp after every rotation of the piston in order to maintain pressure on the piston.

> **Caution**
>
> ◇ Do not twist or cut the rubber piston boot.

☐ 5. Once the piston is fully retracted, reinstall the disc brake pads if necessary.

☐ 6. Reinstall the caliper if necessary.

Clean, Lubricate, and Replace Parking Brake Linkage

Clean and Lubricate Linkage

☐ 1. Safely raise the vehicle.

> **Warning**
>
> ⚠ The vehicle must be raised and supported in a safe manner. Always use approved lifts or jacks and jack stands.

☐ 2. Remove the cable ends if possible.

☐ 3. Spray penetrating oil on both ends of the cable where it enters the sheath.

☐ 4. Allow the penetrating oil to soak for 5 to 10 minutes.

☐ 5. Clean dirt and corrosion from the cable.

☐ 6. Work the cable back and forth in the sheath, applying more penetrating oil as needed.

☐ 7. Check cable operation. If the cable cannot be freed up, it must be replaced.

☐ 8. Apply and release the parking brake.

☐ 9. Test the parking brakes to ensure they operate properly. Readjust the parking brake if necessary.

Replace a Parking Brake Cable

☐ 1. Safely raise the vehicle.

> **Warning**
>
> ⚠ The vehicle must be raised and supported in a safe manner. Always use approved lifts or jacks and jack stands.

☐ 2. Back off the cable adjuster to remove tension from the linkage.

☐ 3. Remove the cable ends.

☐ 4. Loosen and remove any clips or brackets holding the cable to the vehicle.

☐ 5. Compare the old and new cable. Make sure the cable ends and any clips on the new cable exactly match the original cable.

☐ 6. Install the new cable, routing it exactly as the old cable was routed. Be sure the new cable does not contact any moving parts or the exhaust system.

☐ 7. Install the cable ends.

☐ 8. Readjust the parking brake.

☐ 9. Apply and release the parking brake.

☐ 10. Test the parking brakes to ensure they operate properly.

Replace a Parking Brake Lever

☐ 1. Safely raise the vehicle.

> **Warning**
>
> ⚠ The vehicle must be raised and supported in a safe manner. Always use approved lifts or jacks and jack stands.

☐ 2. Back off the cable adjuster to remove tension from the lever.

☐ 3. Remove the cable end(s) from the lever.

☐ 4. Remove any fasteners and remove the lever from the vehicle.

☐ 5. Compare the old and new levers. Make sure they match exactly.

☐ 6. Install the cable(s) on the lever.

☐ 7. Readjust the parking brake.

☐ 8. Apply and release the parking brake.

☐ 9. Test the parking brakes to ensure they operate properly.

☐ 10. Clean the work area and return equipment to storage.

☐ 11. Did you encounter any problems during this procedure? Yes ___ No ___

If Yes, describe the problems: _____

_____

What did you do to correct the problems? _____

_____

☐ 12. Have your instructor check your work and sign this job sheet.

Performance Evaluation—Instructor Use Only

Did the student complete the job in the time allotted? Yes ___ No ___
If No, which steps were not completed?_____
How would you rate this student's overall performance on this job?_____
5–Excellent, 4–Good, 3–Satisfactory, 2–Unsatisfactory, 1–Poor
Comments: _____

_____

_____

INSTRUCTOR'S SIGNATURE _____

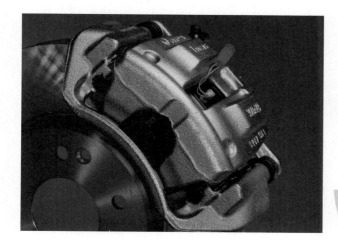

Project

31

Servicing a Brake Hydraulic System

Introduction

Hydraulic lines and hoses deliver hydraulic pressure from the master cylinder to the calipers or wheel cylinders. Valves control the application of this pressure and sometimes activate warning lights. A technician performing brake service must diagnose and service these devices. Common problems are sticking valves, leaks, and hoses that have swelled closed. In Jobs 111 and 112, you will inspect, diagnose, and replace hydraulic valves, lines, and hoses.

Master cylinders and power assists convert the force on the brake pedal into hydraulic pressure. Many brake problems originate in the master cylinder or power assists. Although these units are commonly replaced instead of being overhauled, the technician must know how to diagnose them and inspect their component parts. In Jobs 113 and 114, you will diagnose master cylinder and power assist problems.

Air must be removed from the brake hydraulic system. This is the purpose of bleeding. Bleeding can be done manually or with pressurized bleeding equipment. Many manufacturers recommend occasional flushing of the brake system to remove moisture and overheated brake fluid. The technician performing brake service must be able to carry out these jobs. The technician must also know how to select the proper brake fluid and how to store brake fluid to keep out moisture and other contaminants. In Job 115, you will bleed and flush the brake hydraulic system and properly handle brake fluids.

Project 31 Jobs

- Job 111—Diagnose Brake Hydraulic System Problems
- Job 112—Replace Brake Lines, Hoses, and Valves
- Job 113—Service a Master Cylinder
- Job 114—Service a Power Assist System
- Job 115—Bleed a Brake System

Tools and Materials

The following list contains the tools and materials that may be needed to complete the jobs in this project. The items used will depend on the make and model of vehicle being serviced.

- Vehicle in need of brake service.
- Applicable service information.
- 2500 psi (17,000 kPa) brake pressure gauges (recommend 4000 psi/ 27,000 kPa on ABS systems).
- Brake line bending and flaring tools.
- Line wrenches.
- Brake tools.
- Pressure bleeding equipment.
- Power steering fluid.
- Length of hose (preferably clear).
- Clear jar or bottle.
- Correct type of brake fluid.
- Hand tools as needed.
- Air-operated tools as needed.
- Safety glasses and other protective equipment as needed.

Safety Notice

Before performing this job, review all pertinent safety information in the text, and review safety information with your instructor.

Job 111—Diagnose Brake Hydraulic System Problems

After completing this job, you will be able to inspect and diagnose hydraulic valves, lines, and hoses in base brake systems.

> **Note**
> To service anti-lock brake system valves and lines, see Project 34.

Procedures

☐ 1. Obtain a vehicle to be used in this job. Your instructor may direct you to perform this job on a shop vehicle.

☐ 2. Gather the tools needed to perform this job. Refer to the project's tools and materials list.

Diagnose Line and Hose Problems

☐ 1. Check and adjust tire pressures to ensure that uneven or incorrect pressures do not cause false brake symptoms.

☐ 2. Test drive the vehicle in an area with low traffic density and check for the following problems:
  - Lack or braking (stopping) power.
  - Excessive pedal effort.
  - Pulling to one side as the brakes are applied.
  - Brakes dragging when the brake pedal is not applied.

> **Warning**
> Be sure to select an empty section of road for the test drive. Wait until all other traffic has cleared before beginning the test drive.

Were any problems noted? Yes ___ No ___

If Yes, describe: _____

_____

If Yes, go to step 3.

If No, consult your instructor about how to proceed with this job.

> **Note**
> Hybrid vehicles are equipped with a regenerative braking system. When the brake pedal is depressed, the regenerative braking system switches the vehicle's motor-generator to generation mode. The loading of the generator slows the vehicle, and the friction brake system provides additional stopping torque as needed. A pedal feel emulator, or pedal stroke simulator, is used to provide a consistent pedal feel as braking transitions between regenerative and hydraulic. When the ABS or TCS system is activated, the regenerative braking system is deactivated to allow braking at individual wheels as needed. In a hybrid vehicle, a complaint regarding inconsistent pedal feel may be related to the complex interactions of these systems.

☐ 3. Return to the shop and raise the vehicle.

> ### Warning
> ⚠ The vehicle must be raised and supported in a safe manner. Always use approved lifts or jacks and jack stands.

☐ 4. Inspect the brake system, steering/suspension system, and tires to ensure that any problems noted in step 2 were not caused by problems in the brake friction components, hardware, power booster, or outside factors such as mismatched or defective tires, defective suspension parts, or incorrect wheel alignment.

Were any problems noted? Yes ___ No ___

If Yes, describe: _____

_____

If No, go to step 5.

If Yes, consult your instructor about how to proceed.

☐ 5. Visually inspect lines and hoses for leaks. Check the lines kinks or flattened areas. Check hoses for cracks and bulges, **Figure 111-1**.

Were any defects found? Yes ___ No ___

If Yes, list the defects: _____

_____

☐ 6. Check all junction blocks for leaks, as well as loose fittings or brackets.

Were any defects found? Yes ___ No ___

If Yes, list the defects: _____

_____

> ### Note
> 📄 Steps 7 through 9 should only be performed if other checks indicate that the hose is restricted. Your instructor may have you perform this procedure on a hose that has already been removed from the vehicle.

Figure 111-1. Cracked brake lines, such as the one shown here, should be replaced.

Job 111—Diagnose Brake Hydraulic System Problems (continued)

☐ 7. Remove suspect hose(s).

☐ 8. Attach a short length of vacuum hose or other hose to either end of the brake hose.

☐ 9. Attempt to blow through the hose.

   Is a restriction felt? Yes ___ No ___

   If Yes, the hose is restricted and will cause pulling.

Diagnose Hydraulic Valve Problems

Pressure Differential Valve

☐ 1. Attach a hose to one bleeder screw.

☐ 2. Have an assistant gently press and hold the brake pedal.

☐ 3. Loosen the bleeder screw and allow fluid to exit. Tighten the bleeder screw before having the assistant release the pedal.

   Does the dashboard warning light come on? Yes ___ No ___

   If No, the pressure differential valve is defective or the dashboard light or fuse is defective. Replace the valve after making sure that the problem is not in the electrical components. See Job 138 for warning light checking procedures.

> **Note**
> Newer pressure differential valves will automatically re-center when the bleeder screw is closed. If the valve is installed on an older vehicle, consult the appropriate service literature to determine what steps must be taken to re-center the valve.

Metering Valve

☐ 1. Loosen one *front* brake bleeder screw.

☐ 2. Have an assistant gently press on the brake pedal as you observe the bleeder screw.

   Does fluid exit from the bleeder as soon as the pedal is pressed? Yes ___ No ___

   If Yes, the metering valve is stuck open and should be replaced.

   If No, proceed to step 3.

☐ 3. Have your assistant continue pressing on the brake pedal with sufficient force.

   Does fluid exit the bleeder at some point? Yes ___ No ___

   If No, the metering valve is stuck closed and should be replaced.

☐ 4. Tighten the bleeder screw before having the assistant release the pedal.

Proportioning Valve

> **Note**
> Two pressure gauges are needed to test a proportioning valve.

☐ 1. Attach pressure gauges so that one gauge measures hydraulic pressure entering the proportioning valve and the other gauge measures pressure leaving the valve.

☐ 2. Press on the brake pedal while observing the gauges.

Does the pressure leaving the valve stop rising after the cutoff pressure is reached?
Yes ___ No ___

If No, the proportioning valve is defective and should be replaced.

☐ 3. If the vehicle uses two proportioning valves, repeat steps 1 and 2 on the other valve.

Is the other valve defective? Yes ___ No ___

Combination Valve

> **Note**
>
> The combination valve is made up of two or three of the valves previously discussed. After determining which valve needs to be tested, you can follow the procedures given previously for testing specific valve types.

☐ 1. Determine which valves are contained in the combination valve on the vehicle being serviced.

Valves contained in the combination valve:

1._____

2._____

3._____

☐ 2. Determine which valves will be diagnosed.

Valve(s) to be diagnosed:

1._____

2._____

3._____

☐ 3. Check valve operation according to the type of valve being diagnosed.

Operation(s) performed: _____

_____

Results: _____

_____

Adjust a Height-Compensating Proportioning Valve

> **Note**
>
> Make sure the proportioning valve is not internally defective before adjusting the valve. See the previous section for proportioning valve pressure checking procedures.

☐ 1. Inspect proportioning valve linkage and other parts for damage.

☐ 2. Place the vehicle on a level surface, such as a flat concrete floor or an alignment rack.

> **Caution**
>
> Do not raise the vehicle on a frame lift. This will cause the suspension to drop and affect the proportioning valve adjustment.

Job 111—Diagnose Brake Hydraulic System Problems (continued)

☐ 3. Adjust the proportioning valve for normal operation. There are many methods of adjusting the proportioning valve, depending on the manufacturer. Consult the proper service literature for adjusting procedures.

Adjusting device:_____

Adjustment method: _____

Was the proportioning valve properly adjusted? Yes ____ No ____

☐ 4. Clean the work area and return any equipment to storage.

☐ 5. Did you encounter any problems during this procedure? Yes ____ No ____

If Yes, describe the problems: _____

_____

What did you do to correct the problems? _____

_____

☐ 6. Have your instructor check your work and sign this job sheet.

Performance Evaluation—Instructor Use Only

Did the student complete the job in the time allotted? Yes ____ No ____

If No, which steps were not completed?_____

How would you rate this student's overall performance on this job?_____

5–Excellent, 4–Good, 3–Satisfactory, 2–Unsatisfactory, 1–Poor

Comments: _____

_____

_____

INSTRUCTOR'S SIGNATURE _____

Notes

Name _____ Date_____

Instructor _____ Period _____

Job 112—Replace Brake Lines, Hoses, and Valves

After completing this job, you will be able to replace hydraulic valves, lines, and hoses in base brake systems.

> **Note**
>
> To service anti-lock brake system valves and lines, see Project 34.

Procedures

☐ 1. Obtain a vehicle to be used in this job. Your instructor may direct you to perform this job on a shop vehicle.

☐ 2. Gather the tools needed to perform this job. Refer to the project's tools and materials list.

Replace a Line

☐ 1. Locate the defective line on the vehicle.

☐ 2. Loosen and remove the line fittings.

> **Caution**
>
> Use line wrenches to prevent fitting damage.

☐ 3. Remove any fasteners, brackets, or hose supports holding the line to the vehicle's body.

☐ 4. Remove the line from the vehicle.

☐ 5. If the replacement line is already bent to the proper shape, compare the length, diameter and fittings of the old and new lines, then go to step 7. If the replacement line must be bent to shape, go to step 6.

> **Note**
>
> Replacement lines are usually flared. Compare the flares to ensure that they are compatible with the attaching fittings. If the lines are not flared, flare them as explained later in this job.

☐ 6. If the line must be bent to the proper shape, take the following steps:
  a. Make sure the replacement line will be the proper length when finished. Using a tape measure, trace the old line along its entire length to obtain an overall length; then match the new line to this length. The replacement line can be slightly longer than the original line, but it cannot be shorter.
  b. Compare the diameter and fittings of the new line to those of the old line. If they do not match, do not attempt to use the new line.
  c. Using the old line as a guide, bend the new line to the proper shape.

> **Caution**
>
> Always use a tubing bender to form brake lines. Do not attempt to bend lines by hand; it will kink the lines.

☐ 7. Install the new line on the vehicle.

☐ 8. Install any line fasteners, brackets, or supports.

☐ 9. Install and tighten the line fittings.

☐ 10. Bleed the brake system. See Job 115.

☐ 11. Road test the vehicle.

Double Flare

> ### Warning
>
> ⚠ Always double flare or ISO flare brake lines.

☐ 1. Cut the line to the proper length using a tubing cutter. Do not cut the tubing with a hacksaw.

> ### Caution
>
> ◇ Make sure that the cut is at a 90° angle to the line.

☐ 2. Remove burrs from the inside of the line with a reamer. Most tubing cutters are equipped with a reamer blade.

☐ 3. Lightly chamfer the outside of the line with a file or a grinding wheel.

☐ 4. Install the correct fitting nut on the line.

☐ 5. Insert the line in the proper size opening in the flaring bar. The line should extend out the proper distance as measured by placing the double flare adapter on the flaring bar as shown in **Figure 112-1**.

☐ 6. Tighten the flaring bar around the line.

☐ 7. Insert the end of the double flare adapter over the line.

☐ 8. Place the flaring tool over the adapter and tighten it against the adapter. See **Figure 112-2**.

Figure 112-1. Be sure that the line to be flared extends out the proper distance from the flaring bar.

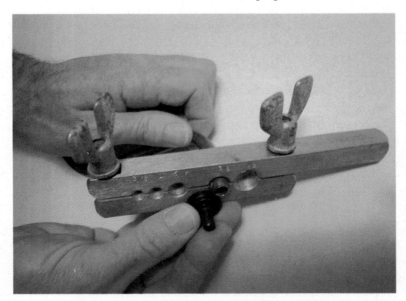

Job 112—Replace Brake Lines, Hoses, and Valves (continued)

Figure 112-2. Use the flaring tool and adapter to make the first flare.

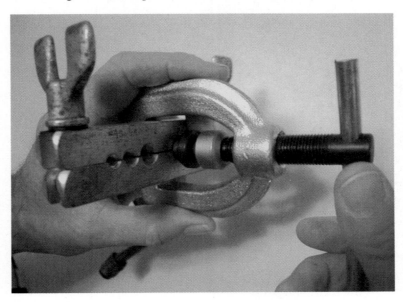

☐   9.   Continue tightening the flaring tool until the adapter is flush against the flaring bar.

☐ 10. Remove the flaring tool and adapter and inspect the line.

            Is the end of the line bell shaped as in **Figure 112-3**? Yes ___ No ___

            If No, repeat steps 1 through 10. If Yes, proceed to step 11.

☐ 11. Reinstall the flaring tool over the flaring bar.

☐ 12. Tighten the flaring tool until the bell shaped part of the line folds in on itself as in **Figure 112-4**.

Figure 112-3. After the first part of the flaring operation, the end of the line should be bell shaped. If you have any doubt as to the quality of this first flare, remove the line from the flaring bar and closely inspect it.

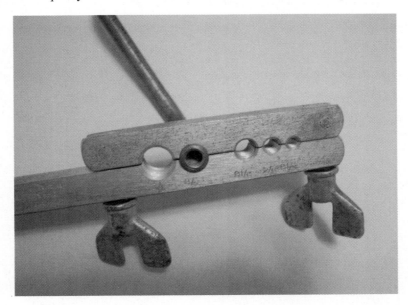

Figure 112-4. In the second part of the flaring process, the flare is folded in on itself. This process is similar to making a single flare.

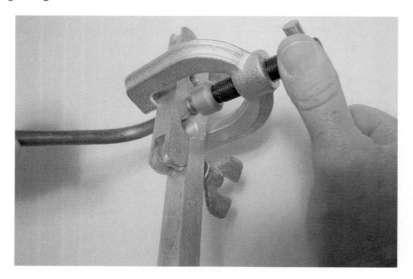

13. Remove the flaring tool and inspect the flare. The flare should look like the flare in **Figure 112-5**.

    Is the flare even all around? Yes ___ No ___

    Is the flare free from cracks and splits? Yes ___ No ___

    If the answer to both of the above questions is Yes, install the line.

    If the answer to either question is No, repeat steps 1 through 13.

ISO Flare

1. Cut the line to the proper length using a tubing cutter. Do not cut the line with a hacksaw.

 Caution

   Make sure that the cut is at a 90° angle to the line.

2. Remove burrs from the inside of the line with a reamer. Most tubing cutters are equipped with a reamer blade.

3. Install the correct fitting nut on the line.

Figure 112-5. If the flare is even all around and has no cracks and splits, the flaring process is complete.

Job 112—Replace Brake Lines, Hoses, and Valves (continued)

☐ 4. Place the line in the proper size forming tool collet and tighten the collet. Make sure the line extends about 3/4″ (19 mm) from the collet.

☐ 5. Turn the forcing screw until it bottoms out in the collet. Do not continue tightening once the screw bottoms out.

☐ 6. Disassemble the collet and make sure the flare is properly formed.

Is the flare properly formed? Yes ___ No ___

If Yes, install the line.

If No, repeat steps 1 through 6.

Replace a Brake Hose

☐ 1. Locate the defective hose on the vehicle.

☐ 2. Loosen and remove the hose fittings.

> **Caution**
> ⚠ Use line wrenches to avoid fitting damage.

☐ 3. Remove any fasteners, brackets, or hose supports holding the hose to the vehicle body.

☐ 4. Remove the hose from the vehicle.

☐ 5. Compare the length and fittings of the old and new hoses.

☐ 6. Install the new hose on the vehicle.

☐ 7. Install and tighten the hose fittings.

☐ 8. Install any hose fasteners, brackets, or hose supports.

☐ 9. Bleed the brake system. See Job 115.

☐ 10. Road test the vehicle.

Replace a Valve

☐ 1. Locate the defective valve on the vehicle.

Valve to be changed: _____

Valve location: _____

☐ 2. Remove any electrical connectors.

☐ 3. Loosen and remove all hydraulic lines.

☐ 4. Remove any fastener brackets.

☐ 5. Remove the valve from the vehicle.

☐ 6. Compare the old and new valves.

☐ 7. Install the new valve and fasteners.

☐ 8. Install and tighten all lines.

☐ 9. Install any electrical connectors.

☐ 10. Bleed the brake system.

☐ 11. Road test the vehicle.

☐ 12. Return tools and test equipment to storage.

☐ 13. Did you encounter any problems during this procedure? Yes ___ No ___

If Yes, describe the problems: _____

_____

What did you do to correct the problems? _____

_____

☐ 14. Have your instructor check your work and sign this job sheet.

Performance Evaluation—Instructor Use Only

Did the student complete the job in the time allotted? Yes ___ No ___
If No, which steps were not completed?_____
How would you rate this student's overall performance on this job?_____
5–Excellent, 4–Good, 3–Satisfactory, 2–Unsatisfactory, 1–Poor
Comments: _____

_____

_____

INSTRUCTOR'S SIGNATURE _____

Job 113—Service a Master Cylinder

After completing this job, you will be able to replace a master cylinder.

Procedures

☐ 1. Obtain a vehicle to be used in this job. Your instructor may direct you to perform this job on a shop vehicle.

☐ 2. Gather the tools needed to perform this job. Refer to the tools and materials list at the beginning of the project.

Remove a Master Cylinder

☐ 1. Place a drip pan under the master cylinder.

☐ 2. Remove any electrical connectors.

☐ 3. Loosen and remove all hydraulic lines.

☐ 4. Cap the hydraulic lines to keep out dirt and moisture.

☐ 5. Remove any fastener brackets.

☐ 6. Remove the nuts holding the master cylinder to the firewall or power booster.

☐ 7. Remove the master cylinder from the vehicle.

Bench Bleed a Master Cylinder

☐ 1. Lightly clamp the master cylinder in a vise. Do not overtighten the vise, or it may damage the master cylinder.

☐ 2. Remove the master cylinder reservoir cover.

☐ 3. Attach clear plastic tubing lines to the master cylinder outlet ports and run them to the reservoir as shown in **Figure 113-1**.

☐ 4. Add the proper type of brake fluid to the reservoir. The required fluid type is usually stamped on the reservoir cover. Do not overfill the reservoir.

Type of brake fluid used: DOT #_____

Figure 113-1. The setup for bench bleeding a master cylinder is shown here.

☐ 5. Push the master cylinder piston inward and release it. Repeat this several times. Fluid should begin to appear in the plastic lines.

Does fluid appear in the lines? Yes ___ No ___

☐ 6. Continue to push and release the master cylinder piston until no air bubbles are present in the fluid moving through the plastic lines.

☐ 7. Top off the brake fluid in the reservoir as needed.

☐ 8. Remove the lines and cap the master cylinder outlet ports.

☐ 9. Reinstall the reservoir cover.

Install a Master Cylinder

☐ 1. Place the master cylinder in position on the vehicle.

☐ 2. Loosely install the nuts holding the master cylinder to the firewall or power booster.

☐ 3. Install brackets if necessary.

☐ 4. Install and tighten the hydraulic lines.

☐ 5. Tighten the nuts holding the master cylinder to the firewall or booster.

☐ 6. Install any electrical connectors.

> **Note**
>
> If a pressure bleeder is available, substitute the bleeder instructions for steps 7 through 12.

☐ 7. Have an assistant pump the brakes several times.

☐ 8. Instruct the assistant to hold the pedal in the applied position.

☐ 9. Use a line wrench to loosen one hydraulic line at the master cylinder. Allow air to exit the fitting.

> **Note**
>
> If the master cylinder has bleeder screws, use them to bleed the master cylinder instead of opening the lines.

☐ 10. When only fluid exits from the line, or when the assistant indicates that the pedal has reached the floor, tighten the fitting.

☐ 11. Closely monitor the fluid level in the reservoir, adding fluid as necessary.

☐ 12. Repeat steps 7 through 11 on all hydraulic lines until only fluid exits from the fittings.

☐ 13. Clean the work area and return any equipment to storage.

☐ 14. Did you encounter any problems during this procedure? Yes ___ No ___

If Yes, describe the problems: _____

_____

What did you do to correct the problems? _____

_____

☐ 15. Have your instructor check your work and sign this job sheet.

Job 113—Service a Master Cylinder (continued)

Performance Evaluation—Instructor Use Only

Did the student complete the job in the time allotted? Yes ___ No ___

If No, which steps were not completed?_____

How would you rate this student's overall performance on this job?_____

5–Excellent, 4–Good, 3–Satisfactory, 2–Unsatisfactory, 1–Poor

Comments: _____

_____

_____

INSTRUCTOR'S SIGNATURE _____

Notes

Job 114—Service a Power Assist System

After completing this job, you will be able to diagnose problems in the power assist.

Procedures

1. Obtain a vehicle to be used in this job. Your instructor may direct you to perform this job on a shop vehicle.

2. Gather the tools needed to perform this job. Refer to the tools and materials list at the beginning of the project.

Check Power Booster: Pedal Travel Test

1. With the engine off, pump the brake pedal 10 times if the vehicle has a vacuum power booster and 20 times if the vehicle uses a Hydro-Boost system.

2. Firmly apply the brake pedal.

3. Start the engine.

> **Note**
>
> If the vehicle uses an auxiliary vacuum pump, simply turn the ignition switch to the *on* position. It is not necessary to start the engine.

Does the brake pedal drop an additional amount when the engine starts? Yes ___ No ___
If Yes, the system is operating properly.

If No, check the following:

Vacuum Power Booster
- Engine condition. ___
- Vacuum hose, **Figure 114-1**. ___
- Booster assembly. ___
- Auxiliary vacuum pump (if used). ___

Figure 114-1. When troubleshooting a vacuum booster, be sure to check the condition of the vacuum lines.

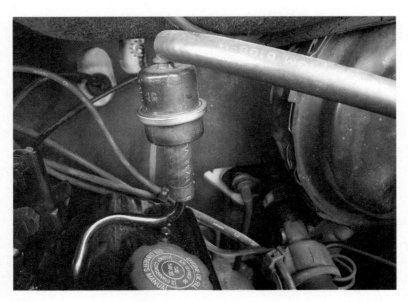

Hydro-Boost System
- Power steering pump and belt. ___
- Pump-to-booster hoses. ___
- Control valve assembly. ___
- Accumulator. ___

Measure and Adjust Booster Push Rod Length

☐ 1. Obtain needed push rod measuring gauge.

☐ 2. Remove the master cylinder from the vacuum or hydraulic booster.

☐ 3. Place the gauge over push rod. Does the gauge indicate the proper push rod length?
Yes ___ No ___

☐ 4. If the answer to the question in step 3 is No, use the proper wrenches to adjust push rod length.

☐ 5. Use the gauge to recheck push rod length.
Is the push rod length now correct? Yes ___ No ___
If No, repeat step 4 as necessary until reading is correct.

☐ 6. Reinstall the master cylinder and check brake operation.

Diagnose a Vacuum Power Booster

☐ 1. Start the engine if it is not already running.

☐ 2. Remove the vacuum line and check valve from the booster.

☐ 3. Place your thumb over the check valve.
Is vacuum present? Yes ___ No ___

☐ 4. Remove the check valve from the vacuum line.

☐ 5. Attach a vacuum gauge to the inlet hose of the vacuum booster and measure the engine vacuum at idle.
Engine vacuum at idle:_____

☐ 6. Reinstall the check valve in the vacuum line and reinstall the valve in the booster.

☐ 7. Stop the engine after allowing it to run for about 60 seconds to establish a vacuum in the booster.

☐ 8. Allow the vehicle to sit for about 5 minutes.

☐ 9. Apply the brake.
Is there the sound of escaping air when the pedal is depressed? Yes ___ No ___
If Yes, the power booster is holding vacuum, go to step 10.
If No, go to step 11.

Note

As part of step 9, you should also feel power assist for at least 3 brake pedal applications. If the pedal is hard (no power assist), this also indicates that the booster is leaking.

☐ 10. Remove the check valve and attempt to blow through it from the hose side.
Does the check valve allow air to pass through? Yes ___ No ___
If Yes, the valve is defective and should be replaced.

☐ 11. With the check valve removed, use a hand vacuum pump to apply vacuum to the booster vacuum inlet. Use fittings and adapters that allow a tight seal at the check valve inlet. See **Figure 114-2**.

Job 114—Service a Power Assist System (continued)

Figure 114-2. When installing a vacuum pump to the booster, adapters may be needed to form a proper seal between the booster inlet and the vacuum pump hose.

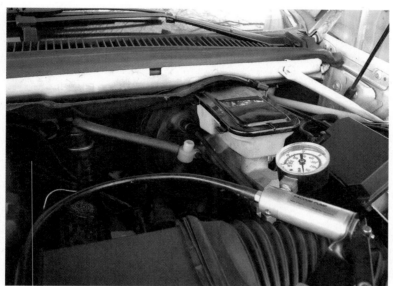

12. Wait about 2 minutes and then check vacuum.

    Does vacuum drop? Yes ___ No ___

    If Yes, the booster is leaking and should be replaced.

Diagnose a Hydro-Boost Power Booster System

1. Check the power steering fluid level.

    Is the level within normal range? Yes ___ No ___

2. Check for leaks at the Hydro-Boost control valve and accumulator.

    Were any leaks noted? Yes ___ No ___

    If Yes, give the location(s): _____

    _____

3. Check belt tension and condition:

    Is the belt properly tensioned? Yes ___ No ___

    Is the belt in good condition? Yes ___ No ___

4. Test the accumulator
    a. Start the engine.
    b. Turn the steering wheel in one direction for five seconds.
    c. Without moving the steering wheel, stop the engine.
    d. Wait about 30 minutes.
    e. Apply the brake. You should feel power assist for two to three brake pedal applications.
    Results of test: _____

5. Based on the results of steps 1 through 4, determine needed service.

    Service needed: _____

    _____

6. With your instructor's approval, make needed repairs.

Warning

⚠️ Pump the brake pedal 20 times to remove all pressure before making repairs.

☐ 7. Recheck system operation.

Do any brake booster problems exist? Yes ___ No ___

If Yes, return to the beginning of this job and recheck the system.

☐ 8. Return tools and test equipment to storage.

☐ 9. Did you encounter any problems during this procedure? Yes ___ No ___

If Yes, describe the problems: _____

_____

What did you do to correct the problems? _____

_____

☐ 10. Have your instructor check your work and sign this job sheet.

Performance Evaluation—Instructor Use Only

Did the student complete the job in the time allotted? Yes ___ No ___
If No, which steps were not completed?_____
How would you rate this student's overall performance on this job?_____
5–Excellent, 4–Good, 3–Satisfactory, 2–Unsatisfactory, 1–Poor
Comments: _____

_____

_____

INSTRUCTOR'S SIGNATURE _____

Job 115—Bleed a Brake System

After completing this job, you will be able to bleed and flush the brake hydraulic system and properly handle brake fluids.

Procedures

☐ 1. Obtain a vehicle to be used in this job. Your instructor may direct you to perform this job on a shop vehicle.

☐ 2. Gather the tools needed to perform this job. Refer to the tools and materials list at the beginning of the project.

Fill Brake Fluid Reservoir

Caution

It is very important to handle and store brake fluid properly. Never reuse brake fluid or mix different types of fluid. Keep brake fluid in its original container until used. Also, make sure the brake fluid container is tightly capped after use to prevent moisture and contaminants from entering the fluid.

☐ 1. Remove the cap on the brake fluid reservoir and observe the fluid level.

or

Observe the fluid level in the clear strip on the side of the reservoir.

☐ 2. If necessary, add fluid:
  a. Observe the reservoir cap or the top of the reservoir to determine what type of fluid should be added.
  b. Obtain the proper type of brake fluid.
  c. Remove the brake fluid container cap.
  d. Remove the reservoir cap.
  e. Add fluid to the reservoir until the full or maximum mark is reached.

☐ 3. Reinstall the reservoir cap.

☐ 4. Tightly recap the brake fluid container.

Check Brake Fluid for Moisture Contamination

Caution

Do not add fresh fluid before performing this test.

☐ 1. Check for moisture-contaminated fluid using one of the following methods:
  • Chemically sensitive test strips:
  a. Remove the cap on the brake fluid reservoir.
  b. Insert a test strip into the reservoir.
  c. Remove the strip and compare the strip color with the color chart on the strip container.
  Does the test strip indicate excessive moisture in the brake fluid? Yes ___ No ___

- Electronic brake fluid tester:
  a. Remove the cap on the brake fluid reservoir.
  b. Insert the tester probe into the fluid.
  c. Read the tester screen. The tester will display a reading showing the percentage of moisture in the brake fluid.

  Does the electronic tester indicate excessive moisture in the brake fluid? Yes ___ No ___

☐ 2. Report excessive brake fluid moisture to your instructor. Your instructor may direct you to flush the brake system using the procedures in this job.

Bleed Brakes

☐ 1. Remove the master cylinder reservoir cover.

☐ 2. Check the level of fluid in the reservoir.

   Is the brake fluid level low? Yes ___ No ___

   If Yes, add fluid if necessary.

   Type of brake fluid added: _____

   Amount of brake fluid added: _____

☐ 3. Tightly close the brake fluid container to ensure that moisture or other contaminants do not enter the fluid.

☐ 4. Replace the reservoir cover.

Bleed Brakes Manually

> **Note**
>
> If the vehicle is equipped with anti-lock brakes, follow the procedure in the service manual as needed. Also, refer to the jobs in Project 34.

☐ 1. Have an assistant pump the brakes slowly several times to build pressure in the hydraulic system.

☐ 2. While the assistant holds the pedal in the applied position, loosen one brake line fitting at the master cylinder. Use a line wrench to prevent fitting damage and place rags under the lines to absorb fluid.

> **Note**
>
> If the master cylinder has bleeder screws, use these instead of opening the lines.

☐ 3. Repeat step 2 at all lines on the master cylinder until only fluid exits when the fittings are loosened.

☐ 4. Raise the vehicle to gain access to the brake bleeder screws.

> **Warning**
>
> ⚠ The vehicle must be raised and supported in a safe manner. Always use approved lifts or jacks and jack stands.

☐ 5. Install one end of a tube on the specified bleeder screw and immerse the other end in a jar or bottle one-quarter filled with clean brake fluid.

Job 115—Bleed a Brake System (continued)

> **Note**
>
> Most manufacturers recommend beginning the bleeding process at the wheel farthest from the master cylinder, usually the right rear.

☐ 6. Have your assistant pump the brake pedal slowly several times and then hold the pedal in the applied position.

☐ 7. Loosen the bleeder screw and allow air and fluid to exit, **Figure 115-1**.

☐ 8. When no more air or fluid exits, close the bleeder. Then, have your assistant release the brake pedal.

☐ 9. Repeat steps 5 through 8 for all the wheels until only fluid exits from every bleeder.

> **Caution**
>
> Frequently check and refill the master cylinder reservoir during the bleeding process.

☐ 10. Ask your assistant to frequently report on pedal height and firmness.
Is pedal height and firmness acceptable after all wheels have been bled? Yes ___ No ___
If No, go to step 11.
If Yes, go to step 12.

☐ 11. Repeat steps 5 through 8 until pedal height and firmness are acceptable.

☐ 12. When bleeding is complete, lower the vehicle and perform a road test.

Bleed Brakes Using Pressure Bleeder

☐ 1. Check that the pressure bleeder is in operating condition and contains enough fluid for the bleeding operation.
Is the bleeder in proper operating condition? Yes ___ No ___
Is the bleeder fluid level acceptable? Yes ___ No ___

Figure 115-1. When manually bleeding a brake system, place a hose in a jar that contains clean brake fluid.

☐ 2. Add fluid to the bleeder if necessary.

☐ 3. Install the pressure bleeder adapter on the brake fluid reservoir.

☐ 4. If necessary, attach an air hose to the pressure bleeder; then adjust the bleeder pressure.
Pressure was set to _____ psi/kPa.

> **Caution**
>
> Do not set the pressure higher than recommended. Pressures above 10 psi (69 kPa) can damage brake fluid reservoirs.

☐ 5. Raise the vehicle to gain access to the brake bleeder screws.

> **Warning**
>
> The vehicle must be raised and supported in a safe manner. Always use approved lifts or jacks and jack stands.

☐ 6. Install a tube on the bleeder screw and immerse the other end of the tube in a jar or bottle partially filled with clean brake fluid.

> **Note**
>
> Most manufacturers recommend beginning the bleeding process at the wheel farthest from the master cylinder, usually the right rear.

☐ 7. Loosen the bleeder screw and allow air and fluid to exit.

☐ 8. When there is no evidence of air flowing from the hose, close the bleeder.

☐ 9. Repeat steps 6 through 8 for the remaining wheels.

☐ 10. Check pedal height and firmness.
Is pedal height and firmness acceptable? Yes ___ No ___
If No, go to step 11.
If Yes, go to step 12.

☐ 11. If the pedal is soft or low, repeat steps 6 through 9 until height and firmness are acceptable.

☐ 12. When all bleeding is complete, lower the vehicle and perform a road test.

Bleed Brakes Using a Vacuum Bleeder

☐ 1. Raise the vehicle to gain access to the brake bleeder screws.

> **Warning**
>
> The vehicle must be raised and supported in a safe manner. Always use approved lifts or jacks and jack stands.

☐ 2. Attach the vacuum bleeder to the bleeder screw farthest from the master cylinder.

☐ 3. Open the bleeder screw.

☐ 4. Operate the vacuum pump to draw air from the hydraulic system.

☐ 5. Repeat steps 2 through 4 for all wheels until only fluid exits from each bleeder.

> **Caution**
>
> Frequently check and refill the master cylinder reservoir during the bleeding process.

Job 115—Bleed a Brake System (continued)

☐ 6. Check pedal height and firmness.

Is pedal height and firmness acceptable? Yes ___ No ___

If No, go to step 7.

If Yes, go to step 8.

☐ 7. If the pedal is soft or low, repeat steps 2 through 6 until height and firmness are acceptable.

☐ 8. When all bleeding is complete, lower the vehicle and perform a road test.

Flush a Brake System

☐ 1. Check the level of brake fluid in the reservoir.

Is the fluid level low? Yes ___ No ___

☐ 2. Add fluid if necessary and replace the reservoir cover.

Type of brake fluid added: _____

Amount of brake fluid added: _____

> **Note**
>
> A pressure bleeder can be used to perform this job. Attach the bleeder as explained in the previous pressure bleeding section.

☐ 3. Raise the vehicle to gain access to the brake bleeder screws.

> **Warning**
>
> ⚠ The vehicle must be raised and supported in a safe manner. Always use approved lifts or jacks and jack stands.

☐ 4. Install one end of a tube on the bleeder screw and place the other end in a jar or bottle.

☐ 5. Have the assistant pump the brake pedal slowly several times then hold the pedal in the applied position.

☐ 6. Loosen the bleeder screw and allow fluid to exit into the jar or bottle.

> **Note**
>
> Pour fluid out of the jar or bottle as needed to keep it from overflowing.

☐ 7. Repeat steps 4 through 6 on the remaining wheels until clean fluid exits from all bleeders.

> **Caution**
>
> ⚠ Frequently check and refill the master cylinder reservoir during the flushing process.

☐ 8. When the lines to all the wheels have been flushed, lower the vehicle and perform a road test.

☐ 9. Clean the work area and return equipment to storage.

☐ 10. Did you encounter any problems during this procedure? Yes ___ No ___

If Yes, describe the problems: _____

_____

What did you do to correct the problems? _____

_____

☐ 11. Have your instructor check your work and sign this job sheet.

Performance Evaluation—Instructor Use Only

Did the student complete the job in the time allotted? Yes ___ No ___

If No, which steps were not completed?_____

How would you rate this student's overall performance on this job?_____

5–Excellent, 4–Good, 3–Satisfactory, 2–Unsatisfactory, 1–Poor

Comments: _____

_____

_____

INSTRUCTOR'S SIGNATURE _____

Project

Servicing Drum Brakes

32

Introduction

While drum brakes have disappeared from the front of vehicles, they are still commonly used on the rear wheels. Drum brake service usually consists of replacing the shoes. Often the rear drums are machined (turned) and the wheel cylinders are rebuilt or replaced. In Jobs 116 and 117, you will inspect and service drum brake components.

Project 32 Jobs

- Job 116—Service Drum Brakes
- Job 117—Machine a Brake Drum

Tools and Materials

The following list contains the tools and materials that may be needed to complete the jobs in this project. The items used will depend on the make and model of vehicle being serviced.

- Vehicle in need of drum brake service.
- Applicable service information.
- Respirator.
- Closed brake cleaning equipment.
- Spring removal tools.
- Wheel cylinder hone.
- Drum micrometer.
- Shoe-to-drum gauge.

- Brake tools.
- Brake bleeding equipment.
- Brake fluid.
- Hand tools as needed.
- Air-operated tools as needed.
- Safety glasses and other protective equipment as needed.

Safety Notice

Before performing this job, review all pertinent safety information in the text, and review safety information with your instructor.

Notes

Job 116—Service Drum Brakes

After completing this job, you will be able to diagnose and service drum brake systems.

Procedures

☐ 1. Obtain a vehicle to be used in this job. Your instructor may direct you to perform this job on a shop vehicle.

☐ 2. Gather the tools needed to perform this job. Refer to the project's tools and materials list.

☐ 3. Raise the vehicle.

> **Warning**
> ⚠ The vehicle must be raised and supported in a safe manner. Always use approved lifts or jacks and jack stands.

☐ 4. Remove the wheels from the axle to be serviced.

☐ 5. Put on the respirator and remove the brake drums.

What did you have to do to remove the drums? _____

_____

> **Warning**
> ⚠ Brake dust may contain asbestos. Wear a respirator whenever you are working on any brake friction materials.

☐ 6. Clean both brake assemblies using a closed cleaning system (liquid or HEPA vacuum).

> **Note**
> 📄 If you are unsure about the placement of brake parts, disassemble one side at a time. Refer to the other side as needed.

Disassemble Brakes

☐ 1. Remove the brake return springs, **Figure 116-1**.

☐ 2. Remove the brake hold-down spring or springs and any related cables or linkage. See **Figure 116-2**.

☐ 3. Remove the parking brake linkage.

☐ 4. Remove the brake shoes from the vehicle.

☐ 5. Remove the parking brake link from the backing plate. Identify the shoe type.

Servo. ___

Non-servo. ___

☐ 6. Place the brake shoes on a clean workbench.

☐ 7. If necessary, remove the spring holding the shoes and star wheel adjuster together and disassemble the brakes.

Figure 116-1. A special tool is needed to remove most return springs.

Figure 116-2. These brake shoes are held in position by a single large hold-down wire. Other shoes are held in place by small coil springs.

Inspect Brakes

☐ 1. Inspect the brake shoes for wear, cracks, and evidence of oil or brake-fluid soaking.

Was any damage found? Yes ____ No ____

If Yes, explain: _____

_____

☐ 2. Clean all springs and other hardware, and check them for bends, breaks, and signs of overheating.

Was any damage found? Yes ____ No ____

If Yes, explain: _____

_____

Job 116—Service Drum Brakes (continued)

3. Return to the vehicle and pull the dust boots away from the wheel cylinder. A slightly damp appearance is acceptable, but *any* liquid is grounds for overhaul or replacement. See **Figure 116-3**.

   Is brake fluid present? Yes ____ No ____

   If Yes, consult your instructor about rebuilding or replacing the wheel cylinder.

4. Inspect the backing plate and attaching hardware.

   Is the backing plate bent or damaged? Yes ____ No ____

   Is the attaching hardware loose or damaged? Yes ____ No ____

5. Check the brake drums for wear, glazing, scoring, and overheating.

   Visual Inspection

	Left	Right
Worn	____	____
Glazed	____	____
Scored	____	____
Overheated	____	____

   Micrometer Readings

	Left	Right
Maximum diameter	_____	_____
Out-of-round	_____	_____
Taper	_____	_____

   Can the drum(s) be turned and reused? Yes ____ No ____

   If No, explain why?_____

   _____

6. Determine which parts are needed to successfully repair the brakes. List them here.

   _____

Figure 116-3. Gently pull back the wheel cylinder dust boots to check for fluid leaks.

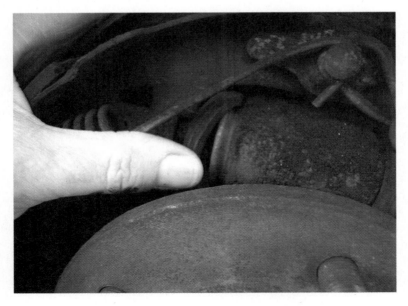

> **Note**
>
> See Job 117 for brake drum turning procedures.

Reassemble Brakes

☐ 1. Obtain new shoes and other parts as necessary.

☐ 2. Reinstall the wheel cylinder if it was removed.

   Fastener torque (if applicable):_____

☐ 3. Apply a small amount of high-temperature lube to the shoe contact pads.

> **Warning**
>
> ⚠ Be careful not to apply too much lube to the contact pads. Excessive lube may get on the brake shoe's friction material and cause brake failure.

☐ 4. Reassemble the shoes and star wheel on the bench if necessary.

☐ 5. Replace the brake link or self-adjuster on the backing plate if needed.

☐ 6. Place the shoe assembly on the backing plate and install the hold-down spring or springs.

> **Note**
>
> If any of the self-adjuster or parking brake parts are located under the hold-down springs, install them before installing the springs.

☐ 7. Reattach the parking brake linkage (rear brakes).

☐ 8. Make sure the shoes engage the wheel cylinder pistons or apply pins.

☐ 9. Install the shoe retractor springs and any related cables or links.

☐ 10. Make sure the self-adjuster operates properly.

☐ 11. Using a shoe-to-drum gauge, adjust the brake shoe clearance.

☐ 12. Reinstall the drum.

☐ 13. Repeat steps 1 through 12 for the other brake assembly.

☐ 14. Bleed the brakes if the wheel cylinders were removed. Refer to Job 115.

☐ 15. Reinstall the wheels and torque the fasteners to the proper values.

   Fastener torque:_____

☐ 16. Road test the vehicle.

☐ 17. Readjust the brake shoe clearance if necessary.
   a. Locate and remove the adjuster plug.
   b. Insert a small screwdriver or awl through the adjuster opening to disengage the self-adjuster lever.
   c. Insert an adjusting spoon into the adjuster opening and turn the star wheel.
   d. Adjust the star wheel until there is a small amount of drag when the wheel is turned.
   e. Remove the screwdriver and adjusting spoon and replace the plug.
   f. Repeat as necessary on other wheels.

☐ 18. Clean the work area and return any equipment to storage.

Job 116—Service Drum Brakes (continued)

☐ 19. Did you encounter any problems during this procedure? Yes ___ No ___

If Yes, describe the problems: _____

_____

What did you do to correct the problems? _____

_____

☐ 20. Have your instructor check your work and sign this job sheet.

Performance Evaluation—Instructor Use Only

Did the student complete the job in the time allotted? Yes ___ No ___

If No, which steps were not completed?_____

How would you rate this student's overall performance on this job?_____

5–Excellent, 4–Good, 3–Satisfactory, 2–Unsatisfactory, 1–Poor

Comments: _____

_____

_____

INSTRUCTOR'S SIGNATURE _____

Notes

Job 117—Machine a Brake Drum

After completing this job, you will be able to machine (turn) brake drums.

Procedures

☐ 1. Put on your safety glasses and respirator.

> **Warning**
>
> ⚠ Rotor or drum machining produces metal chips and brake dust. Wear safety glasses and a respirator at all times while performing this job.

> **Note**
>
> 📄 Actual machining procedures vary between lathe manufacturers. If there are any differences between the procedures given below and the manufacturer's procedures, follow the manufacturer's procedures.

☐ 2. Clean the drum interior.

> **Warning**
>
> ⚠ Use a vacuum or wet cleaner to remove brake dust. Do not remove dust with compressed air.

☐ 3. Remove all grease, dust, and metal debris from the brake lathe arbor.

☐ 4. Check the condition of the cutting bits and make sure the boring bar is tightly installed.

☐ 5. Select the proper adapters and mount the drum on the arbor.

☐ 6. Install the silencer band around the drum.

☐ 7. Make sure the lathe's automatic feed mechanism is disengaged.

☐ 8. Start the lathe and ensure the drum revolves evenly, without excessive wobbling. If the drum wobbles, reposition the drum and adapters as necessary.

☐ 9. Loosen the locknut on the cutting depth adjuster knob and turn the depth adjuster to a position that ensures the cutting bit does not contact the inside of the drum.

☐ 10. Turn the automatic feed mechanism wheel to move the bit to the inside of the drum. Make sure the cutting tip does not contact the rotating drum.

☐ 11. Use the depth adjuster knob to move the cutting bit into slight contact with the inner surface of the drum.

☐ 12. Tighten the lock on the cutting depth adjuster knob.

☐ 13. Turn the movable collar on the cutting bit adjuster to zero.

☐ 14. Use the manual feed to position the cutting bit at the rear of the drum, **Figure 117-1**.

☐ 15. Loosen the locknut and turn the cutting depth adjuster knob to the desired cut.

First cut depth:_____

☐ 16. Set the speed of the automatic feed.

Speed:_____

Is this a rough cut or a finish cut?_____

Figure 117-1. A brake drum is being machined in this image. The drum should machine evenly with no wobble or chattering. (Hunter Engineering)

☐ 17. Engage the automatic feed.

☐ 18. Observe the drum as it is being cut.

☐ 19. When the cutting bit exits the drum, turn off the automatic feed.

☐ 20. If necessary, repeat steps 15 through 20 until the entire drum surface is cut, with no grooves or shiny spots on the surface.

How many cuts were needed to clean up the drum?_____

What was the total cut depth?_____

> **Note**
>
> Be sure to make the final cut at a low speed to produce a smooth finished surface.

☐ 21. Turn off the lathe.

☐ 22. Check the drum diameter using a drum micrometer.

Micrometer reading:_____

Minimum thickness specification:_____

Can the drum be reused? Yes ___ No ___

☐ 23. Remove the silencer band from the drum.

☐ 24. Remove the drum from the lathe arbor.

☐ 25. Repeat steps 2 through 25 for the other drum.

☐ 26. Clean the work area and return tools and equipment to storage.

☐ 27. Did you encounter any problems during this procedure? Yes ___ No ___

If Yes, describe the problems: _____

_____

What did you do to correct the problems? _____

_____

☐ 28. Have your instructor check your work and sign this job sheet.

Job 117—Machine a Brake Drum (continued)

Performance Evaluation—Instructor Use Only

Did the student complete the job in the time allotted? Yes ____ No ____

If No, which steps were not completed?_____

How would you rate this student's overall performance on this job?_____

5–Excellent, 4–Good, 3–Satisfactory, 2–Unsatisfactory, 1–Poor

Comments: _____

_____

_____

INSTRUCTOR'S SIGNATURE _____

Notes

Project

Servicing
Disc Brakes

33

Introduction

Disc brakes are always used on the front wheels of modern vehicles. They are also commonly used on the rear wheels of many vehicles. Disc brake service generally consists of replacing the pads. Sometimes the rotors are turned or replaced. Occasionally the calipers must be rebuilt, although they are usually replaced with new or rebuilt units. In Jobs 118 through 120, you will inspect and service disc brake components.

Project 33 Jobs

- Job 118—Service Disc Brakes
- Job 119—Overhaul a Disc Brake Caliper
- Job 120—Machine a Brake Rotor

Tools and Materials

The following list contains the tools and materials that may be needed to complete the jobs in this project. The items used will depend on the make and model of vehicle being serviced.

- Vehicle in need of disc brake service.
- Applicable service information.
- Closed cleaning equipment.
- Brake hone.
- Brake bleeding equipment.
- Brake fluid.
- Micrometer.
- Dial indicators.
- Hand tools as needed.
- Air-operated tools as needed.
- Safety glasses and other protective equipment as needed.

Safety Notice

Before performing this job, review all pertinent safety information in the text, and review safety information with your instructor.

Notes

Job 118—Service Disc Brakes

After completing this job, you will be able to inspect and replace disc brake components.

Procedures

- [ ] 1. Obtain a vehicle to be used in this job. Your instructor may direct you to perform this job on a shop vehicle.
- [ ] 2. Gather the tools needed to perform this job. Refer to the project's tools and materials list.

Disassemble Disc Brakes

- [ ] 1. Raise the vehicle.

> **Warning**
> ⚠ Raise and support the vehicle in a safe manner. Always use approved lifts or jacks and jack stands.

- [ ] 2. Remove wheels as needed to access the brakes to be serviced.

> **Warning**
> ⚠ Brake pads may contain asbestos. Clean the brake assembly using a closed liquid cleaning system or HEPA vacuum.

- [ ] 3. Place a drip pan under the brake caliper to be serviced.
- [ ] 4. Open the caliper's bleeder valve.
- [ ] 5. Push the caliper piston(s) away from the rotor and into the caliper body.
- [ ] 6. Close the bleeder valve.
- [ ] 7. Repeat steps 3 through 6 on the other wheel.
- [ ] 8. Remove the caliper fasteners.
- [ ] 9. Remove the calipers from the rotors.

> **Caution**
> ◇ To avoid damaging the brake hoses, hang the calipers on wire hooks. Do not allow them to hang freely by their hoses.

- [ ] 10. Remove the pads from the calipers or caliper brackets.
- [ ] 11. Remove the rotor from the hub.

> **Note**
> Most rotors can be pulled from the wheel studs once the wheel and caliper are removed. If the rotor is pressed onto the hub, it must be removed with a special puller. Follow the manufacturer's instructions to remove a pressed-on rotor. Do not attempt to hammer the rotor free. Some rotors can be serviced with an on-vehicle lathe, eliminating the need to remove them.

Inspect Disc Brake Components

☐ 1. Inspect the pads. Indicate any obvious pad defects:
Worn/thin. ___
Cracked. ___
Loose on backing plate. ___
Other. ___

☐ 2. Inspect the pad wear indicator.

Mechanical Type

Locate the pad wear indicator.

Note whether the indicator is recessed at least 1/8″ (3 mm) away from the surface of the pad.

If the indicator is not 1/8″ (3 mm) away from the pad, discard the pad.

> **Note**
>
> If the instrument panel pad wear indicator light is on, the pad can be assumed to be worn to the indicator button.

Electrical or Electronic type

Observe the center of the brake pad.

Note whether the wear indicator button in the pad is visible.

If the wear indicator button is visible, replace the pads.

☐ 3. Check the backing plate and attaching hardware:

Backing plate bent or damaged. ___

Attaching hardware loose, stripped, or damaged. ___

Caliper mounting and sliding surfaces or slide pins worn, burred, or dented. ___

☐ 4. Check the rotor for visible signs of damage.

Visual Inspection

	Left	Right
Worn.	___	___
Glazed.	___	___
Scored.	___	___
Overheated.	___	___
Bearing damage (if applicable).	___	___

Was any damage found? Yes ___ No ___

If Yes, explain: _____

☐ 5. Attach a dial indicator to the rotor as shown in **Figure 118-1**. Check the rotor for warping by slowly turning the rotor through several revolutions. It may be necessary to reinstall the wheel nut temporarily to perform this test.

Maximum variation on the left rotor:_____

Maximum variation on the right rotor:_____

☐ 6. Use an outside micrometer to measure rotor thickness at several points around the disc.

Micrometer Readings

	Left	Right
Maximum thickness:	_____	_____
Minimum thickness:	_____	_____
Thickness variation:	_____	_____
Taper:	_____	_____

Name _____ Date _____

Instructor _____ Period _____

Job 118—Service Disc Brakes (continued)

Figure 118-1. A dial indicator is used to check a rotor for warping. Adjust the wheel bearings before making this test.

Can the rotor(s) be turned and reused? Yes ____ No ____

If No, what is the reason? _____

_____

Note

To remove and turn rotors, see Job 120.

☐ 7. If applicable check the wheel bearings and seals:

	Left	Right
Lubricant dirty/missing.	____	____
Bearings worn/damaged.	____	____
Seals leaking.	____	____

Note

See Job 109 for wheel bearing service procedures.

☐ 8. Determine which parts are needed to successfully repair the brakes. List them in the space provided.

_____

_____

_____

_____

Reassemble Disc Brakes

☐   1.   If necessary, reinstall the bearings in the rotor hub. See Job 109.

☐   2.   Reinstall the rotors if they were removed for turning. Also, reattach the caliper brackets if they were removed.

☐   3.   Apply anti-squeal compound or insulation material to the backs of the pads. Consult the service manual for the exact materials and methods for this step.

☐   4.   Lightly lubricate the caliper sliding surfaces with the correct type of high-temperature lubricant.

☐   5.   Place the new pads in the calipers or brackets as necessary.

☐   6.   Reinstall the calipers over the rotors.

☐   7.   Bleed the brake system at the calipers.

> **Caution**
>
> Check the fluid level in the reservoir before bleeding and add fluid as necessary.

☐   8.   Reinstall the wheels and torque the fasteners to the proper values.

      Fastener torque:_____

☐   9.   Road test the vehicle.

> **Caution**
>
> Before beginning the road test, ensure that you have a firm pedal. Follow the manufacturer's recommendations for seating and burnishing the new brake pads. The burnishing process removes roughness and unevenness between the brake mating surfaces, thermally changes the composition of the pad material, and heat cycles the pad. The seating and burnishing procedure is necessary to ensure proper operation and longevity of the brake pads.

☐   10.   Clean the work area and return any equipment to storage.

☐   11.   Did you encounter any problems during this procedure? Yes ___ No ___

      If Yes, describe the problems: _____

      _____

      What did you do to correct the problems? _____

      _____

☐   12.   Have your instructor check your work and sign this job sheet.

Performance Evaluation—Instructor Use Only

Did the student complete the job in the time allotted? Yes ___ No ___

If No, which steps were not completed?_____

How would you rate this student's overall performance on this job?_____

5–Excellent, 4–Good, 3–Satisfactory, 2–Unsatisfactory, 1–Poor

Comments: _____

_____

_____

INSTRUCTOR'S SIGNATURE _____

Job 119—Overhaul a Disc Brake Caliper

After completing this job, you will be able to overhaul a disc brake caliper.

Procedures

☐   1.   Obtain a vehicle to be used in this job. Your instructor may direct you to perform this job on a shop vehicle.

☐   2.   Gather the tools needed to perform this job. Refer to the project's tools and materials list.

> **Caution**
>
> Always rebuild calipers in pairs. If one caliper is leaking or sticking, the other is probably ready to fail.

☐   3.   Remove the caliper(s) as explained in Job 118.

☐   4.   Place the caliper in a vise.

☐   5.   Remove the dust boot(s) from the caliper piston(s).

☐   6.   Place a wood block in the caliper, across from the piston(s).

> **Note**
>
> If the caliper has two pistons, be sure the wood block is thick enough to keep one piston from completely exiting its bore before the other piston has begun to move. Once one piston is out of its bore, it will be impossible to remove the second piston with air pressure.

☐   7.   Cover the caliper piston(s) with a shop towel.

☐   8.   Apply air pressure to the caliper inlet port to force the caliper from its bore.

> **Warning**
>
> Do not place your hand between the piston(s) and the wood block. Your hand could be severely injured when the piston(s) comes out.

☐   9.   Thoroughly clean all caliper parts in nonpetroleum-based cleaner.

☐ 10.   Inspect the caliper bore(s) and piston(s). Indicate any problems found.

Pitting:___

Rust:___

Heavy scoring:___

In your opinion, can the caliper be overhauled? Yes ___ No ___

If No, explain why: _____

_____

☐ 11.   Hone the caliper bore(s).

Did the bore clean up with light honing? Yes ___ No ___

If No, explain why not: _____

> **Note**
>
> Before honing, be sure the caliper can be honed. Aluminum and some cast iron calipers cannot be honed.

☐ 12. Obtain the correct replacement seal kit.

☐ 13. Lightly lubricate the new seals and the caliper bore with fresh brake fluid or silicone spray.

> **Warning**
>
> ⚠ The lubricant used in step 13 must be compatible with the brake system seals. Do not lubricate with petroleum-based products.

☐ 14. Install the new seal(s) in the caliper piston bore(s).

☐ 15. Install the piston(s) in the caliper.

☐ 16. Install the dust boot.

☐ 17. If needed, repeat steps 3 through 17 for the other caliper.

☐ 18. Install the calipers on the vehicle following the procedures given in Job 118.

☐ 19. Bleed the brake system as outlined in Job 115.

☐ 20. Clean the work area and return any equipment to storage.

☐ 21. Did you encounter any problems during this procedure? Yes ___ No ___

If Yes, describe the problems: _____

_____

What did you do to correct the problems? _____

_____

☐ 22. Have your instructor check your work and sign this job sheet.

Performance Evaluation—Instructor Use Only

Did the student complete the job in the time allotted? Yes ___ No ___
If No, which steps were not completed?_____
How would you rate this student's overall performance on this job?_____
5–Excellent, 4–Good, 3–Satisfactory, 2–Unsatisfactory, 1–Poor
Comments: _____

_____

_____

INSTRUCTOR'S SIGNATURE _____

Name _____ Date _____

Instructor _____ Period _____

Job 120—Machine a Brake Rotor

After completing this job, you will be able to machine a brake rotor.

Procedures

1. Obtain a vehicle to be used in this job. Your instructor may direct you to perform this job on a shop vehicle.

2. Gather the tools needed to perform this job. Refer to the tools and materials list at the beginning of the project.

> **Note**
>
> This portion of the job can be performed with either a bench lathe or an on-car lathe. Be sure to use the same lathe to cut both rotors. Do not use a different type of lathe on each rotor or only cut one rotor. If you are using a bench lathe, proceed with the next step. If you are using an on-car lathe, go to the *Using an On-Vehicle Lathe* section of this job.

Using a Bench Lathe

1. Remove the rotors following the procedure in Job 118. If the rotors use serviceable wheel bearings, clean the wheel bearings and set them aside.

   List the steps needed to remove the rotor: _____

   _____

> **Warning**
>
> ⚠ Use a closed HEPA vacuum or wet cleaner to remove brake dust from the brake assembly. Do not remove dust with compressed air.

2. Remove all grease, dust, and metal debris from the brake lathe arbor.
3. Check the condition of the cutting bits and ensure the boring bar is tightly installed.
4. Select the proper adapters and mount the first rotor on the arbor.
5. Install the silencer band around the rotor.
6. Start the lathe and make sure the rotor revolves evenly, without excessive wobbling. If the rotor wobbles, reposition the rotor and adapters as necessary. See **Figure 120-1**.

Using an On-Vehicle Lathe

1. Check for play in the wheel bearings. If play is excessive, correct it before proceeding.
2. If the rotors are on the driving axle, place the vehicle's transmission or transaxle in neutral.
3. Remove all dust and metal debris from the brake lathe attaching fixtures.
4. Check the condition of the cutting bits and make sure they are tightly installed on the fixture.
5. Select the proper adapters and mount the lathe to the hub and rotor.
6. Compensate for rotor runout or shaft wobble as necessary.
7. Install the silencer band around the rotor. Not all on-vehicle lathes require this step.
8. Start the lathe and make sure it does not vibrate or wobble excessively as the rotor turns. If the lathe vibrates or wobbles excessively, readjust the lathe as necessary.

Figure 120-1. This rotor is being turned on a bench-type lathe. Check the service manual and the lathe operation manual if needed. (Hunter Engineering)

Cutting Rotors

> **Note**
>
> This section applies to both on-vehicle lathes and bench lathes.

1. Loosen the locknut on both cutting depth adjuster knobs and turn the knobs to move the cutting bits into light contact with each side of the rotor.

2. Tighten the locks on each cutting depth adjuster knob.

3. Turn the movable collars on each cutting depth adjuster to zero.

4. Use the manual feed to position the cutting bits at the inside of the rotor.

5. Loosen the locks and turn each cutting depth adjuster knob to the desired cut.

   First-cut depth, outboard rotor surface:_____

   First-cut depth, inboard rotor surface:_____

6. Set the speed of the automatic feed.

   Speed:_____

   Is this a rough cut or a finish cut?_____

7. Engage the automatic feed.

8. Observe each side of the rotor as it is being cut.

9. When the cutting bit exits both sides of the rotor, turn off the automatic feed.

10. If necessary, repeat steps 5 through 9 until both rotor surfaces are smooth, with no grooves or shiny spots on either surface.

    Cuts needed to clean up the rotor:_____

    Total cut on inner surface:_____

    Total cut on outer surface:_____

11. Turn off the lathe.

Job 120—Machine a Brake Rotor (continued)

☐ 12. Check the rotor thickness using a micrometer.

Micrometer reading:_____

Minimum thickness specification:_____

Can the rotor be reused? Yes ___ No ___

☐ 13. Using a lathe-mounted or hand-held grinder, swirl grind the rotor to produce a non-directional finish.

> **Note**
>
> Some manufacturers do not recommend swirl grinding the rotor after machining.

☐ 14. Remove the silencer band from the rotor if necessary.

☐ 15. Remove the rotor from the lathe (bench lathe) or the lathe from the rotor (on-vehicle lathe).

☐ 16. Remove all metal chips from the rotor.

☐ 17. Repack the wheel bearings if applicable.

☐ 18. Repeat the cutting procedure on the other rotor.

☐ 19. Check the rotor thickness using a micrometer.

Micrometer reading: _____

Minimum thickness specification: _____

Can the rotor be reused? Yes _____ No _____

☐ 20. Clean the work area and return tools and equipment to storage.

☐ 21. Did you encounter any problems during this procedure? Yes ___ No ___

If Yes, describe the problems: _____

_____

What did you do to correct the problems? _____

_____

☐ 22. Have your instructor check your work and sign this job sheet.

Performance Evaluation—Instructor Use Only

Did the student complete the job in the time allotted? Yes ___ No ___

If No, which steps were not completed?_____

How would you rate this student's overall performance on this job?_____

5–Excellent, 4–Good, 3–Satisfactory, 2–Unsatisfactory, 1–Poor

Comments: _____

_____

_____

INSTRUCTOR'S SIGNATURE _____

Notes

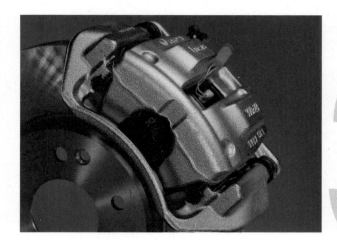

Project

Servicing Anti-lock Braking and Traction Control Systems

Introduction

The anti-lock brake system (ABS) and traction control system (TCS) work with the conventional brake hydraulic and friction elements. The ABS reduces wheel slippage during braking. The TCS reduces slipping during acceleration. On most vehicles with traction control systems, the TCS is interconnected and uses many of the same components as the anti-lock braking system. In Jobs 121 through 124, you will inspect and service ABS and TCS systems.

Project 34 Jobs

- Job 121—Identify and Inspect ABS and TCS Components
- Job 122—Bleed an ABS System
- Job 123—Adjust Speed Sensor Clearance
- Job 124—Diagnose ABS/TCS Problems

Tools and Materials

The following list contains the tools and materials that may be needed to complete the jobs in this project. The items used will depend on the make and model of vehicle being serviced.

- Vehicle in need of anti-lock brake and/or traction control system service.
- Applicable service information.
- Scan tool.
- Test light.
- Multimeter.
- Pressure gauge.
- Hand tools as needed.
- Air-operated tools as needed.
- Safety glasses and other protective equipment as needed.

Safety Notice

Before performing each job, review all pertinent safety information in the text and review safety information with your instructor.

Notes

Job 121—Identify and Inspect ABS and TCS Components

After completing this job, you will be able to inspect anti-lock brake and/or traction control systems for proper operation. You will also be able to identify ABS and TCS problems resulting from component failures due to wear or damage.

Procedures

1. Obtain a vehicle to be used in this job. Your instructor may direct you to perform this job on a shop vehicle.

2. Gather the tools needed to perform this job. Refer to the project's tools and materials list.

3. Obtain the correct service literature for the vehicle being serviced and fill in the following information about the ABS/TCS system being serviced.

   System type: ABS only ___ ABS/TCS ___ 2-wheel system ___ 4-wheel system ___

   Is this an integral system (no separate master cylinder)? Yes ___ No ___

   System manufacturer: _____

4. Check the brake fluid level in the reservoir.

> **Warning**
>
> ⚠ Depressurize integral systems by depressing the brake pedal 40–50 times. Do not turn the ignition on while checking the system fluid.

5. Add fluid to the reservoir as necessary.

   Type of fluid added:_____

> **Warning**
>
> ⚠ Do not use silicone fluid (DOT# 5) in any ABS/TCS system.

6. Start the engine and compare the operation of the dashboard brake, ABS, and TCS warning lights with normal operation as defined in the service literature.

   Did the warning lights come on and then go off after the normal start-up sequence?
   Yes ___ No ___

   If Yes, proceed to step 7.

   If No, refer to Job 124 to diagnose ABS/TCS problems or Job 138 to diagnose brake warning light problems.

7. Raise the vehicle.

> **Warning**
>
> ⚠ The vehicle must be raised and supported in a safe manner. Always use approved lifts or jacks and jack stands.

> **Note**
>
> 📄 If the base brakes are known to be in good condition, or if told to do so by your instructor, skip steps 8 and 9.

☐ 8. Remove all wheels from the vehicle.

☐ 9. Inspect the base brake system and summarize the condition of the hydraulic and friction components (refer to other jobs as necessary).

Describe any defects found in the base-brake hydraulic system: _____

_____

Describe any service needed for the base-brake hydraulic system: _____

_____

Describe any defects found in the base-brake friction system: _____

_____

Describe any service needed for the base-brake friction system: _____

_____

☐ 10. Inspect the wheel speed sensors and rotors (trigger wheels). Look for the following problems:
- Damaged or disconnected sensor.
- Metal shavings on sensor tip. See **Figure 121-1**.
- Debris on rotors. See **Figure 121-2**.
- Damaged rotor.
- Improper sensor-to-rotor clearance.

☐ 11. Describe the condition of each sensor and rotor at the locations listed below. If the ABS/TCS system being serviced does not have the sensor listed, write N/A in the space.

Left-front wheel: _____

Right-front wheel: _____

Left-rear wheel: _____

Right-rear wheel: _____

Transmission-mounted sensor: _____

Differential-mounted sensor: _____

☐ 12. Reinstall the wheels if they were removed. Torque lug nuts to the correct values.

☐ 13. Lower the vehicle.

Figure 121-1. Excessive metal shavings on this wheel speed sensor make it unusable. In many cases, the sensor can be cleaned.

Job 121—Identify and Inspect ABS and TCS Components (continued)

Figure 121-2. Debris on this rotor will affect the signal generated by the sensor.

 14. Locate and note the condition of the following ABS/TCS components. Look for leaks, disconnected or corroded wires, damage, or other defects. If the listed component is not used on the vehicle being serviced, write N/A in the blank.
- Control module: _____
- ABS hydraulic actuator: _____
- Relays:_____
- G-force sensor: _____
- Lateral acceleration sensor:_____
- TCS hydraulic actuator, if used. Put "integral" if the TCS hydraulic actuator is part of the ABS actuator: _____
- TCS disable switch: _____
- Engine torque management control:_____

15. Take the vehicle to a road with little or no traffic and test ABS and/or TCS system operation using the procedures outlined in the service literature.

> **Warning**
>
> ⚠ Be sure to select an empty section of road for these tests. Wait until all other traffic has cleared before beginning the tests. Be sure you have read and understand all the manufacturer's safety instructions before beginning any test.

 16. Consult with your instructor and determine whether further inspection of the ABS/TCS is needed.

Is further testing needed? Yes ___ No ___

> **Note**
>
> Refer to Job 124 for ABS/TCS troubleshooting.

☐ 17. Clean the work area and return any tools and equipment to storage.

☐ 18. Did you encounter any problems during this procedure? Yes ___ No ___

If Yes, describe the problems: _____

_____

What did you do to correct the problems? _____

_____

☐ 19. Have your instructor check your work and sign this job sheet.

Performance Evaluation—Instructor Use Only

Did the student complete the job in the time allotted? Yes ___ No ___

If No, which steps were not completed?_____

How would you rate this student's overall performance on this job?_____

5–Excellent, 4–Good, 3–Satisfactory, 2–Unsatisfactory, 1–Poor

Comments: _____

_____

_____

INSTRUCTOR'S SIGNATURE _____

Job 122—Bleed an ABS System

After completing this job, you will be able to properly bleed an ABS system.

Procedures

☐ 1. Obtain a vehicle to be used in this job. Your instructor may direct you to perform this job on a shop vehicle.

☐ 2. Gather the tools needed to perform this job. Refer to the project's tools and materials list.

☐ 3. Bleed the base brake system according to the procedures given in Job 115.

☐ 4. Attach a scan tool to the data link connector.

☐ 5. Follow the scan tool instructions to bleed the ABS/TCS system.

☐ 6. Check the brake pedal feel and height.

☐ 7. If necessary, repeat steps 3 through 6 until the pedal feel and height are satisfactory.

☐ 8. Clean the work area and return tools and equipment to storage.

☐ 9. Did you encounter any problems during this procedure? Yes ___ No ___

   If Yes, describe the problems: _____

   _____

   What did you do to correct the problems? _____

   _____

☐ 10. Have your instructor check your work and sign this job sheet.

Performance Evaluation—Instructor Use Only

Did the student complete the job in the time allotted? Yes ___ No ___

If No, which steps were not completed?_____

How would you rate this student's overall performance on this job?_____

5–Excellent, 4–Good, 3–Satisfactory, 2–Unsatisfactory, 1–Poor

Comments: _____

_____

_____

INSTRUCTOR'S SIGNATURE _____

Notes

Job 123—Adjust Speed Sensor Clearance

After completing this job, you will be able to adjust wheel speed sensor clearance.

Procedures

☐ 1. Obtain a vehicle to be used in this job. Your instructor may direct you to perform this job on a shop vehicle.

☐ 2. Gather the tools needed to perform this job. Refer to the tools and materials list at the beginning of the project.

☐ 3. Measure the gap between the speed sensor tip and the rotor (trigger wheel) with a feeler gauge. The gauge should be a nonferrous material, such as brass, aluminum, or plastic.

☐ 4. Compare the gap measurement to the manufacturer's specifications.

　　 Is the gap within specifications? Yes ___ No ___

　　 If No, go to step 5.

　　 If Yes, go to step 7.

☐ 5. Adjust the gap until it is within the manufacturer's specifications.

☐ 6. Repeat steps 3 and 4 for all speed sensors.

☐ 7. Clean the work area and return any tools and equipment to storage.

☐ 8. Did you encounter any problems during this procedure? Yes ___ No ___

　　 If Yes, describe the problems: _____

　　 _____

　　 What did you do to correct the problems? _____

　　 _____

☐ 9. Have your instructor check your work and sign this job sheet.

Performance Evaluation—Instructor Use Only

Did the student complete the job in the time allotted? Yes ___ No ___

If No, which steps were not completed?_____

How would you rate this student's overall performance on this job?_____

5–Excellent, 4–Good, 3–Satisfactory, 2–Unsatisfactory, 1–Poor

Comments: _____

_____

_____

INSTRUCTOR'S SIGNATURE _____

Notes

_____

_____

_____

_____

_____

_____

_____

_____

_____

_____

_____

_____

_____

_____

_____

_____

_____

_____

_____

_____

_____

_____

_____

_____

_____

_____

_____

_____

_____

_____

_____

_____

_____

_____

_____

_____

_____

_____

Job 124—Diagnose ABS/TCS Problems

After completing this job, you will be able to diagnose ABS and TCS systems.

Procedures

☐ 1. Obtain a vehicle to be used in this job. Your instructor may direct you to perform this job on a shop vehicle.

☐ 2. Gather the tools needed to perform this job. Refer to the project's tools and materials list.

☐ 3. Obtain a description of the problem from the vehicle operator. Check each problem reported by the vehicle's operator:
   • ABS/TCS lights on or off improperly. ___
   • Poor stopping/no ABS operation. ___
   • Wheel lockup under hard braking. ___
   • Pedal pulsation. ___
   • Other abnormal pedal feel. ___ Explain: _____
   • Noises during hard braking. ___
   • Traction control inoperative. ___
   • Other ABS/TCS problem. ___ Explain: _____

☐ 4. Obtain the correct service literature for the vehicle being serviced and complete the following:

   System type: ABS only ___ ABS/TCS ___ 2-wheel system ___ 4-wheel system ___

   Is this an integral system (no separate master cylinder)? Yes ___ No ___

   System manufacturer: _____

☐ 5. Check for the presence of vehicle modifications that could cause an ABS/TCS problem:
   • Larger or smaller wheels and tires than original equipment. ___
   • Mismatched wheel tire sizes. ___
   • Ride height changed. ___
   • Final drive ratio changed. ___

> **Note**
>
> Sometimes these modifications will be obvious. In other cases, the extent of modification must be determined comparing the manufacturer's specifications with the actual vehicle condition or equipment installed on the vehicle. The owner must be informed that modifications such as the ones above will cause improper ABS/TCS operation.

☐ 6. Check the brake fluid level in the reservoir.

   Is the fluid at correct level? Yes ___ No ___

   If Yes, go to step 8.

   If No, go to step 7.

> **Warning**
>
> ⚠ On some older vehicles the master cylinder reservoir is pressurized. Before opening the reservoir, depressurize the system by depressing the brake pedal 40–50 times. Do not turn the ignition *on* while checking the system fluid.

☐ 7. Add fluid to the reservoir as necessary and check the system for leaks.

Type of fluid added: DOT #_____

Were any leaks found? Yes ___ No ___

If Yes, describe them: _____

> **Warning**
>
> ⚠ Do not use silicone fluid (DOT# 5) in any ABS/TCS system.

☐ 8. Start the engine and compare the operation of the dashboard brake, ABS, and TCS warning lights with normal operation as defined in the service literature.

Do the warning lights come on, then go off after the normal start-up sequence?

Yes ___ No ___

If No, go to step 9.

If Yes, go to step 12.

> **Note**
>
> 📄 See Project 39 to diagnose brake warning light problems.

☐ 9. If the warning lights do not operate properly, determine the nature of the light defect.

The light does not come on when the engine is started. ___ Go to step 10.

The light does not go off after the appropriate time. ___ Go to step 11.

☐ 10. Check the condition of the ABS and TCS system fuses and bulbs. Replace as needed.

> **Caution**
>
> ◇ Do not remove ABS/TCS fuses to check them before retrieving the trouble codes. Removing the fuses will erase the codes.

☐ 11. Connect a scan tool to the vehicle's data link connector. See Job 11 for more information on connecting the scan tool. Follow the scan tool manufacturer's instructions for retrieving codes.

Are codes present? Yes ___ No ___

If Yes, list the stored codes:_____

_____

Do any of the codes apply to the ABS/TCS system? Refer to the appropriate service literature. Yes ___ No ___

If Yes, list the ABS/TCS codes: _____

_____

☐ 12. If the warning lights operate properly, take the vehicle to a road with no other traffic and test ABS and/or TCS system operation using service manual procedures.

> **Caution**
>
> ◇ You must use an empty section of road for these tests. Wait until all other traffic has cleared before beginning the tests. Note all manufacturers' safety instructions before beginning any test.

Does the ABS/TCS system perform normally? Yes ___ No ___

Job 124—Diagnose ABS/TCS Problems (continued)

☐ 13. Based on the results obtained in steps 3 through 12, determine whether a problem exists, and if so, determine the nature of the problem.

Does a problem exist? Yes ____ No ____

Explain your answer:_____

_____

☐ 14. Make further checks as necessary. See **Figure 124-1**. Indicate which checks were made:
- Diagnostic tests using scan tool. ____
- Pressure checks using pressure gauge. ____
- Wheel speed sensors checked (see the following section). ____
- Relays checked. ____
- G-force sensor checked. ____
- Lateral acceleration sensor checked. ____
- TCS hydraulic actuator checked. ____
- TCS disable switch checked. ____
- Engine torque management control checked. ____

List any defective components found: _____

_____

_____

☐ 15. Consult with your instructor about any needed repairs and make any needed repairs according to the service manual and as approved by your instructor.

Repairs made: _____

_____

_____

_____

☐ 16. Recheck ABS/TCS system operation.

Does the system operate properly? Yes ____ No ____

If No, repeat the above steps to isolate the problem.

Figure 124-1. A scan tool can be used to check the operation of various anti-lock brake system components.

Obtain and Interpret Speed Sensor Waveforms

> **Note**
>
> This procedure assumes that the wheel speed sensors have already been checked for obvious damage to the sensor and tone wheel. See Job 121 for information on checking the sensor and tone wheel.

☐ 1. Obtain a waveform meter or oscilloscope.

> **Note**
>
> It may be possible to obtain the waveform readings with a scan tool while driving the vehicle. Consult the vehicle and scan tool manufacturers' instructions.

☐ 2. Set the waveform meter or oscilloscope to record wheel sensor waveform.

☐ 3. Disconnect wheel speed sensor electrical connector and connect the waveform meter or oscilloscope to the wheel speed sensor.

☐ 4. Spin the wheel and observe the waveform meter or oscilloscope screen.

☐ 5. Compare the waveform produced by the wheel speed sensor with the correct waveform as shown in the service literature.

Does the pattern match? Yes ___ No ___

If Yes, go to step 7.

If No, go to step 6.

☐ 6. Consult with your instructor to determine what repair steps are necessary.

Repairs needed: _____

☐ 7. Remove the waveform meter or oscilloscope and reconnect the electrical connector.

☐ 8. Repeat steps 2 through 5 for all suspect wheel speed sensors.

☐ 9. Clean the work area and return any equipment to storage.

☐ 10. Did you encounter any problems during this procedure? Yes ___ No ___

If Yes, describe the problems: _____

_____

What did you do to correct the problems? _____

_____

☐ 11. Have your instructor check your work and sign this job sheet.

Performance Evaluation—Instructor Use Only

Did the student complete the job in the time allotted? Yes ___ No ___

If No, which steps were not completed?_____

How would you rate this student's overall performance on this job?_____

5–Excellent, 4–Good, 3–Satisfactory, 2–Unsatisfactory, 1–Poor

Comments: _____

_____

_____

INSTRUCTOR'S SIGNATURE _____

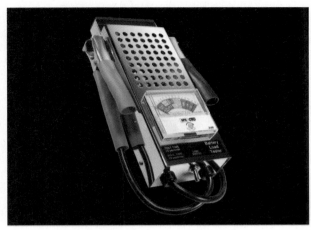

Project

Diagnosing and Servicing a Starting System

36

Introduction

All starting systems are similar and use the same basic components. Although starting systems are relatively simple, they can develop many problems. A technician must diagnose a starting problem carefully because other engine problems can appear to be starting system defects. Also, components to prevent accidental or unauthorized engine cranking have been incorporated into modern starting systems. All of these possible causes should be examined before a starting system component is replaced. In Job 130, you will inspect, test, and diagnose a vehicle's starting system. You will replace a starter relay or solenoid in Job 131, an ignition switch or neutral safety switch in Job 132, and a starter in Job 133.

Project 36 Jobs

- Job 130—Inspect, Test, and Diagnose a Starting System
- Job 131—Replace a Starter Relay or Solenoid
- Job 132—Replace an Ignition Switch or Neutral Safety Switch
- Job 133—Replace a Starter

Tools and Materials

The following list contains the tools and materials that may be needed to complete the jobs in this project. The items used will depend on the make and model of vehicle being serviced.

- Vehicle with a starting system in need of service.
- Applicable service information.
- Multimeter.
- Starting and charging system analyzer.
- Non-powered test light.
- Hand tools as needed.
- Safety glasses and other protective equipment as needed.

Safety Notice

Before performing this job, review all pertinent safety information in the text, and review safety information with your instructor.

Notes

Job 130—Inspect, Test, and Diagnose a Starting System

After completing this job, you will be able to diagnose a starter and related starting system components.

Procedures

☐ 1. Obtain a vehicle to be used in this job. Your instructor may direct you to perform this job on a shop vehicle.

☐ 2. Gather the tools needed to perform the following job. Refer to the project's tools and materials list.

☐ 3. Try to start the vehicle.

   Did the vehicle start? Yes ___ No ___

   Describe any unusual conditions you noticed while attempting to start the vehicle:_____

   _____

   _____

Determine Whether the Starting Problem Is Electrical or Mechanical

☐ 1. Attempt to turn the engine with flywheel turner or socket and breaker bar at the front of the crankshaft.

   Can the engine be turned? Yes ___ No ___

☐ 2. Determine from the actions in step 1 whether the problem is electrical or mechanical.

   Electrical ___ Mechanical ___

   Explain your answer:_____

   _____

   _____

> **Note**
>
> If the engine has a mechanical problem, consult your instructor about what steps to take.

Test the Starter's Electrical Properties

☐ 1. Obtain the needed starting and charging system tester(s).

☐ 2. Obtain the correct electrical specifications for the vehicle you are testing.

☐ 3. Attach the starting and charging system tester to the vehicle according to the manufacturer's instructions. **Figure 130-1** shows how an inductive pick up is attached to read amperage draw.

☐ 4. Check the battery voltage and general battery condition.

   What was the voltage reading?_____

   Were problems found with the battery condition? Yes ___ No ___

   If Yes, describe them: _____

   _____

☐ 5. Disable the ignition or fuel system as necessary to prevent the engine from starting.

☐ 6. Have an assistant crank the engine.

Figure 130-1. In this photograph a multimeter is being used with an inductive pick up to measure amperage. Always clamp the pick up over the negative battery cable.

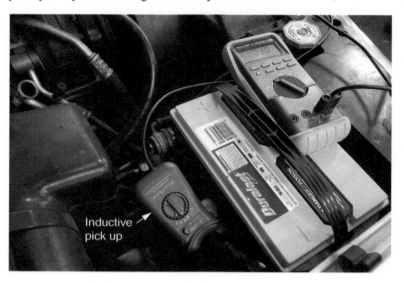

Inductive pick up

7. Using the tester, check the starter current draw and battery voltage as the starter is operated.

Record the amperage draw:_____ amps.

Record the voltage drop:_____ volts. See **Figure 130-2**.

8. Compare the readings with specifications.

Does there appear to be a problem? Yes ___ No ___

If Yes, what is the problem? _____

_____

_____

Use Waveforms to Determine Electrical Problems

> **Note**
>
> Starters produce a distinctive waveform, usually called a sawtooth waveform, and are good candidates for waveform analysis.

Figure 130-2. The cranking voltage shown on this system indicates a good battery and starter. Also check starter amperage draw when possible.

Job 130—Inspect, Test, and Diagnose a Starting System (continued)

1. Obtain a waveform meter and set it to read DC voltage. Make other adjustments as necessary.
2. Attach the positive lead of the waveform meter to a positive battery cable connection.
3. Attach the negative lead of the waveform meter to a metal engine component.
4. Operate the starter and note the waveform.
5. Draw the waveform in the space below.

6. Compare the waveform to a known good waveform. Does the waveform match the known good waveform? Yes _____ No _____

   If No, how does it differ? _____

   _____

   _____

   _____

   What could be causing the incorrect waveform? _____

   _____

   _____

   _____

Make Starting Circuit Voltage Drop Tests

☐ 1. Set a multimeter to read voltage.

☐ 2. Place the leads of the multimeter on either side of the connection to be tested, **Figure 130-3**.

☐ 3. Have an assistant crank the engine.

☐ 4. Observe and record the voltage reading on the multimeter. This reading is the voltage drop.

   Voltage drop:_____

☐ 5. Compare the voltage drop to specifications.

   Is the voltage drop excessive? Yes ___ No ___

   If Yes, what should be done?_____

   _____

Check the Starting System Wiring

☐ 1. Obtain system wiring diagrams.

☐ 2. Using the wiring diagram as a guide, inspect all starting system–related wires. Place a check next to any of the following conditions that you observe:

   • Insulation is discolored and/or swollen (overheating). ___

   • White or green deposits are visible on exposed wires (corrosion). ___

Figure 130-3. Place the multimeter leads on either side of the connection to be tested. Modern multimeters will record positive or negative voltage, so polarity is not important.

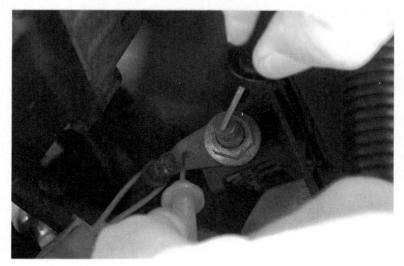

- Connectors are partially melted or their metal is discolored (overheating). ___
- Connectors are loose or disconnected. ___

If any problems were found, describe the location and the corrective action to be taken:

_____

_____

_____

☐ 3. To make resistance checks of a wire or connector, obtain a multimeter and set it to read resistance.

☐ 4. Disconnect the wire or connector to be tested from the vehicle's electrical system.

☐ 5. Place the leads of the multimeter on each end of the wire or connector to be checked and observe the reading.

Reading:_____ ohms.

☐ 6. Compare the reading taken in step 5 to specifications.

Does the wire or connector have excessively high resistance? Yes ___ No ___

If Yes, what should be done next?_____

Test a Solenoid or Relay

☐ 1. Inspect the solenoid or relay for obvious damage.

Did you find any damage? Yes ___ No ___

☐ 2. Obtain a non-powered test light or multimeter. If a multimeter is to be used, set it to read voltage.

☐ 3. Using the test light or multimeter, determine whether voltage is available to the solenoid or relay.

Is voltage available to the solenoid? Yes ___ No ___

If Yes, go to step 7.

If No, go to step 4.

☐ 4. If voltage is not available, probe backward to the spot in the circuit where voltage is available.

Describe the spot where voltage stops: _____

_____

_____

Name _____   Date_____

Instructor _____   Period _____

Job 130—Inspect, Test, and Diagnose a Starting System (continued)

☐ 5. Check the wiring or components at the spot where voltage becomes available. Wiring can be broken internally, so inspect it carefully. Components include connectors, fuses, circuit breakers, and fusible links.

Describe any defects found: _____

_____

_____

☐ 6. Repair the problem to provide power to the relay or solenoid.

☐ 7. If voltage is available to the device, have an assistant operate the circuit. Listen for a clicking noise, or note whether:
- The relay transfers power to the starter. Yes ___ No ___
- The solenoid operates the starter. Yes ___ No ___

If the relay or solenoid now operates, no further checking is needed.

If the relay or solenoid does not operate, obtain a multimeter and set it to read resistance.

☐ 8. Obtain resistance specifications for the relay or solenoid.

Terminals to check:_____ and _____ Resistance range:_____ ohms

Terminals to check:_____ and _____ Resistance range:_____ ohms

Terminals to check:_____ and _____ Resistance range:_____ ohms

☐ 9. Follow the manufacturer's directions to measure the resistance of the relay or solenoid.

Terminals checked:_____ and _____ Resistance reading:_____ ohms

Terminals checked:_____ and _____ Resistance reading:_____ ohms

Terminals checked:_____ and _____ Resistance reading:_____ ohms

Is the resistance reading within specifications. Yes ___ No ___

In your own words, describe what the readings mean:_____

_____

_____

What actions should be taken to correct the problem?_____

_____

_____

Test a Suspected Ignition Switch or Neutral Safety Switch

☐ 1. Use a non-powered test light or multimeter set to read voltage to determine whether voltage is available to the switch.

If voltage is available, go to step 7. If voltage is not available, go to step 2.

☐ 2. If voltage is not available, probe backward to the spot in the circuit where voltage is available.

Describe the spot where voltage stops:_____

_____

_____

☐ 3. At the spot where voltage becomes available, check the wiring or components. Wiring can be broken internally, so inspect it carefully. Components include connectors, fuses, circuit breakers and fusible links.

Describe any defects found: _____

_____

_____

☐ 4. Repair the problem to provide power to the ignition or neutral safety switch.

5. Move the probe or multimeter lead to the output terminal of the switch. Have an assistant operate the circuit while you observe the meter or probe.

Does voltage pass through the switch during operation? Yes ___ No ___

If Yes, no further checking is needed.

If No, obtain a multimeter and set it to read resistance.

> **Note**
>
> An ignition switch is a three- or four-position switch with two or three output terminals. One output terminal should have power when the switch is in the *start* position. Another output terminal should have power when the switch is in the *on* position. The third output terminal should have power when the switch is in the *accessories* position.

6. Obtain resistance specifications for the switch.

Terminals to check:_____ and _____ Resistance range:_____ ohms

Terminals to check:_____ and _____ Resistance range:_____ ohms

Terminals to check:_____ and _____ Resistance range:_____ ohms

7. Follow the manufacturer's directions to measure the resistance of the switch.

Terminals checked:_____ and _____ Resistance reading:_____ ohms

Terminals checked:_____ and _____ Resistance reading:_____ ohms

Terminals checked:_____ and _____ Resistance reading:_____ ohms

Is the resistance reading within specifications? Yes ___ No ___

In your own words, describe what the readings mean:_____

_____

_____

What actions should be taken to correct the problem?_____

_____

_____

8. Clean the work area and return any equipment to storage.

9. Did you encounter any problems during this procedure? Yes ___ No ___

If Yes, describe the problems: _____

_____

What did you do to correct the problems? _____

_____

10. Have your instructor check your work and sign this job sheet.

Performance Evaluation—Instructor Use Only

Did the student complete the job in the time allotted? Yes ___ No ___

If No, which steps were not completed?_____

How would you rate this student's overall performance on this job?_____

5–Excellent, 4–Good, 3–Satisfactory, 2–Unsatisfactory, 1–Poor

Comments: _____

_____

_____

INSTRUCTOR'S SIGNATURE _____

Job 131—Replace a Starter Relay or Solenoid

After completing this job, you will be able to replace a starter relay or solenoid.

Procedures

> **Note**
> Many vehicles have a starter solenoid that is an integral part of the starter unit. In such cases, the entire starter assembly is usually replaced when the solenoid requires service.

- [ ] 1. Obtain a vehicle to be used in this job. Your instructor may direct you to perform this job on a shop vehicle.
- [ ] 2. Gather the tools needed to perform the following job. Refer to the project's tools and materials list.
- [ ] 3. Remove the battery negative cable.
- [ ] 4. Raise the vehicle if necessary to gain access to the relay or solenoid.

> **Warning**
> The vehicle must be raised and supported in a safe manner. Always use approved lifts or jacks and jack stands.

- [ ] 5. Remove the wires attached to the relay or solenoid. Mark the wires for reattachment if necessary.

> **Note**
> It may be necessary to remove the starter at this point.

- [ ] 6. Remove the relay or solenoid fasteners and remove the part from the vehicle, **Figure 131-1**.
- [ ] 7. Compare the old and replacement parts to ensure that the replacement is correct.
- [ ] 8. Place the new relay or solenoid in position and install the fasteners.
- [ ] 9. Reinstall the wires on the device.
- [ ] 10. Lower the vehicle if necessary.
- [ ] 11. Reinstall the battery negative cable.
- [ ] 12. Test the operation of the replacement relay or solenoid.
- [ ] 13. Clean the work area and return any equipment to storage.
- [ ] 14. Did you encounter any problems during this procedure? Yes ___ No ___

    If Yes, describe the problems: _____

    _____

    What did you do to correct the problems? _____

    _____

- [ ] 15. Have your instructor check your work and sign this job sheet.

Figure 131-1. Once the fasteners have been removed, slide the solenoid from the starter plunger and spring assembly.

Performance Evaluation—Instructor Use Only

Did the student complete the job in the time allotted? Yes ___ No ___

If No, which steps were not completed?_____

How would you rate this student's overall performance on this job?_____

5–Excellent, 4–Good, 3–Satisfactory, 2–Unsatisfactory, 1–Poor

Comments: _____

_____

_____

INSTRUCTOR'S SIGNATURE _____

Job 132—Replace an Ignition Switch or Neutral Safety Switch

After completing this job, you will be able to replace a faulty ignition switch or neutral safety switch.

Procedures

☐ 1. Obtain a vehicle to be used in this job. Your instructor may direct you to perform this job on a shop vehicle.

☐ 2. Gather the tools needed to perform the following job. Refer to the project's tools and materials list.

☐ 3. Remove the battery negative cable.

> **Warning**
> ⚠ If the manufacturer's instructions call for disabling the air bag system, do so now. Refer to Job 147.

☐ 4. Remove dashboard or console components as necessary to gain access to the switch.

> **Caution**
> ◇ If you are replacing a neutral safety switch, determine the location of the switch before removing interior parts. The neutral switch on many modern vehicles is located at the manual valve shaft of the transmission or transaxle. It may be necessary to raise the vehicle to reach this type of neutral safety switch.

☐ 5. Remove the wires attached to the switch. Mark the wires for reattachment if necessary.

☐ 6. Remove the switch fasteners and remove the part from the vehicle.

> **Note**
> 📄 Most ignition switch removal procedures call for the use of special tools. Consult the appropriate service literature before attempting to remove the ignition switch.

☐ 7. Compare the old and replacement switches to ensure that the replacement is correct.

☐ 8. Place the new switch in position and install the fasteners.

☐ 9. Reinstall the wires on the switch.

> **Note**
> 📄 Many neutral safety switches must be adjusted to operate in neutral and park only. Consult the appropriate service literature to adjust the switch.

☐ 10. Reinstall dashboard or console components as necessary.

☐ 11. Enable the air bag system. Refer to Job 147.

☐ 12. Reinstall the battery negative cable.

☐ 13. Test the operation of the replacement switch.

☐ 14. Clean the work area and return any equipment to storage.

☐ 15. Did you encounter any problems during this procedure? Yes ___ No ___

If Yes, describe the problems: _____

_____

What did you do to correct the problems? _____

_____

☐ 16. Have your instructor check your work and sign this job sheet.

Performance Evaluation—Instructor Use Only

Did the student complete the job in the time allotted? Yes ___ No ___

If No, which steps were not completed?_____

How would you rate this student's overall performance on this job?_____

5–Excellent, 4–Good, 3–Satisfactory, 2–Unsatisfactory, 1–Poor

Comments: _____

_____

_____

INSTRUCTOR'S SIGNATURE _____

Job 133—Replace a Starter

After completing this job, you will be able to replace a starter.

Procedures

☐ 1. Obtain a vehicle to be used in this job. Your instructor may direct you to perform this job on a shop vehicle.

☐ 2. Gather the tools needed to perform the following job. Refer to the project's tools and materials list.

☐ 3. Remove the battery negative cable.

☐ 4. Raise the vehicle.

> **Warning**
>
> ⚠ The vehicle must be raised and supported in a safe manner. Always use approved lifts or jacks and jack stands.

☐ 5. Remove inspection covers, exhaust components, and skid plates as necessary to gain access to the starter.

☐ 6. Remove the starter wiring, **Figure 133-1**. Depending on the design, you may have to perform this step after completing step 7.

☐ 7. Remove the starter fasteners, **Figure 133-2**.

> **Note**
>
> If adjusting shims are used at the starter, note their position and save them for reinstallation.

☐ 8. Remove the starter from the vehicle.

Figure 133-1. If possible, remove the starter wiring before removing the starter. Most modern starters have only two wires.

Figure 133-2. This starter is held to the engine block with two fasteners. Others are attached to the transmission housing.

☐   9.   Compare the old and replacement starters.

☐ 10. Inspect the flywheel teeth.

         Did you find damage or excessive wear on the flywheel teeth? Yes ___ No ___
         If Yes, the flywheel must be replaced.

☐ 11. Place the starter in position.

☐ 12. Reinstall the wires if this must be done before step 13.

☐ 13. Install any adjusting shims and then install and tighten the starter fasteners. Reinstall any remaining starter wiring.

☐ 14. Reinstall any inspection covers or skid plates that were removed.

☐ 15. Lower the vehicle and reconnect the battery negative cable.

☐ 16. Check starter operation.

> **Note**
>
> If the starter is noisy, and uses adjustment shims, add or remove shims until the starter operates quietly.

☐ 17. Clean the work area and return any equipment to storage.

☐ 18. Did you encounter any problems during this procedure? Yes ___ No ___
         If Yes, describe the problems: _____

         _____

         What did you do to correct the problems? _____

☐ 19. Have your instructor check your work and sign this job sheet.

Job 133—Replace a Starter (continued)

Performance Evaluation—Instructor Use Only

Did the student complete the job in the time allotted? Yes ___ No ___

If No, which steps were not completed?_____

How would you rate this student's overall performance on this job?_____

5–Excellent, 4–Good, 3–Satisfactory, 2–Unsatisfactory, 1–Poor

Comments: _____

_____

_____

INSTRUCTOR'S SIGNATURE _____

Notes

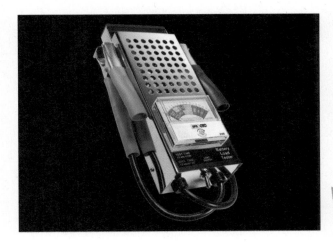

Project

Diagnosing and Servicing a Charging System

37

Introduction

Charging systems are basically simple; they include an alternator and voltage regulator to recharge the battery and provide other electrical needs. However, from vehicle to vehicle, the systems have many small differences that require specialized service. Modern charging systems are usually controlled through the vehicle computer.

Charging system problems are often mistaken for battery and starter problems. It is therefore important to know how to determine whether the charging system is working properly. In Job 134, you will diagnose and service a charging system. You will replace an alternator in Job 135.

Project 37 Jobs

- Job 134—Inspect, Test, and Diagnose a Charging System
- Job 135—Replace an Alternator

Tools and Materials

The following list contains the tools and materials that may be needed to complete the jobs in this project. The items used will depend on the make and model of vehicle being serviced.

- Vehicle with a charging system in need of service.
- Applicable service information.
- Multimeter.
- Starting and charging system analyzer.
- Hand tools as needed.
- Safety glasses and other protective equipment as needed.

Safety Notice

Before performing this job, review all pertinent safety information in the text, and review safety information with your instructor.

Notes

Job 134—Inspect, Test, and Diagnose a Charging System

After completing this job, you will be able to inspect, test, and diagnose a vehicle's charging system.

Procedures

☐   1.   Obtain a vehicle to be used in this job. Your instructor may direct you to perform this job on a shop vehicle.

☐   2.   Gather the tools needed to perform the following job. Refer to the project's tools and materials list.

Check the Operation of the Charging System

☐   1.   Obtain the needed starting and charging system tester.

☐   2.   Obtain the correct electrical specifications for the vehicle that you are testing.

☐   3.   Inspect the alternator drive belt.

Is the belt properly tightened? Yes ___ No ___

Is the belt in good condition? Yes ___ No ___

If you answered No to either of these questions, explain your answer:_____

_____

If the belt is not properly tightened or is damaged, see the belt maintenance information in Job 33.

> **Note**
>
> The belt driving the alternator must be in good condition before the charging system can be checked.

☐   4.   Attach the starting and charging system tester to the vehicle according to the manufacturer's instructions.

☐   5.   Check the battery voltage and general battery condition.

What was the voltage reading?_____

Were problems found with the battery condition or voltage? Yes ___ No ___

If Yes, describe them: _____

_____

> **Caution**
>
> If the battery is discharged or defective, the charging system cannot be adequately tested.

☐   6.   Start the vehicle and check charging system operation. **Figure 134-1** shows idle voltage under load and no load conditions.

What is the charging system voltage at idle?_____ volts.

What is the charging system voltage at 2500 rpm?_____ volts.

What was the maximum charging system amperage?_____ amps.

659

Figure 134-1. Comparing the charging voltage under load and no load conditions gives a good indication of charging system operation.

7. Compare the charging system readings with system specifications.

Are all of the readings within specifications? Yes ___ No ___

If No, list the out-of-specification readings and describe the possible causes: _____

_____

8. Check the alternator stator and diodes (if this test can be made). The tester in **Figure 134-2** indicates whether the diodes are operating properly. Other testers require that you interpret a waveform.

Are the stator and diodes OK? Yes ___ No ___

What possible defects have you identified? _____

_____

Figure 134-2. Testing the diode condition ensures that the charging system is operating properly.

Job 134—Inspect, Test, and Diagnose a Charging System (continued)

☐ 9. If your readings are out of the specified range, make further checks. On some vehicles, it is possible to bypass the voltage regulator. On other vehicles, the charging system can be diagnosed with a scan tool.

Were further checks possible? Yes ___ No ___

If Yes, what did they reveal? _____

_____

_____

_____

Perform Voltage Drop Tests

☐ 1. Obtain a multimeter and set it to read voltage.

☐ 2. Place the leads of the multimeter on either side of the connection to be tested.

☐ 3. Start the engine.

☐ 4. Observe and record the voltage reading on the multimeter. This is the voltage drop.

Voltage drop:_____ volts

☐ 5. Compare the voltage drop to specifications.

Is the voltage drop excessive? Yes ___ No ___

If Yes, what action should be taken?_____

_____

Use Waveforms to Determine Electrical Problems

Note

Alternator stators produce a distinctive waveform and alternators are good candidate for waveform analysis.

1. Obtain a waveform meter and set it to read dc voltage. Make other adjustments as necessary.

2. Attach the positive lead of the waveform meter to the stator terminal of the alternator to be tested.

3. Attach the negative lead of the waveform meter to a metal part on the engine.

4. Start the engine and note the alternator waveform.

5. Draw the waveform in the space below.

☐ 6. Compare the waveform to a known good waveform. Does the waveform match the known good waveform? Yes _____ No _____

If No, how does it differ? _____

_____

_____

_____

What could be causing the incorrect waveform? _____

_____

_____

_____

☐ 7. Clean the work area and return any equipment to storage.

☐ 8. Did you encounter any problems during this procedure? Yes ___ No ___

If Yes, describe the problems: _____

_____

What did you do to correct the problems? _____

_____

☐ 9. Have your instructor check your work and sign this job sheet.

Performance Evaluation—Instructor Use Only

Did the student complete the job in the time allotted? Yes ___ No ___

If No, which steps were not completed?_____

How would you rate this student's overall performance on this job?_____

5–Excellent, 4–Good, 3–Satisfactory, 2–Unsatisfactory, 1–Poor

Comments: _____

_____

_____

INSTRUCTOR'S SIGNATURE _____

Job 135—Replace an Alternator

After completing this job, you will be able to replace a vehicle's alternator.

Procedures

- ☐ 1. Obtain a vehicle to be used in this job. Your instructor may direct you to perform this job on a shop vehicle.
- ☐ 2. Gather the tools needed to perform the following job. Refer to the project's tools and materials list.
- ☐ 3. Remove the battery negative cable.
- ☐ 4. Remove the alternator drive belt. Refer to the belt service information in Job 33.
- ☐ 5. Remove the alternator wiring connectors.
- ☐ 6. Remove the alternator fasteners.
- ☐ 7. Remove the alternator from the vehicle.
- ☐ 8. Compare the old and replacement alternators.
- ☐ 9. Place the alternator in position on the engine, **Figure 135-1**.
- ☐ 10. Install the alternator fasteners and any adjusting shims.
- ☐ 11. Reinstall the alternator wire connectors.
- ☐ 12. Reinstall the alternator drive belt.
- ☐ 13. Adjust the drive belt if necessary and tighten the alternator fasteners.
- ☐ 14. Reconnect the battery negative cable.
- ☐ 15. Start the engine and check alternator operation.
- ☐ 16. Clean the work area and return any equipment to storage.

Figure 135-1. Carefully place the alternator in position.

☐ 17. Did you encounter any problems during this procedure? Yes ___ No ___

If Yes, describe the problems: _____

_____

What did you do to correct the problems? _____

_____

☐ 18. Have your instructor check your work and sign this job sheet.

Performance Evaluation—Instructor Use Only

Did the student complete the job in the time allotted? Yes ___ No ___

If No, which steps were not completed?_____

How would you rate this student's overall performance on this job?_____

5–Excellent, 4–Good, 3–Satisfactory, 2–Unsatisfactory, 1–Poor

Comments: _____

_____

_____

INSTRUCTOR'S SIGNATURE _____

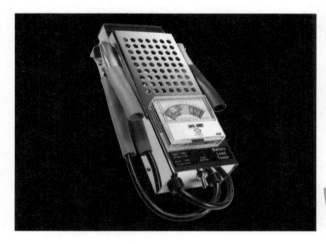

Project
38
Diagnosing and Servicing a Lighting System

Introduction

Some parts of the vehicle lighting system have not changed over many years. Some relatively simple tests and adjustments can be made with electrical testers. The most common problem in the vehicle light circuit is a burned out bulb. Other problems, such as defective switches or high resistance in the circuit, can cause a light to fail to illuminate. In Job 136, you will diagnose a lighting system. You will replace and aim headlights in Job 137.

Project 38 Jobs

- Job 136—Diagnose Lighting System Problems
- Job 137—Replace and Aim Headlights

Tools and Materials

The following list contains the tools and materials that may be needed to complete the jobs in this project. The items used will depend on the make and model of vehicle being serviced.

- Vehicle with lighting system in need of service.
- Applicable service information.
- Multimeter.
- Non-powered test light.
- Hand tools as needed.
- Safety glasses and other protective equipment as needed.

Safety Notice

Before performing this job, review all pertinent safety information in the text, and review safety information with your instructor.

Notes

Job 136—Diagnose Lighting System Problems

After completing this job, you will be able to diagnose problems in a vehicle's lighting systems.

Procedures

☐　1.　Obtain a vehicle to be used in this job. Your instructor may direct you to perform this job on a shop vehicle.

☐　2.　Gather the tools needed to perform the following job. Refer to the project's tools and materials list.

☐　3.　Obtain the needed system wiring diagrams.

General Light Service

☐　1.　Remove the non-functioning bulb and socket from the housing.

☐　2.　Inspect the bulb. Most burned out bulbs will be obviously bad. A defective filament will vibrate when the bulb is lightly shaken.

　　　　Is the bulb burned out? Yes ____ No ____

　　　　If Yes, replace the bulb and recheck its operation.

☐　3.　Visually check lamp sockets for the following:
- Cracked or overheated plastic socket body. ____
- Dented, crushed, or otherwise deformed metal socket body. ____
- Corrosion in metal socket body. ____
- Bent or broken tabs, slots, or other locking devices. ____
- Corroded electrical connectors. ____
- Disconnected electrical connectors. ____

☐　4.　If the bulb and socket appear to be good, use a non-powered test light to check the power and ground circuits at the electrical connector. The control switch should be in the *on* position. **Figure 136-1** shows a typical procedure.

　　　　Are power and ground available at the connector? Yes ____ No ____

　　　　If No, go to step 4.

　　　　If Yes, reconnect the socket and test it for power and ground.

　　　　Are power and a good ground available at the socket? Yes ____ No ____

　　　　If No, replace the socket and recheck its operation.

　　　　If Yes, replace the bulb and recheck operation.

> **Note**
>
> Some older vehicles had a single wire to the bulb socket. The sockets were directly grounded to the vehicle.

Figure 136-1. The electrical connector should be checked to ensure that power is available and that the circuit is grounded. Modern vehicles with many plastic and fiberglass parts usually have a separate ground wire.

☐ 5. Trace the lighting circuit in the proper wiring diagram and answer the following questions: In the light's circuit, what light control switches does the current pass through? _____

_____

In the light's circuit, what relays, circuit breakers, and resistors does power pass through?

_____

What frame or body parts are part of the light's ground circuit? These are the body parts that current passes through on its way back to the battery's negative terminal._____

_____

Is the circuit a parallel, series, or series-parallel type?_____

☐ 6. After referring to the wiring diagram, redraw the circuit, eliminating any wiring that is not directly related to the circuit.

Job 136—Diagnose Lighting System Problems (continued)

☐ 7. Using the wiring diagram, perform checks to identify the wiring or electrical device that caused the light to fail.

Describe the problem and the actions taken to correct it:_____

_____

_____

> **Note**
>
> On older vehicles, a wire often becomes disconnected or a connector plug develops high resistance. This is especially common with taillights and marker lights where the connector is exposed to water and dirt.

Diagnose a Brake Light Problem

☐ 1. Check for the proper operation of the brake light system.

Does the brake light system function properly? Yes ___ No ___

If No, describe the problem: _____

_____

Diagnose the problem using the procedures in one of the following sections.

> **Note**
>
> Brake light problems fall into four categories:
>
> - Some brake lights do not light when the brake pedal is depressed.
> - No brake lights come on when the brake pedal is depressed.
> - The brake lights are on at all times.
> - The front parking lights flash when the brakes are applied.
>
> The diagnosis and service of these problems is discussed in the sections that follow.

Brake Lights Do Not Light

☐ 1. Ensure that the suspected defective light is supposed to come on when the pedal is pressed. Many modern rear light systems use double filament bulbs for the turn signals and brake lights and separate single filament bulbs for the taillights.

☐ 2. Remove the bulb and inspect it as was explained earlier in *General Light Service* section of this job.

Is the bulb burned out? Yes ___ No ___

If Yes, replace the bulb and recheck operation.

> **Note**
>
> The double filament brake light bulbs used on older vehicles may have a burned-out brake light filament but a good taillight filament, or vice versa. If either filament is broken, replace the bulb.

3. If the bulb appears to be good, check for power at the center terminal of the socket using a non-powered test light.

   If power is reaching a good bulb, and it still does not light, go to step 4.

   If power is not reaching the bulb, check the wiring between the switch and light.

4. Inspect the socket for the following conditions. Check all that apply.
   - Socket is properly grounded (check with ohmmeter). ___
   - Bulb contacts are not worn, melted or corroded. ___
   - There is no grease or rust in the bulb socket. ___

5. If the inspection in step 4 turned up any problems, repair or replace the socket as necessary, then recheck bulb operation.

> **Note**
>
> On some older vehicles, a faulty turn signal switch may cause the lights to be off on one side of the vehicle.

All Brake Lights Are Off

> **Note**
>
> When none of the brake lights work, it is unlikely that all of the bulbs have burned out at once. Almost always the defect is in a main electrical device, either the brake light switch or the fuse.

1. Check the fuse controlling the brake lights. If other electrical devices served by this fuse are also inoperative, you can conclude that the fuse is probably blown even before looking at it.

   Is the fuse blown? Yes ___ No ___

   If Yes, replace the fuse and recheck brake light operation.

> **Caution**
>
> If the fuse is blown, you must determine what caused the electrical overload.
> Do not simply replace the fuse and let the vehicle go.

2. If the fuse is OK, disconnect the brake light switch. **Figure 136-2** shows a modern brake light switch. Be sure to disconnect the proper connector.

> **Caution**
>
> Do not confuse the brake light switch with the cruise control release switch.

3. Check the operation of the brake light switch. Use a multimeter or a powered test light.

   Does the switch have infinite resistance (test light off) when the pedal is in the unapplied position? Yes ___ No ___

   Does the switch have low resistance (test light on) when the brake pedal is lightly pressed? Yes ___ No ___

   If the answer is No to either of these questions, adjust or replace the switch as necessary.

   If the answer to both of these questions is Yes, look for disconnected, broken, or shorted wiring.

Job 136—Diagnose Lighting System Problems (continued)

Figure 136-2. The brake light switch on modern vehicles may have inputs to the cruise control and ABS systems. Be sure that you locate the brake light connector before beginning any electrical tests.

Brake Lights On at All Times

☐ 1. Unplug the brake light switch.

Do the lights go off? Yes ___ No ___

If Yes, adjust or replace the brake light switch.

If No, check for shorted wiring.

Front Parking Lights Flash When Brakes Are Applied

☐ 1. Observe the brake light bulbs as an assistant presses on the brake pedal.

Do any bulbs illuminate weakly or not at all? Yes ___ No ___

If Yes, go to step 2.

If No, go to step 4.

☐ 2. Remove and check the bulb.

Is the bulb a dual filament type? Yes ___ No ___

If Yes, go to step 3.

If No, go to step 4.

☐ 3. Closely observe the bulb filaments.

Has one of the filaments broken and fused to the other? Yes ___ No ___

If Yes, this is the source of the problem. Replace the bulb and recheck operation.

If No, go to step 4.

☐ 4. Check for a short between the brake and taillight wiring.

> **Note**
>
> If the vehicle is an older model with dual filament bulbs, disconnect the turn signal harness.

☐ 5. Correct any problems and recheck operation.

Diagnose Problems in Hazard and Turn Signals

☐ 1. Determine which of the following problems is occurring:
- Hazard and/or turn signal lights do not illuminate. ___ Go to step 2.
- Hazard and/or turn signal lights illuminate but do not flash. ___ Go to step 5.
- Hazard and/or turn signal lights flash slowly or only one side flashes. ___ Go to step 8.

☐ 2. If the hazard/turn signal lights do not illuminate, check the fuse.

Is the fuse blown? Yes ___ No ___

If Yes, determine the cause of the overload and correct it. Then, replace the fuse and recheck operation.

If No, reinstall the fuse and go to step 3.

☐ 3. Using wiring diagrams as needed, trace the wiring and determine whether any wires are disconnected or have high resistance.

Were any wiring problems found? Yes ___ No ___

If Yes, correct the problems and recheck operation. Describe the problems found and what you did to correct them. _____

_____

_____

If No, go to step 4.

☐ 4. Check for defective electrical devices such as the flasher unit or control switch.

Were any electrical device problems found? Yes ___ No ___

If Yes, correct as needed and recheck operation.

List the faulty components and describe how the problem was corrected: _____

_____

If No, repeat steps 2 through 4 to locate the problem.

☐ 5. Check the turn signal flasher by substituting a unit known to be good.

Does this solve the problem? Yes ___ No ___

If Yes, replace the flasher and recheck operation.

If No, go to step 6.

☐ 6. Using wiring diagrams as needed, trace the wiring and determine whether any wires are disconnected or have high resistance.

Were any wiring problems found? Yes ___ No ___

If Yes, correct as needed and recheck operation. Describe the problems found and what you did to correct them: _____

_____

_____

No, go to step 7.

☐ 7. Check bulb sockets for corrosion and lack of ground. See **Figure 136-3**.

Were any problems found? Yes ___ No ___

If Yes, correct as needed and recheck operation. Describe the problems found and what you did to correct them: _____

_____

_____

If No, recheck the flashers and wiring until the problem is located.

Job 136—Diagnose Lighting System Problems (continued)

Figure 136-3. If the turn signals do not operate on one side, check the bulb sockets for looseness and corrosion. The burned plastic on this socket may indicate a connection that has overheated.

☐ 8. Using wiring diagrams as needed, determine the following:
- Do any wires have high resistance? Yes ___ No ___
- Are any bulb sockets corroded? Yes ___ No ___
- Are any bulb sockets not properly grounded? Yes ___ No ___

If the answer to any of the above is Yes, correct the problems and recheck turn signal operation. Describe the problems found and what you did to correct them: _____

_____

_____

If the answer to any of the questions is No, go to step 9.

☐ 9. Ensure that the proper flasher is being used.

Is the flasher correct? Yes ___ No ___

If No, install the proper flasher and recheck operation.

If Yes, recheck wiring and wiring sockets until the problem is located.

Reset Maintenance Indicator Light

> **Note**
>
> Indicator light resetting procedures vary between manufacturers, but procedures are almost always given in the owner's manual. For the purposes of this job, a general procedure is outlined below.

☐ 1. Turn the ignition switch to the *on* position.

☐ 2. Operate and hold a dashboard control until the light goes off, or begins blinking.

☐ 3. Turn the ignition switch to the *on* position while continuing to hold the dashboard control.

☐ 4. Turn the ignition switch back to the *on* position and determine whether the light goes off within a few seconds.

Did the light go off? Yes ___ No ___

If No, perform the resetting procedure again.

How did the actual procedure differ from the general procedure given here? _____

_____

_____

☐ 5. Clean the work area and return tools and equipment to storage.

☐ 6. Did you encounter any problems during this procedure? Yes ___ No ___

If Yes, describe the problems: _____

_____

What did you do to correct the problems? _____

_____

☐ 7. Have your instructor check your work and sign this job sheet.

Performance Evaluation—Instructor Use Only

Did the student complete the job in the time allotted? Yes ___ No ___

If No, which steps were not completed?_____

How would you rate this student's overall performance on this job?_____

5–Excellent, 4–Good, 3–Satisfactory, 2–Unsatisfactory, 1–Poor

Comments: _____

_____

_____

INSTRUCTOR'S SIGNATURE _____

Job 137—Replace and Aim Headlights

After completing this job, you will be able to replace and aim a vehicle's headlights.

Procedures

☐ 1. Obtain a vehicle to be used in this job. Your instructor may direct you to perform this job on a shop vehicle.

☐ 2. Gather the tools needed to perform the following job. Refer to the project's tools and materials list.

Replace a Headlight

Sealed Beam Headlights

☐ 1. If necessary, open the vehicle hood to gain access to headlight attachments.

☐ 2. Remove any trim that blocks access to the headlight fasteners.

☐ 3. Remove the fasteners holding the headlight to the headlight housing.

> **Caution**
> Do not move the headlight adjusting screws, this will affect headlight aim.

☐ 4. Carefully remove the headlight from the headlight housing.

☐ 5. Remove the electrical connector.

☐ 6. Compare the old and new headlight to ensure that they are the same size and have the same number of electrical contacts.

☐ 7. Install the electrical connector on the new headlight.

☐ 8. Place the headlight in position in the headlight housing.

☐ 9. Install and tighten the headlight fasteners.

☐ 10. Check headlight operation. If the light does not operate, recheck the electrical connection.

☐ 11. Replace any trim that was removed.

☐ 12. Close the hood if necessary.

Replaceable Bulb Headlights

> **Warning**
> Many modern vehicles have high intensity discharge (HID or Xenon) headlights. The igniters used on high intensity discharge headlights produce potentially hazardous voltages. The bulbs can reach temperatures of over 1300°F (2000°C) when operating. Do not touch a high intensity discharge bulb when it is illuminated. Before working on any modern headlight system, first determine whether it is a high intensity discharge type.

☐ 1. Open the vehicle hood to gain access to the bulb.

☐ 2. Remove any parts that block access to the bulb socket.

> **Note**
>
> If the headlight system is a high intensity discharge type, disconnect the battery
> cable or igniter low-voltage harness and wait several minutes before going to step 3.
> This will ensure that any high voltage has been discharged and that the bulb has
> had time to cool.

3. Remove the bulb:

   Remove the lock ring holding the bulb and socket to the headlight housing, **Figure 137-1**, then pull the bulb from the housing.
   or
   Twist the bulb and socket assembly approximately 1/4 turn counterclockwise, **Figure 137-2**, then pull the bulb from the housing.

4. Remove the bulb, **Figure 137-3**, by pulling it from the socket.

5. Push the new bulb into place, using a clean shop towel on the glass portion.

Figure 137-1. Unscrew the lock ring to release the bulb and socket assembly.

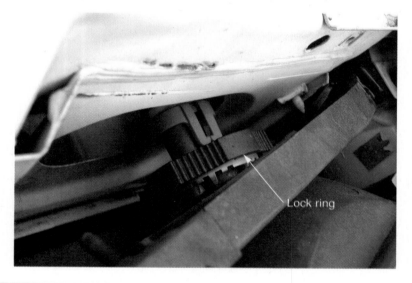

Figure 137-2. Rotate the socket about 1/4 turn to release it from the housing.

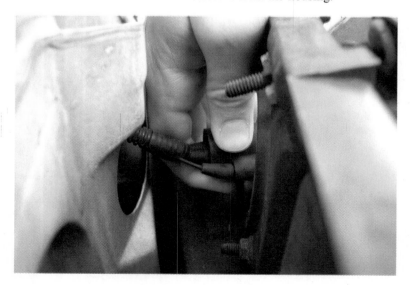

Job 137—Replace and Aim Headlights (continued)

Figure 137-3. Once the bulb is exposed, it can be pulled from the socket. Do not touch the new bulb with your bare hands.

> ⚠ **Caution**
>
> Do not touch the glass of the new bulb. Body oils on the bulb may cause it to shatter when illuminated.

 6. Reinstall the bulb and socket into the housing. Tighten the lock ring if necessary.

☐ 7. Check headlight operation. If the light does not operate, recheck the electrical connection.

☐ 8. Replace any other parts that were removed.

☐ 9. Close the hood if necessary.

Aim the Headlights

☐ 1. Place the vehicle on a level floor.

☐ 2. Test and correct headlight operation on high and low beams.

☐ 3. Make sure all tires are the same size and type and properly inflated. Correct as needed.

☐ 4. Remove any excessive loads from the vehicle trunk or bed.

☐ 5. Locate the headlight adjusters. It may be necessary to open the hood to gain access to the adjusters.

☐ 6. Install the aimers on or in front of the vehicle headlights as needed.

> 📄 **Note**
>
> Some vehicles have built-in headlight aiming bubbles, such as the one in **Figure 137-4**. Follow the manufacturer's instructions to adjust the lights.

 7. Compensate (adjust) the headlight aimers as needed.

☐ 8. Turn on the vehicle headlights. Depending on the manufacturer's instructions, set the headlights to either high or low beam.

Figure 137-4. On some vehicles, bubble-type headlight aiming devices are built into the headlight assembly.

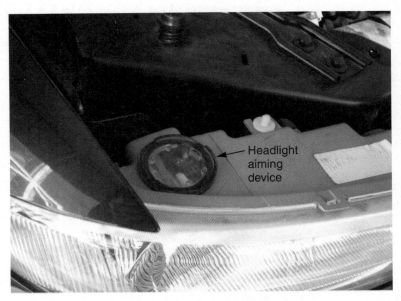

Headlight aiming device

☐ 9. Determine whether the headlights require adjustment.

Do the headlights need adjustment? Yes ____ No ____

If No, go to step 11.

☐ 10. Adjust the headlights as needed and recheck.

Briefly describe what was done to adjust the headlights:_____

_____

☐ 11. Turn off the headlights and close the hood.

☐ 12. Remove the headlight aimers, if necessary.

☐ 13. Clean the work area and return tools and equipment to storage.

☐ 14. Did you encounter any problems during this procedure? Yes ____ No ____

If Yes, describe the problems: _____

_____

What did you do to correct the problems? _____

_____

☐ 15. Have your instructor check your work and sign this job sheet.

Performance Evaluation—Instructor Use Only

Did the student complete the job in the time allotted? Yes ____ No ____

If No, which steps were not completed?_____

How would you rate this student's overall performance on this job?_____

5–Excellent, 4–Good, 3–Satisfactory, 2–Unsatisfactory, 1–Poor

Comments: _____

_____

_____

INSTRUCTOR'S SIGNATURE _____

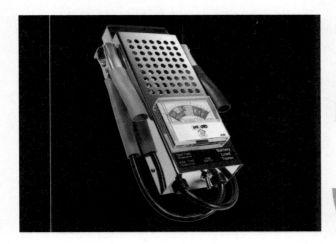

Project

Diagnosing and Servicing Gauges and Warning Lights

39

Introduction

Instrument panels contain many information gauges, such as the speedometer and fuel gauge. Some panels contain engine condition gauges as well. Instrument panels also have many warning lights. These devices sometimes need diagnosis and repair, as do the senders that deliver information to the instrument panel. In Job 138, you will diagnose an instrument panel warning light problem. You will check the operation of electromechanical gauges in Job 139 and electronic gauges in Job 140. In Job 141, you will replace a faulty warning light sensor or gauge sensor.

> **Caution**
>
> This project does not apply to digital gauges or computer controlled warning lights and gauges such as the MIL (check engine light) or the ABS/traction control warning light.

Project 39 Jobs

- Job 138—Diagnose Instrument Panel Warning Light Problems
- Job 139—Check Electromechanical Gauge Operation
- Job 140—Check Electronic Gauge Operation
- Job 141—Replace a Warning Light Sensor or Gauge Sensor

Tools and Materials

The following list contains the tools and materials that may be needed to complete the jobs in this project. The items used will depend on the make and model of vehicle being serviced.

- Vehicle with instrument panel in need of service.
- Applicable service information.
- Jumper wires.
- Multimeter.
- Special sensor testers.
- Wiring diagrams.
- Hand tools as needed.
- Safety glasses and other protective equipment as needed.

Safety Notice

Before performing this job, review all pertinent safety information in the text, and review safety information with your instructor.

Job 138—Diagnose Instrument Panel Warning Light Problems

After completing this job, you will be able to diagnose an instrument panel warning light problem.

Procedures

☐ 1. Obtain a vehicle to be used in this job. Your instructor may direct you to perform this job on a shop vehicle.

☐ 2. Gather the tools needed to perform this job. Refer to the tools and materials list at the beginning of the project.

Diagnose Instrument Panel Warning Light Problems

> **Note**
>
> There are two kinds of warning light problems:
> - The warning light does not light at any time.
> - The warning light is on at all times.
>
> Refer to the appropriate section of this job based on the type of problem you are trying to diagnose.

Warning Light Does Not Light

☐ 1. Check the fuse.

Is the fuse blown? Yes ___ No ___

☐ 2. If the fuse is blown, determine the cause of the electrical overload, correct it, and recheck warning light operation.

☐ 3. If the fuse is good, remove the connector to the electrical switch or sensor that operates the light.

☐ 4. Using a fused jumper wire, ground the switch or sensor connector, **Figure 138-1**.

Does the light come on? Yes ___ No ___

If Yes, the switch or sensor is defective. Adjust or replace as necessary and recheck.

If No, go to step 5.

> **Note**
>
> Most warning light switches and sensors operate by completing a ground through the light. Some switches, such as parking brake switches, can become misadjusted. The electrical connector may be disconnected or the switch or sensor may not be grounded. Pursue all possibilities before deciding that the switch or sensor is defective.

☐ 5. Check the warning light bulb. To reach the bulb, it may be necessary to partially disassemble the dashboard. The bulb may be burned out, or may have fallen out of its socket.

Is the bulb OK? Yes ___ No ___

If Yes, go to step 6.

If No, replace the bulb and recheck warning light operation.

Figure 138-1. Grounding the sensor connector completes the circuit through the instrument panel warning light to ground. If the light circuit is OK, the light will come on.

> **Note**
>
> In some cases, someone has removed the bulb because it was on at all times. Install a new bulb and find out why it is on all the time using the procedure in the *Warning Light Is Always On* section of this job.

6. Obtain the proper wiring diagram and trace the following:
   - Power from the vehicle's positive battery terminal to the warning light fuse.
   - Power from the warning light fuse to the warning light bulb.
   - Power from the warning light bulb to the switch or sensor.
   - Power from the switch or sensor to ground.

7. Using the wiring diagram, draw the warning light circuit, eliminating any wiring that is not directly related to the circuit.

Job 138—Diagnose Instrument Panel Warning Light Problems (continued)

8. Refer to your wiring diagram.

   Could a wiring problem cause the light to fail to illuminate? Yes ___ No ___

   If Yes, check and repair the wiring as necessary and recheck operation.

   If No, recheck the lights, fuses, and sensors as necessary.

Warning Light Is Always On

> **Note**
>
> This procedure assumes that the vehicle has no mechanical problems that would illuminate the light.

1. Determine which vehicle systems operate the light. See **Figure 138-2**.
2. Unplug the connector at the switch or sensor that operates the light, **Figure 138-3**.

   Does the light go out? Yes ___ No ___

   If Yes, go to step 3.

   If No, go to step 4.

Figure 138-2. This light indicates that the windshield washer fluid level is low. If it is on and filling the washer fluid reservoir does not cause the warning light to go out, something is wrong with the light circuit.

Figure 138-3. Unplug the sensor. If the light goes out, the sensor is defective. If the light stays on, the sensor is OK.

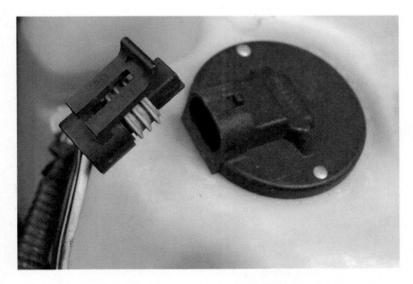

☐ 3. Replace or adjust the switch/sensor as necessary and recheck the warning light operation.

Did replacing the switch/sensor fix the problem? Yes ___ No ___

If Yes, skip to step 5.

☐ 4. If the light stays on, reattach the connector and, using the proper wiring diagram, look for a wire that has grounded against the vehicle body.

Did you locate a problem? Yes ___ No ___

If Yes, repair the wiring and recheck warning light operation.

If No, repeat this step until the problem is located.

☐ 5. Clean the work area and return any equipment to storage.

☐ 6. Did you encounter any problems during this procedure? Yes ___ No ___

If Yes, describe the problems. _____

_____

What did you do to correct the problems? _____

_____

☐ 7. Have your instructor check your work and sign this job sheet.

Performance Evaluation—Instructor Use Only

Did the student complete the job in the time allotted? Yes ___ No ___

If No, which steps were not completed?_____

How would you rate this student's overall performance on this job?_____

5–Excellent, 4–Good, 3–Satisfactory, 2–Unsatisfactory, 1–Poor

Comments: _____

_____

_____

INSTRUCTOR'S SIGNATURE _____

Job 139—Check Electromechanical Gauge Operation

After completing this job, you will be able to check electromechanical gauge operation.

Procedures

☐ 1. Obtain a vehicle to be used in this job. Your instructor may direct you to perform this job on a shop vehicle.

☐ 2. Gather the tools needed to perform this job. Refer to the tools and materials list at the beginning of the project.

> **Caution**
>
> This procedure primarily applies to checking electromechanical gauges. Use Job 140 to check electronic gauges.

☐ 3. Check the gauge fuse.

  Is the fuse OK? Yes ___ No ___

  If Yes, go to step 4.

  If No, go to step 6.

☐ 4. If the fuse is OK, unplug the electrical connection from the related sensor or switch.

  Does this cause a change in the gauge reading? Yes ___ No ___

  If Yes, go to step 6.

☐ 5. Ground the sensor connection to the engine or vehicle body using the following procedure. This procedure is for a temperature circuit, but is representative of all warning gauge checks.

  a. Note the gauge reading, **Figure 139-1**.

  b. Remove the sensor connector and ground the gauge lead to the body. As shown in **Figure 139-2**, some sensors have two connectors, one for the gauge and one for the warning light. Be sure to ground the proper connector.

  c. Observe the gauge reading, **Figure 139-3**.

  Does this cause a change in the gauge reading? Yes ___ No ___

  If No, go to step 8.

Figure 139-1. The gauge is reading minimum temperature. If the engine is at operating temperature, this indicates a problem in the gauge circuit.

Figure 139-2. Grounding the gauge lead to the body. Use a non-powered test light to avoid damaging the gauge.

Figure 139-3. Compare this photo with Figure 139-1. The gauge has moved to the maximum temperature position. This indicates that the gauge is operating.

6. Check the sensor with an ohmmeter or special sensor tester.

   Is the sensor good? Yes ___ No ___

   If Yes, go to step 8.

7. Replace the sensor as explained in Job 141, replace the fuse if necessary, and recheck gauge operation.

8. Check the voltage limiter. The limiter should hold voltage entering the gauge to a steady value, independent of charging voltage.

   Is the limiter OK? Yes ___ No ___

   If Yes, go to step 10.

Job 139—Check Electromechanical Gauge Operation (continued)

> **Note**
>
> If the voltage limiter fails, all gauges in series with the limiter will fail to operate. A defective limiter may also cause gauge readings to appear to be higher or lower than normal.

☐ 9. Replace the voltage limiter and recheck gauge operation.

Did this fix the problem? Yes ___ No ___

If Yes, go to step 11.

☐ 10. Replace the gauge and recheck operation.

Did this fix the problem? Yes ___ No ___

If No, repeat steps 3 through 10.

☐ 11. Clean the work area and return any equipment to storage.

☐ 12. Did you encounter any problems during this procedure? Yes ___ No ___

If Yes, describe the problems: _____

_____

What did you do to correct the problems? _____

_____

☐ 13. Have your instructor check your work and sign this job sheet.

Performance Evaluation—Instructor Use Only

Did the student complete the job in the time allotted? Yes ___ No ___

If No, which steps were not completed?_____

How would you rate this student's overall performance on this job?_____

5–Excellent, 4–Good, 3–Satisfactory, 2–Unsatisfactory, 1–Poor

Comments: _____

_____

_____

INSTRUCTOR'S SIGNATURE _____

Notes

_____
_____
_____
_____
_____
_____
_____
_____
_____
_____
_____
_____
_____
_____
_____
_____
_____
_____
_____
_____
_____
_____
_____
_____
_____
_____
_____
_____
_____
_____
_____
_____
_____
_____
_____
_____
_____
_____
_____
_____
_____
_____
_____
_____

Name _____   Date_____

Instructor _____   Period _____

Job 140—Check Electronic Gauge Operation

After completing this job, you will be able to check electronic gauges for proper operation.

Procedures

☐ 1. Obtain a vehicle to be used in this job. Your instructor may direct you to perform this job on a shop vehicle.

☐ 2. Gather the tools needed to perform this job. Refer to the tools and materials list at the beginning of the project.

> **Note**
>
> The following procedure is a general guide to using the scan tool to diagnose electronic gauges. If the scan tool literature calls for a different procedure or series of steps from what is in this procedure, always perform procedures according to the scan tool literature.
>
> Some electronic gauges cannot be checked with a scan tool. Often the only way to check an electronic gauge is by using a special dedicated tester or substituting a known good sensor. Some electronic gauges and sensors can be checked with a multimeter. If a scan tool cannot be used, proceed directly to step 14.

☐ 3. Obtain the proper scan tool and related service literature.

Type of scan tool: _____

☐ 4. Attach the proper test connector cable and power lead to the scan tool.

☐ 5. Ensure that the ignition switch is in the *off* position.

☐ 6. Locate the correct diagnostic connector and attach the scan tool cable.

Diagnostic connector location: _____

☐ 7. Attach the scan tool power lead to the cigarette lighter or battery.

☐ 8. Ensure that the scan tool is working properly. If the scan tool does not appear to be working, locate and correct any problems before proceeding.

☐ 9. Enter vehicle information as needed to program the scan tool. Many scan tools can be programmed with the proper vehicle information by entering the vehicle identification number (VIN). If the scan tool can read the VIN from the vehicle computer, skip this step.

☐ 10. Turn the ignition switch to the *on* position.

☐ 11. Determine whether the scan tool communicates with the computer module. If the scan tool does not appear to communicate with the computer module, go to step 12. If the scan tool and module are communicating, go to step 13.

☐ 12. Check the instrument panel and module fuses and the wiring between the module and instrument panel and the module and diagnostic connector. Refer to Job 150 as necessary. Correct problems, then proceed to step 13.

> **Note**
>
> If an electronic gauge fails to illuminate, the gauge itself may be defective. If all of the gauges fail to light up, or if the entire instrument panel is dark, the instrument panel may require replacement.

☐ 13. Use the scan tool to access the instrument panel module(s) and check for trouble codes.

Were any codes found? Yes ___ No ___

If Yes, list the codes in the spaces provided:

Code   **Defect**

_____   _____

_____   _____

_____   _____

_____   _____

_____   _____

☐ 14. Turn the ignition switch to the *off* position.

☐ 15. Perform the following actions as directed by the appropriate service literature:
- Use a dedicated tester to determine whether the problem is in the gauge, the control module, or the input sensor.
- Make electrical checks of the gauge, the control module, or the sensor.
- Substitute a known good sensor, gauge, or module and determine whether this corrects the problem.

> **Caution**
>
> ⚠ Never ground an electronic gauge sensor connection. This can damage the electronic circuitry.

Procedure used: _____

_____

Results of the procedure: _____

_____

What actions should be taken to correct the problem? _____

_____

_____

☐ 16. Make repairs or adjustments as directed by your instructor.

☐ 17. Clean the work area and return any equipment to storage.

☐ 18. Did you encounter any problems during this procedure? Yes ___ No ___

If Yes, describe the problems: _____

_____

What did you do to correct the problems? _____

_____

☐ 19. Have your instructor check your work and sign this job sheet.

Performance Evaluation—Instructor Use Only

Did the student complete the job in the time allotted? Yes ___ No ___

If No, which steps were not completed?_____

How would you rate this student's overall performance on this job?_____

5–Excellent, 4–Good, 3–Satisfactory, 2–Unsatisfactory, 1–Poor

Comments: _____

_____

_____

INSTRUCTOR'S SIGNATURE _____

Job 141—Replace a Warning Light Sensor or Gauge Sensor

After completing this job, you will be able to replace a warning light sensor or gauge sensor.

Procedures

☐   1.   Obtain a vehicle to be used in this job. Your instructor may direct you to perform this job on a shop vehicle.

☐   2.   Gather the tools needed to perform this job. Refer to the tools and materials list at the beginning of the project.

☐   3.   Remove the battery negative cable.

☐   4.   Remove the sensor's electrical connector.

☐   5.   If a cooling system sensor is being removed, depressurize and drain the cooling system. Refer to Project 9.

☐   6.   Unscrew the sensor from the attaching part. See **Figure 141-1**.

or

Remove the fastener holding the sensor to the attaching part, **Figure 141-2**.

> ### Note
> Some coolant, oil, or washer fluid level sensors are held in place with a clip. Consult the manufacturer's service information for exact removal procedures.

☐   7.   Compare the old and new sensors.

Do the threads and electrical connectors match? Yes ___ No ___

If No, what should you do next?_____

_____

☐   8.   Apply sealer to the sensor threads, if applicable.

Figure 141-1. Most temperature and pressure sensors are threaded into the engine block. A special socket may be needed to unscrew the sensor.

Figure 141-2. Most speed and RPM sensors are attached to the engine or transmission with a single fastener.

9. Thread the sensor into place and tighten it to the proper torque.

    or

    Place the sensor in position and install the attaching fastener.

10. Reinstall the electrical connector.

11. Reinstall the battery negative cable.

12. Start the engine and check sensor operation.

13. Clean the work area and return any equipment to storage.

14. Did you encounter any problems during this procedure? Yes ____ No ____

    If Yes, describe the problems: _____

    _____

    What did you do to correct the problems? _____

    _____

15. Have your instructor check your work and sign this job sheet.

Performance Evaluation—Instructor Use Only

Did the student complete the job in the time allotted? Yes ____ No ____

If No, which steps were not completed?_____

How would you rate this student's overall performance on this job?_____

5–Excellent, 4–Good, 3–Satisfactory, 2–Unsatisfactory, 1–Poor

Comments: _____

_____

_____

INSTRUCTOR'S SIGNATURE _____

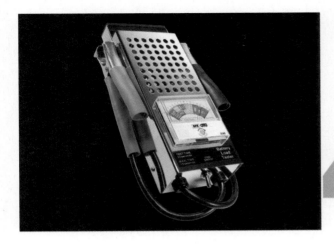

Project

Diagnosing and Servicing Horns and Windshield Washers/Wipers

40

Introduction

Horn circuits are basic devices that use contacts in the steering wheel, a relay, and the horns themselves. Most problems are caused when the steering wheel contacts either do not make contact (an open) or stay in contact continuously (a short). In Job 142, you will diagnose and service a vehicle's horn.

Every vehicle has windshield washers and wipers, and occasionally they need diagnosis and service. Generally, these accessories are relatively simple, but finding the source of a problem can sometimes be complicated. Newer vehicles may be equipped with a rain sensor that energizes the wipers whenever it senses moisture on the windshield. In Job 143, you will diagnose and service windshield washers and wipers.

Project 40 Jobs

- Job 142—Check and Service a Horn
- Job 143—Check and Correct the Operation of Windshield Washers and Wipers

Tools and Materials

The following list contains the tools and materials that may be needed to complete the jobs in this project. The items used will depend on the make and model of vehicle being serviced.

- Vehicle with a windshield wiper/washer system in need of service.
- Vehicle with a horn in need of service.
- Applicable service information.
- Multimeter.
- Fused jumper wire.
- Hand tools as needed.
- Safety glasses and other protective equipment as needed.

Safety Notice

Before performing this job, review all pertinent safety information in the text, and review safety information with your instructor.

Notes

Job 142—Check and Service a Horn

After completing this job, you will be able to diagnose and service a vehicle horn.

Procedures

☐ 1. Obtain a vehicle to be used in this job. Your instructor may direct you to perform this job on a shop vehicle.

☐ 2. Gather the tools needed to perform this job. Refer to the tools and materials list at the beginning of the project.

☐ 3. Obtain the needed system wiring diagrams.

Diagnose a Horn Problem—Horn Does Not Sound

☐ 1. Ensure that power is available to the horn circuit by probing the horn relay with a non-powered test light. Horn relays are usually located in one of the vehicle's fuse boxes, **Figure 142-1**.

Is power available? Yes ___ No ___

If Yes, go to step 2.

If No, trace the circuit until the problem is located. Be sure to check for a blown fuse in the horn circuit.

☐ 2. Turn off all noise-producing accessories.

☐ 3. Press the horn button, or close the contacts, while listening carefully.

Did the horn blow? Yes ___ No ___

Did the horn relay click? Yes ___ No ___

If the horn blows, the horn circuit is operating properly.

If the horn does not blow, but the relay clicks, go to step 4.

If the horn does not blow and the relay does not click, go to step 5.

Figure 142-1. On late-model vehicles, the horn relay is usually located in one of the fuse boxes. Older vehicles may have a separate horn relay located under the hood.

☐ 4. If the relay clicks but the horn does not operate, use a fused jumper wire to directly energize the horn, bypassing the relay.

Did the horn blow? Yes ___ No ___

If Yes, the relay is defective.

If No, the horn is defective or is not properly grounded.

☐ 5. If the relay does not click, directly energize the relay using a fused jumper wire. Most horn relays are grounded through the steering wheel contacts, so grounding the relay contact to the steering wheel should make the horn blow.

Caution

⚠ Always consult the proper wiring diagram before attempting to directly energize a horn relay.

Did grounding the relay contact cause the horn to blow? Yes ___ No ___

If Yes, the relay is good. Go to step 6.

If the horn does not operate and the relay does not click, the relay is defective. Replace the relay and recheck horn operation.

Does the horn function properly? Yes ___ No ___

If Yes, return the vehicle to its owner.

☐ 6. Remove the steering wheel cover and ground the horn wires.

Warning

⚠ If the vehicle is equipped with air bags, disable the air bag system before performing this step. Refer to Job 147.

Does the horn operate? Yes ___ No ___

If Yes, go to step 7.

If No go to step 10.

☐ 7. Obtain a multimeter and set it to read resistance.

☐ 8. Attach the multimeter leads to the horn contact leads at the steering wheel. Some vehicles have only one wire to the contacts, and the leads should be connected between the wire and a good ground.

☐ 9. Operate the steering wheel horn contacts (try to blow the horn) while checking for continuity.

Does the multimeter read close-to-zero resistance when the horn contacts are operated? Yes ___ No ___

If Yes, repeat steps 3 through 9 to determine the problem.

If No, the contacts are defective. Clean or replace the contacts as needed. Refer to the *Clean or Replace Horn Contacts* section of this job.

☐ 10. Perform continuity tests on the wiring between the steering wheel and relay to check for broken or disconnected wires.

Were any problems found? Yes ___ No ___

If Yes, repair the wiring as necessary. Reinstall the steering wheel cover, enable the air bag system, and return the vehicle to its owner.

If No, repeat steps 3 through 10 to determine the problem.

Job 142—Check and Service a Horn (continued)

Diagnose a Horn Problem—Horn Sounds Continuously

☐ 1. Unplug the horn relay.

> **Note**
>
> If the horn continues to sound with the relay unplugged, check for shorted wiring between the relay and horn(s).

☐ 2. Probe the horn relay socket coming from the horn contacts using an ohmmeter or powered test light.

   If the ohmmeter registers low or zero ohms or the powered test light illuminates, the contacts are powering the relay at all times. Disassemble steering column as necessary to repair contacts or related wiring.

> **Warning**
>
> If the vehicle is equipped with air bags, disable the air bag system before disassembling the steering column. Enable the air bag system after completing the repairs. Refer to Job 147.

   If the ohmmeter registers infinite or very high ohms or the powered test light is off, the contacts are good. Replace the relay.

   Are the contacts good? Yes ___ No ___

Clean or Replace Horn Contacts

☐ 1. Remove the battery negative cable.

☐ 2. Refer to the proper service literature and examine the horn contacts to determine whether they can be cleaned.

   Can the contacts be cleaned? Yes ___ No ___

   If Yes, go to step 3.

   If No, go to step 5.

☐ 3. Clean the contacts.

   a. If the vehicle is equipped with air bags, disable the air bag system.
   b. Remove the steering wheel cover to gain access to the contacts.
   c. Lightly spray the contacts with electrical contact or tuner cleaner.
   d. Allow the cleaner to sit for a few seconds.
   e. Wipe off excess cleaner with a clean lint-free rag.
   f. Repeat step c and step d until contacts are clean. Slight dullness or pitting of the contacts is normal.

☐ 4. Reinstall the battery negative cable and retest the contacts.

   Does the horn function correctly? Yes ___ No ___

   If Yes, reinstall the steering wheel cover, enable the air bag system, and return the vehicle to its owner.

   If No, disconnect the battery negative cable. The contacts must be replaced.

☐ 5. Disconnect the electrical connectors at the contacts and remove the contact assembly from the steering wheel.

6. Compare the old contact assembly with the replacement.

   Is the replacement contact assembly identical to the original? Yes ___ No ___

   If No, explain: _____

   _____

7. Reattach the electrical connector to the contact assembly and reinstall the contacts.

8. Reattach the battery negative cable and retest the contacts.

   Does the horn function properly? Yes ___ No ___

   If Yes, reinstall the steering wheel cover, enable the air bag system, and return the vehicle to its owner.

   If No, consult your instructor.

Replace a Horn

1. Remove the battery negative cable.

2. Remove the electrical connectors from the horn.

3. Remove the horn fasteners.

4. Remove the horn from the vehicle and compare it to the replacement.

   Does the replacement horn match the original? Yes ___ No ___

   If No, describe the differences: _____

   _____

   _____

5. Reinstall the horn fasteners.

6. Reinstall the electrical connectors.

7. Reinstall the battery negative cable and check horn operation.

8. Clean the work area and return any equipment to storage.

9. Did you encounter any problems during this procedure? Yes ___ No ___

   If Yes, describe the problems: _____

   _____

   What did you do to correct the problems? _____

   _____

10. Have your instructor check your work and sign this job sheet.

Performance Evaluation—Instructor Use Only

Did the student complete the job in the time allotted? Yes ___ No ___

If No, which steps were not completed?_____

How would you rate this student's overall performance on this job?_____

5–Excellent, 4–Good, 3–Satisfactory, 2–Unsatisfactory, 1–Poor

Comments: _____

_____

_____

INSTRUCTOR'S SIGNATURE _____

Job 143—Check and Correct the Operation of Windshield Washers and Wipers

After completing this job, you will be able to diagnose and service windshield washers and wipers.

Procedures

☐ 1. Obtain a vehicle to be used in this job. Your instructor may direct you to perform this job on a shop vehicle.

☐ 2. Gather the tools needed to perform this job. Refer to the tools and materials list at the beginning of the project.

☐ 3. Obtain the needed system wiring diagrams.

Diagnose Windshield Wiper Motor Problems

> **Note**
>
> For specific windshield wiper problems such as failure to park, incorrect operation at certain speeds only, or failure to stop, consult the manufacturer's service literature.

☐ 1. Ensure that the wiper motor fuse or circuit breaker is not blown or tripped.

   Is the fuse OK? Yes ___ No ___

☐ 2. Disconnect the windshield wiper motor connector. The wiper motor may be located on the engine firewall (bulkhead) or under a cover between the firewall and windshield, **Figure 143-1**. Remove the plastic cover as needed to gain access to the motor.

☐ 3. Turn the ignition switch to the *on* position.

☐ 4. Place the windshield wiper control switch in the *on* position.

Figure 143-1. The wiper motor on many newer vehicles is located under a plastic cover between the firewall and windshield.

5. Connect a non-powered test light to the power wire at the wiper motor's electrical connector.

   Does the test light illuminate? Yes ___ No ___

   If No, trace the wiring between the fuse and motor to isolate the problem.

   If Yes, go to step 6.

6. Check the ground at the motor.

   Is the motor properly grounded? Yes ___ No ___

   If No, repair the ground and retest.

   If Yes, go to step 7.

7. Connect one lead of a fused jumper wire to the input terminal of the motor.

8. Connect the other lead of the jumper wire to the battery positive terminal.

   Does the motor operate? Yes ___ No ___

   If Yes, repeat step 5 to isolate the problem.

   If No, go to step 10.

   If the motor operates but the wipers do not move, check for disconnected linkage or a broken motor shaft.

9. Check the wiper linkage for binding and lack of lubrication.

   Is the linkage clean and lubricated? Yes ___ No ___

   If No, clean and lubricate it until it works freely, then recheck.

   If Yes, go to step 10.

10. Replace the motor and recheck its operation.

Replace a Windshield Wiper Motor

1. Ensure that the wiper motor is in the park position.
2. Disconnect the battery negative cable.
3. Remove the motor's electrical connectors.
4. Remove the motor fasteners.
5. Pull the motor far enough out of the firewall (bulkhead) to reach the shaft nut.
6. Remove the shaft nut and separate the motor shaft and wiper linkage arm.
7. Remove the motor from the vehicle.
8. Compare the old and new motors.
9. Place the new motor in position and install the wiper linkage arm, making sure that the arm is in the proper position.
10. Install and tighten the shaft nut.
11. Place the motor in position on the firewall and install the fasteners.
12. Install the motor's electrical connectors.
13. Reattach the battery negative cable and check motor and wiper linkage operation.

Replace a Wiper or Blade Insert

1. Turn the ignition switch to the *on* position.
2. Turn the windshield wiper switch to the *on* position.
3. Turn the ignition switch to the *off* position when wiper arms reach a place on the windshield where they can be easily reached.
4. Lift the wiper from the glass and detach the retainer. The retainer type depends on whether the wiper rubber insert or the entire wiper assembly is being replaced.

   Describe the type of retainer: _____

Job 143—Check and Correct the Operation of Windshield Washers and Wipers (continued)

☐ 5. Remove the blade insert or wiper assembly as needed.

☐ 6. Install the replacement part.

☐ 7. Install the retainer and ensure that it is securely latched.

☐ 8. Pour a slight amount of water on the windshield.

☐ 9. Turn the ignition switch to the *on* position and note wiper operation.

☐ 10. Turn the wiper and ignition switches to the *off* position.

Diagnose Windshield Washer Operation

☐ 1. Operate the washer control switch and determine whether the pump runs.

Does the pump run? Yes ___ No ___

If No, go to step 2.

If the pump motor runs, but does not pump fluid, check the following:
- Is there enough fluid in the reservoir? Yes ___ No ___
- Are the hoses connected? Yes ___ No ___
- Are the hoses or jet nozzles clear (not plugged)? Yes ___ No ___

If the answer to any of the above questions is No, correct as necessary and retest pump operation.

If the answer to all of the above questions is Yes, go to step 6.

☐ 2. If the pump circuit is controlled through a fuse, check the fuse condition.

Is the fuse OK? Yes ___ No ___

If No, replace it and recheck washer operation.

If Yes, go to step 3.

☐ 3. If the fuse is good, attach a non-powered test light to the washer input terminal of the wiring harness. **Figure 143-2** shows the connector of a typical reservoir-mounted washer pump.

Figure 143-2. Modern washer pumps are installed on the reservoir tank.

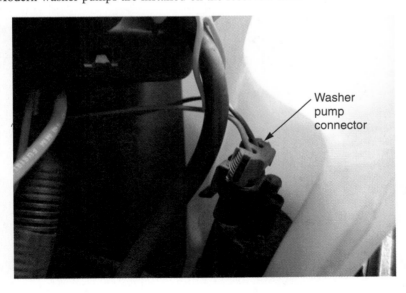

Washer pump connector

Note

SUVs, station wagons, and vans may have an extra washer pump for the rear window. Determine which pump requires testing before proceeding.

☐ 4. Operate the washer switch and note whether the test light illuminates.

Does the test light illuminate? Yes ___ No ___

If Yes, go to step 5.

If No, check the wiring between the fuse and the pump motor.

☐ 5. Check that the pump motor is properly grounded.

Is the motor properly grounded? Yes ___ No ___

If No, repair the ground and retest.

If Yes, go to step 6.

☐ 6. Replace the pump motor and retest.

Replace a Windshield Washer Pump Motor

☐ 1. Remove the battery negative cable.

☐ 2. Remove the pump's electrical connector.

☐ 3. Remove the pump fasteners and remove the pump from the vehicle.

☐ 4. Compare the old and new pumps.

☐ 5. Place the new pump in position and install the fasteners.

☐ 6. Install the pump's electrical connector.

☐ 7. Reattach the battery negative cable and check pump operation.

☐ 8. Clean the work area and return any equipment to storage.

☐ 9. Did you encounter any problems during this procedure? Yes ___ No ___

If Yes, describe the problems: _____

_____

What did you do to correct the problems? _____

_____

☐ 10. Have your instructor check your work and sign this job sheet.

Performance Evaluation—Instructor Use Only

Did the student complete the job in the time allotted? Yes ___ No ___

If No, which steps were not completed?_____

How would you rate this student's overall performance on this job?_____

5–Excellent, 4–Good, 3–Satisfactory, 2–Unsatisfactory, 1–Poor

Comments: _____

_____

_____

INSTRUCTOR'S SIGNATURE _____

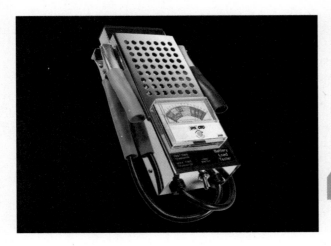

Project

Diagnosing and Servicing Accessories

41

Introduction

Almost every vehicle is now equipped with power windows and door locks. Many tests and adjustments can be made with simple test equipment. You will diagnose and service power windows and power door locks in Job 144.

Other popular accessories on modern automobiles include cruise control and sound systems. You will diagnose and service electric and vacuum-operated cruise control systems in Job 145. You will diagnose problems with a sound system in Job 146.

All modern vehicles have supplemental inflatable restraints (SIR), usually called air bags. SIR systems must be disconnected to service other vehicle components. Always disable the air bag system before attempting to service any component on, or near, the air bag system. These areas include the steering column and dashboard, and sometimes the engine compartment. In Job 147, you will disable and enable an air bag system. You will check an air bag system for defects in Job 148.

Many electrical devices on modern vehicles can be diagnosed with a scan tool, including the following:

- Air bag system.
- Air conditioner and heater.
- Anti-theft and keyless entry systems.
- Daytime running lights.
- Dimmer control.
- Exterior lights.
- Fluid level–sensing system.
- Instrument panel.
- Interior lights.
- Memory seats.
- Power door locks.
- Power sunroof.
- Power tailgate.
- Power windows.
- Radio and sound system.
- Tire pressure monitoring system.
- Vehicle message center.
- Windshield wipers.
- Warning chimes.

The scan tool may be able to pinpoint the actual problem, or it may point the technician to the most likely problem area. The scan tool can also isolate a defective body computer. In Job 149, you will use a scan tool to locate the cause of a vehicle body electrical device malfunction.

Newer vehicles with OBD II systems may contain several on-board modules (computers) to control the engine, transmission/transaxle, anti-lock brakes, and various body systems such as keyless entry and theft protection. A computer that controls body systems is called a body control module. The OBD II system controls communication between these modules. Problems occurring in the modules or the associated wiring can be difficult to track down. A scan tool must be used to begin diagnosing these problems. In Job 150, you will diagnose body control module problems and communication errors between various vehicle modules.

Anti-theft and keyless entry systems are increasingly common. Although they are not as vital as the drive train parts, a failure can be at least an annoyance and possibly a great inconvenience to the owner. In Job 151, you will check the operation of the two types of anti-theft and keyless entry systems and diagnose problems in those systems.

Project 41 Jobs

- Job 144—Diagnose and Service Power Windows and Power Door Locks
- Job 145—Diagnose a Cruise Control Problem
- Job 146—Diagnose Problems with a Radio/Sound System
- Job 147—Disable and Enable an Air Bag System
- Job 148—Diagnose an Air Bag System
- Job 149—Use a Scan Tool to Determine a Body Electrical Problem
- Job 150—Diagnose a Body Control Module and Module Communication Errors
- Job 151—Check the Operation of Anti-Theft and Keyless Entry Systems

Tools and Materials

The following list contains the tools and materials that may be needed to complete the jobs in this project. The items used will depend on the make and model of vehicle being serviced.

- Vehicle with power windows and door locks in need of service.
- Vehicle with an air bag system.
- Vehicle in need of body control module diagnosis.
- Vehicle with anti-theft and keyless entry system in need of service.
- Applicable service information.
- Multimeter.
- Test light.
- Appropriate lubricant.
- Hand tools as needed.
- Scan tool and adapters.
- Scan tool operating instructions.
- Safety glasses and other protective equipment as needed.

Safety Notice

Before performing this job, review all pertinent safety information in the text, and review safety information with your instructor.

Job 144—Diagnose and Service Power Windows and Power Door Locks

After completing this job, you will be able to diagnose and service power windows and power door locks.

Procedures

Diagnose a Power Window Problem

☐ 1. Obtain a vehicle to be used in this job. Your instructor may direct you to perform this job on a shop vehicle.

☐ 2. Gather the tools needed to perform this job. Refer to the tools and materials list at the beginning of the project.

☐ 3. Obtain the proper system wiring diagrams.

☐ 4. Ensure that power is available to the window circuit by removing the window control switch and probing the connectors with the ignition switch in the *on* position. The test light should illuminate on at least one connector.

   Is power available? Yes ___ No ___

   If Yes, go to step 5.

   If No, trace the circuit until the problem is located.

☐ 5. Operate the switch while probing the output terminal(s) with the test light. See **Figure 144-1**.

   Is the switch delivering electricity to the output wires? Yes ___ No ___

   If Yes, go to step 6.

   If No, the switch is defective.

☐ 6. Remove the door panel as explained later in this job.

☐ 7. Use the test light to determine whether power is reaching the motor. See **Figure 144-2**.

   Does the motor have power? Yes ___ No ___

   If Yes, go to step 8.

   If No, check the wiring between the switch and motor.

Figure 144-1. Use a non-powered test light to probe for voltage. If power enters the switch but does not leave it when the switch is operated, the switch is defective.

Figure 144-2. Probe for voltage at the motor's electrical connector. The power connection for a window motor is shown here.

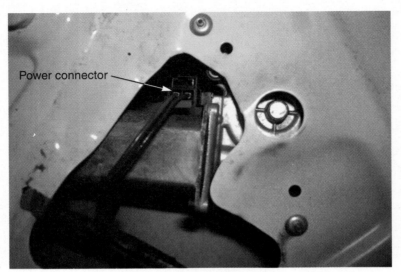

8. Check the glass channel to ensure that the window glass can move freely with no binding.

   Does the glass move freely? Yes ___ No ___

   If No, adjust or lubricate it as necessary and recheck operation.

   If Yes, go to step 9.

9. Check that the motor is properly grounded.

   Is the motor grounded? Yes ___ No ___

   If No, correct the ground and recheck motor operation.

   If Yes, go to step 10.

10. Replace the window motor and recheck window motor operation.

11. Reinstall the door panel, ensuring that all electrical connectors are in place and all fasteners are properly reinstalled. Refer to the *Remove a Door Panel and Replace a Component* section of this job as needed.

Diagnose a Power Door Lock Problem

1. Obtain system wiring diagrams.

2. Ensure that power is available to the door lock circuit by removing the lock control switch and probing the connectors with the ignition switch in the *on* position. The test light should illuminate on at least one connector. See **Figure 144-3**.

   Is power available? Yes ___ No ___

   If power is available at the switch, go to step 3.

   If power is not available, trace the circuit until the problem is located.

3. Operate the switch while probing the output terminal(s) with the test light.

   Is the switch delivering electricity to the output wires? Yes ___ No ___

   If Yes, go to step 4.

   If No, the switch is defective.

4. Remove the door panel as explained in the next section of this job.

5. Use the test light to determine whether power is reaching the door lock solenoid.

   Does the solenoid have power? Yes ___ No ___

   If Yes, go to step 6.

   If No, check the wiring between the switch and solenoid.

Job 144—Diagnose and Service Power Windows and Power Door Locks (continued)

Figure 144-3. In this photo, power is available at the door switch, causing the test light to illuminate.

☐ 6. Check the lock mechanism and related linkage for sticking, misaligned, or bent linkage. See **Figure 144-4**.

Is the linkage properly aligned and lubricated? Yes ___ No ___

If No, adjust or lubricate it as necessary and recheck operation.

If Yes, go to step 7.

☐ 7. Check that the solenoid is properly grounded.

Is the solenoid grounded? Yes ___ No ___

If No, correct the ground and recheck solenoid operation.

If Yes, go to step 8.

Figure 144-4. If the door lock linkage is stuck or bent, the automatic door locks may not function properly.

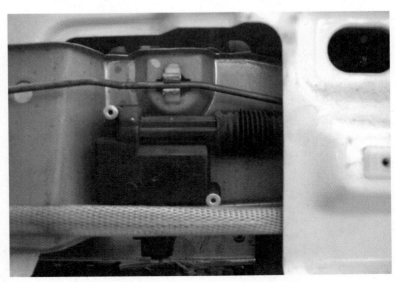

8. Replace the solenoid and recheck window operation. Refer to the next section of this job as needed.

9. Reinstall the door panel, ensuring that all electrical connectors are in place and all fasteners are properly reinstalled. Refer to the next section of this job as needed.

Remove a Door Panel and Replace a Component

1. Remove the battery negative cable.

2. Remove the door panel armrest. If the armrest is part of the panel, skip this step.

3. Remove the pull strap, if used.

4. Remove the electrical switches controlling the windows, door locks, and power mirrors, if the vehicle is so equipped.

> **Note**
>
> Some switches can be left attached to the wiring harness and turned to pass through the switch opening in the door panel.

5. Remove any remaining hardware such as door or window handles.

6. Remove the fasteners holding the door panel to the door frame.

7. Carefully pull the door panel from the door frame. Some door panels must be lifted to disengage them from the window sill or internal slots. See **Figure 144-5**.

8. Remove the inner vapor barrier.

9. Locate the electrical device to be replaced.

Describe the part being replaced: _____

_____

_____

10. Remove the electrical connector(s) at the part to be replaced.

11. Remove the fasteners holding the part.

12. Remove the part from the vehicle.

Figure 144-5. After removing all connectors and fasteners, lift the door panel up to remove it.

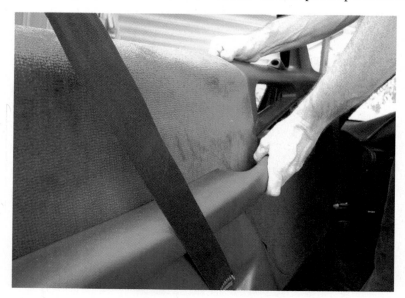

Job 144—Diagnose and Service Power Windows and Power Door Locks (continued)

Warning

⚠ If a window motor is being replaced, it may be necessary to block the window to keep it from falling to the bottom of the door. You may need to take precautions to safely maintain spring pressure when removing the window motor from the regulator assembly. Always consult the manufacturer's service literature before removing any window motor.

☐ 13. Compare the original and replacement parts.

Do the parts match? Yes ___ No ___

If No, explain: _____

_____

☐ 14. Place the new part in position and reinstall the fasteners.

☐ 15. Reinstall the electrical connector(s).

Note

📄 Before installing the door panel, make sure that the window operates properly. Temporarily reconnect the window control switches and the battery negative cable and operate the window. Ensure that no bare electrical connections touch any metal body parts during this procedure. Disconnect the battery, and switches as necessary, before continuing with reassembly.

☐ 16. Reinstall the battery negative cable and check operation.

☐ 17. Reinstall the vapor barrier.

☐ 18. Place the door panel in position and install the fasteners.

☐ 19. Reinstall the door hardware and electrical switches.

☐ 20. Reinstall the pull straps and armrest if necessary.

☐ 21. Clean the work area and return any equipment to storage.

☐ 22. Did you encounter any problems during this procedure? Yes ___ No ___

If Yes, describe the problems: _____

_____

What did you do to correct the problems? _____

_____

☐ 23. Have your instructor check your work and sign this job sheet.

Performance Evaluation—Instructor Use Only

Did the student complete the job in the time allotted? Yes ____ No ____

If No, which steps were not completed?_____

How would you rate this student's overall performance on this job?_____

5–Excellent, 4–Good, 3–Satisfactory, 2–Unsatisfactory, 1–Poor

Comments: _____

_____

_____

INSTRUCTOR'S SIGNATURE _____

Job 145—Diagnose a Cruise Control Problem

After completing this job, you will be able to diagnose vacuum and electronic cruise control systems.

Procedures

> **Note**
>
> The following procedure is a general guide only. Always refer to the manufacturer's service literature for specific procedures.

☐ 1. Obtain a vehicle to be used in this job. Your instructor may direct you to perform this job on a shop vehicle.

☐ 2. Gather the tools needed to perform this job. Refer to the tools and materials list at the beginning of the project.

☐ 3. Determine which type of cruise control is used on the vehicle. Older cruise controls have a vacuum-operated diaphragm and newer cruise controls use an electronically controlled motor. The type can be determined by looking at the underhood servo that operates the throttle, **Figure 145-1**.

☐ 4. Obtain needed system wiring diagrams.

Diagnose a Vacuum-Operated Cruise Control System

☐ 1. Determine whether the speedometer is working.

Is the speedometer working correctly? Yes ___ No ___

If Yes, go to step 2.

If No, locate and correct the speedometer problem and recheck cruise control operation.

Figure 145-1. The two types of cruise control servos are shown here. A—This servo is vacuum operated. Note the vacuum lines attached to the servo body. B—This servo is an electronic cruise control. Vacuum lines are not used.

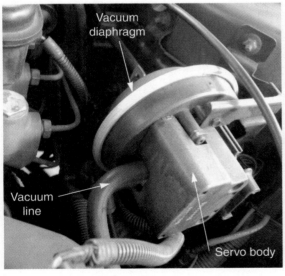

A   B

☐ 2. Check the cruise control fuse condition.

Is the fuse blown? Yes ____ No ____

If No, go to step 3.

If Yes, replace the fuse and recheck cruise control operation.

> **Caution**
>
> If the fuse is blown, you must determine what caused the electrical overload. Do not simply replace the fuse and let the vehicle go.

☐ 3. Check the brake switch.

Is the brake switch connected and properly adjusted? Yes ____ No ____

If Yes, go to step 4.

If No, service the brake switch as needed and recheck cruise control operation.

☐ 4. Check throttle cable adjustment.

Is the throttle cable undamaged and properly adjusted? Yes ____ No ____

If Yes, go to step 5.

If No, adjust or replace the throttle cable as needed and recheck cruise control operation.

☐ 5. Use service literature, wiring diagrams, and a non-powered test light to make electrical checks of the following components:
- Control unit (usually on the turn signal lever).
- Power servo.
- Cruise control wiring.

Do the electrical checks indicate a problem? Yes ____ No ____

If Yes, describe the problem:_____

_____

_____

If No, repeat the above steps until the problem is isolated.

What should you do to correct this problem?_____

_____

_____

> **Note**
>
> Some servos can be checked for vacuum leaks using a vacuum pump, **Figure 145-2**. Refer to the manufacturer's service literature for this procedure.

☐ 6. Repair or replace cruise control components as necessary and recheck cruise control operation.

Diagnose an Electrically Operated Cruise Control System

☐ 1. Determine whether the speedometer is working.

Is the speedometer working correctly? Yes ____ No ____

If Yes, go to step 2.

If No, locate and correct the speedometer problem and recheck cruise control operation.

☐ 2. Check for blown fuse.

Is the fuse blown? Yes ____ No ____

If No, go to step 3.

If Yes, replace the fuse and recheck cruise control operation.

Job 145—Diagnose a Cruise Control Problem (continued)

Figure 145-2. Many vacuum servos can be checked with a vacuum pump. The servo does not need to be removed from the vehicle for this check. If the servo does not move when vacuum is applied, it is usually defective. Some servos have a vacuum bleed, and may not hold vacuum.

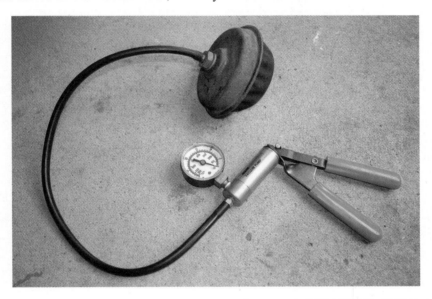

> **Caution**
>
> If the fuse is blown, you must determine what caused the electrical overload. Do not simply replace the fuse and let the vehicle go.

 3.   If possible, attach a scan tool to the vehicle diagnostic connector and retrieve trouble codes. If the cruise control is not connected to the vehicle computer, skip this step.

Did any of the retrieved trouble codes apply to the cruise control system? Yes ___ No ___

If No, proceed to step 4.

If Yes, record the trouble codes and the faults they indicate: _____

_____

Proceed to the appropriate step based on scan tool findings.

4.   Check throttle cable adjustment.

Is the throttle cable undamaged and properly adjusted? Yes ___ No ___

If Yes, go to step 5.

If No, adjust or replace the throttle cable as needed and recheck cruise control operation.

5.   Use service literature, wiring diagrams, and a non-powered test light to make electrical checks at the following components:

- Control unit (usually on the turn signal lever).
- Power servo.
- Cruise control wiring.

Do the electrical checks indicate a problem? Yes ___ No ___

If Yes, describe the problem: _____

_____

If No, repeat the above steps until the problem is isolated.

What should you do to correct this problem?

Project 41—Diagnosing and Servicing Accessories

☐ 6. Clean the work area and return any equipment to storage.

☐ 7. Did you encounter any problems during this procedure? Yes ___ No ___

If Yes, describe the problems: _____

_____

What did you do to correct the problems? _____

_____

☐ 8. Have your instructor check your work and sign this job sheet.

Performance Evaluation—Instructor Use Only

Did the student complete the job in the time allotted? Yes ___ No ___
If No, which steps were not completed?_____
How would you rate this student's overall performance on this job?_____
5–Excellent, 4–Good, 3–Satisfactory, 2–Unsatisfactory, 1–Poor
Comments: _____

_____

_____

INSTRUCTOR'S SIGNATURE _____

Job 146—Diagnose Problems with a Radio/Sound System

After completing this job, you will be able to perform radio and sound system diagnosis.

Procedures

☐ 1. Obtain a vehicle to be used in this job. Your instructor may direct you to perform this job on a shop vehicle.

☐ 2. Gather the tools needed to perform this job. Refer to the tools and materials list at the beginning of the project.

Note

Several hundred types of factory and aftermarket radios and sound systems are available. In addition, almost every vehicle make, model, and year has a different method of mounting components. The following procedures are therefore very general. For exact information on diagnosis and service, always consult the correct manufacturer's manual.

Check an Inoperable Radio and/or Sound System

☐ 1. Turn the ignition and radio/sound system switches to the *on* position.

☐ 2. Turn the volume control up.

Does the device operate? Yes ___ No ___

If Yes, proceed to diagnose other radio complaint(s). Refer to the *Diagnose Radio/Sound System Sound Quality Problem* section of this job.

If No, go to step 3.

☐ 3. Check the system fuse. Use the legend on the fuse box cover to locate the proper fuse, **Figure 146-1**.

Is the fuse blown? Yes ___ No ___

If Yes, replace the fuse and trace the source of the electrical overload that caused the fuse to fail.

If No, go to step 4.

☐ 4. Check for power in the system circuit using a non-powered test light.

Is power present at the radio? Yes ___ No ___

If Yes, proceed to step 5.

If No, locate the wiring problem that causes a lack of power.

☐ 5. Check for a missing system ground. Loss of ground is common in modern vehicles with many plastic dashboard components.

Is the radio/sound system properly grounded? Yes ___ No ___

If Yes, the device is defective and must be removed for service.

If No, locate the missing ground, repair or substitute another ground, and recheck.

Figure 146-1. The fuse will be located in a fuse block with many other fuses. Be sure to remove the right fuse.

Diagnose Radio/Sound System Sound Quality Problem

☐ 1. Determine the exact complaint.

Briefly describe the complaint here: _____

_____

_____

☐ 2. Duplicate the complaint.

Do your observations match the original complaint? Yes ___ No ___

If Yes, proceed to step 3.

If No, talk to the driver or discuss the nature of the complaint with your instructor.

☐ 3. Disconnect and bypass any recently added aftermarket electrical equipment such as amplifiers, speakers, or anti-theft devices.

If there are no aftermarket devices, go to step 5.

List the device(s) disconnected: _____

_____

_____

☐ 4. Recheck system operation.

Has the sound quality been restored to normal? Yes ___ No ___

If Yes, the aftermarket equipment is producing static that is affecting sound quality. Repair or replace the equipment as necessary.

If No, proceed to step 5.

☐ 5. Check radio/sound system operation with the engine off and with the engine running.

When does the problem occur?
- Only with the engine running. ___ Go to step 6.
- Only with the engine off and the ignition switch in the *run* position. ___ Go to step 10.
- Only with the ignition switch in the *accessory* position. ___ Go to step 10.
- All of the time. ___ Go to step 10.

Job 146—Diagnose Problems with a Radio/Sound System (continued)

☐ 6. If the sound quality problem occurs only when the engine is running, identify the type of noise produced.
- A whining noise. ___ Go to step 7.
- A popping or snapping noise. ___ Go to step 9.

☐ 7. Stop the engine and unplug the alternator field wire.

☐ 8. Restart the engine and determine whether the sound quality problem still exists.

Is the noise still present? Yes ___ No ___

If Yes, go to step 9.

If No, the alternator is defective. Repair or replace the alternator as needed and recheck sound quality.

☐ 9. Check the ignition system for defective components. Pay particular attention to secondary ignition components, such as the spark plugs, wires, coil, and distributor cap and rotor when used. The following are typical causes of ignition noises:
- Corrosion at the wire terminals where they connect to the coil(s) or spark plugs.
- Carbon tracking at the coil(s) or spark plugs.
- Arcing at the spark plug wire terminals where they connect to the coil(s) or spark plugs.
- Incorrect routing of the plug wires.

Were any problems found? Yes ___ No ___

If No, proceed to step 10.

If Yes, repair the problem or replace parts as needed and recheck sound quality. Describe any problems found and the corrective actions taken: _____

_____

_____

Did these repairs correct the problem? Yes ___ No ___

If No, recheck your repairs and then proceed to step 10.

☐ 10. If the sound quality problem occurs with the engine off and the ignition switch in the *run* or *accessory* position or occurs all of the time, remove fuses one at a time until the sound quality returns to normal.

Was sound quality restored when a certain fuse was removed? Yes ___ No ___

If No, consult your instructor about how to proceed.

If Yes, the removed fuse was part of the faulty circuit. Identify the fuse that was removed to restore sound quality: _____

☐ 11. Using the appropriate wiring diagram, isolate the faulty component in the circuit.

☐ 12. Check the component grounds.

Is the device properly grounded? Yes ___ No ___

If No, repair the ground circuit and recheck sound quality.

If Yes, go to step 13.

☐ 13. Repair or replace components as needed, then recheck sound quality.

Note

Sometimes further diagnosis procedures, such as checking speaker condition and the antenna ground are needed. Consult the correct service literature for detailed checking procedures.

☐ 14. Clean the work area and return any equipment to storage.

☐ 15. Did you encounter any problems during this procedure? Yes ___ No ___
    If Yes, describe the problems: _____
    _____
    What did you do to correct the problems? _____
    _____

☐ 16. Have your instructor check your work and sign this job sheet.

Performance Evaluation—Instructor Use Only

Did the student complete the job in the time allotted? Yes ___ No ___
If No, which steps were not completed?_____
How would you rate this student's overall performance on this job?_____
5–Excellent, 4–Good, 3–Satisfactory, 2–Unsatisfactory, 1–Poor
Comments: _____
_____
_____

INSTRUCTOR'S SIGNATURE _____

Job 147—Disable and Enable an Air Bag System

After completing this job, you will be able to safely disable and enable an air bag system.

Procedures

☐ 1. Before beginning this job, review the following important air bag safety rules:
- Do *not* place the air bag module where temperatures will exceed 175°F (79.4°C).
- If you drop any part of the air bag system, replace the part.
- When carrying a live air bag module, point the bag and trim cover away from you.
- Do *not* carry an air bag by the connecting wires.
- Place a live air bag module on a bench with the bag and trim cover facing up.
- Do *not* test an air bag with electrical test equipment unless the manufacturer's instructions clearly call for such testing.

> **Warning**
> ⚠ Always obtain the proper service literature and follow manufacturer's procedures exactly before attempting to disable any air bag system.

☐ 2. Obtain a vehicle to be used in this job. Your instructor may direct you to perform this job on a shop vehicle.

☐ 3. Gather the tools needed to perform this job. Refer to the tools and materials list at the beginning of the project.

Disable an Air Bag System

☐ 1. Turn the steering wheel to the straight-ahead position.

☐ 2. Lock the ignition and remove the ignition key.

☐ 3. Locate and remove the air bag fuses. Check the service information, as there may be more than one system fuse.

☐ 4. Locate the air bag connectors under the dashboard. Check the service manual for the exact connector locations. Newer connectors are yellow, but older systems may use another color for the connectors. **Figure 147-1** shows a common air bag connector location.

☐ 5. Disconnect the air bag connectors. To access the connectors, it may be necessary to remove the trim parts from the lower dashboard or center console. Some air bag connectors have a locking pin to ensure that the connector does not come apart accidentally. This pin must be removed to separate the connector. Use the service literature to confirm that you are disconnecting the proper connectors.

> **Warning**
> ⚠ Although the air bags are disabled, they can still inflate without warning. Follow all safety rules.

Enable Air Bag Systems

☐ 1. Make sure that the key is removed from the ignition switch.

☐ 2. Reconnect the air bag connectors and reinstall any trim pieces that were removed.

☐ 3. Reinstall the air bag system fuses.

Figure 147-1. The location of one typical air bag connector is shown here. Note the yellow color of the connector and wires.

4. Position yourself in the passenger compartment so that you are away from the air bags. This will keep you from being injured if the system accidentally deploys.

5. Turn the ignition switch to the *on* position.

6. Allow the air bag system self-diagnostic program to run. This will ensure that the system is operating correctly.

7. Clean the work area and return any equipment to storage.

8. Did you encounter any problems during this procedure? Yes ___ No ___

   If Yes, describe the problems: _____

   _____

   What did you do to correct the problems? _____

   _____

9. Have your instructor check your work and sign this job sheet.

Performance Evaluation—Instructor Use Only

Did the student complete the job in the time allotted? Yes ___ No ___

If No, which steps were not completed?_____

How would you rate this student's overall performance on this job?_____

5–Excellent, 4–Good, 3–Satisfactory, 2–Unsatisfactory, 1–Poor

Comments: _____

_____

_____

INSTRUCTOR'S SIGNATURE _____

Job 148—Diagnose an Air Bag System

After completing this job, you will be able to diagnose an air bag system.

Procedures

☐ 1. Before beginning this job, review the following important air bag safety rules.
   - Do not place the air bag module where temperatures will exceed 175°F (79.4°C).
   - If you drop any part of the air bag system, replace the part.
   - When carrying a live air bag module, point the bag and trim cover away from you.
   - Do not carry an air bag by the connecting wires.
   - Place a live air bag module on a bench with the bag and trim cover facing up.
   - Do not test an air bag with electrical test equipment unless the manufacturer's instructions clearly call for such testing.

> **Warning**
> ⚠ Always obtain the proper service literature and follow manufacturer's procedures exactly before attempting to service any air bag system.

☐ 2. Obtain a vehicle to be used in this job. Your instructor may direct you to perform this job on a shop vehicle.

☐ 3. Gather the tools needed to perform this job. Refer to the tools and materials list at the beginning of the project.

Diagnose an Air Bag System

☐ 1. Attach a scan tool to the vehicle diagnostic connector. Refer to Job 11 and Job 149 as necessary.

☐ 2. Retrieve trouble codes and list them in the space provided.

   Trouble codes: _____

☐ 3. Use the scan tool literature or factory service manual to separate body codes from engine and power train codes. List the body system codes in the following chart:

> **Note**
> 📄 For the purposes of this job, ignore any engine and power train codes.

Code	Defect
_____	_____
_____	_____
_____	_____
_____	_____
_____	_____

Do any of the trouble codes apply to the air bag system? Yes ___ No ___

If Yes, list the air bag–related trouble codes in the following chart:

Code	Defect
_____	_____
_____	_____
_____	_____
_____	_____
_____	_____

☐ 4. Turn the ignition switch to the *off* position.

☐ 5. Clear all trouble codes using the scan tool.

☐ 6. Turn the ignition switch to the *on* position.

☐ 7. Determine whether any trouble codes reset.

Did any codes reset? Yes ___ No ___

If Yes, do any apply to the air bag system? Yes ___ No ___

If Yes, list the air bag codes in the following chart.

Code   **Defect**

_____   _____

_____   _____

_____   _____

_____   _____

_____   _____

☐ 8. Based on the scan tool information, make further checks to isolate the problem(s) revealed by the trouble codes.

Warning

⚠ When performing pinpoint tests on the air bag system, use extreme care and follow the testing procedures described in the service literature precisely.

☐ 9. If your instructor directs, repair the air bag system problem(s).

Repairs performed: _____

☐ 10. Clean the work area and return any equipment to storage.

☐ 11. Did you encounter any problems during this procedure? Yes ___ No ___

If Yes, describe the problems: _____

_____

What did you do to correct the problems? _____

_____

☐ 12. Have your instructor check your work and sign this job sheet.

Performance Evaluation—Instructor Use Only

Did the student complete the job in the time allotted? Yes ___ No ___

If No, which steps were not completed?_____

How would you rate this student's overall performance on this job?_____

5–Excellent, 4–Good, 3–Satisfactory, 2–Unsatisfactory, 1–Poor

Comments: _____

_____

_____

INSTRUCTOR'S SIGNATURE _____

Job 149—Use a Scan Tool to Determine a Body Electrical Problem

After completing this job, you will be able to use a scan tool to access a vehicle body computer for diagnosis purposes.

Procedures

☐ 1. Obtain a vehicle to be used in this job. Your instructor may direct you to perform this job on a shop vehicle.

☐ 2. Gather the tools needed to perform this job. Refer to the tools and materials list at the beginning of the project.   .

> **Note**
>
> There are many kinds of scan tools. The following procedure is a general guide to scan tool use. Always obtain the service instructions for the scan tool that you are using. If the scan tool literature calls for a different procedure or series of steps from what is in this procedure, always perform procedures according to the scan tool literature.

Accessing Vehicle Computer

☐ 1. Use the appropriate service literature to determine whether the vehicle has an OBD I or OBD II computer control system.
   - OBD I system. ___
   - OBD II system. ___

☐ 2. Obtain the appropriate scan tool and related service literature for the vehicle selected and the computer control system determined in step 1.

   Type of scan tool: _____

☐ 3. Attach the proper test connector cable and power lead to the scan tool.

☐ 4. Ensure that the vehicle's ignition switch is in the *off* position.

☐ 5. Locate the correct diagnostic connector.

   Describe the diagnostic connector location: _____

   _____

   - OBD I system connector. ___
   - OBD II system connector. ___

☐ 6. Attach the scan tool test connector cable to the diagnostic connector, using the proper adapter if necessary.

☐ 7. Attach the scan tool power lead to the cigarette lighter or battery terminals if necessary.

> **Note**
>
> OBD II scan tools are powered from terminal 16 of the diagnostic connector.

☐ 8. Observe the scan tool screen to ensure that it is working properly.

☐ 9. Enter vehicle information as needed to program the scan tool. Many scan tools can be programmed with the proper vehicle information by entering the vehicle identification number (VIN).

☐ 10. Turn the ignition switch to the *on* position.

☐ 11. Retrieve any trouble codes and list them in the space provided.

Trouble codes: _____

> **Note**
>
> Modern scan tools can provide additional information about various body electrical systems. Your instructor may ask you to perform other subtasks based on the ability of the scan tool to retrieve information.

☐ 12. Use the scan tool literature or factory service manual to separate body codes from engine and power train codes. List the body system codes in the following chart.

> **Note**
>
> For the purposes of this job, ignore any engine and power train codes.

Code	Defect
_____	_____
_____	_____
_____	_____
_____	_____
_____	_____

☐ 13. Turn the ignition switch to the *off* position.

☐ 14. Disconnect an electrical connector of a body electrical device that is controlled by the body control computer, or remove a fuse to a device controlled by the body control computer. Consult your instructor to determine which connector to disconnect.

☐ 15. Turn the ignition switch to the *on* position.

☐ 16. Retrieve trouble codes and list them in the space provided.

Trouble codes: _____

> **Note**
>
> If no codes were set, the body control computer may be defective. Go to step 24.

☐ 17. Determine whether the new body code(s) correspond to the circuit containing the removed connector or fuse.

Do the codes correspond? Yes ____ No ____

If Yes, go to step 18.

If No, go to step 24.

☐ 18. Turn the ignition switch to the *off* position.

☐ 19. Reinstall all fuses or connectors removed in step 14.

Job 149—Use a Scan Tool to Determine a Body Electrical Problem (continued)

☐ 20. Turn the ignition switch to the *on* position.

☐ 21. Clear all trouble codes using the scan tool.

☐ 22. Turn the ignition switch to the *off* position then back to the *on* position.

☐ 23. Determine whether any trouble codes reset.

Did any codes reset? Yes ___ No ___

If Yes, list them in the following chart:

Code	Defect
_____	_____
_____	_____
_____	_____
_____	_____
_____	_____

☐ 24. If your instructor directs, make further checks to isolate the problem(s) revealed by the trouble codes.

☐ 25. If your instructor directs, repair the problem(s) as necessary. Refer to other job sheets as directed.

☐ 26. Clean the work area and return any equipment to storage.

☐ 27. Did you encounter any problems during this procedure? Yes ___ No ___

If Yes, describe the problems: _____

_____

What did you do to correct the problems? _____

_____

☐ 28. Have your instructor check your work and sign this job sheet.

Performance Evaluation—Instructor Use Only

Did the student complete the job in the time allotted? Yes ___ No ___

If No, which steps were not completed?_____

How would you rate this student's overall performance on this job?_____

5–Excellent, 4–Good, 3–Satisfactory, 2–Unsatisfactory, 1–Poor

Comments: _____

_____

_____

INSTRUCTOR'S SIGNATURE _____

Notes

Job 150—Diagnose a Body Control Module and Module Communication Errors

After completing this job, you will be able to diagnose body control module problems and inter-module communication errors.

Procedures

> **Note**
>
> There are many kinds of scan tools. The following procedure is a general guide to scan tool use. Always obtain the service instructions for the scan tool that you are using. If the scan tool literature calls for a different procedure or series of steps from what is in this procedure, always perform procedures according to the scan tool literature.

☐ 1. Obtain a vehicle to be used in this job. Your instructor may direct you to perform this job on a shop vehicle.

☐ 2. Gather the tools needed to perform this job. Refer to the tools and materials list at the beginning of the project.

☐ 3. Obtain a scan tool and related service literature for the vehicle selected in step 1.

Type of scan tool: _____

☐ 4. Attach the proper test connector cable and power lead to the scan tool.

☐ 5. Ensure that the vehicle's ignition switch is in the *off* position.

☐ 6. Locate the correct diagnostic connector.

Describe the diagnostic connector location: _____

_____

☐ 7. Attach the scan tool test connector cable to the diagnostic connector. If necessary, install the proper adapter.

☐ 8. Attach the scan tool power lead to the cigarette lighter or battery terminals as needed.

> **Note**
>
> OBD II scan tools are powered from terminal 16 of the diagnostic connector.

☐ 9. Observe the scan tool screen to ensure that the scan tool is working properly.

☐ 10. Enter vehicle information as needed to program the scan tool. Many scan tools can be programmed with the proper vehicle information by entering the vehicle identification number (VIN). If the scan tool can read the VIN from the vehicle computer, skip this step.

☐ 11. Turn the vehicle's ignition switch to the *on* position.

☐ 12. Determine whether the scan tool communicates with the computer module. If the scan tool pulls up a screen similar to the one shown in **Figure 150-1**, the scan tool and module are communicating.

Do the module and scan tool communicate? Yes ___ No ___

If Yes, go to step 13.

If No, go to step 17.

Figure 150-1. The scan tool displays information from the body control module. The screen shown here includes internal module timing information and the status of automatic headlight control system.

13. Retrieve trouble codes and list them in the space provided.

   Trouble codes: _____

> **Note**
>
> If more than one code is produced, check with your instructor before continuing. The module and vehicle may have defects that are beyond the scope of this job.

14. Turn the ignition switch to the *off* position.

15. Determine the meaning of the codes by referring to one of the following sources:
   - **Generic OBD II codes:** Refer to a listing of generic OBD II trouble codes. These codes will always have a zero following the U, such as U0XXX.
   - **Manufacturer-specific codes:** Refer to the manufacturer's service information. These codes will always have a number other than zero following the U, such as U1XXX, U2XXX.

   Do any of the codes indicate a loss of communication between modules? Yes ___ No ___
   If Yes, list them in the space provided, and then go to step 16: _____

   _____

   If No, go to step 17.

   If the trouble code cannot be found in a list of generic trouble codes or in the manufacturer's service information, there are two possible reasons:
   - The trouble code has been added since the list was compiled.
   - An internal vehicle computer problem is causing a nonexistent code.

> **Note**
>
> Class 2 serial data codes will always begin with the letter U.

Job 150—Diagnose a Body Control Module and Module Communication Errors (continued)

Code   **Defect**

U_____   _____

U_____   _____

U_____   _____

U_____   _____

U_____   _____

☐ 16. Use the appropriate service literature to determine which steps to take to deal with a particular class 2 serial data trouble code. These steps may include the following:
- Removing the battery negative terminal and checking various wire and module resistance readings with a multimeter.
- Inspecting fuses.
- Checking for battery voltage on input wires.
- Checking module grounds.
- Checking for module wiring that is disconnected or has high resistance.

What problems were found? _____

_____

_____

What should you do to correct them? _____

_____

_____

☐ 17. Use the scan tool to attempt to communicate with other vehicle modules.

Do the various modules communicate? Yes ___ No ___

If Yes, go to step 23.

If No, go to step 18.

☐ 18. Turn the vehicle's ignition switch to the *on* position.

☐ 19. Using a test light, ensure that power is available to both sides of all module fuses. Place a check mark next to the results of your tests.
- Test light lights on both sides of all fuses. ___ Go to step 20.
- Test light lights on one side of a fuse. ___ Go to step 21.
- Test light does not light on either side of a fuse. ___ Go to step 22.

☐ 20. Using a test light, check for power at the module input wire.

Is power available at the module? Yes ___ No ___

If Yes, go to step 23.

If No, repair the wire between the fuse and the module and recheck module operation. Go to step 25 after verifying the problem has been resolved.

☐ 21. Replace the fuse and recheck module operation. Go to step 25 after verifying the problem has been resolved.

Caution

⚠ If the fuse is blown, you must determine what caused the electrical overload. Do not simply replace the fuse and let the vehicle go.

☐ 22. Back probe the system wiring until the reason for the lack of power is determined. Repair the wire or connector between the battery and fuse and recheck operation. Go to step 25 after verifying the problem has been resolved.

☐ 23. Check the ground at the module.

Is the module properly grounded? Yes ____ No ____

If Yes, go to step 24.

If No, repair the ground wire(s) as needed and recheck module operation. Go to step 25 after verifying the problem has been resolved.

☐ 24. Replace the module and recheck its operation.

> **Caution**
>
> Ensure that the ignition switch is in the *off* position and the battery negative cable is removed before replacing the module. Follow all static electricity precautions outlined by the vehicle manufacturer.

☐ 25. Clean the work area and return any equipment to storage.

☐ 26. Did you encounter any problems during this procedure? Yes ____ No ____

If Yes, describe the problems: _____

_____

What did you do to correct the problems? _____

_____

☐ 27. Have your instructor check your work and sign this job sheet.

Performance Evaluation—Instructor Use Only

Did the student complete the job in the time allotted? Yes ____ No ____

If No, which steps were not completed?_____

How would you rate this student's overall performance on this job?_____

5–Excellent, 4–Good, 3–Satisfactory, 2–Unsatisfactory, 1–Poor

Comments: _____

_____

_____

INSTRUCTOR'S SIGNATURE _____

Job 151—Check the Operation of Anti-Theft and Keyless Entry Systems

After completing this job, you will be able to diagnose anti-theft and keyless entry systems.

Procedures

☐ 1. Obtain a vehicle to be used in this job. Your instructor may direct you to perform this job on a shop vehicle.

☐ 2. Gather the tools needed to perform this job. Refer to the tools and materials list at the beginning of the project.

☐ 3. Obtain wiring diagrams or other service literature as needed.

Diagnose an Anti-Theft System That Uses a Key Resistor

> **Note**
>
> Security system problems may or may not illuminate the security light. Security system problems usually will be noticed as failure of the vehicle to start. Later systems can be diagnosed by attaching a scan tool to the vehicle diagnostic connector and accessing trouble codes.

☐ 1. Disconnect the security system wires at the base of the steering column.

☐ 2. Insert the key in the ignition switch.

☐ 3. Obtain a multimeter and set it to the resistance (ohms) position.

☐ 4. Turn the ignition switch to the *on* position and determine which security system wires complete the circuit through the ignition switch and key.

☐ 5. Use the multimeter to check continuity and resistance values through the ignition switch and key.

Resistance reading:_____

Resistance value specified in the service literature:_____

☐ 6. Repeat step 5 while wiggling the key and turning it between the *on* and *start* positions.

Does the resistance value remain constant? Yes ___ No ___

If the multimeter shows no continuity (infinite resistance) or intermittent continuity, the switch contacts or the key resistor are probably bad. No continuity can be caused by a bad key resistor or contacts. Intermittent continuity is most likely caused by loose or corroded contacts. Make repairs as needed, reconnect the wiring, and recheck system operation.

Diagnose a Remote Keyless Entry System

☐ 1. Check that the power door lock system operates properly from the door controls.

Are the power door locks operating properly? Yes ___ No ___

If Yes, go to step 2.

If No, correct the door lock system by referring to Job 144. Then, recheck keyless entry system operation.

☐ 2. Check for radio interference from nearby electrical devices or signal blockage from other vehicles, trees, shrubbery, or buildings.

3. If the power door locks are working properly and there is no radio interference, try to operate the system using two different transmitters.

   If the system operates from one transmitter only, go to step 4.

   If the system will not operate from either transmitter, go to step 6.

4. Replace the battery of the suspect transmitter, **Figure 151-1**, and recheck operation.

   Does the transmitter now operate the door locks? Yes ___ No ___

   If Yes, go to step 9.

   If No, go to step 5.

5. Replace the transmitter and recheck.

   Does the transmitter now operate the door locks? Yes ___ No ___

   If Yes, go to step 9.

   If No, go to step 6.

6. Check the control module fuse.

   Is the fuse OK? Yes ___ No ___

   If Yes, go to step 7.

   If No, replace the fuse and recheck keyless entry system operation.

> **Caution**
>
> If the fuse is blown, you must determine what caused the electrical overload. Do not simply replace the fuse and let the vehicle go.

7. Check the control module wiring for defects and loose connectors, **Figure 151-2**.

   Were problems found in the wiring? Yes ___ No ___

   If No, go to step 8.

   If Yes, repair the wires as needed and recheck keyless entry system operation.

8. Replace the control module and recheck keyless entry system operation. When replacing the control module, make sure that the FCC model number on the original and replacement are the same.

Figure 151-1. Most transmitter batteries can be replaced by prying the case apart to expose the battery. Be careful not to damage the electronic circuitry while changing the battery.

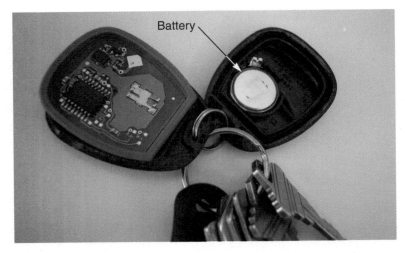

Battery

Job 151—Check the Operation of Anti-Theft and Keyless Entry Systems (continued)

Figure 151-2. A typical control module located under a side trim panel. Check the service information to determine the location of the control module on the vehicle that you are working on.

> ### Note
> It may be necessary to program the new transmitter. Ground the programming connector located near the control module, and push any button on the transmitter. The locks will cycle once to indicate that the transmitter has been programmed.

☐ 9. Clean the work area and return any equipment to storage.

☐ 10. Did you encounter any problems during this procedure? Yes ___ No ___

If Yes, describe the problems. _____

_____

What did you do to correct the problems? _____

☐ 11. Have your instructor check your work and sign this job sheet.

Performance Evaluation—Instructor Use Only

Did the student complete the job in the time allotted? Yes ___ No ___
If No, which steps were not completed?_____
How would you rate this student's overall performance on this job?_____
5–Excellent, 4–Good, 3–Satisfactory, 2–Unsatisfactory, 1–Poor
Comments: _____

_____

INSTRUCTOR'S SIGNATURE _____

733

Notes

Project

Diagnosing and Servicing an Air Conditioning System

42

Introduction

Most modern vehicles now have air conditioning systems. If an air conditioning problem occurs, or if the vehicle has been in service for a long time, the refrigeration system should be checked and the problem determined before any repairs are attempted. This involves making visual and operational checks of the air conditioning system and checking system pressures. An important part of this inspection is testing for leaks, the most common refrigeration system defect. In Job 152, you will inspect a vehicle refrigeration system and determine if further checking is needed. You will also perform a refrigerant leak test. Technicians will also commonly encounter problems in the heating and air conditioning system control components and systems. In Job 153, you will diagnose heating and air conditioning control system problems.

Another common job performed on vehicle air conditioning systems is to recover refrigerant, evacuate the system, and recharge the system with refrigerant. This may be done to allow replacement of refrigeration system components, to remove contaminants from the system, or to ensure that the proper charge has been installed. Procedures are relatively simple, but the possibility of personal injury is always present when dealing with liquid refrigerants. In addition, care must be taken to prevent the release of excessive amounts of refrigerant into the atmosphere. In Job 154, you will recover, evacuate, and recharge a vehicle refrigeration system.

Compressors fail after long service or when they are operated with insufficient refrigerant or oil. Compressor clutches can fail from normal wear on high-mileage vehicles, or because excessive pressures in the refrigeration system have placed excessive strain on the clutch faces, causing slippage and overheating. If the clutch clearance is incorrect, the clutch will fail prematurely. In Job 155, you will remove and replace a compressor and compressor clutch.

In addition to the compressor, refrigeration system components include evaporators, condensers, expansion valves, orifice tubes, accumulators, receiver-driers, and connecting hoses and lines. Aside from the compressor (covered in Job 155) and expansion valves, most modern refrigeration system components have no moving parts. Refrigeration system parts are usually replaced because they are leaking or restricted. Accumulators and receiver-driers are often replaced because the desiccant that they contain is saturated with moisture. Refrigeration system part replacement is usually a matter of removing obstructing parts and the component lines and fasteners, and reversing the procedure to install the new part. In Job 156, you will remove and replace refrigeration system components.

Proper refrigerant handling is also an important part of a technician's job. In the past, the only available refrigerant was R-12, and it was handled thoughtlessly. Usually, the technician simply vented the refrigerant to the atmosphere when repairs had to be done to the refrigeration system. Today, however, several refrigerants are available and federal law (and economic sense) requires that refrigerant be recovered, tested, and reused whenever possible. In Job 157, you will label and store refrigerant, test recycled refrigerant for noncondensable gases, and perform maintenance on refrigerant handling equipment.

Project 42 Jobs

- Job 152—Diagnose an Air Conditioning System
- Job 153—Diagnose Heating and Air Conditioning Control Systems
- Job 154—Recover Refrigerant, and Evacuate and Recharge an Air Conditioning System
- Job 155—Remove and Replace a Compressor and Compressor Clutch
- Job 156—Remove and Replace Air Conditioning System Components
- Job 157—Manage Refrigerant and Maintain Refrigerant Handling Equipment

Tools and Materials

The following list contains the tools and materials that may be needed to complete the jobs in this project. The items used will depend on the make and model of vehicle being serviced.

- Vehicle(s) in need of service.
- Applicable service information.
- Belt tension gauge.
- Compressor service tools.
- Compressor service parts.
- Refrigerant cylinder or one pound cans as needed.
- Refrigerant service center or gauge manifold and vacuum pump.
- Refrigerant recovery/recycling equipment (may be part of service center).
- Refrigerant identifier.
- Temperature gauge.
- Leak detector.
- Non-powered test light or multimeter.
- High-impedance multimeter.
- Hand-operated vacuum pump.
- Hand tools as needed.
- Air-powered tools as needed.
- Safety glasses and other protective equipment as needed.

Safety Notice

Before performing this job, review all pertinent safety information in the text, and review safety information with your instructor. *Wear safety glasses whenever you perform any work on the refrigeration system.*

Job 152—Diagnose an Air Conditioning System

After completing this job, you will be able to inspect a vehicle refrigeration system and test the system for leaks.

Procedures

☐ 1. Obtain a vehicle to be used in this job. Your instructor may direct you to perform this job on a shop vehicle.

☐ 2. Gather the tools needed to perform the following job. Refer to the project's tools and materials list.

☐ 3. Obtain the proper service information and determine the type of refrigeration system used:
- Cycling clutch and orifice tube ___
- Capacity control compressor and orifice tube ___
- Expansion valve system ___
- Expansion valve system with evaporator pressure control ___

☐ 4. Determine the system pressure specifications in the service manual. Write the pressures in the spaces provided.
Low side:_____
High side:_____

> **Warning**
> ⚠ The air conditioning system on a hybrid vehicle operates on the same principles as the air conditioning system on a conventional vehicle. The air conditioner compressor in early model hybrids was typically belt-driven off the engine. In later model hybrids, the compressor is usually powered by a high-voltage electric motor. High-voltage wiring has orange insulation. Before performing any service on a hybrid air conditioning compressor or near high-voltage wiring, the technician must first disable the high-voltage system and disconnect the 12-volt battery.

Visually Inspect the Air Conditioning System

☐ 1. Open the vehicle hood.

☐ 2. Visually check the refrigeration system for obvious refrigerant leaks. Leaks are indicated by oil on refrigeration system parts.
Did you find any leaks? Yes ___ No ___
If Yes, describe the leak locations: _____
_____

☐ 3. Examine the air conditioner compressor's drive belt.
Is the compressor drive belt tension acceptable? Yes ___ No ___
If No, and the belt is a V-belt, tighten the belt. Refer to Job 33 for information on setting belt tension. If the belt is a serpentine belt, check the condition of the tensioner and mounting hardware and tighten or replace as necessary.
Is the compressor drive belt condition acceptable? Yes ___ No ___
If No, describe the damage: _____
_____

☐ 4. Examine the refrigeration hoses and fittings.

Is frayed rubber evident on any hoses? Yes ___ No ___

Are cuts evident on any hoses? Yes ___ No ___

Did you find evidence of leaks on any of the hoses or at any fittings? Yes ___ No ___

Are any of the refrigerator hose fittings damaged? Yes ___ No ___

If you answered Yes to any of these questions, describe the defects:_____

_____

☐ 5. Check the other accessible air conditioning system components for defects.

Does the compressor clutch appear to be in good condition? Yes ___ No ___

If No, describe the defects: _____

_____

Does the compressor clutch have excessive clearance? Yes ___ No ___

If Yes, consult your instructor.

Do you see any dirt or other debris on the condenser? Yes ___ No ___

If Yes, carefully clean the condenser.

Do you see any visible damage to condenser or accumulator/receiver-drier?
Yes ___ No ___

If Yes, describe the damage: _____

_____

Is the evaporator drain obstructed? Yes ___ No ___

> **Note**
> If condensation is dripping from under the vehicle, it can be assumed that the drain is open. However, if slushing noises are heard or the front carpet of the vehicle is wet, make a visual check of the drain hose.

If Yes, consult your instructor.

Is refrigerant oil visible? Yes ___ No ___

If Yes, consult your instructor.

☐ 6. Operate the air conditioning and heater.

• Does the air smell musty or moldy? Yes ___ No ___

If Yes, there is likely microbial growth in the evaporator case. Clean and disinfect the evaporator case. Refer to Job 156.

• Does the air smell like coolant? Yes ___ No ___

If Yes, the heater core may be leaking. Refer to Job 158.

• Does the air smell oily? Yes ___ No ___

If Yes, there may be a massive refrigerant leak in the evaporator. Refer to Job 156.

Report any unusual smells to your instructor.

Determine the Type of Refrigerant Used

☐ 1. Consult the appropriate service literature and check the vehicle's refrigerant identification sticker to answer the following questions.

Identify the original refrigerant used in the vehicle:_____

Does the vehicle have a retrofit label? Yes ___ No ___

If Yes, what refrigerant was used to replace the original refrigerant? _____

Job 152—Diagnose an Air Conditioning System (continued)

☐ 2. Obtain a refrigerant identifier and calibrate it according to manufacturer's instructions. See **Figure 152-1**.

Type of refrigerant identifier used: _____

☐ 3. Remove the service valve caps and install a refrigerant identifier on one service port.

☐ 4. Turn on the refrigerant identifier and record the type(s) of refrigerant in the system. If there are multiple types of refrigerant present, record the percentage of each:

Refrigerant Type　　　　　**Percentage**

_____　　　_____

_____　　　_____

_____　　　_____

_____　　　_____

Are any unknown refrigerants or contaminants present? Yes ___ No ___

If Yes, record the amount (percentage) of unknown refrigerants or contaminants: _____

Does the refrigerant identified agree with the manufacturer's label or the retrofit label? Yes ___ No ___

> **Caution**
>
> If the type of refrigerant is incorrect or the refrigerant appears to be contaminated, consult your instructor before continuing.

☐ 5. Remove the refrigerant identifier from the system.

Troubleshoot the System

☐ 1. Attach manifold gauges or a refrigerant service center to the refrigeration system, **Figure 152-2**.

Describe the equipment used: _____

_____

Figure 152-1. A popular make of refrigerant identifier is shown here.

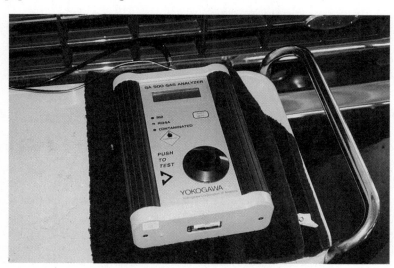

Figure 152-2. Make connections to the refrigeration system carefully. Note the shutoff valves that prevent refrigerant loss.

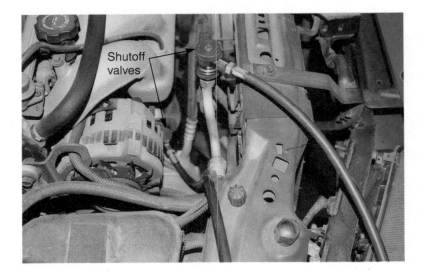

> **Note**
>
> If the hoses do not have shutoff valves, purge the hoses by briefly cracking the hose fittings at the gauge manifold or service center inlet.

2. Check the refrigeration system's static pressure.

   Low side:_____

   High side:_____

> **Note**
>
> If the system shows no static pressure, consult your instructor before performing the next steps.

3. Start the engine and set the throttle to 1500 rpm.

4. Turn the air conditioning system on and set the controls to maximum cooling. Place the blower switch in its highest speed position.

5. Check that the compressor clutch engages and turns the compressor.

   Does the compressor clutch engage? Yes ____ No ____

   If Yes, go to step 7.

   If No, go to step 6.

6. If the compressor clutch does not engage, check for electrical power at the clutch:

   a. Set the air conditioner controls for maximum cooling.

   b. Probe the clutch power connector with a non-powered test light or multimeter.

   Is electrical power reaching the clutch? Yes ____ No ____

   If Yes, the problem is in the clutch electromagnet. Make ohmmeter checks of the electromagnet to determine if the winding is open or shorted and ensure that the electromagnet is properly grounded.

   If No, the control system is faulty or a wire is damaged or unplugged. Continue to substep c.

Job 152—Diagnose an Air Conditioning System (continued)

c. If power is not reaching the clutch, check the following clutch control components as applicable. The use of these devices varies between vehicles, so consult the appropriate service literature before proceeding.
- Compressor relay and control system.
- Low pressure cycling switch.
- High pressure cutout switch.
- Body computer.
- Thermal limiter (older vehicles only).
- Associated wiring.

> **Note**
>
> For more information on the compressor clutch control system, see Job 153.

☐ 7. Install a temperature gauge in the outlet vent nearest the evaporator.

☐ 8. Check the refrigeration system's operating pressure.

Low side:_____

High side:_____

> **Caution**
>
> If the cooling system fans are not operating, or if the high side pressure exceeds 325 psi (2467 kPa), stop the performance test immediately and determine the cause.

How do the pressures compare to the specified pressures listed in step 2? Allow for ambient air temperature when deciding whether system pressures are correct._____
_____

If system pressures are not correct, what could be the cause? _____
_____
_____

☐ 9. Check the air conditioning controls for the following conditions:
- Inoperative blower speeds.
- Inoperative temperature and/or mode controls.
- Temperature changes in response to controls.
- Temperature lowers sufficiently in A/C mode.
- Temperature rises sufficiently in heater mode.
- Interior airflow correct in all modes.

If any of the controls appear to be inoperative, consult your instructor and refer to Job 153.

☐ 10. Observe the refrigeration system to determine if other problems are occurring:

Is the evaporator icing? Yes ___ No ___

Is there any frost on the high side of the system (indicating a high side restriction)? Yes ___ No ___

Do you hear any unusual noises coming from the compressor? Yes ___ No ___

If Yes, describe the noises: _____

11. With the blower in the highest speed, note any unusual odors coming from the ducts.

Is there an odor of oil, indicating an evaporator leak? Yes _____ No _____

Is there an odor of coolant, indicating a heater core leak? Yes _____ No _____

Is there a musty odor, indicating mold growing on the evaporator case? Yes _____ No ____

Are there any other odors? Yes _____ No _____

If Yes, describe _____

12. Place all air conditioner controls in the *off* position.

13. Turn off the engine.

14. Remove gauges or service center hoses from the refrigeration system. Do not replace the service valve caps at this time.

Check for Refrigerant Leaks

1. Obtain a leak detector, **Figure 152-3**.

What type of detector are you working with?_____

> **Warning**
> ⚠ Halide (flame) leak detectors are not recommended. If using a halide leak detector, work in an open area and be extremely careful not to breathe the fumes. Make sure that the flame does not come into contact with flammable components or fuel vapors.

2. Set up and calibrate the leak detector as necessary.

Describe the calibration method: _____

_____

3. Pass the leak detector probe around and under suspected leak areas while observing the detector face.

> **Note**
> 📄 The most obvious places to begin looking for refrigerant leaks are wherever oil residue can be seen.

Were any leaks discovered? Yes ___ No ___

If Yes, where were the leaks located? _____

_____

4. Replace the service valve caps. From the observations made during this job, determine the refrigeration system problem.

Describe the problem: _____

_____

_____

5. Consult your instructor and determine what service and repair steps should be taken.

List the needed service and/or repair steps here: _____

_____

_____

6. With your instructor's permission, take necessary service and/or repair steps.

Service and repairs performed: _____

_____

_____

Job 152—Diagnose an Air Conditioning System (continued)

Figure 152-3. A common make of leak detector is shown here.

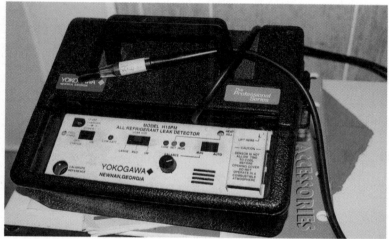

> **Note**
>
> Refer to other jobs in this project for repair and replacement procedures.

☐ 7. Recheck refrigeration system operation.
   Is the problem corrected? Yes ___ No ___
   If Yes, proceed to the next step.
   If No, repeat the job as necessary.

☐ 8. Clean the work area and return tools and equipment to storage. Be sure to turn off the leak detector before storing it.

☐ 9. Did you encounter any problems during this procedure? Yes ___ No ___
   If Yes, describe the problems: _____

   _____

   What did you do to correct the problems? _____

   _____

☐ 10. Have your instructor check your work and sign this job sheet.

Performance Evaluation—Instructor Use Only

Did the student complete the job in the time allotted? Yes ___ No ___
If No, which steps were not completed?_____
How would you rate this student's overall performance on this job?_____
5–Excellent, 4–Good, 3–Satisfactory, 2–Unsatisfactory, 1–Poor
Comments: _____

_____

_____

INSTRUCTOR'S SIGNATURE _____

Notes

Job 153—Diagnose Heating and Air Conditioning Control Systems

After completing this job, you will be able to diagnose a refrigeration control system problem.

Note

This job covers diagnosis of mechanical, vacuum, and electrical heating and air conditioning control system components commonly used on manually controlled systems.

Procedures

☐ 1. Obtain a vehicle to be used in this job. Your instructor may direct you to perform this job on a shop vehicle.

☐ 2. Gather the tools needed to perform the following job. Refer to the project's tools and materials list.

☐ 3. Inspect the cooling system to ensure that it is filled with coolant and that there are no obvious cooling system defects.

   Is the system filled with coolant and in good condition? Yes ___ No ___

☐ 4. Use a refrigerant identifier to ensure that the refrigeration system contains the proper refrigerant and no contaminants.

☐ 5. Attach pressure gauges to the refrigeration system and check system static pressure to ensure that the system is fully charged. See Job 152 if necessary.

   Is the system fully charged? Yes ___ No ___

Caution

The vehicle must be raised and supported in a safe manner. Always use approved lifts or jacks and jack stands. See Figure 6-1.

☐ 6. With the gauges attached, turn the air conditioning system to the full cool position and ensure that the compressor clutch applies and the refrigeration system is operating properly. See Job 152 if necessary.

☐ 7. Install a temperature gauge in the outlet vent nearest the evaporator.

☐ 8. With the system operating and the engine running, check the heating and air conditioning controls for the following conditions:
   - Inoperative blower speeds.
   - Inoperative temperature and/or mode controls.
   - Temperature changes in response to controls.
   - Temperature lowers sufficiently in A/C mode.
   - Temperature rises sufficiently in heater mode.
   - Interior airflow correct in all modes.

   Describe any problems found:_____

   _____

   _____

9. Determine what heating and air conditioning component failure could cause the problem(s) found in step 8.

Possible causes: _____

_____

If the problem could be caused by an electrical component, continue on to the *Check Electrical Component* section.

If the problem could be caused by vacuum or mechanical component, skip to the *Check Mechanical or Vacuum Component* section.

Check Electrical Component

1. Check the fuse that controls the inoperative electrical component. Use the legend on the fuse box cover to locate the proper fuse.

Is the fuse blown? Yes ___ No ___

If Yes, replace the fuse and trace the source of the electrical overload that caused the fuse to fail.

If No, go to step 2.

2. Check for power in the system circuit using a non-powered test light. See **Figure 153-1**.

> **Note**
> If there are problems with multiple accessories, systematically check the wiring ahead of the fuse box in the circuit.

Is there power at the suspect air handling component? Yes ___ No ___

If Yes, proceed to step 3.

If No, locate and check for power at the component's switch or controller in the control panel.

Is there power at the component's switch or controller? Yes ___ No ___

If No, check the wiring between the fuse box and the controller and correct any problems found.

If Yes, the problem is in the controller or the wiring between the controller and the component. Check the wiring for continuity. If the wiring has continuity, the controller is faulty.

Figure 153-1. In this photo, the technician is checking for power at a heating and air conditioning terminal.

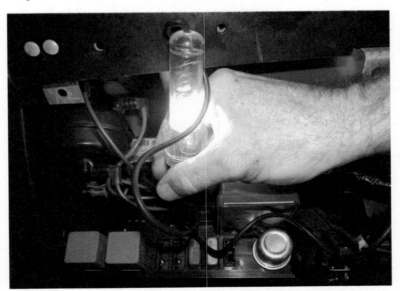

Job 153—Diagnose Heating and Air Conditioning Control Systems (continued)

3. Check for a missing system ground. Loss of ground is common in modern vehicles with many plastic dashboard components.

    Is the heating and air conditioning control panel properly grounded? Yes ___ No ___

    If Yes, the component is probably defective. Proceed to step 4.

    If No, locate the missing ground, repair or substitute another ground, and recheck.

4. Test the system component(s) identified in step 3. Use a high-impedance multimeter to make voltage and resistance tests.

> **Caution**
>
> Do not make any electrical tests of a solid state component unless specifically recommended by the service information. Careless testing can destroy a component.

Component Tested	Type of Test	Reading/Result
_____	_____	_____
_____	_____	_____
_____	_____	_____
_____	_____	_____

5. Based on your test results, write a brief description of the system or component defect(s) and what should be done to return the system to proper operation. After writing the description, skip to the *Make Repairs* section of this job.

    _____

    _____

    _____

    _____

Check Automatic or Semiautomatic Control System

> **Note**
>
> Modern automatic and semiautomatic air conditioning and heating control systems can be accessed with a scan tool. Some older systems can be diagnosed by pushing the system control buttons in a certain sequence.

1. Note any obvious control system failures.

    Is there a failure to change fan speeds? Yes _____No _____

    Is there a failure to change modes? Yes _____ No _____

    Is there a failure to increase or decrease temperature outputs? Yes _____ No _____

    Is there a general failure to respond to controls? Yes _____ No _____

    Does the control panel fail to illuminate? Yes _____ No _____

2. Determine from the service information how the air conditioning and heating control system can be accessed for diagnosis.

    Using a scan tool _____

    Other method _____ Describe the method: _____

3. Use the scan tool or other method to locate any trouble codes. List the codes below.

   _____

   _____

   _____

   _____

4. Determine whether any of the trouble codes apply to the air conditioning and heating control system. List the codes below.

   _____

   _____

   _____

5. Make further control system checks based on the information gathered in steps 1 through 4. Common air conditioning and heating control system defects include blown fuses, bent or disconnected vacuum door linkage, defective blend door motor or diaphragm, defective fan relay, and defective control head.

6. Based on your test results, write a brief description of the system or component defect(s) and what should be done to return the system to proper operation. After writing the description, skip to the *Make Repairs* section of this job.

   _____

   _____

   _____

Check Mechanical or Vacuum Component

1. Remove any components covering the mechanical or vacuum component to be checked.

2. Attempt to operate the component from the control panel while watching the component linkage. See **Figure 153-2**. The engine should be running to test a vacuum-operated component.

   Does the component linkage move? Yes ____ No ____

   If Yes, go to step 4.

   If No, go to step 3.

Figure 153-2. Note that this control assembly contains three cables. Check each cable for free movement and for cracks and wear at the mounting tabs.

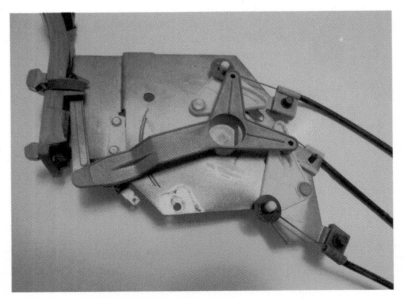

Job 153—Diagnose Heating and Air Conditioning Control Systems (continued)

3. Check the component by one of the following methods, according to the type of component:
   - **Cable- or linkage-operated component:** Disconnect the linkage from the component and attempt to operate the component by hand. If the component does not move, it is defective. If it moves easily, the cable or linkage is defective or requires lubrication and/or adjustment.
   - **Vacuum-operated component:** Use a hand-operated vacuum pump to apply vacuum to the vacuum diaphragm of the component, **Figure 153-3**. If the component will not hold vacuum, it is defective. Note that a few vacuum diaphragms have calibrated leaks and will not hold vacuum. This is acceptable as long as the diaphragm moves the linkage. If the dia-phragm is not leaking but does not move the linkage, disconnect the linkage and retry. If the diaphragm now moves, the linkage is defective or requires lubrication and/or adjustment.

4. Based on your test results, write a brief description of the system or component defect(s) and what should be done to return the system to proper operation.

   _____

   _____

   _____

Make Repairs

1. Consult with your instructor about what repairs are to be made.

2. Replace defective electrical components or lubricate and adjust linkage as necessary.

> **Note**
>
> If lubricating or adjusting the linkage or cable does not solve the problem, the component should be replaced.

3. Recheck heating and air conditioning control system operation.

   Does the system function properly? Yes ___ No ___

Figure 153-3. Apply vacuum to the diaphragm and note whether the linkage moves.

☐ 4. Reinstall any parts that were removed to gain access to the heating and air conditioning control system components.

☐ 5. Clean the work area and return tools and equipment to storage.

☐ 6. Did you encounter any problems during this procedure? Yes ___ No ___

If Yes, describe the problems: _____

_____

What did you do to correct the problems? _____

_____

☐ 7. Have your instructor check your work and sign this job sheet.

Performance Evaluation—Instructor Use Only

Did the student complete the job in the time allotted? Yes ___ No ___

If No, which steps were not completed?_____

How would you rate this student's overall performance on this job?_____

5–Excellent, 4–Good, 3–Satisfactory, 2–Unsatisfactory, 1–Poor

Comments: _____

_____

_____

INSTRUCTOR'S SIGNATURE _____

Job 154—Recover Refrigerant, and Evacuate and Recharge an Air Conditioning System

After completing this job, you will be able to recover refrigerant and evacuate and recharge a vehicle's refrigeration system.

Procedures

☐ 1. Obtain a vehicle to be used in this job. Your instructor may direct you to perform this job on a shop vehicle.

☐ 2. Gather the tools needed to perform the following job. Refer to the project's tools and materials list.

☐ 3. Connect a refrigerant identifier to the vehicle and measure the type and purity of the refrigerant in the system. This was covered in more detail in Job 152.

Refrigerant type:_____

Percentage of this refrigerant:_____

Percentage of other/unknown refrigerants or contaminants:_____

Does the refrigerant identified agree with the manufacturer's label or, if present, the retrofit label? Yes ___ No ___

> **Caution**
> If the type of refrigerant is incorrect or the refrigerant appears to be contaminated, consult with your instructor before continuing.

☐ 4. Remove the refrigerant identifier.

> **Note**
> If you are using a refrigerant service center, proceed to the next section. If you are using a gauge manifold, skip to the *Use a Gauge Manifold and Vacuum Pump to Evacuate and Recharge a System* section of this job.

Use a Refrigerant Service Center to Evacuate and Recharge a System

☐ 1. Attach the service center's hoses to the refrigeration system. A typical service center connection is shown in **Figure 154-1**.

☐ 2. Using the service center controls, recover the system refrigerant.

> **Note**
> If this job is being performed in combination with a refrigeration system component replacement job, perform the component replacement at this time.

☐ 3. Using the service center controls, set the evacuation time as directed by your instructor.

> **Note**
> Skip this step if the service center does not have a timer feature.

Figure 154-1. Service center connections to a typical vehicle are shown here. Connection locations vary from vehicle to vehicle. Refer to the vehicle's service literature before connecting the service center.

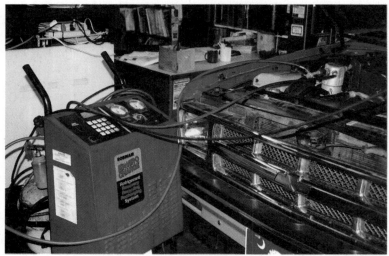

☐ 4. Start the vacuum pump.

☐ 5. When the service center completes evacuation, close the valves on the service center's hoses. Turn off the vacuum pump if necessary.

> **Caution**
>
> ⚠ Do not disconnect the hoses from the service fittings.

What was the total evacuation time?_____

What was the final vacuum reading? _____

☐ 6. Monitor the system vacuum for about 10 minutes.

Did the vacuum decrease? Yes ___ No ___

If Yes, by how much?_____

What do you think is the cause? _____

_____

> **Caution**
>
> ⚠ If the vacuum decrease is excessive (more than 1"–2" Hg), consult your instructor before continuing.

☐ 7. Use the service center controls to set the amount of refrigerant charge according to the system label or as directed by your instructor.

Type of refrigerant:_____

Amount charged into the system:_____

☐ 8. Open the valves on the service center hoses.

☐ 9. Charge the system with the service center.

☐ 10. When charging is complete, start the engine, and set the air conditioning system to *Max A/C*.

☐ 11. Install a temperature gauge in the outlet vent nearest the evaporator.

Job 154—Recover Refrigerant, and Evacuate and Recharge an Air Conditioning System (continued)

☐ 12. Check refrigeration system pressures.

Low side:_____

High side:_____

Are these pressures acceptable? Yes ___ No ___

If No, consult with your instructor about further steps to take.

☐ 13. Record the temperature from the vent:_____

☐ 14. Turn off HVAC system and stop the engine.

☐ 15. Disconnect and remove the service center hoses.

☐ 16. Depending on your instructor's directions, skip ahead to the *Complete the Job* section of this job or repeat the job with a gauge manifold and vacuum pump as described in the following section.

Use a Gauge Manifold and Vacuum Pump to Evacuate and Recharge a System

☐ 1. Attach the gauge manifold to the refrigeration system as shown in **Figure 154-2**. Purge the hoses as necessary.

☐ 2. Recover the refrigerant charge as directed by your instructor. Use the shop equipment as necessary.

Equipment used: _____

Briefly describe the recovery procedure using this equipment: _____

_____

_____

_____

_____

Figure 154-2. Carefully attach the gauge manifold fittings to the refrigeration system. R-134a fittings are push-on types, which are easily attached. Older R-12 fittings must be threaded to the system fittings. An R-134a fitting is shown here.

> **Note**
> If this job is being performed in combination with a refrigeration system component replacement job, perform the component replacement at this time.

☐ 3. Attach the vacuum pump to the gauge manifold service (center) hose, **Figure 154-3**.

☐ 4. Start the vacuum pump and open both manifold valves.

> **Caution**
> Allow 15 minutes minimum time for evacuation.

☐ 5. After the proper time has elapsed, close the valves on the gauge manifold, turn off the vacuum pump, and observe the gauges, **Figure 154-4**.

What was the final vacuum reading? _____

Figure 154-3. Attach the center hose of the gauge manifold to the vacuum pump.

Figure 154-4. Observe the gauges. Vacuum should be indicated on both gauges.

Job 154—Recover Refrigerant, and Evacuate and Recharge an Air Conditioning System (continued)

> **Caution**
>
> ⚠ Do not disconnect the hoses from the service fittings.

☐ 6. Monitor the system vacuum for about 10 minutes.

Did the vacuum decrease? Yes ___ No ___

If Yes, by how much?_____

What do you think is the cause? _____

_____

> **Caution**
>
> ⚠ If the vacuum decrease is excessive (more than 1″–2″ Hg), consult your instructor before continuing.

☐ 7. Attach the gauge manifold to a refrigerant service cart or a cylinder containing the proper refrigerant.

Refrigerant used:_____

☐ 8. Determine the amount of refrigerant charge to be used by referring to the underhood system label or as directed by your instructor.

Amount of refrigerant to be used:_____

☐ 9. If a refrigerant cylinder and an electronic refrigerant scale are to be used to recharge the system, set the cylinder on the scale and note the weight of the cylinder.

Cylinder weight:_____

or

If individual cans are to be used, calculate how many cans and partial cans will be needed to fill the system.

Number of cans needed:_____

☐ 10. Install a temperature gauge in the outlet vent nearest the evaporator.

☐ 11. Start the engine and set the air conditioner controls to maximum.

☐ 12. Open the low side manifold valve and allow the system to charge.

> **Warning**
>
> ⚠ Only charge refrigerant to the low side. Charging refrigerant to the high side while the engine is running could cause the refrigerant can or cylinder to burst.

☐ 13. Continue charging the system until you have added the proper weight of refrigerant or the proper number of refrigerant cans to the system.

☐ 14. Check refrigeration system pressures.

Low side:_____

High side:_____

Are these pressures acceptable? Yes ___ No ___

If No, consult with your instructor about further steps to take.

☐ 15. Add more refrigerant as necessary.

Amount added:_____

☐ 16. Record the temperature from the vent.

Outlet temperature:_____

> **Caution**
>
> Check with your instructor before performing step 16.

☐ 17. Turn the air conditioning system controls to *off* and stop the engine.

☐ 18. Remove the gauge manifold hoses.

☐ 19. Replace the service valve caps.

> **Note**
>
> Adding oil to individual components with the system discharged is explained in the Jobs covering those components.

> **Note**
>
> Your instructor may assign this task using either of the following methods.

Adding Oil to a Charged Refrigeration System using High-Side Pressure

> **Caution**
>
> A special refrigeration oil is used in hybrid vehicles equipped with a high-voltage air conditioning compressor motor. This special refrigeration oil has a much higher electrical resistance than conventional refrigeration oils. Always follow the manufacturer's recommendations when replacing refrigeration oil.

☐ 1. Fill the oil injection canister with the proper system lubricant.

☐ 2. Attach an injection canister to the low-side hose of gauge manifold.

☐ 3. Attach the gauge manifold to the vehicle.

☐ 4. Start the engine and turn on the air conditioner system.

☐ 5. Once the high-side pressure rises above low-side pressure, slowly turn the hand wheels to crack open the manifold valves and allow a small amount of high-side pressure to enter the low side through the gauge manifold. This will push the oil into the low side of the system.

> **Warning**
>
> Oil must slowly enter the low side of the system to prevent compressor valve damage. Do not open the manifold valve hand wheel more than a fraction of a turn during this operation.

Job 154—Recover Refrigerant, and Evacuate and Recharge an Air Conditioning System (continued)

☐   6.   After allowing sufficient time for the oil to enter the low side of the system, turn the hand wheels to close the valves, then allow the low-side and high-side pressures to stabilize.

☐   7.   Turn off the air conditioner and engine.

☐   8.   Remove the gauge manifold from the system.

Adding Oil to a Charged Refrigeration System using a Special Oil Injection Device

☐   1.   Be sure the engine cannot be started.

☐   2.   Obtain an oil injection device that contains the correct refrigeration oil for the system.

☐   3.   Determine the type of hose threads used on the oil injection device.

☐   4.   Determine which system fitting should be removed to attach the hose threads of the oil injection device.

☐   5.   Remove the fitting cap and attach the oil injection device.

☐   6.   Turn the top of the oil injection device with a wrench to push the oil into the system.

☐   7.   Remove the oil injection device and reinstall the fitting cap.

☐   8.   Start the engine.

☐   9.   Turn on the air conditioning system and ensure that the air conditioner compressor has engaged.

☐   10.   Allow the air conditioner to operate for several seconds and then turn the air conditioner off.

☐   11.   Repeat the above steps 9 and 10 to ensure that oil enters the compressor in small amounts. This reduces the chances of the injected oil damaging the compressor valves.

☐   12.   Turn off the air conditioner and engine.

Complete the Job

☐   1.   Check for leaks as directed by your instructor. See Job 152.

☐   2.   Clean the work area and return tools and equipment to storage.

☐   3.   Did you encounter any problems during this procedure? Yes ___ No ___

If Yes, describe the problems: _____

_____

What did you do to correct the problems? _____

_____

☐   4.   Have your instructor check your work and sign this job sheet.

Performance Evaluation—Instructor Use Only

Did the student complete the job in the time allotted? Yes ____ No ____

If No, which steps were not completed?_____

How would you rate this student's overall performance on this job?_____

5–Excellent, 4–Good, 3–Satisfactory, 2–Unsatisfactory, 1–Poor

Comments: _____

_____

_____

INSTRUCTOR'S SIGNATURE _____

Job 155—Remove and Replace a Compressor and Compressor Clutch

After completing this job, you will be able to remove and replace a compressor and compressor clutch.

Procedures

- [ ] 1. Obtain a vehicle to be used in this job. Your instructor may direct you to perform this job on a shop vehicle.
- [ ] 2. Gather the tools needed to perform the following job. Refer to the project's tools and materials list.

> **Warning**
>
> ⚠ Before performing any service on a hybrid air conditioning compressor, the technician must first disable the high-voltage system and disconnect the 12-volt battery.

Replace the Compressor Clutch

- [ ] 1. Remove the compressor drive belt.
- [ ] 2. Remove the nut holding the clutch plate to the shaft, if necessary. See **Figure 155-1**.
- [ ] 3. Remove the clutch plate. Usually a special tool is needed. See **Figure 155-2**.
- [ ] 4. Remove the snap ring holding the pulley to the compressor.
- [ ] 5. Remove the pulley from the compressor. A puller may be needed.
- [ ] 6. Examine the clutch faces for signs of scoring and overheating.

     Does the clutch require replacement? Yes ___ No ___

Figure 155-1. Some clutch plates are held to the compressor shaft by a nut. This nut must be removed before the clutch plate can be removed.

Figure 155-2. A special puller is needed to remove the clutch plate from this compressor.

7. If the electromagnet is to be replaced, perform the following steps:
   a. Detach the electrical connector.
   b. Remove the snap ring or bolts holding the electromagnet.
   c. Remove the electromagnet from the compressor.
   d. Ensure that the new electromagnet is the correct replacement.
   e. Place the new electromagnet in position.
   f. Reinstall the snap ring or bolts as applicable.
   g. Reattach the electrical connector.

8. If the pulley bearing is to be replaced, perform the following steps:
   a. Remove the snap rings holding the bearing in the pulley.
   b. Remove the bearing from the pulley.
   c. Clean the bearing surface of the pulley.
   d. Ensure that the new bearing is the correct replacement.
   e. Install the new bearing in the pulley.
   f. Reinstall the snap ring.

9. Install the new pulley assembly on the compressor.

10. Install the pulley snap ring.

11. Install the clutch plate on the compressor shaft.

12. Using necessary special tools, pull the clutch plate into position while monitoring clutch gap. See **Figure 155-3**.

    Final clutch gap:_____

13. Install the nut holding the clutch plate to the shaft if necessary.

14. Reinstall the drive belt.

15. Start the engine, turn on the air conditioner, and check the compressor clutch operation.

Replace the Compressor

1. Identify and recover the refrigerant. Refer to Job 154 for exact procedures.

Caution

 If the type of refrigerant is incorrect, or the refrigerant appears to be contaminated, consult with your instructor before continuing.

Job 155—Remove and Replace a Compressor and Compressor Clutch (continued)

Figure 155-3. Using the special installation tool, slowly pull the clutch plate into position while checking clutch gap.

☐ 2. Ensure that all pressure has been removed from the refrigeration system.

☐ 3. Remove the compressor drive belt.

☐ 4. Remove any parts preventing removal of the compressor.

☐ 5. Remove the compressor lines and cap the open line fittings.

Type of fitting(s):_____

☐ 6. Remove fasteners holding the compressor to the engine and remove the compressor.

☐ 7. Drain and measure the oil from the compressor.

How much oil was drained from the compressor? _____

Is this the amount specified in the service information? Yes ___ No ___

☐ 8. Examine the oil for the following abnormal conditions:
- Dirty.
- Milky.
- Contaminated with metal particles.
- Contaminated with other debris.

Is the oil clean? Yes ___ No ___

If No, describe the condition of the oil: _____

_____

> **Note**
> If the compressor contained an incorrect quantity of oil, or if the oil was not clean, consult your instructor. Other refrigeration system service operations may be necessary.

☐ 9. Place the compressor on a clean bench.

> **Note**
>
> If a compressor holding fixture is available, use it to secure the compressor.

☐ 10. Remove any parts such as pressure switches that will be transferred to the new compressor.

☐ 11. Compare the old and new compressors.

Do they match? Yes ___ No ___

If No, consult with your instructor about obtaining the correct replacement compressor.

☐ 12. Determine the type of refrigeration oil needed.

What source did you use to determine the proper type of oil?_____

☐ 13. Fill the compressor with the proper type and amount of new oil. Check the service information for the proper amount of oil needed. The amount of oil needed may be different from the amount drained.

How much oil was added to the compressor?_____

> **Caution**
>
> To prevent compressor failure, the correct amount of oil must be added to the compressor. Amounts vary widely, from 5 ounces to over 11 ounces. Many new and remanufactured compressors come filled with oil. It is a good practice to drain and measure the new oil, and add more if necessary.

☐ 14. Install the new compressor on the engine.

☐ 15. Remove the line caps and install the lines on the compressor.

> **Caution**
>
> Replace any O-rings or gaskets that were removed. Lubricate the new O-rings or gaskets with the correct type of refrigerant oil.

☐ 16. Replace the accumulator or receiver-drier if necessary.

> **Note**
>
> Most technicians prefer to replace the accumulator or receiver-drier if there is any evidence that the desiccant in the old unit has absorbed excess moisture.

☐ 17. Turn the compressor through at least 10 revolutions by hand to remove any excess oil from the cylinders.

☐ 18. Reinstall the drive belt and set the proper belt tension.

Job 155—Remove and Replace a Compressor and Compressor Clutch (continued)

Add a Refrigeration System Filter If Needed

☐ 1. Before deciding to install a filter, answer the following questions concerning the vehicle refrigeration system.
- Has the refrigeration system suffered a catastrophic compressor failure? Yes ___ No ___
- Has flushing been unsuccessful in removing all debris from the refrigeration system? Yes ___ No ___
- Has the expansion valve or orifice tube become blocked during prior refrigeration system operation? Yes ___ No ___

If the answer to any of the above questions is Yes, do you think that the system would benefit from the installation of a filter? Yes ___ No ___

Discuss your answer with your instructor before continuing.

☐ 2. If the decision is made to install a filter, evacuate and recover the system refrigerant.

☐ 3. Determine a location in the high-pressure line with sufficient clearance to install the filter.

☐ 4. Using a tubing cutter, cut the high-pressure line.

> **Note**
>
> Cut out just enough of the high-pressure line to allow the filter to be installed.

☐ 5. Lubricate all fittings and O-rings as necessary.

☐ 6. Place the high-pressure filter in the line, ensuring that it fits securely away from moving parts.

☐ 7. Tighten the filter fittings.

☐ 8. Install any filter brackets as necessary.

Complete the Service

☐ 1. Evacuate the system and check for leaks.

☐ 2. Recharge the system and check its performance.

☐ 3. Clean the work area and return tools and equipment to storage.

☐ 4. Did you encounter any problems during this procedure? Yes ___ No ___

If Yes, describe the problems: _____

_____

What did you do to correct the problems? _____

_____

☐ 5. Have your instructor check your work and sign this job sheet.

Performance Evaluation—Instructor Use Only

Did the student complete the job in the time allotted? Yes ＿＿ No ＿＿
If No, which steps were not completed?＿＿＿＿＿＿＿＿＿＿＿＿
How would you rate this student's overall performance on this job?＿＿＿＿
5–Excellent, 4–Good, 3–Satisfactory, 2–Unsatisfactory, 1–Poor

Comments: ＿＿＿＿＿＿＿＿＿＿＿＿＿＿＿＿＿＿＿＿＿＿＿＿＿＿＿＿＿＿＿＿＿
＿＿＿＿＿＿＿＿＿＿＿＿＿＿＿＿＿＿＿＿＿＿＿＿＿＿＿＿＿＿＿＿＿＿＿＿＿＿＿
＿＿＿＿＿＿＿＿＿＿＿＿＿＿＿＿＿＿＿＿＿＿＿＿＿＿＿＿＿＿＿＿＿＿＿＿＿＿＿

INSTRUCTOR'S SIGNATURE ＿＿＿＿＿＿＿＿＿＿＿＿＿＿＿＿＿＿＿＿＿＿＿＿＿＿

Job 156—Remove and Replace Air Conditioning System Components

After completing this job, you will be able to remove and replace refrigeration system components.

Procedures

> **Caution**
>
> Before beginning, check with your instructor to determine whether the vehicle battery should be disconnected for this procedure.

1. Obtain a vehicle to be used in this job. Your instructor may direct you to perform this job on a shop vehicle.
2. Gather the tools needed to perform the following job. Refer to the project's tools and materials list.
3. Identify the refrigeration system component(s) to be replaced.

   List the components to be replaced in this job: _____

   _____
4. Identify and recover the refrigerant. Refer to Job 154 for exact procedures.

> **Caution**
>
> If the refrigerant is an incorrect type or appears to be contaminated, consult with your instructor before continuing.

5. Ensure that all pressure has been removed from the refrigeration system.
6. Remove any brackets or covers blocking access to the component. If an expansion valve is being replaced, remove the sensing bulb from the evaporator outlet line.
7. Remove the component's line fittings. Fittings may be located under the cowl, **Figure 156-1**. In the figure, the fittings are held to the evaporator with easily removable studs and nuts. However, other fittings may require special tools for removal.

   Types of fitting(s) used:_____

   Special tools needed to remove lines: _____

   Fitting seals: _____
8. Cap any open fittings.
9. Remove the brackets or other fasteners holding the component to the vehicle.
10. Remove the component from the vehicle.

> **Note**
>
> To replace the evaporator and expansion valve on most modern vehicles, the evaporator case must be removed from the vehicle and split or otherwise disassembled. See **Figure 156-2**.

Figure 156-1. Many evaporator fittings are located under the vehicle cowl and may require careful removal to avoid damage.

Figure 156-2. Once the case is removed from the vehicle, it must be split to expose the evaporator and expansion valve.

☐ 11. To confirm your original diagnosis, check the removed components as follows.

Do the condenser or evaporator leak or clog? Yes _____ No _____

Are there metal particles or excessive debris in the expansion valve or orifice tube screen? Yes _____ No _____

Is there excess moisture in the accumulator or receiver-dryer? Yes _____ No _____

Are there leaks, restrictions, or swelled sections in the hoses? Yes _____ No _____

Are there leaks, kinked sections, or clogging in the lines? Yes _____ No _____

If the answer to any of the above is Yes, consult your instructor to determine whether the system should be flushed or other parts replaced.

☐ 12. Compare the old and new components.

Job 156—Remove and Replace Air Conditioning System Components (continued)

> **Note**
>
> Consult your instructor if the new component does not seem to be an exact replacement for the old unit.

☐ 13. Drain the oil from the old component into a container and measure the amount of oil removed.

How much oil was drained from the component? _____

Is this the amount specified in the service information? Yes ___ No ___

☐ 14. Examine the oil for the following abnormal conditions:
- Dirty.
- Milky.
- Contaminated with metal particles.
- Contaminated with other debris.

Is the oil clean? Yes ___ No ___

If No, describe the condition of the oil: _____

_____

> **Note**
>
> If the component contained an incorrect quantity of oil, or if the oil was not clean, consult your instructor. Other refrigeration system service operations may be necessary.

☐ 15. Determine the type of refrigeration oil needed.

What source did you use to determine the proper type of oil?_____

☐ 16. Fill the component with the proper type and amount of new oil. Check the service information, as the amount of oil needed may be different from the amount drained. If specifications are not available, follow these general guidelines:
- Accumulator—1 oz.
- Receiver-drier—1 oz.
- Condenser—1 oz.
- Evaporator:
  Ford or GM—3 oz.
  Chrysler—2 oz.
  Imports—1 oz.

How much oil was added to the component? _____

> **Note**
>
> It is usually not necessary to add oil to a replacement hose or line. If the system appears to have lost considerable oil because of a leak, your instructor may direct you to add an additional one or two ounces of oil.

☐ 17. Obtain new seals and gaskets for all parts that were removed. Do not reuse old seals or gaskets.

☐ 18. Lubricate the new seals and gaskets with the same type of refrigerant oil used in the system.

Caution

If an evaporator is being replaced, ensure that the evaporator drain in the case is not restricted and that the case is free of debris and mold.

☐ 19. Place the new component in position and loosely install the mounting brackets or fasteners.

Note

If the expansion valve is being replaced, place the sensing bulb in position on the evaporator outlet and wrap the bulb in insulating tape. See **Figure 156-3**.

☐ 20. Reassemble the evaporator case as necessary, **Figure 156-4**.

Figure 156-3. Carefully place the new parts in position. If you are installing an expansion valve, wrap the expansion valve with insulating tape before assembling the two case halves.

Figure 156-4. Reassemble the case after the internal parts are replaced.

Job 156—Remove and Replace Air Conditioning System Components (continued)

☐ 21. Remove the fitting caps and install the fittings using new seals and/or gaskets.

☐ 22. Tighten all mounting brackets or fasteners.

☐ 23. If necessary, tighten the fittings to the recommended torque.

☐ 24. Repeat steps 6 through 23 for any other refrigeration system components to be replaced.

Note

Most technicians prefer to replace the accumulator or receiver-drier when another part is replaced, or if there is any evidence that the desiccant in the old unit has absorbed excess moisture.

☐ 25. Evacuate the system and ensure that the system holds vacuum.

☐ 26. Recharge the system and check performance.

☐ 27. Check for leaks, concentrating on fittings that were dissembled as part of this job.

☐ 28. Clean the work area and return tools and equipment to storage.

☐ 29. Did you encounter any problems during this procedure? Yes ___ No ___

If Yes, describe the problems: _____

_____

What did you do to correct the problems? _____

_____

☐ 30. Have your instructor check your work and sign this job sheet.

Performance Evaluation—Instructor Use Only

Did the student complete the job in the time allotted? Yes ___ No ___

If No, which steps were not completed?_____

How would you rate this student's overall performance on this job?_____

5–Excellent, 4–Good, 3–Satisfactory, 2–Unsatisfactory, 1–Poor

Comments: _____

_____

_____

INSTRUCTOR'S SIGNATURE _____

Notes

Job 157—Manage Refrigerant and Maintain Refrigerant Handling Equipment

After completing this job, you will be able to label and store refrigerant, test recycled refrigerant for noncondensable gases, and maintain refrigerant handling equipment.

Procedures

☐ 1. Obtain a vehicle to be used in this job. Your instructor may direct you to perform this job on a shop vehicle.

☐ 2. Gather the tools needed to perform the following job. Refer to the project's tools and materials list.

Label and Store Refrigerant

☐ 1. Open the vehicle hood and determine the original type of refrigerant used in the vehicle by examining the factory refrigerant label.

Refrigerant type:_____

☐ 2. Look for the presence of a retrofit label.

Did you find a retrofit label? Yes ___ No ___

If Yes, what refrigerant was used to replace the original refrigerant? _____

☐ 3. Use a refrigerant identifier to identify the type of refrigerant(s) used in the system. Refer to Job 152 for specific instructions.

Type of refrigerant identifier used: _____

Refrigerant Type	Percentage
_____	_____
_____	_____
_____	_____
_____	_____

Are any unknown refrigerants or contaminants present? Yes ___ No ___

If Yes, record the amount (percentage) of unknown refrigerants or contaminants: _____

Does the refrigerant identified agree with the manufacturer's label or, if present, the retrofit label? Yes ___ No ___

> **Caution**
>
> If the type of refrigerant is incorrect or the refrigerant appears to be contaminated, consult your instructor before continuing.

☐ 4. Remove the refrigerant identifier from the system.

☐ 5. Attach the recovery equipment to the refrigeration system.

Type of equipment used: _____

☐ 6. Determine what sort of receptacle should be used to recover the equipment.
- **R12.** White tank or internal tank in recovery/recycling equipment. ___
- **R134a.** Blue tank or internal tank in recovery/recycling equipment. ___
- **Contaminated or unknown refrigerant.** Gray tank with yellow top. ___
- **Other type of receptacle.** Explain: _____

☐ 7. Use the recovery equipment controls to recover the system refrigerant.

☐ 8. If the recycled refrigerant contains noncondensable gases, purge the receptacle by placing it upright and cracking the valve until refrigerant exits.

> **Caution**
>
> Do not allow excessive amounts of refrigerant to escape.

☐ 9. If necessary, label the refrigerant receptacle.
☐ 10. Check the refrigerant receptacle for leaks.
☐ 11. Store the refrigerant receptacle in a cool place.

Maintain Refrigerant Handling Equipment

☐ 1. Have your instructor identify a piece of equipment on which to perform the maintenance.
☐ 2. Ensure that the equipment is not attached to a vehicle's refrigeration system.

Replace Service Center Filter-Drier

☐ 1. Locate the service center's filter-drier, **Figure 157-1**.
☐ 2. Follow the manufacturer's instructions to recover any refrigerant from the filter tubing.
☐ 3. Remove the filter bracket.
☐ 4. Loosen and remove the filter fittings.
☐ 5. Place the new filter in position and install the fittings.
☐ 6. Reinstall the filter bracket.
☐ 7. Evacuate the filter tubing.
☐ 8. Recharge the service center and check for leaks.

Check Service Center for Leaks

☐ 1. Obtain a leak detector.

Detector type: _____

Figure 157-1. Most service center filter-driers are located where they can be easily serviced.

Job 157—Manage Refrigerant and Maintain Refrigerant Handling Equipment (continued)

> **Warning**
>
> ⚠ Halide (flame) leak detectors are not recommended. If using a halide leak detector, work in an open area and be extremely careful not to breathe the fumes. Make sure that the flame does not come into contact with flammable components or fuel vapors.

☐ 2. Set up and calibrate the leak detector as necessary. Refer to Job 152.

☐ 3. Pass the leak detector probe around and under suspected service center leak areas while observing the detector face.

Were any leaks discovered? Yes ___ No ___

If Yes, where were the leaks located? _____

Change Vacuum Pump Oil

☐ 1. Place the vacuum pump over a drain pan.

☐ 2. Open the drain valve and allow the oil to drain into the drain pan.

☐ 3. Obtain the proper type of vacuum pump oil.

> **Caution**
>
> ⟁ Obtain and use the correct type of vacuum pump oil. Do not use any other kind of oil.

☐ 4. Close the drain valve and remove the oil fill plug.

☐ 5. Add oil until the oil level reaches the proper level in the sight glass, **Figure 157-2**.

☐ 6. Reinstall the oil fill plug.

☐ 7. Clean the work area and return tools and equipment to storage. Be sure to turn off the leak detector before storing it.

Figure 157-2. The most common type of oil sight glass used on vacuum pumps is shown here.

8. Did you encounter any problems during this procedure? Yes ___ No ___
   If Yes, describe the problems: _____

   _____

   What did you do to correct the problems? _____

   _____

9. Have your instructor check your work and sign this job sheet.

Performance Evaluation—Instructor Use Only

Did the student complete the job in the time allotted? Yes ___ No ___
If No, which steps were not completed?_____
How would you rate this student's overall performance on this job?_____
5–Excellent, 4–Good, 3–Satisfactory, 2–Unsatisfactory, 1–Poor
Comments: _____

_____

_____

INSTRUCTOR'S SIGNATURE _____

Project

Diagnosing and Servicing a Heating and Ventilation System

43

Introduction

Basically, heater cores have not changed in many years, although modern cores are often made of aluminum and plastic instead of copper. They function as small radiators, allowing heat from the cooling system to enter the passenger compartment. Heater cores may leak or become clogged, in which case they are replaced instead of being repaired. Related heater components include the heater hoses and may include a heater shutoff valve. In Job 158, you will inspect and service a vehicle's heater components.

Heating and air conditioning systems depend on the air handling components (blower fan and motor, air ducts and hoses, outlet vents, cabin filters) for delivery of the heated or cooled air. A heating and air conditioning system cannot be successfully serviced unless the air handling components are in good working order. Ducts and hoses can be improperly positioned or disconnected. Outlets can be stuck closed or incapable of positioned correctly. Cabin filters can become clogged and restrict airflow. In Job 159, you will inspect and, if necessary, service the air handling components of a heating and air conditioning system.

Project 43 Jobs

- Job 158—Inspect and Service Heater Components
- Job 159—Inspect and Service Air Handling Components

Tools and Materials

The following list contains the tools and materials that may be needed to complete the jobs in this project. The items used will depend on the make and model of vehicle being serviced.

- Vehicle in need of service.
- Applicable service information.
- Test light or multimeter.
- Drain pan(s).
- Correct type of antifreeze.

- Air-powered tools as needed.
- Hand tools as needed.
- Safety glasses and other protective equipment as needed.

Safety Notice

Before performing this job, review all pertinent safety information in the text, and review safety information with your instructor. *Wear safety glasses whenever you perform any work on the refrigeration system.*

Job 158—Inspect and Service Heater Components

After completing this job, you will be able to replace a heater core and restore the cooling system to proper operation.

Procedures

☐ 1. Obtain a vehicle to be used in this job. Your instructor may direct you to perform this job on a shop vehicle.

☐ 2. Gather the tools needed to perform the following job. Refer to the project's tools and materials list.

Check the Heater, Heater Hoses, and Heater Shutoff Valve

☐ 1. Open the hood and locate the heater hoses and the heater shutoff valve, if used. The valve may be located on the engine or near the firewall, **Figure 158-1**.

> **Note**
>
> A few vehicles use an electrically operated pump to move coolant through the heater system. Consult the appropriate service information to inspect and service this pump.

☐ 2. Make a visual check for leaks at the hoses, hose connections, and heater shutoff valve. If there is any evidence of a leak, pressure check the cooling system according to the procedures in Job 31.

Did you locate any leaks? Yes ___ No ___

If Yes, where? _____

☐ 3. Start the engine and check the shutoff valve operation. The shutoff valve should move when vacuum is applied or removed, for instance when changing between the air conditioning and heater modes or when the temperature lever is moved from full heating to full cooling.

Does the shutoff valve react to control changes? Yes ___ No ___

Replace a Heater Shutoff Valve

☐ 1. Ensure that the engine and cooling system are sufficiently cool to allow the radiator cap to be safely removed.

☐ 2. Remove the radiator cap, then locate the heater shutoff valve.

☐ 3. Place a suitable container under the heater shutoff valve.

☐ 4. Remove the heater hoses.

> **Warning**
>
> ⚠ Clean up any coolant spills. Antifreeze is poisonous to people and animals.

☐ 5. Remove the vacuum line from the shutoff valve vacuum diaphragm and remove the shutoff valve.

Figure 158-1. Heater shutoff valves are located in one of the heater hose lines, usually the inlet. A—Linkage-type shutoff valve located in a heater hose. B—Plunger-type shutoff valve located on the engine.

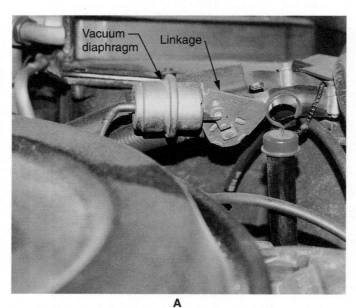

A

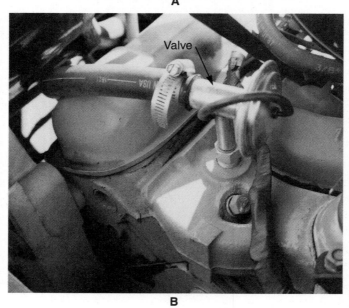

B

6. Compare the old and new valves. Some shutoff valves are fairly complex, **Figure 158-2**, and should be carefully inspected. Test that the replacement valve is correct by blowing through both the old and new valves. If you can blow through one valve and not the other, the replacement is incorrect.

7. Install the shutoff valve on the heater hoses. Use sealer if necessary.

8. Install and tighten the heater hose clamps.

9. Install the vacuum line on the diaphragm.

10. Add the proper kind of coolant as needed. Refer to Job 32.

   What type of coolant is used on this vehicle?_____

Job 158—Inspect and Service Heater Components (continued)

Figure 158-2. This heater shutoff valve has three heater hose connections and three vacuum hose connections. The replacement valve must match the old valve exactly.

> **Caution**
>
> Many modern vehicles use special long-life coolant. Do not mix coolant types.

 11. Start the engine and check for leaks and proper heater operation.

> **Note**
>
> On some vehicles, it will be necessary to bleed the heater core to remove trapped air. Loosen the highest heater hose and allow air to bleed from the heater core and hoses as the engine operates.

 12. Ensure that the cooling system is full and install the radiator cap.

Replace a Heater Core

☐ 1. Ensure that the engine and cooling system are sufficiently cool to allow the radiator cap to be safely removed.

☐ 2. Remove the radiator cap, then locate the radiator drain plug.

☐ 3. Place a suitable container under the drain plug.

☐ 4. Open the radiator drain plug. Ensure that coolant drains into the container.

> **Warning**
>
> Clean up any coolant spills. Antifreeze is poisonous to people and animals.

 5. Allow as much coolant as possible to drain out.

☐ 6. Remove the heater hoses.

7. If the heater core will be removed through the passenger compartment, blow through the heater core to remove as much coolant as possible.

8. Remove housings or covers as necessary to gain access to the heater core, **Figure 158-3**.

9. If necessary, remove the heater case or module containing the heater core. **Figure 158-4** shows a heater case removed from the vehicle.

Note

On some vehicles, it may be necessary to discharge the refrigeration system and disconnect the evaporator before removing the heater core. Refer to Job 154 for details about evacuating a refrigeration system and Job 156 for details about disconnecting an evaporator.

Figure 158-3. As shown here, heater core removal sometimes requires that the instrument panel be partially removed.

Figure 158-4. Once the heater case or module containing the heater core has been removed from the vehicle, it can be disassembled to remove the heater core. Be sure to clean up spilled antifreeze.

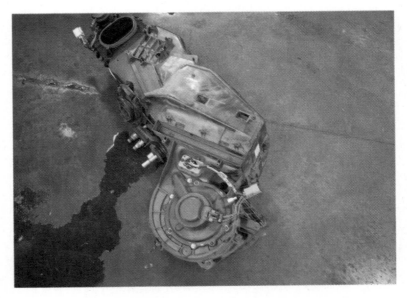

Job 158—Inspect and Service Heater Components (continued)

☐ 10. Remove the fasteners holding the heater core to the case or module, if used.

☐ 11. Lift the heater core from the case or module.

☐ 12. Compare the original and replacement heater cores.

Do the sizes match? Yes ___ No ___

Do the shapes match? Yes ___ No ___

Are the hose nipples in the same position? Yes ___ No ___

If you answered No to any of the preceding questions, describe the differences and inform your instructor: _____

_____

☐ 13. Place the heater core in position in the case or module, making sure that it is properly positioned.

☐ 14. Reinstall the attaching brackets if used.

☐ 15. Reinstall the case or module as necessary.

☐ 16. Reinstall any housings or covers that were removed.

☐ 17. If necessary, reinstall the evaporator connections and recharge the refrigeration system.

☐ 18. Install the heater hoses. Use sealer if necessary.

☐ 19. Install and tighten the heater hose clamps.

☐ 20. Close the drain plug, then refill the system with the proper kind of coolant. Refer to Job 32.

What type of coolant is used on this vehicle?_____

> **Caution**
> Many modern vehicles use special long-life coolant. Do not mix coolant types.

☐ 21. Start the engine and check for leaks and proper heater operation.

> **Note**
> On some vehicles, it will be necessary to bleed the heater core to remove trapped air. Loosen the highest heater hose and allow air to bleed from the heater core and hoses as the engine operates.

☐ 22. Ensure that the cooling system is full and install the radiator cap.

☐ 23. Clean the work area, including any coolant spills, and return any equipment to storage.

☐ 24. Did you encounter any problems during this procedure? Yes ___ No ___

If Yes, describe the problems: _____

_____

What did you do to correct the problems? _____

_____

☐ 25. Have your instructor check your work and sign this job sheet.

Performance Evaluation—Instructor Use Only

Did the student complete the job in the time allotted? Yes ___ No ___

If No, which steps were not completed?_____

How would you rate this student's overall performance on this job?_____

5–Excellent, 4–Good, 3–Satisfactory, 2–Unsatisfactory, 1–Poor

Comments: _____

_____

_____

INSTRUCTOR'S SIGNATURE _____

Job 159—Inspect and Service Air Handling Components

After completing this job, you will be able to inspect and service the air handling components of a vehicle's heating and air conditioning system.

Procedures

☐ 1. Obtain a vehicle to be used in this job. Your instructor may direct you to perform this job on a shop vehicle.

☐ 2. Gather the tools needed to perform the following job. Refer to the project's tools and materials list.

Inspect Ducts, Hoses, and Outlet Vents

☐ 1. Remove instrument panel components or cover panels as necessary to access the air handling system ductwork and hoses.

☐ 2. Make a visual inspection of the ducts and hoses. Check carefully for misaligned or disconnected ducts, torn hoses, and loose connections.

   Were any problems noted? Yes ___ No ___

   If Yes, describe the problems: _____

   _____

☐ 3. Inspect the air handling system outlet vents for the following problems:
   • Stuck or broken shutoff flaps.
   • Stuck outlets.
   • Stuck or broken outlet vent fins.
   • Other outlet vent damage.

   Were any problems noted? Yes ___ No ___

   If Yes, describe the problems: _____

   _____

☐ 4. Consult with your instructor about what repair steps to take.

   Repairs to be made: _____

   _____

☐ 5. Make repairs as necessary.

☐ 6. Recheck airflow and overall HVAC system operation.

Inspect/Replace Cabin Filter

Note

The cabin filter should be changed every 12,000–15,000 miles (19,000–24,000 km) or whenever the system airflow becomes sluggish.

☐ 1. Determine the cabin filter location. **Figure 159-1** shows the most common cabin filter locations. The location may be shown in the owners' manual or a parts catalog.

   Where is the cabin filter located? _____

Figure 159-1. Common cabin air filter locations are shown here. A—Inside the glove box. B—Under the hood. C—Under the dash.

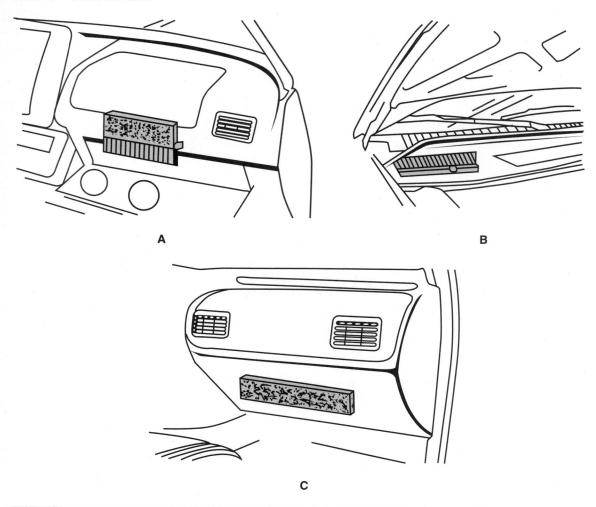

A

B

C

> **Note**
>
> The following two sections are arranged by the filter location. Follow the procedure in the section that is appropriate for the filter you are replacing. Disregard the other procedure.

Cabin Filter Located in the Passenger Compartment

 1. If the cabin filter is located in the passenger compartment under the glove compartment, remove the kick panel covering the filter housing then loosen the housing. If the cabin filter is located behind the glove compartment, open the glove compartment and the filter access doors.

 2. Remove the cabin filter.

> **Note**
>
> Some vehicles have two filters installed in the housing. Both filters should be changed. Remove the top filter by pulling it forward, then pull the second filter upward into the space previously occupied by the first filter and remove it.

Job 159—Inspect and Service Air Handling Components (continued)

☐ 3. Observe the cabin filter condition and describe it in the space provided.

_____

_____

_____

> **Caution**
>
> ⚠ This type of filter usually cannot be successfully cleaned. If the filter is dirty, it should be replaced.

☐ 4. Install the new filter(s) or reinstall the inspected filter(s).

> **Note**
>
> 📄 If the housing has two filters, slide the first new filter into the place occupied by the top filter, and then push it downward into place. Next, install the other new filter in the top position.

☐ 5. Reinstall the kick panel or close the access door and glove compartment door.

Cabin Filter Located in Cowl

☐ 1. If necessary, place the windshield wipers in the straight up position.

☐ 2. Open the hood and remove the right side cowl cover to expose the cabin air filter.

☐ 3. Pull on the removal tab to remove the filter from the recess.

☐ 4. Install the new filter and make sure that it is fully seated in the recess.

☐ 5. Reinstall the cowl cover.

☐ 6. Close the hood.

☐ 7. Park the windshield wipers.

☐ 8. Recheck airflow and overall system operation.

Diagnose a Blower Motor Problem

☐ 1. Check blower operation by the following procedure:
   a. Turn the ignition switch to the *on* position.
   b. Turn the heater and air conditioner controls to any operating position.
   c. Operate the blower motor through the various speed settings.

   Does the blower operate in all speed ranges? Yes ___ No ___

   If No, what speeds are inoperative? _____

☐ 2. If all or any speeds are inoperative check the related components:
   • Blower fuses.
   • Blower relay.
   • Blower speed control switch.
   • Blower resistor assembly, **Figure 159-2**.
   • Blower power module (when used).
   • Blower shaft free movement.
   • Blower motor windings.
   • Related wiring.

Figure 159-2. Printed circuit–type resistor assembly. Note the burned area. This vehicle had no low blower speed.

Burned area

Describe the results of the tests: _____

_____

_____

3. After consulting your instructor, make repairs as needed.

4. Recheck system operation.

Does the blower operate properly in all speeds? Yes ___ No ___

If Yes, go to step 5. If not, repeat steps 2 through 4 as necessary.

5. Clean the work area, including any coolant spills, and return any equipment to storage.

6. Did you encounter any problems during this procedure? Yes ___ No ___

If Yes, describe the problems: _____

_____

What did you do to correct the problems? _____

_____

7. Have your instructor check your work and sign this job sheet.

Performance Evaluation—Instructor Use Only

Did the student complete the job in the time allotted? Yes ___ No ___

If No, which steps were not completed?_____

How would you rate this student's overall performance on this job?_____

5–Excellent, 4–Good, 3–Satisfactory, 2–Unsatisfactory, 1–Poor

Comments: _____

_____

_____

INSTRUCTOR'S SIGNATURE _____

Project

Diagnosing General Engine Concerns

44

Introduction

The most important part of engine performance service is determining what is wrong. It is almost impossible to service modern vehicles without knowing how to diagnose them. Scan tools are an important diagnostic instrument. Every vehicle system is controlled by, or sends input to, the on-board computer. This information can only be accessed with a scan tool. In Job 160, you will learn how to diagnose engine performance problems. You will learn how to access and reprogram a vehicle computer in Job 161.

Exhaust gas analyzers are used to measure the gases exiting the engine, and therefore the air-fuel ratio and engine condition. Exhaust gas analyzers are invaluable for diagnosing performance, drivability, and emissions problems. In many states and cities, the technician must be able to use an exhaust gas analyzer to perform state emissions inspections. In Job 162, you will use an exhaust gas analyzer to check exhaust gases and interpret the results.

Noise, vibration, and harshness are commonly referred to as NVH and are a frequent source of complaints. Modern vehicles use more lightweight materials, which allows more NVH to reach the passenger compartment. All of the drivetrain components of front-wheel drive vehicles are located at the front of the vehicle, near the passenger compartment. Vans, pickup trucks, and SUVs have large sheet metal stampings that amplify any NVH developed by the moving parts. In Job 163, you will locate and correct NVH complaints.

Project 44 Jobs

- Job 160—Diagnosing Engine Performance Problems
- Job 161—Use a Scan Tool to Access and Reprogram a Vehicle Computer
- Job 162—Use a 4- or 5-Gas Analyzer to Check Emission System Operation
- Job 163—Diagnose an NVH Problem

Tools and Materials

The following list contains the tools and materials that may be needed to complete the jobs in this project. The items used will depend on the make and model of vehicle being serviced.

- Vehicle in need of performance diagnosis.
- Vehicle in need of emission testing.
- Vehicle with an NVH problem.
- Applicable service literature.
- 4- or 5-gas analyzer.
- Scan tool and adapters.
- Scan tool operating instructions.
- Multimeter.
- Hand tools as needed.
- Air-powered tools as needed.
- Safety glasses and other protective equipment as needed.

Safety Notice

Before performing this job, review all pertinent safety information in the text, and review safety information with your instructor.

Job 160—Diagnose Engine Performance Problems

After completing this job, you will be able to diagnose an engine performance problem.

Procedures

☐ 1. Obtain a vehicle to be used in this job. Your instructor may direct you to perform this job on a shop vehicle.

☐ 2. Gather the tools needed to perform this job. Refer to the tools and materials list at the beginning of the project.

☐ 3. Determine the exact concern by performing the following tasks:
- Question the vehicle driver. ___
- Consult previous service records. ___
- Road test the vehicle. ___
- Consult service information. ___

Does your description of the concern agree with the driver's description of the concern?
Yes ___ No ___

If No, explain why: _____

_____

Does the service information address the specific concern? Yes ___ No ___

☐ 4. Check for obvious problems by performing the following checks:

a. Make a visual inspection of the engine compartment.

List the problems found: _____

_____

b. Make visual inspections at other applicable points on the vehicle.

List the other areas checked and any problems found:_____

_____

c. Retrieve any trouble codes that may have been set. Refer to Job 11 as needed.

List the codes: _____

_____

d. Perform basic electrical checks of the charging and starting system. Refer to Job 130 for the starting system and Job 134 for the charging system.

Describe the battery's state of charge:_____

Record the starter amperage draw:_____

Record the voltage drop during cranking:_____

Record the charging system output:_____

Record the charging system voltage at idle:_____

Record the charging system voltage at 2500 rpm:_____

Were any obvious problems found? Yes ___ No ___

If Yes, explain:_____

_____

Do you think that this could be the cause of the concern determined in step 3?
Yes ___ No ___

Explain your answer:_____

_____

Note

In cases where an obvious cause is located before all steps are completed, the instructor may approve the elimination of some steps.

☐ 5. Determine which component or system is causing the concern by performing the following tasks:
   - Use the proper service information to compare the concern with the trouble codes retrieved.
   - Determine whether the indicated problems could be the cause of the driver complaint.
   - Use the appropriate test equipment to make additional checks to determine which system or part is causing the problem.

   List any abnormal conditions found:_____

   _____

   List any readings that are out of specifications: _____

   Do you think your findings have isolated the cause of the concern? Yes ___ No ___

   Explain your answer:_____

   _____

☐ 6. Eliminate any duplicate causes of the concern by checking components that could cause the findings found in step 5.
   - Make physical checks.

   Describe the checks made and list any defective components or systems detected: _____

   _____

   - Make electrical checks.

   Describe the checks made and list any defective components or systems detected: _____

   _____

   Do you think these defects could be the cause of the original concern? Yes ___ No ___

   Explain your answer:_____

   _____

☐ 7. Isolate and recheck causes of the concern by performing the following:
   - Recheck damaged components.
   - Repeat electrical checks.

   Did the rechecking procedure reveal any new problems or establish that the suspected components were in fact good? Yes ___ No ___

   If Yes, explain: _____

   _____

   If there were any differences with the conclusions in step 6, what was the reason? _____

   _____

☐ 8. Correct the defect by making necessary repairs or adjustments.

   Describe the services performed: _____

   _____

☐ 9. Recheck system operation by performing one or more of the following:
   - Make checks to engine and vehicle using test equipment. ___
   - Check emissions using an exhaust gas analyzer. ___
   - Road test the vehicle. ___

   Did the services in step 8 correct the concern? Yes ___ No ___

   If No, return to step 3 and repeat procedures until the cause of the concern is isolated.

☐ 10. Clean the work area and return any equipment to storage.

Job 160—Diagnose Engine Performance Problems (continued)

☐ 11. Did you encounter any problems during this procedure? Yes ___ No ___

If Yes, describe the problems: _____

_____

What did you do to correct the problems? _____

_____

☐ 12. Have your instructor check your work and sign this job sheet.

Performance Evaluation—Instructor Use Only

Did the student complete the job in the time allotted? Yes ___ No ___

If No, which steps were not completed?_____

How would you rate this student's overall performance on this job?_____

5–Excellent, 4–Good, 3–Satisfactory, 2–Unsatisfactory, 1–Poor

Comments: _____

_____

_____

INSTRUCTOR'S SIGNATURE _____

Notes

_____
_____
_____
_____
_____
_____
_____
_____
_____
_____
_____
_____
_____
_____
_____
_____
_____
_____
_____
_____
_____
_____
_____
_____
_____
_____
_____
_____
_____
_____
_____
_____
_____
_____
_____
_____
_____
_____
_____
_____

Job 161—Use a Scan Tool to Access and Reprogram a Vehicle Computer

After completing this job, you will be able to use a scan tool to access a vehicle engine or power train computer for diagnosis purposes.

Procedures

☐ 1. Obtain a vehicle to be used in this job. Your instructor may direct you to perform this job on a shop vehicle.

☐ 2. Gather the tools needed to perform this job. Refer to the tools and materials list at the beginning of the project.

☐ 3. Obtain a scan tool and related service literature for the vehicle selected in step 1.

Type of scan tool: _____

> **Note**
>
> There are many kinds of scan tools. The following procedure is a general guide to scan tool use. Always obtain the service instructions for the scan tool that you are using. If the scan tool literature calls for a different procedure or series of steps from those in this procedure, always perform procedures according to the scan tool literature.

Accessing the Vehicle Computer

☐ 1. Attach the proper test connector cable and power lead to the scan tool.

☐ 2. Ensure that the ignition switch is in the *off* position.

☐ 3. Locate the correct diagnostic connector.

Diagnostic connector location: _____

☐ 4. Attach the scan tool test connector cable to the diagnostic connector, using the proper adapter if necessary.

☐ 5. Attach the scan tool power lead to the cigarette lighter or battery terminals if necessary.

> **Note**
>
> OBD II scan tools are powered from terminal 16 of the diagnostic connector.

☐ 6. Observe the scan tool screen to ensure that it is working properly.

☐ 7. Enter vehicle information as needed to program the scan tool. Many scan tools can be programmed with the proper vehicle information by entering the vehicle identification number (VIN).

☐ 8. Turn the ignition key to the *on* position.

☐ 9. Retrieve trouble codes and list them below.

Trouble codes: _____

> **Note**
>
> Vehicle computer systems in good condition will produce a single code to indicate that the engine is not running. If more than one code is produced, check with your instructor before continuing. The computer system and vehicle may have defects that are beyond the scope of this job.

☐ 10. Use the scan tool literature or factory service manual to determine the meaning of the codes.

Code #   **Defect**

_____ _____

_____ _____

_____ _____

_____ _____

_____ _____

☐ 11. Start the engine.

☐ 12. Have another student drive the vehicle while you record the following information. If a particular reading is not available, write "NA" in the space.
- Idle RPM:_____

Is the idle speed within specifications? Yes ___ No ___
- Intake air temperature:_____

Is the air intake temperature within specifications? Yes ___ No ___
- MAF sensor reading:_____

Is the MAF sensor reading within specifications? Yes ___ No ___
- MAP sensor reading:_____

Is the MAP sensor reading within specifications? Yes ___ No ___
- Fuel trim counts at idle:_____ At cruising speed:_____

Are the fuel trim counts within specifications? Yes ___ No ___
- Does the torque converter lockup clutch apply at cruising speed? Yes ___ No ___
- Speeds at which the shift solenoids are energized:

First:_____

Second:_____

Third:_____

Fourth:_____
- Coolant temperature at the end of the test:_____°F or °C (circle one)
- Other readings as required by your instructor.

Type of Reading	Value (Include Units of Measure)	Within Specifications?
_____	_____	Yes ___ No ___
_____	_____	Yes ___ No ___
_____	_____	Yes ___ No ___
_____	_____	Yes ___ No ___

☐ 13. If a drivability problem is encountered based on the above steps, use the scan tool freeze frame feature to capture readings that occur at the time the malfunction occurs.
- Set the scan tool to freeze frame.
- Drive the vehicle until the malfunction occurs.
- Observe the freeze frame data.

Does any freeze frame reading indicate a possible cause of the malfunction?

Yes _____ No _____

Job 161—Use a Scan Tool to Access and Reprogram a Vehicle Computer (continued)

If Yes, describe: _____

_____

_____

☐ 14. When the testing is complete, turn the ignition switch to the *off* position.

☐ 15. If your instructor directs, make further checks to isolate the problem(s) revealed by the trouble codes or other readings.

☐ 16. If your instructor directs, repair the problem(s) as necessary. Refer to other jobs as directed.

> **Note**
> If the vehicle has an OBD II system, do not erase codes after repairs are completed. Erasing codes will also remove any monitors that have been run, which may cause the vehicle to fail its next emissions test.

17. After all repairs are complete, operate the vehicle to run the monitors again. If any monitors fail, a trouble code will set, indicating the need for further service. When attempting to set the monitors, keep several points in mind.
   - Some monitors are designed to run under certain ambient temperature conditions. In some parts of the United States, these temperatures may not be reached for many months, or ever.
   - Monitors may not reset a trouble code until they have run several times. Therefore, it is not possible to determine the status of repairs without keeping the vehicle for an excessive amount of time.
   - The vehicle may have to be returned to the owner before all monitors have reset.
   - Most state emission inspections require that a certain percentage, usually about 80%, of the monitors to have been run to pass inspection.

Reprogram the Computer

☐ 1. Determine the method to be used to reprogram the vehicle computer.

What reprogramming method will be used? Upload-download ___ Pass through ___

Each of the methods for reprogramming the computer is covered in one of the following two sections of the job. Follow the procedure in the appropriate section and disregard the other procedure.

Upload-Download Method

> **Note**
> This method is sometimes referred to as indirect programming.

☐ 1. If not done previously, ensure that the ignition switch is in the *Off* position and connect the scan tool to the vehicle diagnostic connector.

☐ 2. Observe the scan tool screen and ensure that the tool is operating properly.

☐ 4. If necessary, use the scan tool keypad to enter vehicle information.

☐ 5. Turn the ignition switch to the *On* position if necessary.

6. Download the existing ECM programming into the scan tool.

   Briefly describe the steps taken to download the programming. Be sure to include menus selected and scan tool keys operated. _____

   _____

   _____

7. When downloading is complete, turn the ignition switch to the *Off* position and disconnect the scan tool from the vehicle diagnostic connector.

8. Take the scan tool to the Internet connection, modem, or programming computer. See **Figure 161-1**.

9. Access the programming database.

   Database name: _____

10. Download the updated programming into the scan tool.

    Briefly describe the steps taken to begin the downloading process: _____

    _____

    _____

    _____

> **Note**
>
> Several minutes may be needed to overwrite the existing programming. The computer and/or scan tool will indicate the finish of the downloading process.

11. When the new in programming has been installed in the scan tool, unplug the tool and take it to the vehicle.

12. Plug the scan tool into the diagnostic connector and, if necessary, turn the ignition switch to the *On* position.

Figure 161-1. Once the scan tool is attached to the programming computer, use the computer keyboard and mouse to check the vehicle programming and download updated programming as necessary.

Job 161—Use a Scan Tool to Access and Reprogram a Vehicle Computer (continued)

☐ 13. Operate the scan tool as necessary to install the new programming into the vehicle ECM.

Briefly describe the steps taken to install the new programming, including scan tool menus selected:_____

_____

_____

_____

☐ 14. After reprogramming is complete, start the engine and test drive the vehicle to ensure that the new programming has been properly entered.

Does operation of the ECM and vehicle indicate that the new programming has been properly installed? Yes ___ No ___

If No, describe the problem (s): _____

_____

What could be the reason for the problem(s)? _____

_____

_____

> **Note**
>
> After reprogramming, many ECM's must go through a relearn procedure to properly control idle speed and other vehicle functions. This may take several minutes of driving. Some manufacturers require the technician to perform a specific series of relearning steps. Consult the manufacturer's service information for these procedures.

☐ 15. Park the vehicle and turn the ignition switch to the *Off* position.

☐ 16. Unplug the scan tool from the diagnostic connector.

Pass-Through Method

> **Note**
>
> This method is sometimes referred to as direct programming, flash reprogramming, or J2534 programming.

☐ 1. Park the vehicle at a place in the shop where it can be attached to the Internet connection using the available connecting cables.

☐ 2. If not done previously, ensure that the ignition switch is in the *Off* position.

☐ 3. Connect the scan tool or reprogramming tool to the vehicle diagnostic connector.

☐ 4. Attach the scan tool or reprogramming tool to the Internet connection using correct cables. This connection is usually made through an Internet-ready computer.

☐ 5. Turn the ignition switch to the *On* position if necessary.

☐ 6. Use the computer (and scan tool if required) to access the remote reprogramming website.

Name of website:_____

☐ 7. Follow website instructions to reprogram the vehicle ECM.

Briefly describe the steps taken to reprogram the ECM: _____

_____

> **Note**
>
> Several minutes may be needed to overwrite the existing programming. The website will indicate when the program has finished downloading.

☐ 8. After reprogramming is complete, exit the website and turn the ignition switch to the *Off* position.

☐ 9. Unplug the scan tool or reprogramming tool from the vehicle diagnostic connector and remove it from the vehicle.

☐ 10. Start the engine and test drive the vehicle to ensure that the new programming has been properly entered.

Does operation of the ECM and vehicle indicate that the new programming has been properly installed? Yes ___ No ___

If No, describe the problem(s): _____

_____

_____

What could be the reason for the problem(s)? _____

_____

_____

> **Note**
>
> After reprogramming, many ECM's must go through a relearn procedure to properly control idle speed and other vehicle functions. This may take several minutes of driving. Some manufacturers require the technician to perform a specific series of relearning steps. Consult the manufacturer's service information for these procedures.

☐ 11. If necessary, disconnect the scan tool or reprogramming tool from the computer.

☐ 12. Turn off the computer and return any cables to storage.

Job Wrap-Up

☐ 1. Return the scan tool to storage.

☐ 2. Clean the work area and return any equipment to storage.

☐ 3. Did you encounter any problems during this procedure? Yes ___ No ___

If Yes, describe the problems: _____

_____

What did you do to correct the problems? _____

☐ 4. Have your instructor check your work and sign this job sheet.

Performance Evaluation—Instructor Use Only

Did the student complete the job in the time allotted? Yes ___ No ___

If No, which steps were not completed?_____

How would you rate this student's overall performance on this job?_____

5–Excellent, 4–Good, 3–Satisfactory, 2–Unsatisfactory, 1–Poor

Comments: _____

_____

_____

INSTRUCTOR'S SIGNATURE _____

Job 162—Use a 4- or 5-Gas Analyzer to Check Emission System Operation

After completing this job, you should be able to use a 4- or 5-gas analyzer to check and interpret exhaust gas readings.

Procedures

☐ 1. Obtain a vehicle to be used in this job. Your instructor may direct you to perform this job on a shop vehicle.

☐ 2. Gather the tools needed to perform this job. Refer to the tools and materials list at the beginning of the project.

☐ 3. Obtain the exhaust gas analyzer and start it. Many exhaust gas analyzers require several minutes to reach operating temperature. During this time, other preparatory steps can be performed.

   Type of analyzer:_____

   Number of gases analyzed:_____

☐ 4. Start the engine and allow it to reach operating temperature.

☐ 5. Check engine temperature using a temperature tester.

   Type of tester used:_____

> **Caution**
>
> Do not rely on the vehicle temperature gauge; they are not reliable for exact temperature measurements.

If the engine temperature is below normal, allow the engine to warm further. Ensure that the thermostat is not stuck open or missing.

If the engine temperature is above normal, check cooling system condition as outlined in Job 31. Correct as necessary before proceeding with the test.

☐ 6. Calibrate the exhaust gas analyzer.

☐ 7. Inspect the vehicle for obvious problems that will affect emissions readings, such as disconnected wiring or hoses; damaged or missing components; and fuel, oil, or coolant leaks. Refer to other jobs as needed and correct obvious problems before proceeding.

☐ 8. Place the exhaust gas analyzer hose in the exhaust pipe.

☐ 9. Make other engine connections as needed.

☐ 10. Observe exhaust readings at different engine speeds and write the readings in the following chart. Indicate whether each reading is a percentage or a grams per mile figure. Write NA in the space if the gas cannot be read by this analyzer.

	Idle	1000 RPM*	2500 RPM*	Other Speed* (list)_____
HC	_____	_____	_____	_____
CO	_____	_____	_____	_____
NO_x	_____	_____	_____	_____
CO_2	_____	_____	_____	_____
O_2	_____	_____	_____	_____

*The equipment or vehicle manufacturer or state regulations may require a different engine speed for testing.

☐ 11. If the exhaust gas analyzer has a printout feature, print a copy of the exhaust gas readings.

Do the analyzer readings indicate that the emission controls and engine are operating properly? Yes ___ No ___

If No, which readings are incorrect?_____

What could cause these readings?_____

_____

_____

☐ 12. Turn off the engine.

☐ 13. Remove the analyzer hose from the tailpipe and remove other engine connections as applicable.

☐ 14. Make other tests to the engine and emission control systems as necessary.

What were the results of the tests? _____

_____

_____

☐ 15. After consulting with your instructor, make any needed repairs to the engine and/or emissions control systems.

List the repairs made: _____

_____

_____

☐ 16. When repairs are complete, repeat steps 5 through 12 to recheck the exhaust gas readings. Record the readings in the following chart. Indicate whether each reading is a percentage or a grams per mile figure. Write NA in the space if the gas cannot be read by this analyzer.

	Idle	1000 RPM*	2500 RPM*	Other Speed* (list)_____
HC	_____	_____	_____	_____
CO	_____	_____	_____	_____
NO_x	_____	_____	_____	_____
CO_2	_____	_____	_____	_____
O_2	_____	_____	_____	_____

*The equipment or vehicle manufacturer or state regulations may require a different engine speed for testing.

☐ 17. Compare the readings in steps 10 and 16.

Do the readings indicate that any problems have been corrected? Yes ___ No ___

If No, repeat steps 4 through 16 to isolate the problem.

☐ 18. Clean the work area and return any equipment to storage.

☐ 19. Did you encounter any problems during this procedure? Yes ___ No ___

If Yes, describe the problems: _____

_____

What did you do to correct the problems? _____

_____

☐ 20. Have your instructor check your work and sign this job sheet.

Job 162—Use a 4- or 5-Gas Analyzer to Check Emission System Operation (continued)

Performance Evaluation—Instructor Use Only

Did the student complete the job in the time allotted? Yes ___ No ___

If No, which steps were not completed?_____

How would you rate this student's overall performance on this job?_____

5–Excellent, 4–Good, 3–Satisfactory, 2–Unsatisfactory, 1–Poor

Comments: _____

_____

_____

INSTRUCTOR'S SIGNATURE _____

Notes

Job 163—Diagnose an NVH Problem

After completing this job, you will be able to follow logical troubleshooting procedures to locate causes of vehicle noise, vibration, and harshness.

Procedures

☐ 1. Obtain a vehicle to be used in this job. Your instructor may direct you to perform this job on a shop vehicle.

☐ 2. Gather the tools needed to perform this job. Refer to the tools and materials list at the beginning of the project.

☐ 3. Familiarize yourself with the common sources of NVH complaints. **Figure 163-1** shows some typical causes of NVH complaints on a modern vehicle.

☐ 4. Determine the exact NVH problem by performing the following tasks:
   - Question the vehicle driver. _____
   - Consult previous service records. _____
   - Road test the vehicle. _____

   Briefly describe the problem: _____

   _____

   _____

   _____

   Does your description of the problem agree with the driver's original description of the problem? Yes ____ No ____

   If No, explain: _____

   _____

   _____

☐ 5. Check for obvious problems by performing the following checks:
   - Make a visual underhood inspection. ____
   - Retrieve trouble codes. ____
   - Make basic electrical and mechanical checks. ____

   Describe any obvious problems found:_____

   _____

   _____

   Do you think that this could be the cause of the problem determined in step one?
   Yes ____ No ____

   Explain your answer:_____

   _____

   _____

Figure 163-1. Common areas of NVH complaints in vehicles with different power train arrangements. A—Rear-wheel drive. B—Front-wheel drive. (*continued*)

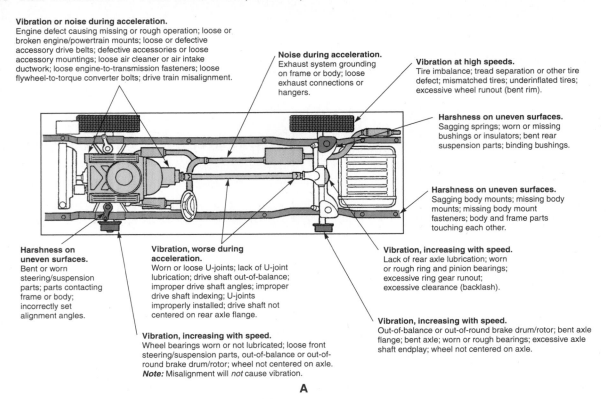

Vibration or noise during acceleration.
Engine defect causing missing or rough operation; loose or broken engine/powertrain mounts; loose or defective accessory drive belts; defective accessories or loose accessory mountings; loose air cleaner or air intake ductwork; loose engine-to-transmission fasteners; loose flywheel-to-torque converter bolts; drive train misalignment.

Noise during acceleration.
Exhaust system grounding on frame or body; loose exhaust connections or hangers.

Vibration at high speeds.
Tire imbalance; tread separation or other tire defect; mismatched tires; underinflated tires; excessive wheel runout (bent rim).

Harshness on uneven surfaces.
Sagging springs; worn or missing bushings or insulators; bent rear suspension parts; binding bushings.

Harshness on uneven surfaces.
Sagging body mounts; missing body mounts; missing body mount fasteners; body and frame parts touching each other.

Harshness on uneven surfaces.
Bent or worn steering/suspension parts; parts contacting frame or body; incorrectly set alignment angles.

Vibration, worse during acceleration.
Worn or loose U-joints; lack of U-joint lubrication; drive shaft out-of-balance; improper drive shaft angles; improper drive shaft indexing; U-joints improperly installed; drive shaft not centered on rear axle flange.

Vibration, increasing with speed.
Lack of rear axle lubrication; worn or rough ring and pinion bearings; excessive ring gear runout; excessive clearance (backlash).

Vibration, increasing with speed.
Out-of-balance or out-of-round brake drum/rotor; bent axle flange; bent axle; worn or rough bearings; excessive axle shaft endplay; wheel not centered on axle.

Vibration, increasing with speed.
Wheel bearings worn or not lubricated; loose front steering/suspension parts, out-of-balance or out-of-round brake drum/rotor; wheel not centered on axle. *Note:* Misalignment will *not* cause vibration.

A

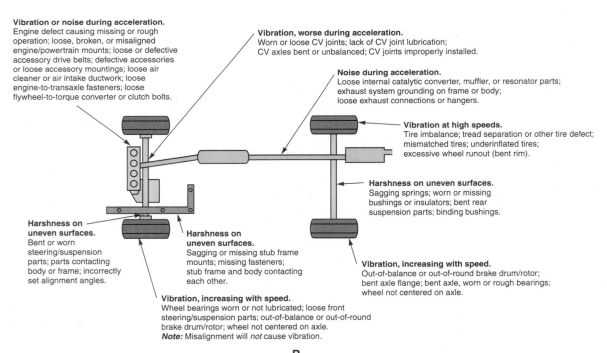

Vibration or noise during acceleration.
Engine defect causing missing or rough operation; loose, broken, or misaligned engine/powertrain mounts; loose or defective accessory drive belts; defective accessories or loose accessory mountings; loose air cleaner or air intake ductwork; loose engine-to-transaxle fasteners; loose flywheel-to-torque converter or clutch bolts.

Vibration, worse during acceleration.
Worn or loose CV joints; lack of CV joint lubrication; CV axles bent or unbalanced; CV joints improperly installed.

Noise during acceleration.
Loose internal catalytic converter, muffler, or resonator parts; exhaust system grounding on frame or body; loose exhaust connections or hangers.

Vibration at high speeds.
Tire imbalance; tread separation or other tire defect; mismatched tires; underinflated tires; excessive wheel runout (bent rim).

Harshness on uneven surfaces.
Sagging springs; worn or missing bushings or insulators; bent rear suspension parts; binding bushings.

Harshness on uneven surfaces.
Bent or worn steering/suspension parts; parts contacting body or frame; incorrectly set alignment angles.

Harshness on uneven surfaces.
Sagging or missing stub frame mounts; missing fasteners; stub frame and body contacting each other.

Vibration, increasing with speed.
Wheel bearings worn or not lubricated; loose front steering/suspension parts; out-of-balance or out-of-round brake drum/rotor; wheel not centered on axle. *Note:* Misalignment will *not* cause vibration.

Vibration, increasing with speed.
Out-of-balance or out-of-round brake drum/rotor; bent axle flange; bent axle, worn or rough bearings; wheel not centered on axle.

B

Job 163—Diagnose an NVH Problem (continued)

Figure 163-1 (continued). C—Four-wheel drive.

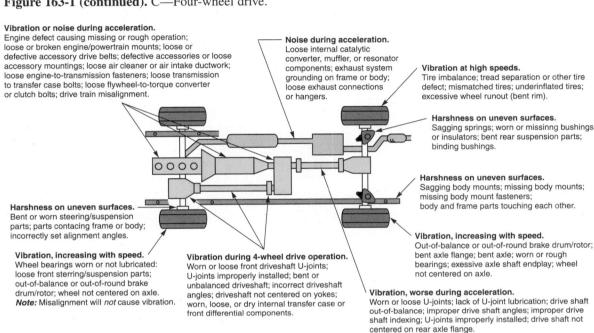

Vibration or noise during acceleration.
Engine defect causing missing or rough operation; loose or broken engine/powertrain mounts; loose or defective accessory drive belts; defective accessories or loose accessory mountings; loose air cleaner or air intake ductwork; loose engine-to-transmission fasteners; loose transmission to transfer case bolts; loose flywheel-to-torque converter or clutch bolts; drive train misalignment.

Noise during acceleration.
Loose internal catalytic converter, muffler, or resonator components; exhaust system grounding on frame or body; loose exhaust connections or hangers.

Vibration at high speeds.
Tire imbalance; tread separation or other tire defect; mismatched tires; underinflated tires; excessive wheel runout (bent rim).

Harshness on uneven surfaces.
Sagging springs; worn or missinng bushings or insulators; bent rear suspension parts; binding bushings.

Harshness on uneven surfaces.
Sagging body mounts; missing body mounts; missing body mount fasteners; body and frame parts touching each other.

Harshness on uneven surfaces.
Bent or worn steering/suspension parts; parts contacing frame or body; incorrectly set alignment angles.

Vibration, increasing with speed.
Wheel bearings worn or not lubricated: loose front sterring/suspension parts; out-of-balance or out-of-round brake drum/rotor; wheel not centered on axle.
Note: Misalignment will *not* cause vibration.

Vibration during 4-wheel drive operation.
Worn or loose front driveshaft U-joints; U-joints improperly installed; bent or unbalanced driveshaft; incorrect driveshaft angles; driveshaft not centered on yokes; worn, loose, or dry internal transfer case or front differential components.

Vibration, increasing with speed.
Out-of-balance or out-of-round brake drum/rotor; bent axle flange; bent axle; worn or rough bearings; exessive axle shaft endplay; wheel not centered on axle.

Vibration, worse during acceleration.
Worn or loose U-joints; lack of U-joint lubrication; drive shaft out-of-balance; improper drive shaft angles; improper drive shaft indexing; U-joints improperly installed; drive shaft not centered on rear axle flange.

C

6. Determine which component or system is causing the problem by performing the following:

> **Note**
>
> In cases where an obvious problem is located before all steps are completed, the instructor may approve the elimination of some steps.

- Obtain the proper service manual and other service information. ___
- Use the service information to compare the problem with any trouble codes retrieved. ___
- Obtain the needed test equipment. ___
- Check engine systems. ___
- Check the exhaust system. ___
- Check drive train systems. ___
- Check the heating/air conditioning system. ___

List any abnormal conditions or readings that were not within specifications: _____

_____

_____

_____

Do you think this could be the cause of the problem? Yes ___ No ___

Explain your answer:_____

_____

_____

☐ 7. Eliminate causes of problem by checking components that could cause the problems found in step 6.
- Make visual checks. ___
- Make physical checks for loose parts. ___
- Make electrical checks. ___

List any defective components or systems that you located: _____

_____

_____

_____

Do you think these defects could be the cause of the original problem? Yes ___ No ___

Explain your answer:_____

_____

_____

☐ 8. Isolate and recheck the causes of the problem. Recheck damaged components and repeat the electrical checks.

Did the rechecking procedure reveal any new problems or establish that suspected components were in fact good? Yes ___ No ___

If there were any differences with the conclusions in step 6, what was the reason? _____

_____

_____

_____

☐ 9. Correct the defect by making necessary repairs or adjustments.

Briefly describe the services performed: _____

_____

_____

☐ 10. Recheck system operation by performing either or both of the following:
- Make checks using test equipment. ___
- Road test the vehicle. ___

Did the services in step 9 correct the problem? Yes ___ No ___

If No, what steps should you take now?_____

_____

_____

_____

☐ 11. Clean the work area and return any equipment to storage.

☐ 12. Did you encounter any problems during this procedure? Yes ___ No ___

If Yes, describe the problems: _____

_____

What did you do to correct the problems? _____

_____

☐ 13. Have your instructor check your work and sign this job sheet.

Job 163—Diagnose an NVH Problem (continued)

Performance Evaluation—Instructor Use Only

Did the student complete the job in the time allotted? Yes ____ No ____

If No, which steps were not completed?_____

How would you rate this student's overall performance on this job?_____

5–Excellent, 4–Good, 3–Satisfactory, 2–Unsatisfactory, 1–Poor

Comments: _____

_____

_____

INSTRUCTOR'S SIGNATURE _____

Notes

Project
Diagnosing and Repairing an Ignition System

45

Introduction

Modern vehicles use many types of ignition systems. Many engines continue to use distributors. The pick up coil is usually located in the distributor, although a few systems have a distributor and a crankshaft-mounted pick up. Engines without a distributor have crankshaft pick ups. Some systems have a second pick up on the camshaft. Many distributorless ignitions use the waste spark system with the coil firing two plugs at once, one on the compression stroke and the other on the exhaust stroke. A later variation of this system has two plugs per cylinder. The latest ignition systems have one ignition coil per cylinder. The coil near plug system uses coil wires, while the direct, or coil on plug, version places the coil directly over the spark plug. You will inspect and test an ignition system's secondary circuit components in Job 163, and its primary circuit components in Job 164.

Project 45 Jobs

- Job 164—Inspect and Test an Ignition System's Secondary Circuit Components and Wiring
- Job 165—Inspect and Test an Ignition System's Primary Circuit Components and Wiring

Tools and Materials

- One or more vehicles in need of service.
- Applicable service information.
- Spark tester.
- Equipment capable of reading waveforms.
- Multimeter or ohmmeter.
- Timing light with advance meter (optional).
- Hand tools as needed.
- Safety glasses and other protective equipment as needed.

Safety Notice

Before performing this job, review all pertinent safety information in the text, and review safety information with your instructor.

Notes

Job 164—Inspect and Test an Ignition System's Secondary Circuit Components and Wiring

After completing this job, you will be able to inspect and service a vehicle's secondary ignition circuit.

Procedures

1. Obtain a vehicle to be used in this job. Your instructor may direct you to perform this job on a shop vehicle.

2. Gather the tools needed to perform this job. Refer to the tools and materials list at the beginning of the project.

3. Determine the type of ignition system to be tested. See **Figure 164-1** for some typical ignition systems.
   - Distributorless (waste spark). ____
   - Distributorless with two plugs per cylinder. ____
   - Coil-near-plug (uses coil wires). ____
   - Coil-on-plug (direct, no plug wires). ____
   - Distributor with internal pick up. ____
   - Distributor without internal pick up. ____

4. Locate service information for the ignition system to be tested.

Figure 164-1. Four common types of ignition systems are shown here. A—Distributorless ignition. B—Coil-on-plug ignition. C—Dual ignition (in this design, a combination of coil-over-plug and coil-near-plug systems). D—Distributor-type ignition.

A

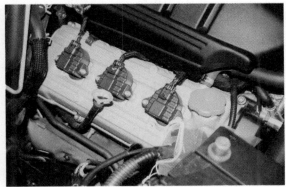

B

C

D

Test for a Spark from the Secondary System

☐ 1. Detach coil or plug wire or remove one coil from its plug.

☐ 2. Attach a spark tester to the wire or coil as applicable.

☐ 3. Have an assistant crank the engine as you observe the spark. Describe the spark produced:
- No spark is produced. ___
- The spark is weak. Describe: _____
- The spark is acceptable. Describe: _____

☐ 4. If there is no spark or the spark is weak, make further checks to isolate the problem to the primary or secondary ignition system components, as outlined in the following sections of the job.

Read and Interpret Ignition System Waveforms

☐ 1. Obtain one of the following testers, capable of reading ignition system waveforms.
- Engine Analyzer. ___
- Oscilloscope. ___
- Waveform Meter. ___

☐ 2. Attach the tester leads to the ignition system connections.

Attachment method(s): _____

_____

☐ 3. Set the tester controls to match the ignition system type.

☐ 4. Start the engine and observe the oscilloscope patterns.

☐ 5. Draw the secondary oscilloscope pattern.

☐ 6. Draw the primary oscilloscope pattern.

Do the primary and secondary patterns indicate that the ignition system is operating correctly? Yes ___ No ___

If No, what could be the problem? _____

_____

☐ 7. Stop the engine and remove the tester leads.

Job 164—Inspect and Test an Ignition System's Secondary Circuit Components and Wiring (continued)

Check the Spark Plugs

☐ 1. Remove the spark plugs from the cylinder head.

☐ 2. Visually inspect the spark plugs for excessive wear, fouling, or other signs of damage. List the defects found by cylinder number.

Plug from Cylinder #   **Defect**

Plug from Cylinder #	Defect
_____	_____
_____	_____
_____	_____
_____	_____
_____	_____
_____	_____
_____	_____
_____	_____
_____	_____
_____	_____
_____	_____
_____	_____

☐ 3. Determine the cause of the plug's condition, replace defective plugs, and retest the ignition system.

Check the Spark Plug Wires

☐ 1. Visually inspect the wires for the following:
- Chafing or other damage. ___
- Cracked or split wire insulation. ___
- Evidence of flashover or shorting. ___
- Damaged boots. ___

Describe any defects:_____

☐ 2. Use an ohmmeter to check wire resistance. List any defects found by cylinder number.

Wire from Cylinder #   **Defect**

Wire from Cylinder #	Defect
_____	_____
_____	_____
_____	_____
_____	_____
_____	_____
_____	_____
_____	_____
_____	_____
_____	_____
_____	_____
_____	_____

☐ 3. After consulting with your instructor, replace defective plug wires and recheck the ignition system.

Check the Distributor Cap and Rotor

☐ 1. Remove the cap from the distributor body.

☐ 2. Visually inspect the cap and rotor for carbon tracks and flashover.

☐ 3. Check the cap and rotor for cracks or other evidence of mechanical damage.

☐ 4. After consulting with your instructor, replace defective components as necessary and recheck the ignition system.

☐ 5. Clean the work area and return any equipment to storage.

☐ 6. Did you encounter any problems during this procedure? Yes ___ No ___

If Yes, describe the problems: _____

_____

What did you do to correct the problems? _____

_____

☐ 7. Have your instructor check your work and sign this job sheet.

Performance Evaluation—Instructor Use Only

Did the student complete the job in the time allotted? Yes ___ No ___

If No, which steps were not completed?_____

How would you rate this student's overall performance on this job?_____

5–Excellent, 4–Good, 3–Satisfactory, 2–Unsatisfactory, 1–Poor

Comments: _____

_____

_____

INSTRUCTOR'S SIGNATURE _____

Job 165—Inspect and Test an Ignition System's Primary Circuit Components and Wiring

After completing this job, you will be able to inspect and service a vehicle's secondary ignition circuit.

Procedures

☐ 1. Obtain a vehicle to be used in this job. Your instructor may direct you to perform this job on a shop vehicle.

☐ 2. Gather the tools needed to perform this job. Refer to the tools and materials list at the beginning of the project.

☐ 3. Determine the type of ignition system to be tested.

Type of ignition system: _____

Visually Inspect Primary Circuit Wiring

☐ 1. Obtain system wiring diagrams.

☐ 2. Using the wiring diagram as a guide, inspect all ignition system wiring for the following conditions. Indicate any defects found:

> **Note**
>
> Wiring and connectors are often located in inaccessible locations, such as the ignition primary connectors shown in **Figure 165-1**. It is important to check all connections, even when it involves removing other parts to gain access to them.

- Insulation discolored and/or swelled (overheating). ___
- White or green deposits on exposed wires (corrosion). ___
- Connectors melted or metal discolored (overheating). ___
- Connectors loose, disconnected. ___

Figure 165-1. The primary connectors on this coil pack and module assembly are hidden behind vacuum hoses and the automatic transaxle dipstick.

Partially obstructed connector

If any problems were found, describe the problems and the correct action needed to correct them: _____

_____

_____

Make Resistance Checks of Wires and/or Connectors

☐ 1. Obtain a multimeter and set it to read resistance.

☐ 2. Disconnect the wire or connector to be tested from the vehicle's electrical system.

☐ 3. Place the leads of the multimeter on each end of the wire or connector to be checked and observe the reading.

Reading:_____ ohms.

☐ 4. Compare this reading to specifications. Most wires and connectors should show very low resistance, usually less than .5 ohm.

Does the wire or connector have excessively high resistance? Yes ___ No ___

If Yes, what should be done next?_____

☐ 5. After consulting with your instructor, make repairs as necessary and recheck.

Inspect Solid State Components

☐ 1. Remove parts as necessary to gain access to the solid state ignition components that are to be inspected.

☐ 2. Inspect and describe the condition of the pickup coils and/or speed sensors and their related toothed wheels or shutters. Look for the following conditions:

- Sensor damaged or disconnected.
- Metal shavings on sensor tip.
- Debris on wheel/shutter.
- Damaged wheel/shutter.
- Improper sensor clearance.

Record the type of sensors being inspected and any defects they may have in the following chart. If an inspected sensor has no defects, write OK in the Defects blank.

Type of Sensor	Defects
_____	_____
_____	_____
_____	_____
_____	_____
_____	_____
_____	_____
_____	_____

☐ 3. Locate and note the condition of the following ignition system components. Look for, and record, damage such as cracks, damaged connectors, obvious overheating, disconnected or corroded wires, or other defects.

Ignition module (if separate from the ECM): _____

_____

ECM: _____

Relays when used:_____

Other components (list): _____

_____

_____

Job 165—Inspect and Test an Ignition System's Primary Circuit Components and Wiring (continued)

> **Caution**
>
> Perform steps 4 through 8 only when specifically recommended by the applicable service information.

☐ 4. To make resistance checks of a solid-state component, obtain the service literature listing the resistance specifications and test terminals for the component.

☐ 5. Obtain a multimeter and set it to read resistance.

☐ 6. Disconnect the component to be tested from the vehicle's electrical system.

☐ 7. Place the leads of the multimeter on the proper terminals of the component and observe the reading.

   Reading:_____ ohms.

☐ 8. Compare this reading to specifications.

   Does the component show excessively high or low resistance? Yes ___ No ___

   If Yes, what should be done next?_____

   _____

☐ 9. After consulting with your instructor, make repairs as necessary and recheck.

Test the Ignition Module and/or Engine Control Module

☐ 1. Locate and note the condition of the ignition module (if separate from the ECM) and ECM. Look for, and record, damage such as cracks, damaged connectors, obvious overheating, disconnected or corroded wires, or other defects.

   _____

   _____

   _____

   _____

☐ 2. Turn the vehicle's ignition switch to the *on* position.

☐ 3. Using a test light, ensure that power is available to both sides of all module fuses. Place a check mark next to the results of your tests.
   - Test light lights on both sides of all fuses. _____ Go to step 4.
   - Test light lights on one side of a fuse. _____ Replace the fuse and recheck module operation.
   - Test light does not light on either side of a fuse. _____ Locate and repair the wiring problem between fuse and ignition switch or battery, then go to step 4.

> **Caution**
>
> If the fuse is blown, you must determine what caused the electrical overload. Do not simply replace the fuse and let the vehicle go.

☐ 4. Using a test light, check for power at the module input wire.

   Is power available at the module? Yes _____ No _____

   If Yes, go to step 5.

   If No, repair the wire between the fuse and the module and recheck module operation.

☐ 5. Back probe the system wiring until the reason for the lack of power is determined. Repair the wire or connector between the battery and fuse and recheck operation.

6. Check the ground at the module.

   Is the module properly grounded? Yes _____ No _____

   If Yes, go to step 7.

   If No, repair the ground wire(s) as needed and recheck module operation.

> **Caution**
>
> Perform steps 7 through 11 only when specifically recommended by the applicable service information.

7. Obtain service information listing the resistance specifications and test terminals for the component.

8. Obtain a multimeter and set it to read resistance.

9. Disconnect the component to be tested from the vehicle's electrical system.

10. Place the leads of the multimeter on the proper terminals of the component and observe the reading.

    Reading: _____ ohms.

11. Compare this reading to specifications.

    Does the component show excessively high or low resistance? Yes _____ No _____

    If Yes, go to step 12.

    If No, what should be done next? _____

    _____

12. Replace the module and recheck its operation.

> **Caution**
>
> Ensure that the ignition switch is in the *off* position and the battery negative cable is removed before replacing any module. Follow all static electricity precautions outlined by the vehicle manufacturer.

Test Ignition Coil(s)

1. Make a visual inspection of the suspect coil(s). Indicate any defects found.
   - Cracks. ___
   - Flashover/carbon tracks. ___
   - Damaged terminals. ___

2. Obtain resistance specifications for the coil(s).

   Terminals to check:_____ and _____ Resistance range:_____ ohms

   Terminals to check:_____ and _____ Resistance range:_____ ohms

   Terminals to check:_____ and _____ Resistance range:_____ ohms

3. Follow manufacturer's directions to measure the resistance of the coil(s).

   Terminals checked:_____ and _____ Resistance range:_____ ohms

   Terminals checked:_____ and _____ Resistance range:_____ ohms

   Terminals checked:_____ and _____ Resistance range:_____ ohms

4. Are the resistance readings within specifications? Yes ___ No ___

   If No, describe the problem(s) revealed by out-of-specification readings: _____

   _____

5. After consulting with your instructor, replace the defective coil(s) and recheck.

Job 165—Inspect and Test an Ignition System's Primary Circuit Components and Wiring (continued)

Check and Set the Initial Timing

> **Note**
>
> The following procedure can only be performed on some vehicles with distributor ignition systems. On most modern vehicles, the timing can be read from a scan tool, but not adjusted.

☐ 1. Locate timing procedures and specifications on underhood emissions label.

Write the timing specification here:

Degrees:_____ before TDC / after TDC (circle one).

☐ 2. Clean and highlight the timing marks as necessary.

☐ 3. Detach electrical connectors or vacuum hoses as directed by the emissions label. See **Figure 165-2**.

Describe the actions taken: _____

_____

☐ 4. Attach a timing light to the engine. If the timing light has an advance meter, turn the meter to the 0 degrees position.

☐ 5. Start the engine and ensure that the idle speed is correct.

☐ 6. Point the timing light at the timing marks and ensure that the marks are illuminated enough to make accurate readings.

☐ 7. Observe the ignition timing. Write the observed timing in the space provided:

Degrees:_____ before TDC / after TDC (circle one).

Compare timing with specifications. Is the timing correct? Yes ___ No ___

If No, loosen distributor hold down bolt and turn distributor until timing is correct. Retighten the hold down bolt and recheck the timing.

Figure 165-2. Many modern vehicles have an electrical connector that must be disengaged to set the initial timing. On older vehicles, a vacuum line may need to be disconnected.

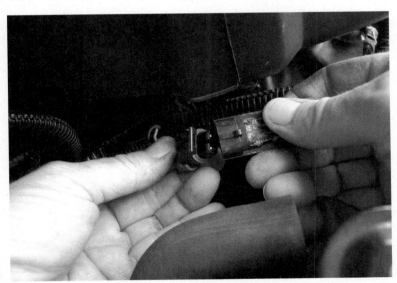

☐ 8. Turn off the engine.

☐ 9. Reattach any removed hoses/connectors.

Check the Timing Advance

☐ 1. Restart the engine with the timing light attached.

☐ 2. Point the timing light at the timing marks.

☐ 3. Raise engine speed and observe the ignition timing.

Does the timing advance as engine speed is increased? Yes ___ No ___

> **Note**
>
> If exact specifications for timing advance at a specific rpm are not available, as a general rule, timing should advance 35° to 45° at 2500 rpm.

If No, what could be the problem? _____

_____

Check Knock Sensor Operation

☐ 1. Strike the engine lightly with a wrench or small hammer and observe ignition timing.

Does the timing retard when the engine is struck? Yes ___ No ___

If No, what could be the problem? _____

☐ 2. Turn off engine.

☐ 3. Disconnect the timing light.

☐ 4. After consulting with your instructor, make repairs as necessary and recheck the timing.

☐ 5. Clean the work area and return any equipment to storage.

☐ 6. Did you encounter any problems during this procedure? Yes ___ No ___

If Yes, describe the problems: _____

_____

What did you do to correct the problems? _____

_____

☐ 7. Have your instructor check your work and sign this job sheet.

Performance Evaluation—Instructor Use Only

Did the student complete the job in the time allotted? Yes ___ No ___

If No, which steps were not completed?_____

How would you rate this student's overall performance on this job?_____

5–Excellent, 4–Good, 3–Satisfactory, 2–Unsatisfactory, 1–Poor

Comments: _____

_____

_____

INSTRUCTOR'S SIGNATURE _____

Project

Diagnosing and Servicing Fuel and Air Induction Systems

46

Introduction

Fuel must be supplied to the injectors (or carburetor on older engines) or the engine will not run. Major components of the fuel supply system are the fuel pump (some engines have two), fuel lines and hoses, and fuel filters. Failure in any of these components can cause the engine to run poorly or stop running. In Job 166, you will test fuel quality. In Job 167, you will test and replace fuel supply components.

The throttle body houses the throttle valve, which responds to driver input to regulate the flow of air into the intake manifold. The throttle body also houses the injector(s) in a throttle body injection system. In Job 168, you will inspect and clean a throttle body assembly. Idle speed controls keep the engine idling smoothly despite changes in load and temperature. In Job 169, you will check an idle speed motor or an idle air control valve.

A modern engine may have one of several types of fuel systems. Most modern engines are equipped with multiport, or multipoint, fuel injection, with an injector for each cylinder. Older vehicles may have central fuel injection with one or two injectors located at the top of the intake manifold. There are a variety of tests available to help a technician diagnose a faulty fuel injection system.

Incorrect injector waveforms can alert you to injector failure and help you diagnose the cause of that failure. An injector balance test measures the amount of fuel delivered by the injectors over a specified period of time. This test is useful in detecting clogged injectors or injectors that are failing to fully open. In Job 170, you will check fuel injector waveforms and perform an injector balance test.

Once the source of the fuel system problem has been determined, it is necessary to correct the problem. This frequently involves replacing worn or faulty injectors. In Job 171, you will replace a faulty fuel injector.

In Job 172, you will check the operation of a turbocharger or supercharger.

Project 46 Jobs

- Job 166—Test Fuel Quality
- Job 167—Test and Replace Fuel Supply System Components
- Job 168—Inspect and Service a Throttle Body
- Job 169—Check Idle Speed Motor or Idle Air Control
- Job 170—Check Fuel Injector Waveforms and Perform a Fuel Injector Balance Test
- Job 171—Remove and Replace a Fuel Injector
- Job 172—Check Supercharger and Turbocharger Operation

Tools and Materials

The following list contains the tools and materials that may be needed to complete the jobs in this project. The items used will depend on the make and model of vehicle being serviced.

- Vehicle in need of fuel system service.
- Vehicle in need of exhaust service.
- Exhaust gas analyzer.
- Fuel pressure tester.
- Equipment capable of reading waveforms.
- Dedicated injector tester or scan tool and adapters.
- Scan tool operating instructions.
- Pressure gauge.
- Fire extinguisher.
- Hand tools as needed.
- Air-powered tools as needed.
- Safety glasses and other protective equipment as needed.
- Service information as needed.
- Clear glass or plastic container.
- Graduated cylinder.
- Cetane hydrometer.
- Vacuum gauge.

Safety Notice

Before performing this job, review all pertinent safety information in the text, and review safety information with your instructor.

Job 166—Test Fuel Quality

After completing this job, you will be able to verify gasoline and diesel fuel quality.

Note

Fuel quality and/or contamination is often overlooked as a cause of driveability problems. Low-quality fuel can be the cause of stalling and poor cold engine operation, detonation (pinging) under load, and generally sluggish operation. If a driveability problem cannot be isolated, sometimes switching grades or brands of gasoline or diesel fuel may solve the problem. In this job, you will verify the quality of gasoline and diesel fuel.

Procedures

Inspect Gasoline Sample for Contamination

☐ 1. Place a drip pan under the vehicle and have a fire extinguisher handy.

☐ 2. Disconnect the fuel line between the fuel tank and the filter.

Note

Dirt in the fuel filter is normal. Do not decide that the fuel is contaminated based on the contents of the fuel filter.

☐ 3. Drain about 6 oz (200 ml) of gasoline into a glass or plastic container.

☐ 4. Visually inspect for dirt, rust, or other particles suspended in the gasoline. Are particles present? Yes _____ No _____

☐ 5. Observe whether the sample is cloudy, indicating water or other contamination. Is the sample cloudy? Yes _____ No _____

☐ 6. Allow the sample to sit for 5–10 minutes. Water will settle to the bottom and be easily observed. Is any water present? Yes _____ No _____

☐ 7. If the answer to any of the above questions is Yes, the gasoline is contaminated. After consulting your instructor, clean the fuel tank and related lines. In some cases, a badly contaminated tank must be replaced.

☐ 8. After all tests and, if necessary, repairs, are completed, reconnect the fuel line between the tank and the filter.

Perform Alcohol in Gasoline Test

Note

This test checks for excess alcohol (more than 10%) in the gasoline sample. Excess alcohol can cause hard starting, stalling and other performance problems when the engine is fully warmed up.

☐ 1. Obtain a 3 oz (110 ml) graduated cylinder.

☐ 2. Pour 2.7 oz (90 ml) of sample gasoline into the cylinder.

☐ 3. Pour 0.3 oz (10 ml) of water into the cylinder.

☐ 4. Seal the top of the cylinder and shake the cylinder vigorously.

☐ 5. Allow the cylinder to sit for 5–10 minutes.

☐ 6. Observe the combined water and alcohol in the bottom of the cylinder. If the combined water in the bottom of the cylinder is more than 0.6 oz (20 ml), the sample contains more than 10% alcohol. Does the sample contain more than 10% alcohol? Yes _____ No _____

If Yes, suggest that the vehicle owner switch gasoline brands.

Inspect Diesel Fuel Sample for Contamination

☐ 1. Place a drip pan under the vehicle and have a fire extinguisher handy.

☐ 2. Disconnect the fuel line between the fuel tank and the filter.

> **Note**
>
> Dirt in the fuel filter is normal. Do not decide that the fuel is contaminated based on the contents of the fuel filter.

☐ 3. Drain about 6 oz (200 ml) of diesel fuel into a glass or plastic container.

☐ 4. Visually inspect for particles suspended in the diesel fuel. This is usually evidence of water contamination, which allows microorganisms to grow. Does the sample contain suspended particles? Yes _____ No _____

☐ 5. Allow the sample to sit for 5–10 minutes. Water will settle to the bottom and be easily observed. Is any water present? Yes _____ No _____

☐ 6. If the answer to any of the above questions is Yes, the diesel fuel is contaminated. After consulting your instructor, clean the fuel tank and related lines. In some cases, a badly contaminated tank must be replaced.

☐ 7. After all tests and, if necessary, repairs, are completed, reconnect the fuel line between the tank and the filter.

Measure Diesel Fuel Cetane Rating

☐ 1. Obtain a cetane hydrometer.

> **Note**
>
> A cetane hydrometer will provide a quick and relatively accurate cetane reading. If your shop uses another type of cetane tester, follow the tester instructions to check cetane.

☐ 2. Ensure that the temperature of the diesel fuel to be sampled is between 75 and 95°F (24–35°C). If necessary, place the sample in a spot where it can reach this temperature range.

☐ 3. Draw a sample of diesel fuel into the hydrometer.

☐ 4. Be sure that the float is not touching the bottom of the cylinder, then read the number on the float. Diesel fuel should have a minimum cetane number of 40. Does the fuel have a 40 rating? Yes _____ No _____

If No, suggest that the vehicle owner switch fuel brands.

Name _____ Date_____

Instructor _____ Period _____

Job 166—Test Fuel Quality (continued)

Check for Diesel Fuel Waxing

> **Note**
> Waxing occurs in cold weather and can clog filters and injectors, keeping the engine from starting or causing it to stall.

☐ 1. Place a sample of diesel fuel in a cold place. This can be done by placing the sample outside in cold weather (below 32°F or 0°C) or placing it in the freezer section of a refrigerator.

☐ 2. Wait several hours to allow the sample to cool down completely.

☐ 3. Observe the fuel. Wax will form easily observable particles in the fuel. Are wax particles present? Yes _____ No _____

If Yes, suggest that the vehicle owner switch fuel brands.

☐ 4. Clean the work area and return any equipment to storage.

Did you encounter any problems during this procedure? Yes _____ No _____

If Yes, describe the problems: _____

What did you do to correct them? _____

☐ 5. Have your instructor check your work and sign this job sheet.

Performance Evaluation—Instructor Use Only

Did the student complete the job in the time allotted? Yes ___ No ___

If No, which steps were not completed?_____

How would you rate this student's overall performance on this job?_____

5–Excellent, 4–Good, 3–Satisfactory, 2–Unsatisfactory, 1–Poor

Comments: _____

_____

_____

INSTRUCTOR'S SIGNATURE _____

Notes

Job 167—Test and Replace Fuel Supply System Components

After completing this job, you will be able to test and replace fuel supply components.

> **Warning**
> ⚠ Gasoline is extremely flammable, and can be ignited by the smallest spark. Even a spark caused by dropping a tool on the shop floor is sufficient to ignite gasoline. Be sure that a fire extinguisher is nearby and that you know how to use it before starting any procedure that requires removing fuel line connections.

Procedures

☐ 1. Obtain a vehicle to be used in this job. Your instructor may direct you to perform this job on a shop vehicle.

☐ 2. Gather the tools needed to perform this job. Refer to the tools and materials list at the beginning of the project.

☐ 3. Verify that your fire extinguisher is handy and properly charged.

☐ 4. Visually inspect the fuel lines and hoses for the following defects:
- Leaks at connections.
- Kinked lines or hoses.
- Collapsed lines or hoses.
- Swelled hoses.
- Loose or missing clips or brackets.

Describe any fuel line and hose problems found:_____

_____

Describe what should be done to correct the problems:_____

_____

Remove, Replace, and Test Fuel Filters

☐ 1. Locate the fuel filter on the vehicle.

☐ 2. Relieve fuel system pressure.

☐ 3. Remove the fuel filter bracket fasteners.

☐ 4. Place a drip pan under the filter.

☐ 5. Remove the fuel line fittings and allow excess fuel to drip into the pan.

> **Warning**
> ⚠ Some gasoline will spill as the fittings are removed. Be sure to clean up any spilled gasoline before proceeding.

☐ 6. Check the filter condition by one of the following methods:

Drain the fuel from the inlet side of the filter and observe the fuel's condition. If the fuel is contaminated, the filter must be replaced.

or

Attempt to blow through the filter. Restriction will make the filter difficult to blow through, indicating that the filter must be replaced.

> **Warning**
>
> ⚠ Avoid breathing gasoline vapors.

☐ 7. Compare the old and new filters to ensure that the new filter is correct. Check for correct size, fittings, and number of outlets.

Do the filters match? Yes ___ No ___

If No, what could account for the difference? _____

_____

☐ 8. Install the new filter.

☐ 9. Tighten the fittings.

☐ 10. Install the bracket fasteners.

☐ 11. Start the engine and check for fuel leaks.

Test Fuel Pressure (Electric Pump)

> **Note**
>
> 📋 This procedure also tests the operation of the fuel pressure regulator.

☐ 1. Attach a fuel pressure gauge to the fuel system. See **Figure 167-1**.

> **Warning**
>
> ⚠ Some gasoline will spill as the test connections are made. Be sure to clean up any spilled gasoline before proceeding.

☐ 2. Turn the ignition switch to the *on* position.

☐ 3. Observe the fuel system pressure.

Is the fuel pressure acceptable? Yes ___ No ___

If No, describe the possible causes: _____

_____

_____

Figure 167-1. Fuel pressure gauge attachment methods are shown here. A—Some systems, especially those using central fuel injectors, require an adapter to attach the gauge to the test port. B—An adapter is *not* needed to connect the pressure gauge to most multi-port fuel injection system test ports.

A   B

Job 167—Test and Replace Fuel Supply System Components (continued)

☐ 4. With the pressure gauge attached, start the engine.

☐ 5. Observe the fuel system pressure.

Did the fuel pressure drop slightly when the engine was started? Yes ___ No ___

Does the pressure remain steady as the engine operates? Yes ___ No ___

If No, describe possible causes: _____

_____

_____

☐ 6. Consult your instructor for any corrective actions to be taken.

Test Electric Fuel Pump Relays

☐ 1. Allow the engine to stand for at least 5 minutes.

☐ 2. Locate the fuel pump relay.

☐ 3. Place your hand on the relay and have an assistant turn the ignition switch to the *on* position.

Does the relay click? Yes ___ No ___

If Yes, skip to the *Test Electric Fuel Pump Amperage* section of this job.

If No, go to step 4.

☐ 4. Make other electrical checks as necessary to isolate the relay problem.

Test the Electric Fuel Pump's Amperage Draw

☐ 1. Obtain the proper amperage draw specifications from the appropriate service literature. If amperage draw specifications are not available, determine the type of fuel system and use the general guidelines given here.
   - **Multi-port fuel injection:** 4–6 amps.
   - **Throttle body injection:** 3–5 amps.
   - **Central point injection:** 8–10 amps.

☐ 2. Obtain an amperage tester or multimeter with amperage testing capabilities and set it to a minimum range of 0–20 amps.

☐ 3. Attach the tester to the fuel pump connectors as necessary. If the tester uses an inductive clamp, place it around the pump's power wire.

☐ 4. Turn the ignition switch to the *on* position and make sure that all other electrical devices are turned off.

☐ 5. Measure pump amperage.

Amperage:_____

Is the pump amperage within specifications? Yes ___ No ___

If the pump amperage is low, the pump brushes are probably worn and are not contacting the armature properly. If the amperage is high, the pump bearings are worn and are allowing the armature to drag against the field windings. In either case, the pump must be replaced.

Test Electric Fuel Pump Waveforms

☐ 1. Obtain an oscilloscope or waveform meter and attach the scope leads as necessary. Set the controls to measure the pump waveform.

☐ 2. Turn the ignition switch to the *on* position and make sure that all other electrical devices are turned off.

3. Observe the waveform. A pump in good condition will produce an even series of voltage spikes (sometimes called a sawtooth pattern). The amperage peaks should be near the rated amperage. Poor brush contact or open commutator windings in the pump will show up as uneven or missing voltage spikes.

   Does the waveform indicate a pump in good condition? Yes ___ No ___

Replace an Inline Electric Fuel Pump

1. Remove the battery negative cable.

2. Locate the pump on the vehicle, **Figure 167-2**. If the pump is not visible under the hood, raise the vehicle and follow the fuel line.

   Where is the pump located? _____

   _____

Warning

⚠ The vehicle must be raised and supported in a safe manner. Always use approved lifts or jacks and jack stands.

---

3. Relieve the fuel pressure.
4. Disconnect the electrical connector at the pump.
5. Place a drip pan under the pump.
6. Loosen the pump fittings.
7. Allow excess fuel to drip into the pan, then remove the pump fittings.

Warning

⚠ If fuel continues to drip from the lines, plug them as necessary.

---

8. Remove the pump bracket fasteners.
9. Remove the pump from the vehicle.

Figure 167-2. A typical location for an inline electric fuel pump is shown here. The bracket holding the pump and filter must be removed to gain access to the pump.

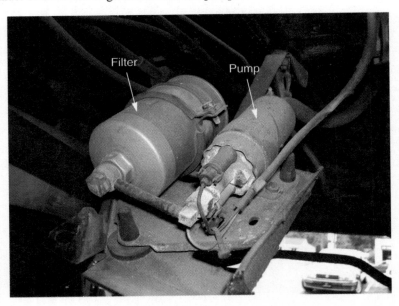

Job 167—Test and Replace Fuel Supply System Components (continued)

☐ 10. Compare the old and new pumps to ensure that the new pump is correct.

☐ 11. Place the new pump in position on the vehicle.

☐ 12. Install but do not tighten the bracket fasteners.

☐ 13. Install and tighten the fuel line fittings.

☐ 14. Tighten the pump bracket fasteners.

☐ 15. Install the electrical connector.

☐ 16. Reinstall the battery negative cable.

☐ 17. Turn the ignition switch to the *on* position and check pump operation.

> **Warning**
>
> ⚠ Make sure that the pump fittings do not leak.

Replacing an In-Tank Electric Fuel Pump

☐ 1. Remove the battery negative cable.

☐ 2. Drain the fuel from the tank into an approved fuel collecting device. When the fuel is done draining, move the fuel and container out of the work area.

> **Warning**
>
> ⚠ Do not allow spilled fuel or vapors to collect. Wipe up spills and ensure that the area is properly ventilated.

☐ 3. Disconnect the electrical connectors, fuel lines, and hoses from the tank assembly.

☐ 4. Place a transmission jack or other lowering device under the fuel tank. Raise it into position so that it will support the tank.

☐ 5. Remove the fuel tank fasteners.

> **Note**
>
> 📄 You may need to disconnect the fuel tank filler neck before lowering the tank from the vehicle.

☐ 6. Lower the fuel tank from the vehicle and place it on a bench.

☐ 7. Locate the cam lock that holds the fuel pump assembly in place.

☐ 8. Locate the tabs that hold the cam lock in place and bend them out enough to allow the cam to be turned.

☐ 9. Turn the cam lock using a special tool.

> **Note**
>
> 📄 Some tanks use a snap ring or bolts and do not require a special tool.

☐ 10. Lift the assembly from the tank. See **Figure 167-3**.

Figure 167-3. A typical tank-mounted electric fuel pump assembly is shown here. Always check the strainer (sock) at the pump inlet for contamination.

11. Remove the tank-to-assembly O-ring and discard it.

12. Wipe up all spilled fuel, and then bring the assembly to a clean workbench.

13. Remove the pump from the pickup tube and sender as needed.

14. Remove the strainer (often called the sock) from the pump assembly and inspect it closely. If the strainer appears to be clogged or has any tears or holes, replace it.

15. Compare the old and new pumps to ensure that the new pump is correct.

16. Install the pump strainer.

17. Reattach the pickup tube and sender to the pump, if applicable.

18. Install a new O-ring in the tank's O-ring groove.

> **Note**
>
> Some manufacturers recommend lubricating O-rings before installation.

19. Lower the pump into the tank opening.

> **Caution**
>
> Watch carefully to ensure that the strainer is not damaged or folded over enough to restrict fuel flow.

20. Install the cam lock.

21. Tap the lock fully into position, and reposition the holding tabs.

22. Move the tank under the vehicle and raise it into position.

23. Reinstall the tank in the vehicle. Reconnect any electrical connectors, hoses, and lines that were disconnected during tank removal.

24. Return the fuel to the tank.

25. Reinstall the battery negative cable.

Job 167—Test and Replace Fuel Supply System Components (continued)

- [ ] 26. Turn the ignition switch to the *on* position and check that the pump operates correctly.

Test Mechanical Fuel Pump Pressure

- [ ] 1. Attach a fuel pressure gauge to the fuel system.

> **Warning**
>
> ⚠ Some gasoline will spill as the test connections are made. Be sure to clean up any spilled gasoline before proceeding.

- [ ] 2. Start the engine.
- [ ] 3. Record the fuel system pressure.

  Fuel system pressure:_____

  Is the system pressure within specifications? Yes ___ No ___

  If No, the fuel pump may be faulty; there may be a leak in the pump's intake line, allowing air to be drawn in; or there may be a restriction in the system, such as a collapsed fuel line or clogged strainer. Consult your instructor for further diagnostic or corrective actions that should be taken.

Replace a Mechanical Fuel Pump

- [ ] 1. Locate the fuel pump on the engine.
- [ ] 2. Relieve pressure from the fuel system.
- [ ] 3. Place a drip pan under the pump.
- [ ] 4. Remove the fuel line fittings and allow excess fuel to drip into the pan.

> **Warning**
>
> ⚠ If fuel continues to drip from the lines, plug them as necessary.

- [ ] 5. Remove the fasteners holding the pump to the engine, **Figure 167-4**.
- [ ] 6. Remove the pump from the engine.

> **Note**
>
> 📑 Some fuel pumps are operated by a push rod, which could fall free from the block when the pump is removed. If you are replacing a push rod–type pump, remove the pump slowly to avoid dropping the push rod.

- [ ] 7. Scrape off all old gasket material from the pump mounting surface.
- [ ] 8. Compare the old and new pumps.

  Do the pumps match? Yes ___ No ___

  If No, what could be the reason? _____

  _____

- [ ] 9. Install a new fuel pump gasket on the engine, using sealer as necessary.
- [ ] 10. Place the new pump on the engine, being sure to align the pump arm (or push rod) with the camshaft.

Figure 167-4. Most mechanical fuel pumps are attached to the engine with two fasteners. The fasteners can be removed with standard hand tools.

> **Note**
>
> If you are installing a push rod–type pump, you can use grease to hold the push rod in position while you install the pump.

11. Install and tighten the fuel pump fasteners.
12. Install the fuel line fittings.
13. Start the engine and check that the pump operates correctly.
14. Clean the work area and return any equipment to storage.
15. Did you encounter any problems during this procedure? Yes ___ No ___
    If Yes, describe the problems: _____
    _____
    What did you do to correct the problems? _____
16. Have your instructor check your work and sign this job sheet.

Performance Evaluation—Instructor Use Only

Did the student complete the job in the time allotted? Yes ___ No ___
If No, which steps were not completed?_____
How would you rate this student's overall performance on this job?_____
5–Excellent, 4–Good, 3–Satisfactory, 2–Unsatisfactory, 1–Poor
Comments: _____
_____
_____

INSTRUCTOR'S SIGNATURE _____

Job 168—Inspect and Service a Throttle Body

After completing this job, you will be able to check injector waveforms for indications of failure.

Procedures

- [ ] 1. Obtain a vehicle to be used in this job. Your instructor may direct you to perform this job on a shop vehicle.
- [ ] 2. Gather the tools needed to perform this job.
- [ ] 3. Obtain the proper service manual and other service information.

Check for Loose or Damaged Air Intake System Ducts or Hoses

- [ ] 1. Open the vehicle hood and remove any air ducts or engine covers as necessary.
- [ ] 2. Locate the air filter housing.
- [ ] 3. Remove the fasteners holding the air filter housing.
- [ ] 4. Remove the housing to gain access to the filter.
- [ ] 5. Remove the filter and inspect the filter element for excessive dirt buildup.
- [ ] 6. Observe the air intake system for obvious damage or missing parts.
- [ ] 7. Remove necessary ductwork to expose the throttle body. Look for broken fittings, damaged gaskets, loose fasteners, and a worn throttle shaft.
- [ ] 8. Check for disconnected, cracked, and split vacuum lines.

     Were any vacuum line problems found? Yes ___ No ___

     If Yes, explain: _____

     _____
- [ ] 9. Inspect the intake manifold for broken fittings and damaged gaskets.
- [ ] 10. Check ductwork for splits, cracks, and loose or missing fittings.

     Were any ductwork problems found? Yes ___ No ___

     If Yes, explain: _____

     _____

Check the Spray Patterns in a Throttle Body Injection (TBI) System

- [ ] 1. Attach a timing light to any spark plug wire.
- [ ] 2. Remove the air cleaner.
- [ ] 3. Start the engine.
- [ ] 4. Point the timing light and observe the injector spray pattern, **Figure 168-1**.

     Does the pattern appear to be steady and evenly cone-shaped? Yes ___ No ___

     If No, what could be the problem? _____

     _____
- [ ] 5. Turn off the engine.
- [ ] 6. Disconnect the timing light.

Clean the Throttle Body

- [ ] 1. Remove the air cleaner or intake duct as applicable.
- [ ] 2. Inspect the throttle body for deposits.

Figure 168-1. Point the timing light at the fuel injector and observe the spray pattern. Raise the idle to get a more consistent flash.

3. Remove any sensors or solenoids that could be damaged by solvent.

4. Spray solvent into throttle body and allow the solvent to soak for several minutes.

5. If necessary, use a toothbrush to scrub the throttle body assembly, **Figure 168-2**.

Warning

⚠ Do not allow toothbrush bristles to enter the intake system.

---

6. Flush deposits from the throttle body using more spray solvent.

7. Allow the throttle body to air dry for several minutes.

8. Reinstall the air cleaner or intake ducts and any sensors or solenoids that were removed.

9. Start the engine and check for leaks and proper engine operation.

Figure 168-2. Carefully scrub throttle body deposits using an old toothbrush or other soft brush.

Job 168—Inspect and Service a Throttle Body (continued)

☐ 10. Clean the work area and return any equipment to storage.

☐ 11. Did you encounter any problems during this procedure? Yes ___ No ___

If Yes, describe the problems: _____

_____

What did you do to correct the problems? _____

_____

☐ 12. Have your instructor check your work and sign this job sheet.

Performance Evaluation—Instructor Use Only

Did the student complete the job in the time allotted? Yes ___ No ___

If No, which steps were not completed?_____

How would you rate this student's overall performance on this job?_____

5–Excellent, 4–Good, 3–Satisfactory, 2–Unsatisfactory, 1–Poor

Comments: _____

_____

_____

INSTRUCTOR'S SIGNATURE _____

Notes

Job 169—Check Idle Speed Motor or Idle Air Control

After completing this job, you will be able to check the operation of the idle speed motor or idle air control.

Procedures

- ☐ 1. Obtain a vehicle to be used in this job. Your instructor may direct you to perform this job on a shop vehicle.
- ☐ 2. Gather the tools needed to perform this job. Refer to the tools and materials list at the beginning of the project.

Check Idle Speed Motor or Idle Air Control

> **Note**
>
> Do not move the throttle during this procedure.

- ☐ 1. Attach a tachometer to the engine.
- ☐ 2. Start the engine.
- ☐ 3. Observe the engine speed (engine should be warm).
- ☐ 4. Place the transmission in gear (parking brake on) or turn on the air conditioner.
  Does the idle speed motor compensate for the increased load and maintain idle speed?
  Yes ___ No ___
  If Yes, skip to step 6.
  If No, what could be the problem? _____
  _____
  _____
- ☐ 5. Consult your instructor for the corrective action to be taken.
- ☐ 6. Stop the engine.
- ☐ 7. Remove the tachometer.

Remove and Replace an Idle Speed Motor

- ☐ 1. If necessary, remove the air cleaner to gain access to the idle speed motor.
- ☐ 2. Determine the idle speed motor design, **Figure 169-1**. Some motors are threaded into the throttle body; others are held in place with fasteners.
- ☐ 3. Select the proper tools and remove the idle speed motor.
- ☐ 4. Compare the idle speed motor with the replacement part.
  Do the replacement idle speed motor and the replacement match? Yes ___ No ___
  If No, consult your instructor.
- ☐ 5. Install the replacement idle speed motor.
- ☐ 6. Reinstall the air cleaner if necessary.
- ☐ 7. Start the engine and ensure that the replacement idle speed motor is operating properly.

Figure 169-1. Two common types of idle speed motors are shown here. A—This idle speed motor is threaded into the throttle body. B—This motor is held in place with fasteners.

A   B

8. Clean the work area and return any equipment to storage.

9. Did you encounter any problems during this procedure? Yes ___ No ___

If Yes, describe the problems: _____

_____

What did you do to correct the problems:_____

_____

10. Have your instructor check your work and sign this job sheet.

Performance Evaluation—Instructor Use Only

Did the student complete the job in the time allotted? Yes ___ No ___

If No, which steps were not completed?_____

How would you rate this student's overall performance on this job?_____

5–Excellent, 4–Good, 3–Satisfactory, 2–Unsatisfactory, 1–Poor

Comments: _____

_____

_____

INSTRUCTOR'S SIGNATURE _____

Job 170—Check Fuel Injector Waveforms and Perform a Fuel Injector Balance Test

After completing this job, you will be able to check injector waveforms for indications of failure. You will also be able to perform a fuel injector balance test.

Procedures

- [ ] 1. Obtain a vehicle to be used in this job. Your instructor may direct you to perform this job on a shop vehicle.

- [ ] 2. Gather the tools needed to perform this job. Refer to the tools and materials list at the beginning of the project.

- [ ] 3. Obtain the proper service manual and other service information.

Check Injector Waveforms

- [ ] 1. Obtain one of the following testers capable of reading injector waveforms:
    - Engine Analyzer. ____
    - Oscilloscope. ____
    - Waveform Meter. ____

- [ ] 2. Set the tester to read the waveform of the injection system being tested.

- [ ] 3. Attach the analyzer or meter leads to the injector connections of one injector. The tester must be connected to the injector in a way that allows the injector to function, or the waveform will not be usable.

    Attachment method(s): _____

    _____

    _____

> **Note**
>
> To check a cold start injector, **Figure 170-1**, the engine must be allowed to cool down.

- [ ] 4. Start the engine and observe the waveform. **Figure 170-2** shows correct and incorrect waveforms for a common type of injector.

    Is the observed waveform correct for this particular injector? Yes ____ No ____

    If No, what are some possible causes? _____

    _____

    _____

- [ ] 5. Repeat steps 3 and 4 for each injector to be tested.

    List any injector that does not have the proper waveform and write a brief description of the possible cause:_____

    _____

    _____

- [ ] 6. Turn off the engine and disconnect the meter, analyzer, or oscilloscope.

Figure 170-1. Some older vehicles use a cold start injector. It can be checked like any other injector. However, you must work quickly since the computer turns this injector off when the engine warms up.

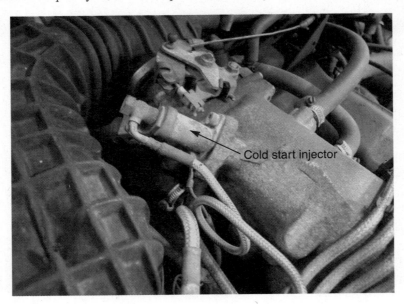

Cold start injector

Perform an Injector Balance Test

☐ 1. Ensure that the vehicle ignition switch is in the *off* position.

☐ 2. Connect a fuel pressure gauge to the fuel rail.

☐ 3. Attach the tester to the first injector, or install the scan tool in the diagnostic connector.

☐ 4. Turn the ignition switch to the *on* position.

☐ 5. Bleed all air from the gauge lines.

☐ 6. After bleeding is finished, wait ten seconds and then record the gauge pressure.

   Pressure with the injector closed:_____

☐ 7. Turn the injector on, using the tester switch or the scan tool.

☐ 8. Note the pressure as soon as the gauge needle reaches its lowest point.

   Pressure with the injector open:_____

☐ 9. Repeat steps 3 through 8 for all injectors and record the readings by injector (cylinder) number.

Injector #	Closed	Open
2	_____	_____
3	_____	_____
4	_____	_____
5	_____	_____
6	_____	_____
7	_____	_____
8	_____	_____
9	_____	_____
10	_____	_____

☐ 10. Retest injectors that appear to be defective.

Job 170—Check Fuel Injector Waveforms and Perform a Fuel Injector Balance Test (continued)

Figure 170-2. Note the difference between the correct and incorrect waveforms shown here for a common injection system. A—Correct waveform. B—Incorrect waveform indicating a shorted injector. (Autonerdz.com/AES)

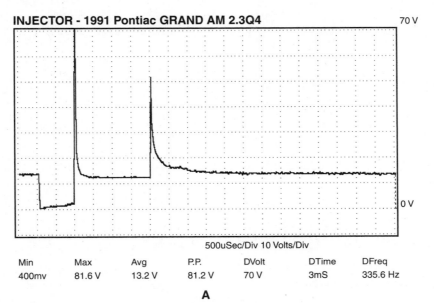

INJECTOR - 1991 Pontiac GRAND AM 2.3Q4

70 V

0 V

500uSec/Div 10 Volts/Div

Min	Max	Avg	P.P.	DVolt	DTime	DFreq
400mv	81.6 V	13.2 V	81.2 V	70 V	3mS	335.6 Hz

A

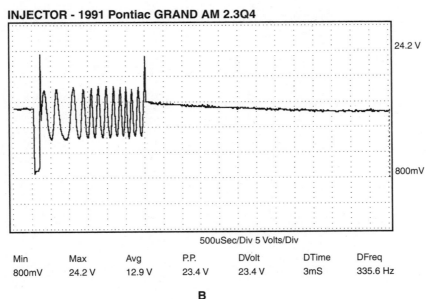

INJECTOR - 1991 Pontiac GRAND AM 2.3Q4

24.2 V

800mV

500uSec/Div 5 Volts/Div

Min	Max	Avg	P.P.	DVolt	DTime	DFreq
800mV	24.2 V	12.9 V	23.4 V	23.4 V	3mS	335.6 Hz

B

☐ 11. Once all injectors have been tested, average all of the readings using the following process:
    a. Subtract the low reading (injector open) from the high reading (injector closed). This gives the pressure drop for each injector.
    b. Add all of the injector drop readings.
    c. Divide the above figure by the number of injectors. This gives the average injector drop.

☐ 12. Compare the drop of each injector with the average.

843

> **Note**
>
> Any injector with a pressure drop that deviates from the average by 10 psi (69 kPa) or more is suspect.

Do any injectors deviate from the average drop? Yes ___ No ___

What conclusions can you draw from these readings? _____

_____

_____

_____

☐ 13. Clean the work area and return any equipment to storage.

☐ 14. Did you encounter any problems during this procedure? Yes ___ No ___

If Yes, describe the problems: _____

_____

What did you do to correct the problems? _____

_____

☐ 15. Have your instructor check your work and sign this job sheet.

Performance Evaluation—Instructor Use Only

Did the student complete the job in the time allotted? Yes ___ No ___

If No, which steps were not completed? _____

How would you rate this student's overall performance on this job? _____

5–Excellent, 4–Good, 3–Satisfactory, 2–Unsatisfactory, 1–Poor

Comments: _____

_____

_____

INSTRUCTOR'S SIGNATURE _____

Job 171—Remove and Replace a Fuel Injector

After completing this job, you will be able to remove and replace a fuel injector.

Procedures

- [ ] 1. Obtain a vehicle to be used in this job. Your instructor may direct you to perform this job on a shop vehicle.
- [ ] 2. Gather the tools needed to perform this job. Refer to the tools and materials list at the beginning of the project.
- [ ] 3. Obtain the proper service manual and other service information.

Remove and Replace a Fuel Injector

- [ ] 1. Relieve fuel system pressure.
- [ ] 2. Remove the intake plenum (if applicable) or other parts that block access to the fuel rail or injector.
- [ ] 3. Remove the injector's electrical connector.
- [ ] 4. Remove the fuel rail if applicable. Note in **Figure 171-1** how the fuel rail is constructed. Attempting to remove any of the injectors without removing the rail first will damage the rail.

> **Warning**
>
> Clean up any spilled gasoline.

- [ ] 5. Pull the injector from the manifold or throttle body.
- [ ] 6. Install the new injector.

Figure 171-1. The fuel rail (shown here installed in the intake manifold) must be removed before removing any of the injectors. Most fuel rails have an internal coating that will crack and flake off if the rail is bent.

☐   7.   Reinstall the fuel rail if necessary.

☐   8.   Tighten all fittings.

☐   9.   Reinstall the injector's electrical connector.

☐ 10. Reinstall the intake plenum (if applicable) and any other components that were removed.

☐ 11. Start the engine and check for leaks and proper engine operation.

☐ 12. Clean the work area and return any equipment to storage.

☐ 13. Did you encounter any problems during this procedure? Yes ___ No ___

        If Yes, describe the problems: _____

        _____

        What did you do to correct the problems? _____

        _____

☐ 14. Have your instructor check your work and sign this job sheet.

Performance Evaluation—Instructor Use Only

Did the student complete the job in the time allotted? Yes ___ No ___

If No, which steps were not completed?_____

How would you rate this student's overall performance on this job?_____

5–Excellent, 4–Good, 3–Satisfactory, 2–Unsatisfactory, 1–Poor

Comments: _____

_____

_____

INSTRUCTOR'S SIGNATURE _____

Job 172—Check Supercharger and Turbocharger Operation

After completing this job, you will be able to check a supercharger or turbocharger for proper operation.

Procedures

☐ 1. Obtain a vehicle to be used in this job. Your instructor may direct you to perform this job on a shop vehicle.

☐ 2. Gather the tools needed to perform this job. Refer to the tools and materials list at the beginning of the project.

☐ 3. Obtain the proper supercharger/turbocharger service information. If boost pressures and manifold vacuum specifications are given, list them in the spaces provided. If no specifications are available, go to the next section.

Engine rpm	Boost pressure	Manifold vacuum
_____	_____ psi/kPa	_____ in Hg/mm Hg
_____	_____ psi/kPa	_____ in Hg/mm Hg
_____	_____ psi/kPa	_____ in Hg/mm Hg
_____	_____ psi/kPa	_____ in Hg/mm Hg

☐ 4. Attach pressure and vacuum gauges to the engine.

☐ 5. Measure pressure and vacuum at the correct engine rpm. Record the readings.

Engine rpm	Boost pressure	Manifold vacuum
_____	_____ psi/kPa	_____ in Hg/mm Hg
_____	_____ psi/kPa	_____ in Hg/mm Hg
_____	_____ psi/kPa	_____ in Hg/mm Hg
_____	_____ psi/kPa	_____ in Hg/mm Hg

☐ 6. Compare the actual boost pressures and vacuum readings to the specifications.

Do actual readings match the specifications? Yes ___ No ___

If No, what defects are indicated?_____

_____

_____

Many external defects can make it appear as if the supercharger or turbocharger is defective:
- A clogged catalytic converter or other exhaust system restriction can cause high supercharger or turbocharger pressures.
- A clogged air filter can cause low supercharger or turbocharger pressures.
- Engine ignition, fuel, and compression problems can appear to be a supercharger or turbocharger problem.

Check the exhaust system, air filter, and general engine condition before condemning the supercharger or turbocharger.

Check a Supercharger

☐ 1. Visually check the supercharger, **Figure 172-1**, for the following:
- Oil leaks. ___
- Disconnected linkage or hoses. ___
- Loose fasteners or damaged gaskets. ___
- Obvious mechanical damage, such as a cracked case. ___

Are any problems found? Yes ___ No ___

If Yes, list: _____

_____

☐ 2. If internal damage is suspected, remove the belt and turn the supercharger by hand. The rotors should freely turn with no bearing roughness or scraping sounds.

Are any defects present? Yes ___ No ___

If Yes, explain: _____

_____

☐ 3. Check the condition of the belt and pulley.

Are the belt and pulley in good condition? Yes ___ No ___

If No, explain: _____

_____

☐ 4. Check that the supercharger's magnetic clutch (if used) engages when boost is needed.

Does the clutch engage when the throttle plate is quickly opened beyond half throttle? Yes ___ No ___

Check a Turbocharger

☐ 1. Remove the air inlet hose from the turbocharger.

☐ 2. Visually check the impeller blades for problems, **Figure 172-2**. Identify any problems found.
- Cracks. ___
- Bent or broken blades. ___
- Blade wear. ___

Are defects found? Yes ___ No ___

If Yes, explain: _____

_____

Figure 172-1. Check the supercharger at all points. It may be necessary to remove engine covers to observe all parts of the supercharger. Many superchargers develop oil leaks at the pulley end seals. This can be spotted by the presence of oil on the pulley and under the vehicle hood directly over the pulley.

Job 172—Check Supercharger and Turbocharger Operation (continued)

Figure 172-2. Carefully inspect the impeller blades for damage.

☐ 3. Check the inlet for excessive oil, which indicates a leaking compressor seal.

Is excessive oil found? Yes ____ No ____

☐ 4. Carefully turn the compressor wheel (impeller). There should be no bearing roughness or scraping sounds.

Are any defects noted? Yes ____ No ____

If Yes, explain: _____

_____

☐ 5. Push and pull on the impeller while turning. There will be a minimal amount of radial (back-and-forth) and axial (in-and-out) play, but there should be no contact between the impeller/turbine wheels and the turbocharger housing.

Does there appear to be excessive radial or axial play? Yes ____ No ____

If Yes, explain: _____

_____

> **Note**
>
> Some manufacturers specify maximum radial or axial play. This check can be made with a dial indicator.

☐ 6. If steps 4 and 5 indicate the possibility of damage to the turbine blades, remove exhaust components as needed and check the turbine blades for cracks, bent or broken blades, and wear. Also, check the turbine and chamber for excessive carbon buildup, which indicates a leaking compressor seal.

Are any defects present? Yes ____ No ____

If Yes, explain: _____

_____

7. If the engine uses two turbochargers, repeat steps 1 through 6 on the other turbocharger.

8. When inspection is finished, replace the exhaust system components and the air inlet hose(s).

Check Turbocharger Waste Gate and Actuator Diaphragm Operation

> **Caution**
>
> Some waste gates can be adjusted to vary boost pressure. Do not raise the boost pressure over the manufacturer's established maximums. Severe engine damage can result.

1. With the engine off, attempt to move the waste gate linkage. The linkage should freely move. It should move toward the closed position when the linkage is released.

    Does the waste gate operate properly? Yes ___ No ___

> **Note**
>
> The linkage may not fully move to the closed position when the linkage is released. This is acceptable as long as the linkage freely moves.

2. Check the condition of the vacuum actuator by one of the following methods:
    - Connect a vacuum pump to the waste gate actuator diaphragm and apply 5–10 inches of vacuum to the diaphragm.
    - Start the engine with the manifold hose connected to the waste gate actuator diaphragm.

    Does the waste gate close? Yes ___ No ___

    If No, the diaphragm is defective.

3. If the engine uses two turbochargers, repeat steps 1 and 2 and the other turbocharger.

Check an Intercooler

> **Note**
>
> Not all supercharged and turbocharged engines have an intercooler.

1. Remove any shrouds or vehicle parts covering the intercooler.

2. Check the intercooler for leaks, especially at any seals.

3. Check for dirt buildup on the outer and inner fins.

4. Check for mechanical damage, such as cracks and broken tubing.

5. Replace the shrouds and parts covering the intercooler.

6. After consulting your instructor, decide what repairs are necessary.

    Needed repairs: _____

    _____

    _____

7. Clean the work area and return any equipment to storage.

Job 172—Check Supercharger and Turbocharger Operation (continued)

☐ 8. Did you encounter any problems during this procedure? Yes ___ No ___

If Yes, describe the problems: _____

_____

What did you do to correct the problems? _____

_____

☐ 9. Have your instructor check your work and sign this job sheet.

Performance Evaluation—Instructor Use Only

Did the student complete the job in the time allotted? Yes ___ No ___

If No, which steps were not completed?_____

How would you rate this student's overall performance on this job?_____

5–Excellent, 4–Good, 3–Satisfactory, 2–Unsatisfactory, 1–Poor

Comments: _____

_____

_____

INSTRUCTOR'S SIGNATURE _____

Notes

Project
Diagnosing and Servicing an Exhaust System

47

Introduction

Every technician will eventually have to diagnose and correct an exhaust system problem. Most exhaust system problems consist of leaks and restrictions. All modern exhaust systems have one or more catalytic converters and are connected to other emission control devices such as exhaust gas recirculation (EGR) systems and oxygen sensors. In Job 173, you will troubleshoot an exhaust system.

Most exhaust system repair consists of replacing rusted out or damaged parts. Exhaust systems replacement procedures remain relatively simple. Modern exhaust systems are connected to such emission control devices as catalytic converters, exhaust gas recirculation (EGR) systems, and oxygen sensors. The technician must therefore be careful when replacing exhaust system parts so that these components are not damaged. In Job 174, you will replace exhaust system components.

Project 47 Jobs

- Job 173—Inspect and Test an Exhaust System
- Job 174—Replace Exhaust System Components

Tools and Materials

The following list contains the tools and materials that may be needed to complete the jobs in this project. The items used will depend on the make and model of vehicle being serviced.

- Vehicle in need of exhaust system service.
- Air chisel with adapters.
- Pipe cone.
- Pipe expander.
- Vacuum gauge or pressure gauge.
- Hand tools as needed.
- Air-powered tools as needed.
- Safety glasses and other protective equipment as needed.

Safety Notice

Before performing this job, review all pertinent safety information in the text, and review safety information with your instructor.

Notes

Job 173—Inspect and Test an Exhaust System

After completing this job, you will be able to examine and test exhaust systems.

Procedures

☐   1.   Obtain a vehicle to be used in this job. Your instructor may direct you to perform this job on a shop vehicle.

☐   2.   Gather the tools needed to perform this job. Refer to the tools and materials list at the beginning of the project.

Check and Refill Diesel Exhaust Fluid (Diesel Vehicles Only)

☐   1.   If a diesel vehicle is being serviced, check the diesel exhaust fluid (DEF) tank level. Some tanks have a gauge, while others require that the cap be removed and the level observed.

       Is the level adequate? Yes ___ No ___

☐   2.   If DEF is needed, remove the filler cap if not already removed, and pour DEF into the tank as needed.

☐   3.   Ensure that the instrument panel LOW DEF warning light goes off when the engine is started.

> **Note**
> DEF is not dangerous. Small amounts of DEF, if spilled, can be washed away.

Check for Exhaust Leaks and Other Exhaust System Damage

☐   1.   Start the engine and set the speed to fast idle.

☐   2.   Raise the vehicle on a lift and determine the exhaust system configuration.

       What type of exhaust system does the vehicle have? Single ___ Dual ___

       How many catalytic converters does the vehicle have?_____

       Does the vehicle have one or more resonators? Yes ___ No ___

> **Warning**
> ⚠ The vehicle must be raised and supported in a safe manner. Always use approved lifts or jacks and jack stands.

☐   3.   Look and listen for leaks. Leaks can be heard, and rusted out components are usually visible.

       Were any leaks found? Yes ___ No ___

       If Yes, list: _____

> **Warning**
> ⚠ Exhaust system parts become hot quickly. Do not touch any exhaust parts with your bare hands.

☐ 4. Turn off the engine. If necessary, lower the vehicle to turn the engine off and then raise the vehicle again.

☐ 5. Visually inspect the exhaust system.
   - Are there broken, missing, or corroded hangers? Yes ___ No ___
   If Yes, describe: _____

   _____

   - Are there bent or disconnected hangers? Yes ___ No ___
   If Yes, describe: _____

   _____

   - Is exhaust hanger or bracket attaching hardware loose or missing? Yes ___ No ___
   If Yes, describe: _____

   _____

   - Are there missing, loose, or badly corroded heat shields? (Some corrosion is acceptable, but there must be no missing sections or holes in the shields.) Yes ___ No ___
   If Yes, describe: _____

   _____

   - Are any heat shield–attaching fasteners loose or missing? Yes ___ No ___
   If Yes, describe: _____

   _____

☐ 6. Lower the vehicle.

☐ 7. Return the engine to its normal idle speed and turn it off.

Perform a Backpressure Test to Check for Exhaust Restriction

> **Note**
>
> Your instructor will direct you to use one or both of the following methods to check for exhaust restriction. If your instructor directs you to use only one of the following methods, skip the procedure that does not apply.

Method(s) used: _____

Vacuum Gauge Method

☐ 1. Attach a vacuum gauge to the intake manifold and a tachometer to the engine ignition system.

☐ 2. Start the engine.

☐ 3. Record manifold vacuum as the engine idles.
   Vacuum:_____

☐ 4. Raise the engine speed to 2500 rpm to check for low vacuum at high engine speed.
   Vacuum:_____

   If the vacuum reading at 2500 rpm is about the same or higher than at idle, the exhaust system is not restricted. If the vacuum reading at 2500 rpm is less than the reading at idle, the system is restricted. Many manufacturers consider a 3 psi (21 kPa) drop to be grounds for further investigation and part replacement.

☐ 5. Return the engine to its normal idle speed and turn it off.

☐ 6. Remove the vacuum gauge from the intake manifold.

Job 173—Inspect and Test an Exhaust System (continued)

Pressure Gauge Method

☐ 1. Attach a pressure gauge to the exhaust manifold by removing an oxygen sensor and threading the pressure gauge into the opening. If the engine uses a air injection pump, the check valve can be removed and the pressure gauge installed in the opening.

  If you are testing for a clogged catalytic converter, remove the oxygen sensor nearest to the front of the converter and install the pressure gauge in the opening.

☐ 2. Start the engine.

☐ 3. Raise engine speed to 2500 rpm and record the exhaust pressure.

  Pressure:_____

☐ 4. Compare the back pressure to specifications. If back pressure specifications are not available, the back pressure should be no more than 4 to 5 psi on a single exhaust system or 1/2 to 2 psi on one side of a dual exhaust system.

  Does the exhaust appear to be restricted? Yes ___ No ___

  If Yes, what could be the cause? _____

  _____

  If you are testing for a clogged converter on a dual exhaust system, repeat steps 1 through 4 for the other converter.

☐ 5. Clean the work area and return any equipment to storage.

☐ 6. Did you encounter any problems during this procedure? Yes ___ No ___

  If Yes, describe the problems: _____

  _____

  What did you do to correct the problems? _____

  _____

☐ 7. Have your instructor check your work and sign this job sheet.

Performance Evaluation—Instructor Use Only

Did the student complete the job in the time allotted? Yes ___ No ___

If No, which steps were not completed?_____

How would you rate this student's overall performance on this job?_____

5–Excellent, 4–Good, 3–Satisfactory, 2–Unsatisfactory, 1–Poor

Comments: _____

_____

_____

INSTRUCTOR'S SIGNATURE _____

Notes

_____

_____

_____

_____

_____

_____

_____

_____

_____

_____

_____

_____

_____

_____

_____

_____

_____

_____

_____

_____

_____

_____

_____

_____

_____

_____

_____

_____

_____

_____

_____

_____

_____

_____

_____

_____

_____

_____

_____

_____

_____

Job 174—Replace Exhaust System Components

After completing this job, you will be able to replace exhaust system components, such as mufflers, exhaust pipes, and catalytic converters.

Procedures

☐ 1. Obtain a vehicle to be used in this job. Your instructor may direct you to perform this job on a shop vehicle.

☐ 2. Gather the tools needed to perform this job. Refer to the tools and materials list at the beginning of the project.

☐ 3. Raise the vehicle.

> **Warning**
> ⚠ The vehicle must be raised and supported in a safe manner. Always use approved lifts or jacks and jack stands.

☐ 4. Remove any heat shields or skid plates that prevent access to the exhaust system component to be replaced.

☐ 5. Apply penetrating oil to all pipe bracket fasteners and to joint clamps. If an exhaust manifold is being replaced, apply penetrating oil to the manifold-to-engine fasteners.

☐ 6. Remove any parts that are attached to the exhaust component, such as oxygen sensors or EGR tubes.

☐ 7. Remove the clamps and bracket fasteners.

☐ 8. Loosen the outlet joint and pull the pipe free with a chain wrench. If the parts are welded together, cut off the pipe where necessary, ensuring that the remaining pipe can engage the new parts.

> **Note**
> 📄 Carefully remove and handle any pipe or parts that will be reused, such as exhaust doughnuts.

☐ 9. If any parts are joined at flanges, remove the fasteners and separate the flanges.

☐ 10. Remove the components from the vehicle.

☐ 11. After removal, use a pipe-end straightening cone or a pipe expander to straighten the ends of all pipes that will be reused. See **Figure 174-1**.

☐ 12. Clean the inside or outside of the nipples that will be reused. If a nipple is distorted, use a pipe expander to restore it to a perfectly round state.

☐ 13. Inspect all bell-mouth connectors, **Figure 174-2**, and replace any that are damaged.

☐ 14. Determine what parts will be needed to finish the job.

List the parts:_____

_____

_____

Figure 174-1. Carefully straighten the ends of exhaust components that will be reused.

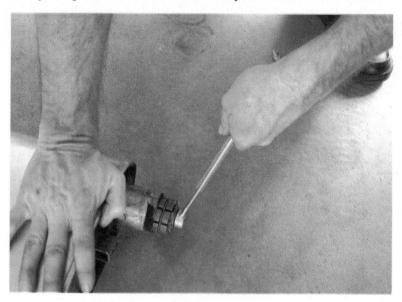

Figure 174-2. Check all bell-mouth connectors for distortion and replace as necessary.

☐ 15. Compare the old and new exhaust system parts.

Do the parts match? Yes ___ No ___

If No, explain why: _____

_____

_____

☐ 16. Scrape old gasket material off of any flanged connectors that will be reused, such as the one in **Figure 174-3**.

☐ 17. Apply a coating of exhaust system sealant to any section of a component that will be installed inside of another pipe.

☐ 18. Slide the pipes and nipples together.

Job 174—Replace Exhaust System Components (continued)

Figure 174-3. A typical flanged connector is shown here.

> **Caution**
>
> Make certain that the depth is correct.

 19. Slide the clamps into position and tighten them lightly.

☐ 20. Install and tighten any flanged connections, making sure that gaskets are used where necessary.

> **Caution**
>
> Make sure that all parts are installed in the correct position and are facing in the proper direction. Make sure that the exhaust system parts are aligned with each other and do not contact any other part of the vehicle.

 21. Tighten all clamps and brackets.

> **Caution**
>
> Do not over tighten the clamps.

 22. Install any related parts, such as oxygen sensors, EGR tubes, or heat shields.

☐ 23. Start the engine and check for exhaust leaks.

☐ 24. Clean the work area and return any equipment to storage.

☐ 25. Did you encounter any problems during this procedure? Yes ___ No ___

If Yes, describe the problems: _____

_____

What did you do to correct the problems? _____

_____

☐ 26. Have your instructor check your work and sign this job sheet.

Performance Evaluation—Instructor Use Only

Did the student complete the job in the time allotted? Yes ___ No ___

If No, which steps were not completed? _____

How would you rate this student's overall performance on this job? _____

5–Excellent, 4–Good, 3–Satisfactory, 2–Unsatisfactory, 1–Poor

Comments: _____

_____

_____

INSTRUCTOR'S SIGNATURE _____

Project

Diagnosing and Servicing Emissions Control Systems

48

Introduction

This project covers diagnosis and service of the exhaust gas recirculation (EGR), evaporative emissions, and exhaust treatment (air injection and catalytic converter) systems. Any of the three systems can develop problems that affect emissions, engine performance, and driveability. Some of these systems, such as the air injection system and catalytic converter, work together to clean the exhaust gases, and a problem in one can cause a malfunction in the other. A defect in the EGR system can cause an apparently unrelated problem such as detonation or missing. As with the other two types of emissions control systems covered in this project, an inoperative PCV or evaporative emissions system can cause driveability and emissions problems. Lack of crankcase ventilation can also damage the engine.

You will test and service an EGR valve in Job 175 and an air pump and air injection system in Job 176. In Job 177, you will inspect, test, and replace a catalytic converter. Finally, you will inspect and service positive crankcase ventilation and evaporative emissions control systems in Job 178.

Project 48 Jobs

- Job 175—Test and Service an EGR Valve
- Job 176—Test and Service an Air Pump and Air Injection System
- Job 177—Test and Service a Catalytic Converter
- Job 178—Test and Service Positive Crankcase Ventilation (PCV) and Evaporative Emissions Control Systems

Tools and Materials

The following list contains the tools and materials that may be needed to complete the jobs in this project. The items used will depend on the make and model of vehicle being serviced.

- One or more vehicles requiring emissions service.
- Applicable service information.
- Scan tool.
- Rubber hammer.
- Pressure gauge.
- Smoke generating machine.
- Infrared thermometer.
- Drill motor and cleaning attachments.
- Hand tools as needed.
- Air-operated tools as needed.
- Safety glasses and other protective equipment as needed.

Safety Notice

Before performing this job, review all pertinent safety information in the text, and review safety information with your instructor.

Job 175—Test and Service an EGR Valve

After completing this job, you will be able to test, clean, and remove and replace an EGR valve.

Procedures

☐ 1. Obtain a vehicle to be used in this job. Your instructor may direct you to perform this job on a shop vehicle.

☐ 2. Gather the tools needed to perform this job. Refer to the tools and materials list at the beginning of the project.

☐ 3. Obtain the proper service information for the vehicle being serviced.

☐ 4. Locate the EGR valve.

Where is the EGR valve located on the engine?_____

_____

Is the EGR valve vacuum-operated or solenoid operated? See **Figure 175-1**. _____

_____

Describe the location of EGR control devices:_____

☐ 5. Check EGR valve operation by one of the following procedures, depending on the type of EGR valve being tested.

Vacuum-Operated EGR Valve:
• Observe that the valve moves when the engine is accelerated.

or
• Apply vacuum to the diaphragm and note valve movement. Check any vacuum control solenoids or thermostatic EGR controls, especially if the EGR valve itself appears to be working properly.

Solenoid-Operated EGR Valve:
• Attach a scan tool and check EGR solenoid and position sensor operation. If possible, open and close the solenoid using the scan tool.

or
• Use a test light or multimeter to check EGR power supply and solenoid operation.

Does the valve operate properly? Yes ___ No ___

☐ 6. Determine the EGR valve and control system condition from the tests made in step 5.

Is the EGR valve operating properly? Yes ___ No ___

If No, what could be causing the problem? _____

_____

☐ 7. Remove the EGR valve from the engine. See the *Remove and Replace an EGR Valve* section of this job.

☐ 8. Visually inspect the EGR passages.

Are any passages blocked? Yes ___ No ___

If Yes, the passages must be cleaned using the procedures in the *Clean an EGR Valve and Passages* section of this job.

Figure 175-1. Vacuum-operated and solenoid-operated EGR valves can be distinguished by whether they have vacuum or electrical connections. A—Solenoid-operated EGR valve. B—Vacuum-operated EGR valve.

A

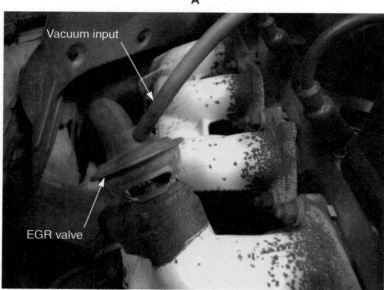

B

> **Note**
>
> One or more clogged EGR passages can cause misfiring by diverting excess exhaust gases to one or two cylinders.

- [ ] 9. Inspect EGR transfer tube or tubes, if used. Check for obvious cracks or breaks in the EGR tube or evidence of exhaust leakage.
- [ ] 10. After consulting your instructor, make repairs as needed and recheck system operation.

Remove and Replace an EGR Valve

- [ ] 1. Remove the EGR valve electrical connector or vacuum line as applicable.
- [ ] 2. Remove the fasteners holding the valve to the engine.

Job 175—Test and Service an EGR Valve (continued)

☐   3.   Compare the old and new valves to ensure that they match and transfer any related components such as EGR position sensors, to the new valve.

☐   4.   Clean the EGR valve mounting surfaces on the engine.

☐   5.   Install the new EGR valve and tighten the fasteners.

☐   6.   Reinstall the electrical connector and/or vacuum line.

☐   7.   Start the engine and test EGR valve operation.

Clean an EGR Valve and Passages

☐   1.   Remove the EGR valve as described in the previous section.

☐   2.   If necessary, clean the valve passages with a wire brush. You can use solvent, but it usually has no effect on hardened carbon deposits. Some EGR valves can be partially disassembled for thorough cleaning, **Figure 175-2**. Use one of the following techniques, depending on the condition of the passages.

Caution

   Do not allow solvent to contact any rubber or electrical parts on the EGR valve.

- **If the passage is not entirely clogged,** attach a circular wire brush or a length of old speedometer cable or stranded wire to an electric drill. Run the brush or wire into the passage while operating the drill at low speed. It is not necessary to remove every fragment of soot. Some soot formation is normal, however, all major deposits should be removed. All loose deposits should be removed from the EGR passages and surrounding area. Do not allow the debris to enter the intake manifold.
- **If the passage is completely clogged,** create an opening by hand turning a drill bit into the passage. Once an opening has been created, finish cleaning the passage as explained in the previous paragraph.

Figure 175-2. On some EGR valves, the pintle seat can be removed, allowing a more thorough cleaning.

Pintle seat

☐ 3. Reinstall the EGR valve.

☐ 4. Start the engine and check EGR and engine operation.

☐ 5. Clean the work area and return any equipment to storage.

☐ 6. Did you encounter any problems during this procedure? Yes ___ No ___

If Yes, describe the problems: _____

_____

What did you do to correct the problems? _____

_____

☐ 7. Have your instructor check your work and sign this job sheet.

Performance Evaluation—Instructor Use Only

Did the student complete the job in the time allotted? Yes ___ No ___

If No, which steps were not completed?_____

How would you rate this student's overall performance on this job?_____

5–Excellent, 4–Good, 3–Satisfactory, 2–Unsatisfactory, 1–Poor

Comments: _____

_____

_____

INSTRUCTOR'S SIGNATURE _____

Job 176—Test and Service an Air Pump and Air Injection System

After completing this job, you will be able to test an air injection system and replace faulty components.

Procedures

- [ ] 1. Obtain a vehicle to be used in this job. Your instructor may direct you to perform this job on a shop vehicle.
- [ ] 2. Gather the tools needed to perform this job. Refer to the tools and materials list at the beginning of the project.
- [ ] 3. Obtain the proper service information for the vehicle being serviced.
- [ ] 4. If the air injection system is electronically controlled, use a scan tool to obtain any air injection system trouble codes.

    Are any codes present? Yes ___ No ___

    If Yes, describe: _____
- [ ] 5. If codes are present, proceed to make further electrical tests as necessary. If no codes are present, proceed to the next section.

Test an Air Pump and Air Injection System Valve Operation

> **Note**
> The air injection systems on most newer vehicles are controlled by the on-board computer and can be checked with a scan tool. Not all systems use the parts described below. Many newer vehicles no longer use an air injection system.

- [ ] 1. Open the vehicle hood and visually check for damaged or disconnected hoses and a missing or loose drive belt.
- [ ] 2. Start the engine. The engine should be cold if possible.

> **Note**
> Perform steps 3 and 4 quickly, before the engine reaches normal operating temperature.

- [ ] 3. Observe the pump output with the engine running.
- [ ] 4. Observe the operation of the switching valve as the engine warms up.
- [ ] 5. Place your hand near the diverter valve outlet as an assistant opens and closes the throttle valve, **Figure 176-1**. You should feel air blowing out of the outlet as the engine is quickly accelerated and decelerated. This tells you that the diverter valve is operating properly.
- [ ] 6. Check the condition of the check valve(s) by removing the inlet hose and starting the engine. There should be little or no exhaust flow from the check valve.

> **Note**
> The check valve can also be tested by removing it from the engine and attempting to blow through it. If you can blow through the inlet side but not the outlet side, the valve is OK. If you can blow through both sides, the valve is not sealing against exhaust gases and should be replaced.

Figure 176-1. Hold your hand near the diverter valve outlet and feel for exhausting air as the throttle is opened and then closed.

☐ 7. From the information gathered in steps 3 through 6, determine whether the air injection system is working properly.

Is the system working properly? Yes ___ No ___

If No, what could be the problem? _____

☐ 8. After consulting your instructor, make repairs as needed and recheck system operation.

Replace an Air Injection Pump

☐ 1. Remove the pump drive belt.

☐ 2. Remove any hoses attached to the pump.

☐ 3. Remove the fasteners holding the pump to the engine and remove the pump.

☐ 4. Compare the old and new pumps.

☐ 5. Transfer parts from the old pump to the new pump as needed.

☐ 6. Place the new pump in position.

☐ 7. Install the pump fasteners.

☐ 8. Install the hoses.

☐ 9. Reinstall and tighten the pump drive belt.

☐ 10. Start the engine and check pump operation.

Replace a Switching or Diverter Valve

☐ 1. Remove any hoses attached to the valve.

☐ 2. If necessary, remove the fasteners holding the valve to the pump.

☐ 3. Compare the old and new valves.

☐ 4. Place the new valve in position.

☐ 5. Install the valve fasteners.

☐ 6. Install the hoses.

☐ 7. Start the engine and check the valve and system operation.

Job 176—Test and Service an Air Pump and Air Injection System (continued)

Replace a Check Valve

☐ 1. Remove the hose from the inlet side of the check valve.

☐ 2. Loosen the fitting holding the check valve to the injector tube assembly.

> **Note**
>
> Use two wrenches to avoid damaging the injector tube assembly. See **Figure 176-2**.

☐ 3. Remove the check valve from the injector tube assembly.

☐ 4. Compare the old and new check valves to ensure that the fittings are correct, **Figure 176-3**.

☐ 5. Install the replacement check valve on the injector tube assembly, **Figure 176-4**.

☐ 6. Reinstall the hose on the check valve and tighten the hose clamp if necessary.

Replace an Injector Tube Assembly

☐ 1. Allow the engine to cool.

☐ 2. Remove the hose from the check valve at the injector tube assembly.

☐ 3. Remove the fittings holding the injector tube assembly to the exhaust manifold.

☐ 4. Carefully pull the injector tube assembly out of the exhaust manifold.

☐ 5. Remove the check valve from the injector tube assembly.

☐ 6. Install the check valve on the replacement injector tube assembly.

☐ 7. Place the injector tube assembly in position on the exhaust manifold.

☐ 8. Tighten the injector tube fittings to the exhaust manifold.

☐ 9. Reinstall the hose on the check valve.

Figure 176-2. Loosen the check valve using two wrenches to avoid damaging the tubing.

Figure 176-3. Compare the old and replacement check valves. The replacement may not look exactly the same as the original, but it should fit and function like the original.

Figure 176-4. Carefully thread the new check valve onto the discharge tubing.

☐ 10. Clean the work area and return any equipment to storage.

☐ 11. Did you encounter any problems during this procedure? Yes ___ No ___
If Yes, describe the problems: _____

_____

What did you do to correct the problems? _____

_____

☐ 12. Have your instructor check your work and sign this job sheet.

Job 176—Test and Service an Air Pump and Air Injection System (continued)

Performance Evaluation—Instructor Use Only

Did the student complete the job in the time allotted? Yes ___ No ___

If No, which steps were not completed?_____

How would you rate this student's overall performance on this job?_____

5–Excellent, 4–Good, 3–Satisfactory, 2–Unsatisfactory, 1–Poor

Comments: _____

_____

_____

INSTRUCTOR'S SIGNATURE _____

Notes

Job 177—Test and Service a Catalytic Converter

After completing this job, you will be able to test a catalytic converter for physical damage and correct operation. You will be able to replace a faulty catalytic converter.

Procedures

☐ 1. Obtain a vehicle to be used in this job. Your instructor may direct you to perform this job on a shop vehicle.

☐ 2. Gather the tools needed to perform this job. Refer to the tools and materials list at the beginning of the project.

☐ 3. Obtain the proper service information for the vehicle being serviced.

Test Catalytic Converter Operation

Note

Catalytic converter operation can be tested with a 4- or 5-gas analyzer. Excessive CO may indicate a defective converter. Note also that a defect in the air injection system can cause improper catalytic converter operation.

In addition to the tests described in this job, a backpressure test can also be performed to check for a clogged catalytic converter. Refer to Job 173 for procedures for performing a backpressure test.

Mechanical Damage Test

☐ 1. Ensure that the catalytic converter is cool.

☐ 2. Strike the converter housing with a rubber hammer.

Does the converter rattle when it is pounded? Yes ___ No ___

If Yes, the converter has internal damage, such as loose catalyst, screens, or baffles. Refer to the *Replace a Catalytic Converter* section in this job.

Scan Tool Test

Note

The following procedure can be performed only on OBD II–equipped vehicles. The scan tool test may uncover other engine and emission control problems.

☐ 1. Obtain a scan tool that is compatible with the vehicle being tested.

☐ 2. Attach a scan tool to the vehicle diagnostic connector.

☐ 3. Retrieve any trouble codes that are set.

☐ 4. Operate the engine and observe the waveform readings from the oxygen sensors upstream and downstream of the converter.

☐ 5. Determine whether the oxygen sensor inputs indicate that the converter is operating properly.

Is the converter working properly? Yes ___ No ___

Explain: _____

What could be the cause? _____

_____

_____

Temperature Test

☐ 1. Obtain an infrared thermometer.

Type of thermometer: _____

☐ 2. Run the vehicle at a high engine speed (about 2500 rpm) for two minutes. This needs to be done to ensure that the catalytic converter has had time to begin operating.

> **Note**
>
> If the engine is started cold, a longer run time may be needed. Leave the engine running during steps 3 and 4.

☐ 3. Quickly aim the thermometer at the front of the converter and note the temperature. See **Figure 177-1**.

Temperature:_____ °F or °C (circle one)

☐ 4. Quickly aim the thermometer at the rear of the converter and note temperature. See **Figure 177-2**.

Temperature _____ °F or °C (circle one)

Figure 177-1. Aim the thermometer at the front of the converter body. Be sure to aim at the body of the converter, not any attached heat shields.

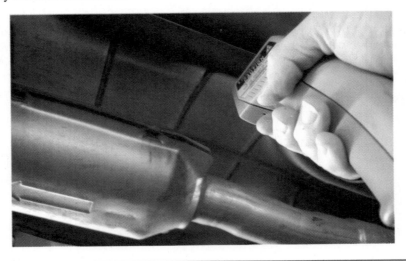

Figure 177-2. Aim the thermometer at the rear of the catalytic converter. Be sure to aim at the body of the converter, not any attached heat shields.

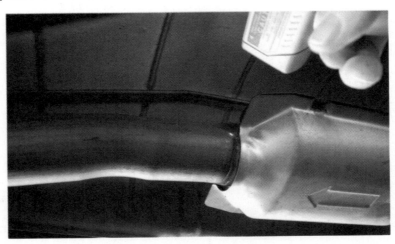

Job 177—Test and Service a Catalytic Converter (continued)

☐ 5. Turn off the engine and compare the readings taken in steps 3 and 4.

Is there an substantial difference between the two temperatures? Yes ___ No ___

Is the rear temperature greater? Yes ___ No ___

If the answer to either question is No, the converter is not working properly.

☐ 6. Consult the manufacturer's service information for exact temperature specifications. If specifications are not available use the following general guide:

- The converter must reach 500–600°F (260–315°C) to begin working. If converter temperatures are below this when the engine is fully warmed up, the catalyst is coated or is worn out.
- Normal converter operating temperatures are 750–1200°F (400–650°C). A temperature higher than this indicates that the converter is overheating, usually because of a rich fuel mixture. At temperatures above 2000°F (1100°C), the converter catalyst is in danger of melting.
- There should be a temperature increase of 40–100°F (5–37°C) between the front and rear of the converter. No increase or a small increase indicates that the catalyst is coated or is worn out.

Replace a Catalytic Converter

☐ 1. Raise the vehicle.

> **Warning**
> ⚠ Raise and support the vehicle in a safe manner. Always use approved lifts or jacks and jack stands.

☐ 2. Remove any heat shields or skid plates that prevent access to the converter.

☐ 3. Remove the air inlet tube from the converter.

☐ 4. Remove any clamps and bracket fasteners.

> **Note**
> 📄 Carefully handle any pipes or other parts that will be reused.

☐ 5. Remove the flange fasteners and separate the flanges.

☐ 6. Remove the converter from the vehicle.

☐ 7. Compare the old and new converters.

Do the converters match? Yes ___ No ___

If No, explain: _____

_____

_____

☐ 8. Scrape old gasket material from flanged connectors that will be reused.

☐ 9. Install and tighten the flange connections, making sure that gaskets are used where necessary.

> **Caution**
>
> ⚠ Make sure that all parts are installed in the correct position and facing in the proper direction. Make sure that all exhaust system parts are aligned with each other and do not contact any other part of the vehicle.

☐ 10. Install and tighten all clamps and brackets.

☐ 11. Install the air inlet tube.

☐ 12. Start the engine and check for exhaust leaks.

☐ 13. Clean the work area and return any equipment to storage.

☐ 14. Did you encounter any problems during this procedure? Yes ___ No ___

If Yes, describe the problems: _____

_____

What did you do to correct the problems? _____

_____

☐ 15. Have your instructor check your work and sign this job sheet.

Performance Evaluation—Instructor Use Only

Did the student complete the job in the time allotted? Yes ___ No ___

If No, which steps were not completed?_____

How would you rate this student's overall performance on this job?_____

5–Excellent, 4–Good, 3–Satisfactory, 2–Unsatisfactory, 1–Poor

Comments: _____

_____

INSTRUCTOR'S SIGNATURE _____

Job 178—Test and Service Positive Crankcase Ventilation (PCV) and Evaporative Emissions Control Systems

After completing this job, you will be able to service PCV and evaporative emissions systems.

Related Trouble Codes

Trouble codes are not specifically addressed in this job. For background information, some common OBD II evaporative emissions trouble codes are listed below. There are usually no trouble codes associated with the PCV system.

P0440	Evaporative emissions system (general).
P0442	Small leak detected.
P0446	System performance.
P0455	Large leak detected.
P0496	Flow during non-purge.
P1441	Flow during non-purge.

Procedures

☐ 1. Obtain a vehicle to be used in this job. Your instructor may direct you to perform this job on a shop vehicle.

☐ 2. Gather the tools needed to perform this job. Refer to the tools and materials list at the beginning of the project.

☐ 3. Obtain the proper service information for the vehicle being serviced.

Test and Service a PCV System

☐ 1. Locate the PCV valve on the engine. Most are located on the valve cover, **Figure 178-1**, while a few are located on the valley cover under the intake manifold.

Describe the location of the PCV valve:_____

_____

☐ 2. Visually inspect the PCV system hoses. Look for hoses that are disconnected, leaking, or collapsed.

Were any hose problems found? Yes ___ No ___

If Yes, what was done to correct them? _____

_____

☐ 3. Check PCV system condition by using one or more of the following procedures:
   • Use a tester to ensure that air is flowing into the crankcase when the engine is running.
   Did the PCV valve pass the test? Yes ___ No ___
   • Remove the PCV valve from its grommet with the engine idling.
   Does the engine speed up, even slightly? Yes ___ No ___

Note

Some late model PCV valves are attached to the valve cover with a cam lock. A special tool is needed to remove and install the cam lock.

Figure 178-1. Most PCV valves are installed on the valve cover. A few are mounted on the valley cover of V-type engines. Their removal may require that other components be removed. Note that the PCV shown here has been removed from its grommet.

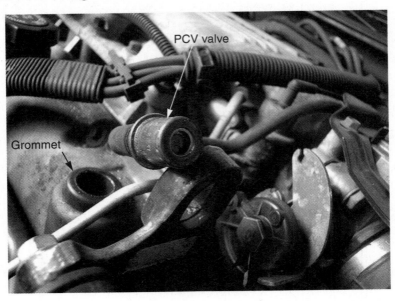

- Remove the PCV valve from its grommet. Place your finger over the PCV valve with the engine running.
  Is vacuum present? Yes ___ No ___
- Stop the engine with the PCV valve removed from its grommet. Shake the PCV valve.
  Does the PCV valve rattle when shaken? Yes ___ No ___

If the answers to any of the preceding questions is No, what could be the problem? _____

_____

_____

_____

What should be done to correct the problem? _____

_____

_____

Test and Service an Evaporative Emissions System

> **Note**
>
> The year and type of test vehicle and the available test equipment will determine which of the following tests can be performed. Consult with your instructor to determine which tests will be performed.

1. Visually inspect the evaporative emissions system hoses. Look for disconnected, leaking, or collapsed hoses.

   Did you find any hose problems? Yes ___ No ___

   If Yes, what was done to correct them? _____

   _____

> **Warning**
>
> Escaping fuel vapors are extremely flammable. Keep the test area well ventilated and make sure that cigarettes, sparking devices, or any other ignition sources are kept well away from the area.

Job 178—Test and Service Positive Crankcase Ventilation (PCV) and Evaporative Emissions Control Systems (continued)

Test Purge Air Flow with Flow Meter

☐ 1. Attach a purge flow meter to the evaporative emissions system according to the flow meter manufacturer's instructions.

☐ 2. Start the vehicle.

☐ 3. Observe the purge air flow with the engine running. Flow should be at least one liter (slightly more than a quart) during the test. Most vehicles with an operating evaporative emissions system will have much more flow, often as much as 25 liters (28 quarts) during the test.

☐ 4. Turn off the engine.

Pressure Test the Evaporative Emissions System

☐ 1. Locate the evaporative emissions system pressure test fitting on the vehicle. See **Figure 178-2**.

 Note

On some vehicles, the scan tool can be used to operate the fuel pump to pressurize the fuel tank. On other vehicles, a special pump is used to create a vacuum on the evaporative emissions system. Consult the proper service manuals for exact procedures.

☐ 2. Attach a nitrogen pressure source and a pressure gauge to the pressure test fitting.

☐ 3. Open the nitrogen tank valve and pressurize the evaporative emissions system to the required pressure.

Figure 178-2. Most evaporative emissions system pressure test fittings will have a green protective cap. It may have an identification and caution label as shown here.

4. Close the nitrogen tank valve and wait the specified time while observing the pressure gauges.

Does the pressure drop? Yes ___ No ___

If Yes, locate the system leak and consult your instructor for the corrective action to be taken.

Locate Evaporative Emissions System Leaks with a Smoke Generating Machine

1. Make sure that the gas tank is between 1/4 and 3/4 full.
2. Use a scan tool to set the evaporative system purge and vent valves to the closed position to seal the evaporative emissions system.
3. Connect the shop air or nitrogen source to the smoke machine.
4. Set the smoke generating machine flow gauge so that the pointer and ball are aligned.
5. Install the smoke generating machine output hose on the appropriate fitting of the evaporative emissions system. Use adapters as needed to ensure a leak proof seal.
6. Start the flow of smoke into the evaporative emissions system and remove the gas cap until smoke begins to exit from the filler neck.
7. Reinstall the gas cap.

Note
Steps 8 and 9 apply to a common type of smoke generation machine. If another type of machine is used, follow the manufacturer's instructions instead.

8. Note the position of the flow gauge ball.

If the ball has moved upward, a leak is present. Go to step 9.

If the ball does not move, or moves downward, the system is not leaking. Go to step 10.

9. If the ball movement indicates a leak, carefully observe the entire evaporative emissions system for signs of smoke. Areas to concentrate on include the gas cap, tank fittings, charcoal canister, and all hose fittings.

Caution
Use a caged droplight or other safe light source to illuminate suspected areas if necessary.

Was smoke observed at a potential leak point? Yes ___ No ___

If Yes, consult your instructor about what steps to take.

If No, recheck the smoke machine connections to determine the actual cause of the pressure drop.

10. Use the scan tool to return the evaporative system valves to their normal position. If no leaks are suspected in the evaporative emissions system, skip to step 13.
11. If approved by your instructor, correct the leak.

Briefly explain what you did to make repairs. _____

_____

_____

_____

Job 178—Test and Service Positive Crankcase Ventilation (PCV) and Evaporative Emissions Control Systems (continued)

☐ 12. Recheck the evaporative system for leaks.

Were any leaks found? Yes ___ No ___

If Yes, recheck the system for additional leaks and retest.

☐ 13. Clean the work area and return any equipment to storage.

☐ 14. Did you encounter any problems during this procedure? Yes ___ No ___

If Yes, describe the problems: _____

_____

What did you do to correct the problems? _____

_____

☐ 15. Have your instructor check your work and sign this job sheet.

Performance Evaluation—Instructor Use Only

Did the student complete the job in the time allotted? Yes ___ No ___

If No, which steps were not completed?_____

How would you rate this student's overall performance on this job?_____

5–Excellent, 4–Good, 3–Satisfactory, 2–Unsatisfactory, 1–Poor

Comments: _____

_____

_____

INSTRUCTOR'S SIGNATURE _____

Notes

NATEF Correlation Chart
Maintenance and Light Repair (MLR)

The following chart correlates the jobs in this *Shop Manual* to the 2013 NATEF Maintenance and Light Repair (MLR) Task List.

ENGINE REPAIR

For every task in this list, the following safety requirement must be strictly enforced:
- Comply with personal and environmental safety practices associated with clothing; eye protection; hand tools; power equipment; proper ventilation; and the handling, storage, and disposal of chemicals/materials in accordance with local, state, and federal safety and environmental regulations.

I. ENGINE REPAIR
A. General

Task Number and Description	Priority	Job #s
1. Research applicable vehicle and service information, vehicle service history, service precautions, and technical service bulletins.	P-1	2
2. Verify operation of the instrument panel engine warning indicators.	P-1	11, 138, 139, 140, 141
3. Inspect engine assembly for fuel, oil, coolant, and other leaks; determine necessary action.	P-1	7
4. Install engine covers using gaskets, seals, and sealers as required.	P-1	3, 4, 27, 28, 34
5. Remove and replace timing belt; verify correct camshaft timing.	P-1	13, 28
6. Perform common fastener and thread repair, to include: remove broken bolt, restore internal and external threads, and repair internal threads with thread insert.	P-1	15
7. Identify hybrid vehicle internal combustion engine service precautions.	P-3	12, 13

I. ENGINE REPAIR
B. Cylinder Head and Valve Train

Task Number and Description	Priority	Job #s
1. Adjust valves (mechanical or hydraulic lifters).	P-1	25

I. ENGINE REPAIR
C. Lubrication and Cooling Systems

Task Number and Description	Priority	Job #s
1. Perform cooling system pressure and dye tests to identify leaks; check coolant condition and level; inspect and test radiator, pressure cap, coolant recovery tank, and heater core and galley plugs; determine necessary action.	P-1	31
2. Inspect, replace, and adjust drive belts, tensioners, and pulleys; check pulley and belt alignment.	P-1	31, 33
3. Remove, inspect, and replace thermostat and gasket/seal.	P-1	36
4. Inspect and test coolant; drain and recover coolant; flush and refill cooling system with recommended coolant; bleed air as required.	P-1	31, 32
5. Perform engine oil and filter change.	P-1	6

AUTOMATIC TRANSMISSION AND TRANSAXLE

II. AUTOMATIC TRANSMISSION AND TRANSAXLE
A. General

Task Number and Description	Priority	Job #s
1. Research applicable vehicle and service information, fluid type, vehicle service history, service precautions, and technical service bulletins.	P-1	2
2. Check fluid level in a transmission or a transaxle equipped with a dipstick.	P-1	39, 44
3. Check fluid level in a transmission or a transaxle not equipped with a dipstick.	P-1	39, 44
4. Check transmission fluid condition; check for leaks.	P-2	39, 44

II. AUTOMATIC TRANSMISSION AND TRANSAXLE
B. In-Vehicle Transmission/Transaxle

Task Number and Description	Priority	Job #s
1. Inspect, adjust, and replace external manual valve shift linkage, transmission range sensor/switch, and park/neutral position switch.	P-2	39, 45
2. Inspect for leakage at external seals, gaskets, and bushings.	P-2	3, 4
3. Inspect, replace, and align power train mounts.	P-2	37, 38
4. Drain and replace fluid and filter(s).	P-1	44

II. AUTOMATIC TRANSMISSION AND TRANSAXLE
C. Off-Vehicle Transmission and Transaxle

Task Number and Description	Priority	Job #s
1. Describe the operational characteristics of a continuously variable transmission (CVT).	P-3	39
2. Describe the operational characteristics of a hybrid vehicle drive train.	P-3	39

MANUAL DRIVE TRAIN AND AXLES

III. MANUAL DRIVE TRAIN AND AXLES
A. General

Task Number and Description	Priority	Job #s
1. Research applicable vehicle and service information, fluid type, vehicle service history, service precautions, and technical service bulletins.	P-1	2
2. Drain and refill manual transmission/transaxle and final drive unit.	P-1	77, 78, 79, 80, 86, 89
3. Check fluid condition; check for leaks.	P-2	60, 65

III. MANUAL DRIVE TRAIN AND AXLES (continued)
B. Clutch

Task Number and Description	Priority	Job #s
1. Check and adjust clutch master cylinder fluid level.	P-1	71, 73
2. Check for system leaks.	P-1	71, 73

III. MANUAL DRIVE TRAIN AND AXLES
C. Transmission/Transaxle

Task Number and Description	Priority	Job #s
1. Describe the operational characteristics of an electronically-controlled manual transmission/transaxle.	P-3	69

III. MANUAL DRIVE TRAIN AND AXLES
D. Drive Shaft, Half Shafts, Universal and Constant-Velocity (CV) Joints

Task Number and Description	Priority	Job #s
1. Inspect, remove, and replace front-wheel drive (FWD) bearings, hubs, and seals.	P-2	68
2. Inspect, service, and replace shafts, yokes, boots, and universal/ CV joints.	P-2	66, 67

III. MANUAL DRIVE TRAIN AND AXLES
E. Differential Case Assembly

Task Number and Description	Priority	Job #s
1. Clean and inspect differential housing; check for leaks; inspect housing vent.	P-2	7, 65, 87
2. Check and adjust differential housing fluid level.	P-1	60
3. Drain and refill differential housing.	P-1	86, 89
E.1 Drive Axles		
1. Inspect and replace drive axle wheel studs.	P-2	108

III. MANUAL DRIVE TRAIN AND AXLES
F. Four-wheel Drive/All-wheel Drive

Task Number and Description	Priority	Job #s
1. Inspect front-wheel bearings and locking hubs.	P-3	67, 68, 83, 92
2. Check for leaks at drive assembly seals; check vents; check lube level.	P-2	65, 81

IV. SUSPENSION AND STEERING SYSTEMS
A. General

Task Number and Description	Priority	Job #s
1. Research applicable vehicle and service information, vehicle service history, service precautions, and technical service bulletins.	P-1	2
2. Disable and enable supplemental restraint system (SRS).	P-1	94, 147

IV. SUSPENSION AND STEERING
B. Related Suspension and Steering Service

Task Number and Description	Priority	Job #s
1. Inspect rack and pinion steering gear inner tie rod ends (sockets) and bellows boots.	P-1	95
2. Determine proper power steering fluid type; inspect fluid level and condition.	P-1	91, 97
3. Flush, fill, and bleed power steering system.	P-2	97
4. Inspect for power steering fluid leakage; determine necessary action.	P-1	97
5. Remove, inspect, replace, and adjust power steering pump drive belt.	P-1	33, 97
6. Inspect and replace power steering hoses and fittings.	P-2	90, 97
7. Inspect pitman arm, relay (center link/intermediate) rod, idler arm and mountings, and steering linkage damper.	P-1	90, 94, 96
8. Inspect tie rod ends (sockets), tie rod sleeves, and clamps.	P-1	90
9. Inspect upper and lower control arms, bushings, and shafts.	P-1	90
10. Inspect and replace rebound and jounce bumpers.	P-1	90, 99
11. Inspect track bar, strut rods/radius arms, and related mounts and bushings.	P-1	90
12. Inspect upper and lower ball joints (with or without wear indicators).	P-1	90
13. Inspect suspension system coil springs and spring insulators (silencers).	P-1	90
14. Inspect suspension system torsion bars and mounts.	P-1	90
15. Inspect and replace front stabilizer bar (sway bar) bushings, brackets, and links.	P-1	90, 98
16. Inspect strut cartridge or assembly.	P-1	90, 101
17. Inspect front strut bearing and mount.	P-1	90, 101
18. Inspect rear suspension system lateral links/arms (track bars), control (trailing) arms.	P-1	90, 99
19. Inspect rear suspension system leaf spring(s), spring insulators (silencers), shackles, brackets, bushings, center pins/bolts, and mounts.	P-1	90, 100

IV. SUSPENSION AND STEERING (continued)
B. Related Suspension and Steering Service

Task Number and Description	Priority	Job #s
20. Inspect, remove, and replace shock absorbers; inspect mounts and bushings.	P-1	90, 102
21. Inspect electric power-assisted steering.	P-3	97
22. Identify hybrid vehicle power steering system electrical circuits and safety precautions.	P-2	93
23. Describe the function of the power steering pressure switch.	P-3	97

IV. SUSPENSION AND STEERING
C. Wheel Alignment

Task Number and Description	Priority	Job #s
1. Perform prealignment inspection and measure vehicle ride height; determine necessary action.	P-1	90, 103

IV. SUSPENSION AND STEERING
D. Wheels and Tires

Task Number and Description	Priority	Job #s
1. Inspect tire condition; identify tire wear patterns; check for correct size and application (load and speed ratings) and adjust air pressure; determine necessary action.	P-1	81, 104
2. Rotate tires according to manufacturer's recommendations.	P-1	106
3. Dismount, inspect, and remount tire on wheel; balance wheel and tire assembly (static and dynamic).	P-1	105, 106
4. Dismount, inspect, and remount tire on wheel equipped with tire pressure monitoring system sensor.	P-2	105, 106
5. Inspect tire and wheel assembly for air loss; perform necessary action.	P-1	104
6. Repair tire using internal patch.	P-1	105
7. Identify and test tire pressure monitoring systems (indirect and direct) for operation; verify operation of instrument panel lamps.	P-2	105
8. Demonstrate knowledge of steps required to remove and replace sensors in a tire pressure monitoring system.	P-2	105

V. BRAKES
A. General

Task Number and Description	Priority	Job #s
1. Research applicable vehicle and service information, vehicle service history, service precautions, and technical service bulletins.	P-1	2
2. Describe procedure for performing a road test to check brake system operation, including an anti-lock brake system (ABS).	P-1	107
3. Install wheel and torque lug nuts.	P-1	105, 116, 118

V. BRAKES
B. Hydraulic System

Task Number and Description	Priority	Job #s
1. Measure brake pedal height, travel, and free play (as applicable); determine necessary action.	P-1	107
2. Check master cylinder for external leaks and proper operation.	P-1	107
3. Inspect brake lines, flexible hoses, and fittings for leaks, dents, kinks, rust, cracks, bulging, wear, loose fittings and supports; determine necessary action.	P-1	107, 111
4. Select, handle, store, and fill brake fluids to proper level.	P-1	113, 115
5. Identify components of brake warning light system.	P-3	111, 136, 138
6. Bleed and/or flush brake system.	P-1	115
7. Test brake fluid for contamination.	P-1	115

V. BRAKES
C. Drum Brakes

Task Number and Description	Priority	Job #s
1. Remove, clean, inspect, and measure brake drum diameter; determine necessary action.	P-1	107, 116
2. Refinish brake drum and measure final drum diameter; compare with specifications.	P-1	117
3. Remove, clean, and inspect brake shoes, springs, pins, clips, levers, adjusters/self-adjusters, other related brake hardware, and backing support plates; lubricate and reassemble.	P-1	116
4. Inspect wheel cylinders for leaks and proper operation; remove and replace as needed.	P-2	116
5. Pre-adjust brake shoes and parking brake; install brake drums or drum/hub assemblies and wheel bearings; make final checks and adjustments.	P-2	116

V. BRAKES (continued)
D. Disc Brakes

Task Number and Description	Priority	Job #s
1. Remove and clean caliper assembly; inspect for leaks and damage/wear to caliper housing; determine necessary action.	P-1	107, 118
2. Clean and inspect caliper mounting and slides/pins for proper operation, wear, and damage; determine necessary action.	P-1	118
3. Remove, inspect, and replace pads and retaining hardware; determine necessary action.	P-1	118
4. Lubricate and reinstall caliper, pads, and related hardware; seat pads and inspect for leaks.	P-1	118
5. Clean and inspect rotor, measure rotor thickness, thickness variation, and lateral runout; determine necessary action.	P-1	118
6. Remove and reinstall rotor.	P-1	118
7. Refinish rotor on vehicle; measure final rotor thickness and compare with specifications.	P-1	120
8. Refinish rotor off vehicle; measure final rotor thickness and compare with specifications.	P-1	120
9. Retract and re-adjust caliper piston on an integral parking brake system.	P-3	110
10. Check brake pad wear indicator; determine necessary action.	P-2	118
11. Describe importance of operating vehicle to burnish/break-in replacement brake pads according to manufacturer's recommendations.	P-1	118

V. BRAKES
E. Power-Assist Units

Task Number and Description	Priority	Job #s
1. Check brake pedal travel with, and without, engine running to verify proper power booster operation.	P-2	114
2. Check vacuum supply (manifold or auxiliary pump) to vacuum-type power booster.	P-1	114

V. BRAKES
F. Miscellaneous (Wheel Bearings, Parking Brakes, Electrical, Etc.)

Task Number and Description	Priority	Job #s
1. Remove, clean, inspect, repack, and install wheel bearings; replace seals; install hub and adjust bearings.	P-1	5, 109
2. Check parking brake cables and components for wear, binding, and corrosion; clean, lubricate, adjust or replace as needed.	P-2	110
3. Check parking brake operation and parking brake indicator light system operation; determine necessary action.	P-1	110

V. BRAKES (continued)
F. Miscellaneous (Wheel Bearings, Parking Brakes, Electrical, Etc.)

Task Number and Description	Priority	Job #s
4. Check operation of brake stop light system.	P-1	136
5. Replace wheel bearing and race.	P-2	5, 109
6. Inspect and replace wheel studs.	P-1	108

V. BRAKES
G. Electronic Brakes, and Traction and Stability Control Systems

Task Number and Description	Priority	Job #s
1. Identify traction control/vehicle stability control system components.	P-3	121
2. Describe the operation of a regenerative braking system.	P-3	111

ELECTRICAL/ELECTRONIC SYSTEMS

VI. ELECTRICAL/ELECTRONIC SYSTEMS
A. General

Task Number and Description	Priority	Job #s
1. Research applicable vehicle and service information, vehicle service history, service precautions, and technical service bulletins.	P-1	2
2. Demonstrate knowledge of electrical/electronic series, parallel, and series-parallel circuits using principles of electricity (Ohm's Law).	P-1	136
3. Use wiring diagrams to trace electrical/electronic circuits.	P-1	130, 136, 138
4. Demonstrate proper use of a digital multimeter (DMM) when measuring source voltage, voltage drop (including grounds), current flow, and resistance.	P-1	130, 136, 142
5. Demonstrate knowledge of the causes and effects from shorts, grounds, opens, and resistance problems in electrical/electronic circuits.	P-2	130, 138, 142
6. Check operation of electrical circuits with a test light.	P-2	130, 136
7. Check operation of electrical circuits with fused jumper wires.	P-2	125, 138
8. Measure key-off battery drain (parasitic draw).	P-1	125
9. Inspect and test fusible links, circuit breakers, and fuses; determine necessary action.	P-1	125, 130, 138, 139
10. Perform solder repair of electrical wiring.	P-1	126
11. Replace electrical connectors and terminal ends.	P-1	126

VI. ELECTRICAL/ELECTRONIC SYSTEMS (continued)
B. Battery Service

Task Number and Description	Priority	Job #s
1. Perform battery state-of-charge test; determine necessary action.	P-1	127
2. Confirm proper battery capacity for vehicle application; perform battery capacity test; determine necessary action.	P-1	127
3. Maintain or restore electronic memory functions.	P-1	125, 128
4. Inspect and clean battery; fill battery cells; check battery cables, connectors, clamps, and hold-downs.	P-1	127, 128
5. Perform slow/fast battery charge according to manufacturer's recommendations.	P-1	128
6. Jump-start vehicle using jumper cables and a booster battery or an auxiliary power supply.	P-1	129
7. Identify high-voltage circuits of electric or hybrid electric vehicle and related safety precautions.	P-3	127, 128
8. Identify electronic modules, security systems, radios, and other accessories that require reinitialization or code entry after reconnecting vehicle battery.	P-1	128
9. Identify hybrid vehicle auxiliary (12v) battery service, repair, and test procedures.	P-3	127, 128

VI. ELECTRICAL/ELECTRONIC SYSTEMS
C. Starting System

Task Number and Description	Priority	Job #s
1. Perform starter current draw test; determine necessary action.	P-1	130
2. Perform starter circuit voltage drop tests; determine necessary action.	P-1	130
3. Inspect and test starter relays and solenoids; determine necessary action.	P-2	130, 131
4. Remove and install starter in a vehicle.	P-1	133
5. Inspect and test switches, connectors, and wires of starter control circuits; determine necessary action.	P-2	130

VI. ELECTRICAL/ELECTRONIC SYSTEMS
D. Charging System

Task Number and Description	Priority	Job #s
1. Perform charging system output test; determine necessary action.	P-1	134
2. Inspect, adjust, or replace generator (alternator) drive belts; check pulleys and tensioners for wear; check pulley and belt alignment.	P-1	33, 134
3. Remove, inspect, and re-install generator (alternator).	P-2	134, 135
4. Perform charging circuit voltage drop tests; determine necessary action.	P-1	134

VI. ELECTRICAL/ELECTRONIC SYSTEMS (continued)
E. Lighting Systems

Task Number and Description	Priority	Job #s
1. Inspect interior and exterior lamps and sockets including headlights and auxiliary lights (fog lights/driving lights); replace as needed.	P-1	136
2. Aim headlights.	P-2	137
3. Identify system voltage and safety precautions associated with high-intensity discharge headlights.	P-2	137

VI. ELECTRICAL/ELECTRONIC SYSTEMS
F. Accessories

Task Number and Description	Priority	Job #s
1. Disable and enable airbag system for vehicle service; verify indicator lamp operation.	P-1	147
2. Remove and reinstall door panel.	P-1	144
3. Describe the operation of keyless entry/remote-start systems.	P-3	151
4. Verify operation of instrument panel gauges and warning/indicator lights; reset maintenance indicators.	P-1	136, 138, 139, 140, 141
5. Verify windshield wiper and washer operation; replace wiper blades.	P-1	143

HEATING AND AIR CONDITIONING

VII. HEATING AND AIR CONDITIONING
A. General

Task Number and Description	Priority	Job #s
1. Research applicable vehicle and service information, vehicle service history, service precautions, and technical service bulletins.	P-1	2

VII. HEATING AND AIR CONDITIONING
B. Refrigeration System Components

Task Number and Description	Priority	Job #s
1. Inspect and replace A/C compressor drive belts, pulleys, and tensioners; determine necessary action.	P-1	33, 152, 155
2. Identify hybrid vehicle A/C system electrical circuits and the service/safety precautions.	P-2	152
3. Inspect A/C condenser for airflow restrictions; determine necessary action.	P-1	152

VII. HEATING AND AIR CONDITIONING
C. Heating, Ventilation, and Engine Cooling Systems

Task Number and Description	Priority	Job #s
1. Inspect engine cooling and heater systems hoses; perform necessary action.	P-1	31, 33, 158

VII. HEATING AND AIR CONDITIONING (continued)
D. Operating Systems and Related Controls

Task Number and Description	Priority	Job #s
1. Inspect A/C-heater ducts, doors, hoses, cabin filters, and outlets; perform necessary action.	P-1	159
2. Identify the source of A/C system odors.	P-2	152

ENGINE PERFORMANCE

VIII. ENGINE PERFORMANCE
A. General

Task Number and Description	Priority	Job #s
1. Research applicable vehicle and service information, vehicle service history, service precautions, and technical service bulletins.	P-1	2
2. Perform engine absolute (vacuum/boost) manifold pressure tests; determine necessary action	P-1	8
3. Perform cylinder power balance test; determine necessary action.	P-2	8
4. Perform cylinder cranking and running compression tests; determine necessary action.	P-1	9
5. Perform cylinder leakage test; determine necessary action.	P-1	9
6. Verify engine operating temperature.	P-1	31
7. Remove and replace spark plugs; inspect secondary ignition components for wear and damage.	P-1	164, 165

VIII. ENGINE PERFORMANCE
B. Computerized Controls

Task Number and Description	Priority	Job #s
1. Retrieve and record diagnostic trouble codes, OBD monitor status, and freeze-frame data; clear codes when applicable.	P-1	11, 161
2. Describe the importance of operating all OBDII monitors for repair verification.	P-1	161

VIII. ENGINE PERFORMANCE
C. Fuel, Air Induction, and Exhaust Systems

Task Number and Description	Priority	Job #s
1. Replace fuel filter(s).	P-1	167
2. Inspect, service, or replace air filters, filter housings, and intake duct work.	P-1	168
3. Inspect integrity of the exhaust manifold, exhaust pipes, muffler(s), catalytic converter(s), resonator(s), tail pipe(s), and heat shields; determine necessary action.	P-1	173, 174, 177

VIII. ENGINE PERFORMANCE (continued)
C. Fuel, Air Induction, and Exhaust Systems

Task Number and Description	Priority	Job #s
4. Inspect condition of exhaust system hangers, brackets, clamps, and heat shields; repair or replace as needed.	P-1	173, 174, 177
5. Check and refill diesel exhaust fluid (DEF).	P-3	173

VIII. ENGINE PERFORMANCE
D. Emissions Control Systems

Task Number and Description	Priority	Job #s
1. Inspect, test, and service positive crankcase ventilation (PCV) filter/breather cap, valve, tubes, orifices, and hoses; perform necessary action.	P-2	178

REQUIRED SUPPLEMENTAL TASKS

Shop and Personal Safety

Task Number and Description	
1. Identify general shop safety rules and procedures.	1
2. Utilize safe procedures for handling of tools and equipment.	1
3. Identify and use proper placement of floor jacks and jack stands.	1
4. Identify and use proper procedures for safe lift operation.	1
5. Utilize proper ventilation procedures for working within the lab/shop area.	1
6. Identify marked safety areas.	1
7. Identify the location and the types of fire extinguishers and other fire safety equipment; demonstrate knowledge of the procedures for using fire extinguishers and other fire safety equipment.	1
8. Identify the location and use of eye wash stations.	1
9. Identify the location of the posted evacuation routes.	1
10. Comply with the required use of safety glasses, ear protection, gloves, and shoes during lab/shop activities.	1
11. Identify and wear appropriate clothing for lab/shop activities.	1
12. Secure hair and jewelry for lab/shop activities.	1
13. Demonstrate awareness of the safety aspects of supplemental restraint systems (SRS), electronic brake control systems, and hybrid vehicle high voltage circuits.	1
14. Demonstrate awareness of the safety aspects of high voltage circuits (such as high intensity discharge (HID) lamps, ignition systems, injection systems, etc.).	1, 137
15. Locate and demonstrate knowledge of material safety data sheets (MSDS).	1, 2

Tools and Equipment

Task Number and Description	
1. Identify tools and their usage in automotive applications.	1, 11
2. Identify standard and metric designation.	2, 28, 31, 40, 172
3. Demonstrate safe handling and use of appropriate tools.	1
4. Demonstrate proper cleaning, storage, and maintenance of tools and equipment.	1
5. Demonstrate proper use of precision measuring tools (i.e. micrometer, dial indicator, dial caliper).	16, 18, 21, 27

Preparing Vehicle for Service

Task Number and Description	
1. Identify information needed and the service requested on a repair order.	2
2. Identify purpose and demonstrate proper use of fender covers, mats.	6
3. Demonstrate use of the three Cs (concern, cause, and correction).	160
4. Review vehicle service history.	2
5. Complete work order to include customer information, vehicle identifying information, customer concern, related service history, cause, and correction.	2

Preparing Vehicle for Customer

Task Number and Description	
1. Ensure vehicle is prepared to return to customer per school/company policy (floor mats, steering wheel cover, etc.).	6

NATEF Correlation Chart
Automobile Service Technology (AST)

The following chart correlates the jobs in this *Shop Manual* to the 2013 NATEF Automobile Service Technology (AST) Task List.

ENGINE REPAIR

For every task in this list, the following safety requirement must be strictly enforced:
- Comply with personal and environmental safety practices associated with clothing; eye protection; hand tools; power equipment; proper ventilation; and the handling, storage, and disposal of chemicals/materials in accordance with local, state, and federal safety and environmental regulations.

I. ENGINE REPAIR
A. General: Engine Diagnosis; Removal and Reinstallation (R & R)

Task Number and Description	Priority	Job #s
1. Complete work order to include customer information, vehicle identifying information, customer concern, related service history, cause, and correction.	P-1	2
2. Research applicable vehicle and service information, such as internal engine operation, vehicle service history, service precautions, and technical service bulletins.	P-1	2
3. Verify operation of the instrument panel engine warning indicators.	P-1	11, 138, 139, 140, 141
4. Inspect engine assembly for fuel, oil, coolant, and other leaks; determine necessary action.	P-1	7
5. Install engine covers using gaskets, seals, and sealers as required.	P-1	3, 4, 27, 28, 34
6. Remove and replace timing belt; verify correct camshaft timing.	P-1	13, 28
7. Perform common fastener and thread repair, to include: remove broken bolt, restore internal and external threads, and repair internal threads with thread insert.	P-1	15
8. Inspect, remove, and replace engine mounts.	P-2	37, 38
9. Identify hybrid vehicle internal combustion engine service precautions.	P-3	12, 13
10. Remove and reinstall engine in an OBDII or newer vehicle; reconnect all attaching components and restore the vehicle to running condition.	P-3	12, 30

I. ENGINE REPAIR
B. Cylinder Head and Valve Train Diagnosis and Repair

Task Number and Description	Priority	Job #s
1. Remove cylinder head; inspect gasket condition; install cylinder head and gasket; tighten according to manufacturer's specifications and procedures.	P-1	13, 28
2. Clean and visually inspect a cylinder head for cracks; check gasket surface areas for warpage and surface finish; check passage condition.	P-1	3, 20

I. ENGINE REPAIR (continued)
B. Cylinder Head and Valve Train Diagnosis and Repair

Task Number and Description	Priority	Job #s
3. Inspect push rods, rocker arms, rocker arm pivots and shafts for wear, bending, cracks, looseness, and blocked oil passages (orifices); determine necessary action.	P-2	21
4. Adjust valves (mechanical or hydraulic lifters).	P-1	25
5. Inspect and replace camshaft and drive belt/chain; includes checking drive gear wear and backlash, end play, sprocket and chain wear, overhead cam drive sprocket(s), drive belt(s), belt tension, tensioners, camshaft reluctor ring/tone-wheel, and valve timing components; verify correct camshaft timing.	P-1	24, 28
6. Establish camshaft position sensor indexing.	P-1	28

I. ENGINE REPAIR
C. Engine Block Assembly Diagnosis and Repair

Task Number and Description	Priority	Job #s
1. Remove, inspect, or replace crankshaft vibration damper (harmonic balancer).	P-2	13, 19, 29

I. ENGINE REPAIR
D. Lubrication and Cooling Systems Diagnosis and Repair

Task Number and Description	Priority	Job #s
1. Perform cooling system pressure and dye tests to identify leaks; check coolant condition and level; inspect and test radiator, pressure cap, coolant recovery tank, and heater core and galley plugs; determine necessary action.	P-1	31
2. Identify causes of engine overheating.	P-1	8, 9, 10, 31
3. Inspect, replace, and adjust drive belts, tensioners, and pulleys; check pulley and belt alignment.	P-1	31, 33
4. Inspect and test coolant; drain and recover coolant; flush and refill cooling system with recommended coolant; bleed air as required.	P-1	13, 32
5. Inspect, remove, and replace water pump.	P-2	31, 34
6. Remove and replace radiator.	P-2	35
7. Remove, inspect, and replace thermostat and gasket/seal.	P-1	36
8. Inspect and test fan(s) (electrical or mechanical), fan clutch, fan shroud, and air dams.	P-1	31
9. Perform oil pressure tests; determine necessary action.	P-1	10
10. Perform engine oil and filter change.	P-1	6
11. Inspect auxiliary coolers; determine necessary action.	P-3	31
12. Inspect, test, and replace oil temperature and pressure switches and sensors.	P-2	139, 141

II. AUTOMATIC TRANSMISSION AND TRANSAXLE
A. General: Transmission and Transaxle Diagnosis

Task Number and Description	Priority	Job #s
1. Identify and interpret transmission/transaxle concern, differentiate between engine performance and transmission/transaxle concerns; determine necessary action.	P-1	39
2. Research applicable vehicle and service information fluid type, vehicle service history, service precautions, and technical service bulletins.	P-1	2
3. Diagnose fluid loss and condition concerns; determine necessary action.	P-1	7, 39, 44, 65
4. Check fluid level in a transmission or a transaxle equipped with a dip-stick.	P-1	39, 44
5. Check fluid level in a transmission or a transaxle not equipped with a dip-stick.	P-1	39, 44
6. Perform stall test; determine necessary action.	P-3	39
7. Perform lock-up converter system tests; determine necessary action.	P-3	39
8. Diagnose transmission/transaxle gear reduction/multiplication concerns using driving, driven, and held member (power flow) principles.	P-1	39
9. Diagnose pressure concerns in a transmission using hydraulic principles (Pascal's Law).	P-2	40

II. AUTOMATIC TRANSMISSION AND TRANSAXLE
B. In-Vehicle Transmission/Transaxle Maintenance and Repair

Task Number and Description	Priority	Job #s
1. Inspect, adjust, and replace external manual valve shift linkage, transmission range sensor/switch, and park/neutral position switch.	P-2	39, 45
2. Inspect for leakage; replace external seals, gaskets, and bushings.	P-2	3, 4
3. Inspect, test, adjust, repair, or replace electrical/electronic components and circuits including computers, solenoids, sensors, relays, terminals, connectors, switches, and harnesses.	P-1	48
4. Drain and replace fluid and filter(s).	P-1	44
5. Inspect, replace, and align power train mounts.	P-2	37, 38

II. AUTOMATIC TRANSMISSION AND TRANSAXLE
C. Off-Vehicle Transmission and Transaxle Repair

Task Number and Description	Priority	Job #s
1. Remove and reinstall transmission/transaxle and torque converter; inspect engine core plugs, rear crankshaft seal, dowel pins, dowel pin holes, and mating surfaces.	P-1	50, 51, 58, 59
2. Inspect, leak test, and flush or replace transmission/transaxle oil cooler, lines, and fittings.	P-1	46

II. AUTOMATIC TRANSMISSION AND TRANSAXLE (continued)
C. Off-Vehicle Transmission and Transaxle Repair

Task Number and Description	Priority	Job #s
3. Inspect converter flex (drive) plate, converter attaching bolts, converter pilot, converter pump drive surfaces, converter end play, and crankshaft pilot bore.	P-2	50, 51, 52, 53, 55
4. Describe the operational characteristics of a continuously variable transmission (CVT).	P-3	39
5. Describe the operational characteristics of a hybrid vehicle drive train.	P-3	39

MANUAL DRIVE TRAIN AND AXLES

III. MANUAL DRIVE TRAIN AND AXLES
A. General: Drive Train Diagnosis

Task Number and Description	Priority	Job #s
1. Identify and interpret drive train concerns; determine necessary action.	P-1	60
2. Research applicable vehicle and service information, fluid type, vehicle service history, service precautions, and technical service bulletins.	P-1	2
3. Check fluid condition; check for leaks; determine necessary action.	P-1	60, 65
4. Drain and refill manual transmission/transaxle and final drive unit.	P-1	77, 78, 79, 80, 86, 89

III. MANUAL DRIVE TRAIN AND AXLES
B. Clutch Diagnosis and Repair

Task Number and Description	Priority	Job #s
1. Diagnose clutch noise, binding, slippage, pulsation, and chatter; determine necessary action.	P-1	71
2. Inspect clutch pedal linkage, cables, automatic adjuster mechanisms, brackets, bushings, pivots, and springs; perform necessary action.	P-1	71, 72
3. Inspect and replace clutch pressure plate assembly, clutch disc, release (throw-out) bearing and linkage, and pilot bearing/bushing (as applicable).	P-1	75
4. Bleed clutch hydraulic system.	P-1	73
5. Check and adjust clutch master cylinder fluid level; check for leaks.	P-1	71, 73
6. Inspect flywheel and ring gear for wear and cracks; determine necessary action.	P-1	75
7. Measure flywheel runout and crankshaft end play; determine necessary action.	P-2	75

III. MANUAL DRIVE TRAIN AND AXLES
C. Transmission/Transaxle Diagnosis and Repair

Task Number and Description	Priority	Job #s
1. Inspect, adjust, and reinstall shift linkages, brackets, bushings, cables, pivots, and levers.	P-2	64, 78, 80

III. MANUAL DRIVE TRAIN AND AXLES (continued)
C. Transmission/Transaxle Diagnosis and Repair

Task Number and Description	Priority	Job #s
2. Describe the operational characteristics of an electronically-controlled manual transmission/transaxle.	P-3	69

III. MANUAL DRIVE TRAIN AND AXLES
D. Drive Shaft and Half Shaft, Universal and Constant-Velocity (CV) Joint Diagnosis and Repair

Task Number and Description	Priority	Job #s
1. Diagnose constant-velocity (CV) joint noise and vibration concerns; determine necessary action.	P-1	67, 68
2. Diagnose universal joint noise and vibration concerns; perform necessary action.	P-2	61, 62, 63, 66
3. Inspect, remove, and replace front-wheel drive (FWD) bearings, hubs, and seals.	P-1	68
4. Inspect, service, and replace shafts, yokes, boots, and universal/CV joints.	P-1	66, 67
5. Check shaft balance and phasing; measure shaft runout; measure and adjust driveline angles.	P-2	61, 62, 63

III. MANUAL DRIVE TRAIN AND AXLES
E. Drive Axle Diagnosis and Repair

E.1 Ring and Pinion Gears and Differential Case Assembly

Task Number and Description	Priority	Job #s
1. Clean and inspect differential housing; check for leaks; inspect housing vent.	P-2	7, 65, 87
2. Check and adjust differential housing fluid level.	P-1	60
3. Drain and refill differential housing.	P-1	86, 89
4. Inspect and replace companion flange and pinion seal; measure companion flange runout.	P-2	66, 87, 88

E.2 Drive Axles

Task Number and Description	Priority	Job #s
1. Inspect and replace drive axle wheel studs.	P-1	108
2. Remove and replace drive axle shafts.	P-1	67, 84, 85
3. Inspect and replace drive axle shaft seals, bearings, and retainers.	P-2	67, 84, 85
4. Measure drive axle flange runout and shaft end play; determine necessary action.	P-2	84

III. MANUAL DRIVE TRAIN AND AXLES (continued)
F. Four-wheel Drive/All-wheel Drive Component Diagnosis and Repair

Task Number and Description	Priority	Job #s
1. Inspect, adjust, and repair shifting controls (mechanical, electrical, and vacuum), bushings, mounts, levers, and brackets.	P-3	83
2. Inspect front-wheel bearings and locking hubs; perform necessary action(s).	P-3	67, 68, 83, 92
3. Check for leaks at drive assembly seals; check vents; check lube level.	P-3	65, 81
4. Identify concerns related to variations in tire circumference and/or final drive ratios.	P-3	81

SUSPENSION AND STEERING

IV. SUSPENSION AND STEERING
A. General: Suspension and Steering Systems

Task Number and Description	Priority	Job #s
1. Research applicable vehicle and service information, vehicle service history, service precautions, and technical service bulletins.	P-1	2

IV. SUSPENSION AND STEERING
B. Steering Systems Diagnosis and Repair

Task Number and Description	Priority	Job #s
1. Disable and enable supplemental restraint system (SRS).	P-1	94, 147
2. Remove and replace steering wheel; center/time supplemental restraint system (SRS) coil (clock spring).	P-1	94, 147
3. Diagnose steering column noises, looseness, and binding concerns (including tilt mechanisms); determine necessary action.	P-2	90, 93
4. Diagnose power steering gear (non-rack and pinion) binding, uneven turning effort, looseness, hard steering, and noise concerns; determine necessary action.	P-2	90, 97
5. Diagnose power steering gear (rack and pinion) binding, uneven turning effort, looseness, hard steering, and noise concerns; determine necessary action.	P-2	90, 97
6. Inspect steering shaft universal-joint(s), flexible coupling(s), collapsible column, lock cylinder mechanism, and steering wheel; perform necessary action.	P-2	93, 94
7. Remove and replace rack and pinion steering gear; inspect mounting bushings and brackets.	P-2	95
8. Inspect rack and pinion steering gear inner tie rod ends (sockets) and bellows boots; replace as needed.	P-2	95
9. Determine proper power steering fluid type; inspect fluid level and condition.	P-1	91, 97

IV. SUSPENSION AND STEERING (continued)
B. Steering Systems Diagnosis and Repair

Task Number and Description	Priority	Job #s
10. Flush, fill, and bleed power steering system.	P-2	97
11. Inspect for power steering fluid leakage; determine necessary action.	P-1	97
12. Remove, inspect, replace, and adjust power steering pump drive belt.	P-1	33, 97
13. Remove and reinstall power steering pump.	P-2	97
14. Remove and reinstall press-fit power steering pump pulley; check pulley and belt alignment.	P-2	97
15. Inspect and replace power steering hoses and fittings.	P-2	90, 97
16. Inspect and replace pitman arm, relay (center link/intermediate) rod, idler arm and mountings, and steering linkage damper.	P-2	90, 94, 96
17. Inspect, replace, and adjust tie rod ends (sockets), tie rod sleeves, and clamps.	P-1	90, 95, 96
18. Identify hybrid vehicle power steering system electrical circuits and safety precautions.	P-2	93
19. Inspect electric power-assisted steering.	P-3	97

IV. SUSPENSION AND STEERING
C. Suspension Systems Diagnosis and Repair

Task Number and Description	Priority	Job #s
1. Diagnose short and long arm suspension system noises, body sway, and uneven ride height concerns; determine necessary action.	P-1	90
2. Diagnose strut suspension system noises, body sway, and uneven ride height concerns; determine necessary action.	P-1	90
3. Inspect, remove, and install upper and lower control arms, bushings, shafts, and rebound bumpers.	P-3	90, 99
4. Inspect, remove, and install strut rods and bushings.	P-3	90, 98
5. Inspect, remove, and install upper and/or lower ball joints (with or without wear indicators).	P-2	90, 99
6. Inspect, remove, and install steering knuckle assemblies.	P-3	90
7. Inspect, remove, and install short and long arm suspension system coil springs and spring insulators.	P-3	90, 100
8. Inspect, remove, and install torsion bars and mounts.	P-3	90
9. Inspect, remove, and install front stabilizer bar (sway bar) bushings, brackets, and links.	P-3	90, 98
10. Inspect, remove, and install strut cartridge or assembly, strut coil spring, insulators (silencers), and upper strut bearing mount.	P-3	90, 101
11. Inspect, remove, and install track bar, strut rods/radius arms, and related mounts and bushings.	P-3	90, 96, 98, 99
12. Inspect rear suspension system leaf spring(s), bushings, center pins/bolts, and mounts.	P-1	90

IV. SUSPENSION AND STEERING (continued)
D. Related Suspension and Steering Service

Task Number and Description	Priority	Job #s
1. Inspect, remove, and replace shock absorbers; inspect mounts and bushings.	P-1	90, 102
2. Remove, inspect, and service or replace front and rear wheel bearings.	P-1	5, 83, 92
3. Describe the function of the power steering pressure switch.	P-3	90

IV. SUSPENSION AND STEERING
E. Wheel Alignment Diagnosis, Adjustment, and Repair

Task Number and Description	Priority	Job #s
1. Diagnose vehicle wander, drift, pull, hard steering, bump steer, memory steer, torque steer, and steering return concerns; determine necessary action.	P-1	103
2. Perform prealignment inspection and measure vehicle ride height; perform necessary action.	P-1	90, 103
3. Prepare vehicle for wheel alignment on alignment machine; perform four-wheel alignment by checking and adjusting front and rear wheel caster, camber; and toe as required; center steering wheel.	P-1	103
4. Check toe-out-on-turns (turning radius); determine necessary action.	P-2	103
5. Check SAI (steering axis inclination) and included angle; determine necessary action.	P-2	103
6. Check rear wheel thrust angle; determine necessary action.	P-1	103
7. Check for front wheel setback; determine necessary action.	P-2	103
8. Check front and/or rear cradle (subframe) alignment; determine necessary action.	P-3	103
9. Reset steering angle sensor.	P-2	103

IV. SUSPENSION AND STEERING
F. Wheels and Tires Diagnosis and Repair

Task Number and Description	Priority	Job #s
1. Inspect tire condition; identify tire wear patterns; check for correct tire size and application (load and speed ratings) and adjust air pressure; determine necessary action.	P-1	81, 104
2. Diagnose wheel/tire vibration, shimmy, and noise; determine necessary action.	P-2	103, 104, 106
3. Rotate tires according to manufacturer's recommendations.	P-1	106
4. Measure wheel, tire, axle flange, and hub runout; determine necessary action.	P-2	84, 104
5. Diagnose tire pull problems; determine necessary action.	P-2	103
6. Dismount, inspect, and remount tire on wheel; balance wheel and tire assembly (static and dynamic).	P-1	105, 106

IV. SUSPENSION AND STEERING (continued)
F. Wheels and Tires Diagnosis and Repair

Task Number and Description	Priority	Job #s
7. Dismount, inspect, and remount tire on wheel equipped with tire pressure monitoring system sensor.	P-2	105, 106
8. Inspect tire and wheel assembly for air loss; perform necessary action.	P-1	104
9. Repair tire using internal patch.	P-1	105
10. Identify and test tire pressure monitoring system (indirect and direct) for operation; verify operation of instrument panel lamps.	P-2	105
11. Demonstrate knowledge of steps required to remove and replace sensors in a tire pressure monitoring system.	P-1	105

BRAKES

V. BRAKES
A. General: Brake Systems Diagnosis

Task Number and Description	Priority	Job #s
1. Identify and interpret brake system concerns; determine necessary action.	P-1	107
2. Research applicable vehicle and service information, vehicle service history, service precautions, and technical service bulletins.	P-1	2
3. Describe procedure for performing a road test to check brake system operation; including an anti-lock brake system (ABS).	P-1	107
4. Install wheel and torque lug nuts.	P-1	105, 116, 118

V. BRAKES
B. Hydraulic System Diagnosis and Repair

Task Number and Description	Priority	Job #s
1. Diagnose pressure concerns in the brake system using hydraulic principles (Pascal's Law).	P-1	107, 111
2. Measure brake pedal height, travel, and free play (as applicable); determine necessary action.	P-1	107
3. Check master cylinder for internal/external leaks and proper operation; determine necessary action.	P-1	107
4. Remove, bench bleed, and reinstall master cylinder.	P-1	113
5. Diagnose poor stopping, pulling or dragging concerns caused by malfunctions in the hydraulic system; determine necessary action.	P-3	111
6. Inspect brake lines, flexible hoses, and fittings for leaks, dents, kinks, rust, cracks, bulging, and wear; check for loose fittings and supports; determine necessary action.	P-1	107, 111
7. Replace brake lines, hoses, fittings, and supports.	P-2	112

V. BRAKES (continued)
B. Hydraulic System Diagnosis and Repair

Task Number and Description	Priority	Job #s
8. Fabricate brake lines using proper material and flaring procedures (double flare and ISO types).	P-2	112
9. Select, handle, store, and fill brake fluids to proper level.	P-1	113, 115
10. Inspect, test, and/or replace components of brake warning light system.	P-3	111, 112, 136, 138, 141
11. Identify components of brake warning light system.	P-2	111, 112, 136, 138
12. Bleed and/or flush brake system.	P-1	115
13. Test brake fluid for contamination.	P-1	115

V. BRAKES
C. Drum Brake Diagnosis and Repair

Task Number and Description	Priority	Job #s
1. Diagnose poor stopping, noise, vibration, pulling, grabbing, dragging or pedal pulsation concerns; determine necessary action.	P-1	107
2. Remove, clean, inspect, and measure brake drum diameter; determine necessary action.	P-1	107, 116
3. Refinish brake drum and measure final drum diameter; compare with specifications.	P-1	117
4. Remove, clean, and inspect brake shoes, springs, pins, clips, levers, adjusters/self-adjusters, other related brake hardware, and backing support plates; lubricate and reassemble.	P-1	116
5. Inspect wheel cylinders for leaks and proper operation; remove and replace as needed.	P-2	116
6. Pre-adjust brake shoes and parking brake; install brake drums or drum/ hub assemblies and wheel bearings; perform final checks and adjustments.	P-2	116

V. BRAKES
D. Disc Brake Diagnosis and Repair

Task Number and Description	Priority	Job #s
1. Diagnose poor stopping, noise, vibration, pulling, grabbing, dragging, or pulsation concerns; determine necessary action.	P-1	107
2. Remove and clean caliper assembly; inspect for leaks and damage/ wear to caliper housing; determine necessary action.	P-1	107, 118
3. Clean and inspect caliper mounting and slides/pins for proper operation, wear, and damage; determine necessary action.	P-1	118
4. Remove, inspect, and replace pads and retaining hardware; determine necessary action.	P-1	118
5. Lubricate and reinstall caliper, pads, and related hardware; seat pads and inspect for leaks.	P-1	118

V. BRAKES (continued)
D. Disc Brake Diagnosis and Repair

Task Number and Description	Priority	Job #s
6. Clean and inspect rotor; measure rotor thickness, thickness variation, and lateral runout; determine necessary action.	P-1	118
7. Remove and reinstall rotor.	P-1	118
8. Refinish rotor on vehicle; measure final rotor thickness and compare with specifications.	P-1	120
9. Refinish rotor off vehicle; measure final rotor thickness and compare with specifications.	P-1	120
10. Retract and re-adjust caliper piston on an integrated parking brake system.	P-3	110
11. Check brake pad wear indicator; determine necessary action.	P-2	118
12. Describe importance of operating vehicle to burnish/break-in replacement brake pads according to manufacturer's recommendations.	P-1	118

V. BRAKES
E. Power-Assist Units Diagnosis and Repair

Task Number and Description	Priority	Job #s
1. Check brake pedal travel with, and without, engine running to verify proper power booster operation.	P-2	114
2. Check vacuum supply (manifold or auxiliary pump) to vacuum-type power booster.	P-1	114
3. Inspect vacuum-type power booster unit for leaks; inspect the check-valve for proper operation; determine necessary action.	P-1	114
4. Inspect and test hydraulically-assisted power brake system for leaks and proper operation; determine necessary action.	P-3	114
5. Measure and adjust master cylinder push rod length.	P-3	114

V. BRAKES
F. Miscellaneous (Wheel Bearings, Parking Brakes, Electrical, Etc.) Diagnosis and Repair

Task Number and Description	Priority	Job #s
1. Diagnose wheel bearing noises, wheel shimmy, and vibration concerns; determine necessary action.	P-3	109
2. Remove, clean, inspect, repack, and install wheel bearings; replace seals; install hub and adjust bearings.	P-1	5, 109
3. Check parking brake cables and components for wear, binding, and corrosion; clean, lubricate, adjust, or replace as needed.	P-2	110
4. Check parking brake operation and parking brake indicator light system operation; determine necessary action.	P-1	110
5. Check operation of brake stop light system.	P-1	136
6. Replace wheel bearing and race.	P-2	5, 109

V. BRAKES (continued)
F. Miscellaneous (Wheel Bearings, Parking Brakes, Electrical, Etc.) Diagnosis and Repair

Task Number and Description	Priority	Job #s
7. Inspect and replace wheel studs.	P-1	108
8. Remove and reinstall sealed wheel bearing assembly.	P-2	109

V. BRAKES
G. Electronic Brake, Traction and Stability Control Systems Diagnosis and Repair

Task Number and Description	Priority	Job #s
1. Identify and inspect electronic brake control system components; determine necessary action.	P-1	121
2. Identify traction control/vehicle stability control system components.	P-3	121
3. Describe the operation of a regenerative braking system.	P-3	111

ELECTRICAL/ELECTRONIC SYSTEMS

VI. ELECTRICAL/ELECTRONIC SYSTEMS
A. General: Electrical System Diagnosis

Task Number and Description	Priority	Job #s
1. Research applicable vehicle and service information, vehicle service history, service precautions, and technical service bulletins.	P-1	2
2. Demonstrate knowledge of electrical/electronic series, parallel, and series-parallel circuits using principles of electricity (Ohm's Law).	P-1	136
3. Demonstrate proper use of a digital multimeter (DMM) when measuring source voltage, voltage drop (including grounds), current flow, and resistance.	P-1	130, 136, 142
4. Demonstrate knowledge of the causes and effects from shorts, grounds, opens, and resistance problems in electrical/electronic circuits.	P-1	130, 138, 142
5. Check operation of electrical circuits with a test light.	P-1	130, 136
6. Check operation of electrical circuits with fused jumper wires.	P-1	125, 138
7. Use wiring diagrams during the diagnosis (troubleshooting) of electrical/electronic circuit problems.	P-1	130, 136, 138
8. Diagnose the cause(s) of excessive key-off battery drain (parasitic draw); determine necessary action.	P-1	125
9. Inspect and test fusible links, circuit breakers, and fuses; determine necessary action.	P-1	125, 130, 138, 139
10. Inspect and test switches, connectors, relays, solenoid solid-state devices, and wires of electrical/electronic circuits; determine necessary action.	P-1	130, 132, 142
11. Replace electrical connectors and terminal ends.	P-1	126

VI. ELECTRICAL/ELECTRONIC SYSTEMS (continued)
A. General: Electrical System Diagnosis

Task Number and Description	Priority	Job #s
12. Repair wiring harness.	P-3	126
13. Perform solder repair of electrical wiring.	P-1	126

VI. ELECTRICAL/ELECTRONIC SYSTEMS
B. Battery Diagnosis and Service

Task Number and Description	Priority	Job #s
1. Perform battery state-of-charge test; determine necessary action.	P-1	127
2. Confirm proper battery capacity for vehicle application; perform battery capacity test; determine necessary action.	P-1	127
3. Maintain or restore electronic memory functions.	P-1	125, 128
4. Inspect and clean battery; fill battery cells; check battery cables, connectors, clamps, and hold-downs.	P-1	127, 128
5. Perform slow/fast battery charge according to manufacturer's recommendations.	P-1	128
6. Jump-start vehicle using jumper cables and a booster battery or an auxiliary power supply.	P-1	129
7. Identify high-voltage circuits of electric or hybrid electric vehicle and related safety precautions.	P-3	126
8. Identify electronic modules, security systems, radios, and other accessories that require reinitialization or code entry after reconnecting vehicle battery.	P-1	128
9. Identify hybrid vehicle auxiliary (12v) battery service, repair, and test procedures.	P-3	127, 128

VI. ELECTRICAL/ELECTRONIC SYSTEMS
C. Starting System Diagnosis and Repair

Task Number and Description	Priority	Job #s
1. Perform starter current draw tests; determine necessary action.	P-1	130
2. Perform starter circuit voltage drop tests; determine necessary action.	P-1	130
3. Inspect and test starter relays and solenoids; determine necessary action.	P-2	130, 131
4. Remove and install starter in a vehicle.	P-1	133
5. Inspect and test switches, connectors, and wires of starter control circuits; determine necessary action.	P-2	130, 132
6. Differentiate between electrical and engine mechanical problems that cause a slow-crank or a no-crank condition.	P-2	130

VI. ELECTRICAL/ELECTRONIC SYSTEMS (continued)
D. Charging System Diagnosis and Repair

Task Number and Description	Priority	Job #s
1. Perform charging system output test; determine necessary action.	P-1	134
2. Diagnose (troubleshoot) charging system for causes of undercharge, no-charge, or overcharge conditions.	P-1	134
3. Inspect, adjust, or replace generator (alternator) drive belts; check pulleys and tensioners for wear; check pulley and belt alignment.	P-1	33, 134
4. Remove, inspect, and re-install generator (alternator).	P-1	134, 135
5. Perform charging circuit voltage drop tests; determine necessary action.	P-1	134

VI. ELECTRICAL/ELECTRONIC SYSTEMS
E. Lighting Systems Diagnosis and Repair

Task Number and Description	Priority	Job #s
1. Diagnose (troubleshoot) the causes of brighter-than-normal, intermittent, dim, or no light operation; determine necessary action.	P-1	136
2. Inspect interior and exterior lamps and sockets including headlights and auxiliary lights (fog lights/driving lights); replace as needed.	P-1	136
3. Aim headlights.	P-2	137
4. Identify system voltage and safety precautions associated with high-intensity discharge headlights.	P-2	137

VI. ELECTRICAL/ELECTRONIC SYSTEMS
F. Gauges, Warning Devices, and Driver Information Systems Diagnosis and Repair

Task Number and Description	Priority	Job #s
1. Inspect and test gauges and gauge sending units for causes of abnormal gauge readings; determine necessary action.	P-2	139, 140
2. Diagnose (troubleshoot) the causes of incorrect operation of warning devices and other driver information systems; determine necessary action.	P-2	138, 139, 140, 141

VI. ELECTRICAL/ELECTRONIC SYSTEMS
G. Horn and Wiper/Washer Diagnosis and Repair

Task Number and Description	Priority	Job #s
1. Diagnose (troubleshoot) causes of incorrect horn operation; perform necessary action.	P-1	142
2. Diagnose (troubleshoot) causes of incorrect wiper operation; diagnose wiper speed control and park problems; perform necessary action.	P-2	143
3. Diagnose (troubleshoot) windshield washer problems; perform necessary action.	P-2	143

VI. ELECTRICAL/ELECTRONIC SYSTEMS (continued)
H. Accessories Diagnosis and Repair

Task Number and Description	Priority	Job #s
1. Diagnose (troubleshoot) incorrect operation of motor-driven accessory circuits; determine necessary action.	P-2	144
2. Diagnose (troubleshoot) incorrect electric lock operation (including remote keyless entry); determine necessary action.	P-2	144, 151
3. Diagnose (troubleshoot) incorrect operation of cruise control systems; determine necessary action.	P-3	145
4. Diagnose (troubleshoot) supplemental restraint system (SRS) problems; determine necessary action.	P-2	148
5. Disable and enable an airbag system for vehicle service; verify indicator lamp operation.	P-1	147
6. Remove and reinstall door panel.	P-1	144
7. Check for module communication errors (including CAN/BUS systems) using a scan tool.	P-2	150
8. Describe the operation of keyless entry/remote-start systems.	P-3	151
9. Verify operation of instrument panel gauges and warning/indicator lights; reset maintenance indicators.	P-1	136, 138, 139, 140, 141
10. Verify windshield wiper and washer operation, replace wiper blades.	P-1	143

HEATING AND AIR CONDITIONING

VII. HEATING AND AIR CONDITIONING
A. General: A/C System Diagnosis and Repair

Task Number and Description	Priority	Job #s
1. Identify and interpret heating and air conditioning problems; determine necessary action.	P-1	152
2. Research applicable vehicle and service information, vehicle service history, service precautions, and technical service bulletins.	P-1	2
3. Performance test A/C system; identify problems.	P-1	152, 153
4. Identify abnormal operating noises in the A/C system; determine necessary action.	P-2	152
5. Identify refrigerant type; select and connect proper gauge set; record temperature and pressure readings.	P-1	152, 153
6. Leak test A/C system; determine necessary action.	P-1	152
7. Inspect condition of refrigerant oil removed from A/C system; determine necessary action.	P-2	155
8. Determine recommended oil and oil capacity for system application.	P-1	155
9. Using a scan tool, observe and record related HVAC data and trouble codes.	P-3	11

VII. HEATING AND AIR CONDITIONING (continued)
B. Refrigeration System Component Diagnosis and Repair

Task Number and Description	Priority	Job #s
1. Inspect and replace A/C compressor drive belts, pulleys, and tensioners; determine necessary action.	P-1	33, 152, 155
2. Inspect, test, service or replace A/C compressor clutch components and/or assembly; check compressor clutch air gap; adjust as needed.	P-2	152, 155
3. Remove, inspect, and reinstall A/C compressor and mountings; determine recommended oil quantity.	P-2	155
4. Identify hybrid vehicle A/C system electrical circuits and service/safety precautions.	P-2	152
5. Determine need for an additional A/C system filter; perform necessary action.	P-3	159
6. Remove and inspect A/C system mufflers, hoses, lines, fittings, O-rings, seals, and service valves; perform necessary action.	P-2	152, 156
7. Inspect A/C condenser for airflow restrictions; perform necessary action.	P-1	152
8. Remove, inspect, and reinstall receiver/drier or accumulator/drier; determine recommended oil quantity.	P-2	155, 156
9. Remove, inspect, and install expansion valve or orifice (expansion) tube.	P-1	156
10. Inspect evaporator housing water drain; perform necessary action.	P-1	152
11. Determine procedure to remove and reinstall evaporator; determine required oil quantity.	P-2	156

VII. HEATING AND AIR CONDITIONING
C. Heating, Ventilation, and Engine Cooling Systems Diagnosis and Repair

Task Number and Description	Priority	Job #s
1. Inspect engine cooling and heater systems hoses; perform necessary action.	P-1	31, 33, 158
2. Inspect and test heater control valve(s); perform necessary action.	P-2	158
3. Determine procedure to remove, inspect, and reinstall heater core.	P-2	158

VII. HEATING AND AIR CONDITIONING
D. Operating Systems and Related Controls Diagnosis and Repair

Task Number and Description	Priority	Job #s
1. Inspect and test A/C-heater blower motors, resistors, switches, relays, wiring, and protection devices; perform necessary action.	P-1	152, 159
2. Diagnose A/C compressor clutch control systems; determine necessary action.	P-2	152
3. Diagnose malfunctions in the vacuum, mechanical, and electrical components and controls of the heating, ventilation, and A/C (HVAC) system; determine necessary action.	P-2	153
4. Inspect and test A/C-heater control panel assembly; determine necessary action.	P-3	153

VII. HEATING AND AIR CONDITIONING (continued)
D. Operating Systems and Related Controls Diagnosis and Repair

Task Number and Description	Priority	Job #s
5. Inspect and test A/C-heater control cables, motors, and linkages; perform necessary action.	P-3	153
6. Inspect A/C-heater ducts, doors, hoses, cabin filters, and outlets; perform necessary action.	P-1	159
7. Identify the source of A/C system odors.	P-2	152
8. Check operation of automatic or semi-automatic heating, ventilation, and air-conditioning (HVAC) control systems; determine necessary action.	P-2	153

VII. HEATING AND AIR CONDITIONING
E. Refrigerant Recovery, Recycling, and Handling

Task Number and Description	Priority	Job #s
1. Perform correct use and maintenance of refrigerant handling equipment according to equipment manufacturer's standards.	P-1	154, 157
2. Identify and recover A/C system refrigerant.	P-1	154
3. Recycle, label, and store refrigerant.	P-1	154, 157
4. Evacuate and charge A/C system; add refrigerant oil as required.	P-1	154

ENGINE PERFORMANCE

VIII. ENGINE PERFORMANCE
A. General: Engine Diagnosis

Task Number and Description	Priority	Job #s
1. Identify and interpret engine performance concerns; determine necessary action.	P-1	160
2. Research applicable vehicle and service information, vehicle service history, service precautions, and technical service bulletins.	P-1	2
3. Diagnose abnormal engine noises or vibration concerns; determine necessary action.	P-3	163
4. Diagnose the cause of excessive oil consumption coolant consumption, unusual exhaust color, odor, and sound; determine necessary action.	P-2	173
5. Perform engine absolute (vacuum/boost) manifold pressure tests; determine necessary action.	P-1	8
6. Perform cylinder power balance test; determine necessary action.	P-2	8
7. Perform cylinder cranking and running compression tests; determine necessary action.	P-1	9
8. Perform cylinder leakage test; determine necessary action.	P-1	9
9. Diagnose engine mechanical, electrical, electronic, fuel, and ignition concerns; determine necessary action.	P-2	8, 9, 10, 130, 134, 160, 164, 167, 170

VIII. ENGINE PERFORMANCE (continued)
A. General: Engine Diagnosis

Task Number and Description	Priority	Job #s
10. Verify engine operating temperature; determine necessary action.	P-1	31
11. Verify correct camshaft timing.	P-1	28

VIII. ENGINE PERFORMANCE
B. Computerized Controls Diagnosis and Repair

Task Number and Description	Priority	Job #s
1. Retrieve and record diagnostic trouble codes, OBD monitor status, and freeze frame data; clear codes when applicable.	P-1	11, 161
2. Access and use service information to perform step-by-step (trouble-shooting) diagnosis.	P-1	2
3. Perform active tests of actuators using a scan tool; determine necessary action.	P-2	170, 178
4. Describe the importance of running all OBDII monitors for repair verification.	P-1	161

VIII. ENGINE PERFORMANCE
C. Ignition System Diagnosis and Repair

Task Number and Description	Priority	Job #s
1. Diagnose (troubleshoot) ignition system related problems such as no-starting, hard starting, engine misfire, poor driveability, spark knock, power loss, poor mileage, and emissions concerns; determine necessary action.	P-2	160, 161, 164, 165
2. Inspect and test crankshaft and camshaft position sensor(s); perform necessary action.	P-1	28, 29, 165
3. Inspect, test, and/or replace ignition control module, powertrain/engine control module; reprogram as necessary.	P-3	165
4. Remove and replace spark plugs; inspect secondary ignition components for wear and damage.	P-1	164, 165

VIII. ENGINE PERFORMANCE
D. Fuel, Air Induction, and Exhaust Systems Diagnosis and Repair

Task Number and Description	Priority	Job #s
1. Check fuel for contaminants; determine necessary action.	P-2	166
2. Inspect and test fuel pumps and pump control systems for pressure, regulation, and volume; perform necessary action.	P-1	167
3. Replace fuel filter(s).	P-1	167
4. Inspect, service, or replace air filters, filter housings, and intake duct work.	P-1	168
5. Inspect throttle body, air induction system, intake manifold and gaskets for vacuum leaks and/or unmetered air.	P-2	168
6. Inspect and test fuel injectors.	P-2	168, 170

VIII. ENGINE PERFORMANCE (continued)
D. Fuel, Air Induction, and Exhaust Systems Diagnosis and Repair

Task Number and Description	Priority	Job #s
7. Verify idle control operation.	P-1	169
8. Inspect integrity of the exhaust manifold, exhaust pipes, muffler(s), catalytic converter(s), resonator(s), tail pipe(s), and heat shields; perform necessary action.	P-1	173, 174, 177
9. Inspect condition of exhaust system hangers, brackets, clamps, and heat shields; repair or replace as needed.	P-1	173, 174, 177
10. Perform exhaust system back-pressure test; determine necessary action.	P-2	173
11. Check and refill diesel exhaust fluid (DEF).	P-3	173

VIII. ENGINE PERFORMANCE
E. Emissions Control Systems Diagnosis and Repair

Task Number and Description	Priority	Job #s
1. Diagnose oil leaks, emissions, and driveability concerns caused by the positive crankcase ventilation (PCV) system; determine necessary action.	P-3	160, 178
2. Inspect, test, and service positive crankcase ventilation (PCV) filter/breather cap, valve, tubes, orifices, and hoses; perform necessary action.	P-2	178
3. Diagnose emissions and driveability concerns caused by the exhaust gas recirculation (EGR) system; determine necessary action.	P-3	160, 175
4. Inspect, test, service, and replace components of the EGR system including tubing, exhaust passages, vacuum/pressure controls, filters, and hoses; perform necessary action.	P-2	175
5. Inspect and test electrical/electronically-operated components and circuits of air injection systems; perform necessary action.	P-3	176
6. Inspect and test catalytic converter efficiency.	P-2	177
7. Inspect and test components and hoses of the evaporative emissions control system; perform necessary action.	P-1	178
8. Interpret diagnostic trouble codes (DTCs) and scan tool data related to the emissions control systems; determine necessary action.	P-3	11, 160, 177, 178

REQUIRED SUPPLEMENTAL TASKS

Shop and Personal Safety

Task Number and Description	
1. Identify general shop safety rules and procedures.	1
2. Utilize safe procedures for handling of tools and equipment.	1
3. Identify and use proper placement of floor jacks and jack stands.	1
4. Identify and use proper procedures for safe lift operation.	1
5. Utilize proper ventilation procedures for working within the lab/shop area.	1
6. Identify marked safety areas.	1
7. Identify the location and the types of fire extinguishers and other fire safety equipment; demonstrate knowledge of the procedures for using fire extinguishers and other fire safety equipment.	1
8. Identify the location and use of eye wash stations.	1
9. Identify the location of the posted evacuation routes.	1
10. Comply with the required use of safety glasses, ear protection, gloves, and shoes during lab/shop activities.	1
11. Identify and wear appropriate clothing for lab/shop activities.	1
12. Secure hair and jewelry for lab/shop activities.	1
13. Demonstrate awareness of the safety aspects of supplemental restraint systems (SRS), electronic brake control systems, and hybrid vehicle high voltage circuits.	1
14. Demonstrate awareness of the safety aspects of high voltage circuits (such as high intensity discharge (HID) lamps, ignition systems, injection systems, etc.).	1, 137
15. Locate and demonstrate knowledge of material safety data sheets (MSDS).	1, 2

Tools and Equipment

Task Number and Description	
1. Identify tools and their usage in automotive applications.	1, 11
2. Identify standard and metric designation.	2, 28, 31, 40, 172
3. Demonstrate safe handling and use of appropriate tools.	1
4. Demonstrate proper cleaning, storage, and maintenance of tools and equipment.	1
5. Demonstrate proper use of precision measuring tools (i.e. micrometer, dial-indicator, dial-caliper).	16, 18, 21, 27

Preparing Vehicle for Service

Task Number and Description	
1. Identify information needed and the service requested on a repair order.	2
2. Identify purpose and demonstrate proper use of fender covers, mats.	6

Preparing Vehicle for Service (continued)	
Task Number and Description	
3. Demonstrate use of the three Cs (concern, cause, and correction).	160
4. Review vehicle service history.	2
5. Complete work order to include customer information, vehicle identifying information, customer concern, related service history, cause, and correction.	2
Preparing Vehicle for Customer	
Task Number and Description	
1. Ensure vehicle is prepared to return to customer per school/company policy (floor mats, steering wheel cover, etc.).	6

NATEF Correlation Chart
Master Automobile Service Technology (MAST)

The following chart correlates the jobs in this *Shop Manual* to the 2013 NATEF Master Automobile Service Technology (MAST) Task List.

ENGINE REPAIR

For every task in this list, the following safety requirement must be strictly enforced:
- Comply with personal and environmental safety practices associated with clothing; eye protection; hand tools; power equipment; proper ventilation; and the handling, storage, and disposal of chemicals/ materials in accordance with local, state, and federal safety and environmental regulations.

I. ENGINE REPAIR
A. General: Engine Diagnosis; Removal and Reinstallation (R & R)

Task Number and Description	Priority	Job #s
1. Complete work order to include customer information, vehicle identifying information, customer concern, related service history, cause, and correction.	P-1	2
2. Research applicable vehicle and service information, such as internal engine operation, vehicle service history, service precautions, and technical service bulletins.	P-1	2
3. Verify operation of the instrument panel engine warning indicators.	P-1	11, 138, 139, 140, 141
4. Inspect engine assembly for fuel, oil, coolant, and other leaks; determine necessary action.	P-1	7
5. Install engine covers using gaskets, seals, and sealers as required.	P-1	3, 4, 27, 28, 34
6. Remove and replace timing belt; verify correct camshaft timing.	P-1	13, 28
7. Perform common fastener and thread repair, to include: remove broken bolt, restore internal and external threads, and repair internal threads with thread insert.	P-1	15
8. Inspect, remove, and replace engine mounts.	P-2	37, 38
9. Identify hybrid vehicle internal combustion engine service precautions.	P-3	12, 13
10. Remove and reinstall engine in an OBDII or newer vehicle; reconnect all attaching components and restore the vehicle to running condition.	P-3	12, 30

I. ENGINE REPAIR
B. Cylinder Head and Valve Train Diagnosis and Repair

Task Number and Description	Priority	Job #s
1. Remove cylinder head; inspect gasket condition; install cylinder head and gasket; tighten according to manufacturer's specifications and procedures.	P-1	13, 28
2. Clean and visually inspect a cylinder head for cracks; check gasket surface areas for warpage and surface finish; check passage condition.	P-1	3, 20

I. ENGINE REPAIR (continued)
B. Cylinder Head and Valve Train Diagnosis and Repair

Task Number and Description	Priority	Job #s
3. Inspect push rods, rocker arms, rocker arm pivots and shafts for wear, bending, cracks, looseness, and blocked oil passages (orifices); determine necessary action.	P-2	21
4. Adjust valves (mechanical or hydraulic lifters).	P-1	25
5. Inspect and replace camshaft and drive belt/chain; includes checking drive gear wear and backlash, end play, sprocket and chain wear, overhead cam drive sprocket(s), drive belt(s), belt tension, tensioners, camshaft reluctor ring/tone-wheel, and valve timing components; verify correct camshaft timing.	P-1	24, 28
6. Establish camshaft position sensor indexing.	P-1	28
7. Inspect valve springs for squareness and free height comparison; determine necessary action.	P-3	21
8. Replace valve stem seals on an assembled engine; inspect valve spring retainers, locks/keepers, and valve lock/keeper grooves; determine necessary action.	P-3	21, 26
9. Inspect valve guides for wear; check valve stem-to-guide clearance; determine necessary action.	P-3	20
10. Inspect valves and valve seats; determine necessary action.	P-3	20
11. Check valve spring assembled height and valve stem height; determine necessary action.	P-3	23
12. Inspect valve lifters; determine necessary action.	P-2	21
13. Inspect and/or measure camshaft for runout, journal wear, and lobe wear.	P-2	21
14. Inspect camshaft bearing surface for wear, damage, out-of-round, and alignment; determine necessary action.	P-3	21

I. ENGINE REPAIR
C. Engine Block Assembly Diagnosis and Repair

Task Number and Description	Priority	Job #s
1. Remove, inspect, or replace crankshaft vibration damper (harmonic balancer).	P-2	13, 19, 29
2. Disassemble engine block; clean and prepare components for inspection and reassembly.	P-1	13
3. Inspect engine block for visible cracks, passage condition, core and gallery plug condition, and surface warpage; determine necessary action.	P-2	14
4. Inspect and measure cylinder walls/sleeves for damage, wear, and ridges; determine necessary action.	P-2	13, 16
5. Deglaze and clean cylinder walls.	P-2	16

I. ENGINE REPAIR (continued)
C. Engine Block Assembly Diagnosis and Repair

Task Number and Description	Priority	Job #s
6. Inspect and measure camshaft bearings for wear, damage, out-of-round, and alignment; determine necessary action.	P-3	21
7. Inspect crankshaft for straightness, journal damage, keyway damage, thrust flange and sealing surface condition, and visual surface cracks; check oil passage condition; measure end play and journal wear; check crankshaft position sensor reluctor ring (where applicable); determine necessary action.	P-1	14, 27
8. Inspect main and connecting rod bearings for damage and wear; determine necessary action.	P-2	14, 27
9. Identify piston and bearing wear patterns that indicate connecting rod alignment and main bearing bore problems; determine necessary action.	P-3	14, 18
10. Inspect and measure piston skirts and ring lands; determine necessary action.	P-2	18
11. Determine piston-to-bore clearance.	P-2	18
12. Inspect, measure, and install piston rings.	P-2	18
13. Inspect auxiliary shaft(s) (balance, intermediate, idler, counterbalance or silencer); inspect shaft(s) and support bearings for damage and wear; determine necessary action; reinstall and time.	P-2	19
14. Assemble engine block.	P-1	27

I. ENGINE REPAIR
D. Lubrication and Cooling Systems Diagnosis and Repair

Task Number and Description	Priority	Job #s
1. Perform cooling system pressure and dye tests to identify leaks; check coolant condition and level; inspect and test radiator, pressure cap, coolant recovery tank, heater core and gallery plugs; determine necessary action.	P-1	31
2. Identify causes of engine overheating.	P-1	8, 9, 10, 31
3. Inspect, replace, and adjust drive belts, tensioners, and pulleys; check pulley and belt alignment.	P-1	31, 33
4. Inspect and test coolant; drain and recover coolant; flush and refill cooling system with recommended coolant; bleed air as required.	P-1	31, 32
5. Inspect, remove, and replace water pump.	P-2	31, 34
6. Remove and replace radiator.	P-2	35
7. Remove, inspect, and replace thermostat and gasket/seal.	P-1	36
8. Inspect and test fan(s) (electrical or mechanical), fan clutch, fan shroud, and air dams.	P-1	31
9. Perform oil pressure tests; determine necessary action.	P-1	10
10. Perform engine oil and filter change.	P-1	6

I. ENGINE REPAIR (continued)
D. Lubrication and Cooling Systems Diagnosis and Repair

Task Number and Description	Priority	Job #s
11. Inspect auxiliary coolers; determine necessary action.	P-3	31
12. Inspect, test, and replace oil temperature and pressure switches and sensors.	P-2	139, 141
13. Inspect oil pump gears or rotors, housing, pressure relief devices, and pump drive; perform necessary action.	P-2	19

AUTOMATIC TRANSMISSION AND TRANSAXLE

II. AUTOMATIC TRANSMISSION AND TRANSAXLE
A. General: Transmission and Transaxle Diagnosis

Task Number and Description	Priority	Job #s
1. Identify and interpret transmission/transaxle concern, differentiate between engine performance and transmission/transaxle concerns; determine necessary action.	P-1	39
2. Research applicable vehicle and service information fluid type, vehicle service history, service precautions, and technical service bulletins.	P-1	2
3. Diagnose fluid loss and condition concerns; determine necessary action.	P-1	7, 39, 44, 65
4. Check fluid level in a transmission or a transaxle equipped with a dip-stick.	P-1	39, 44
5. Check fluid level in a transmission or a transaxle not equipped with a dip-stick.	P-1	39, 44
6. Perform pressure tests (including transmissions/transaxles equipped with electronic pressure control); determine necessary action.	P-1	39, 40
7. Diagnose noise and vibration concerns; determine necessary action.	P-2	37, 39
8. Perform stall test; determine necessary action.	P-3	39
9. Perform lock-up converter system tests; determine necessary action.	P-3	39
10. Diagnose transmission/transaxle gear reduction/multiplication concerns using driving, driven, and held member (power flow) principles.	P-1	39
11. Diagnose electronic transmission/transaxle control systems using appropriate test equipment and service information.	P-1	39
12. Diagnose pressure concerns in a transmission using hydraulic principles (Pascal's Law).	P-2	40

II. AUTOMATIC TRANSMISSION AND TRANSAXLE
B. In-Vehicle Transmission/Transaxle Maintenance and Repair

Task Number and Description	Priority	Job #s
1. Inspect, adjust, and replace external manual valve shift linkage, transmission range sensor/switch, and park/neutral position switch.	P-2	39, 45
2. Inspect for leakage; replace external seals, gaskets, and bushings.	P-2	3, 4

II. AUTOMATIC TRANSMISSION AND TRANSAXLE (continued)
B. In-Vehicle Transmission/Transaxle Maintenance and Repair

Task Number and Description	Priority	Job #s
3. Inspect, test, adjust, repair, or replace electrical/electronic components and circuits including computers, solenoids, sensors, relays, terminals, connectors, switches, and harnesses.	P-1	48
4. Drain and replace fluid and filter(s).	P-1	44
5. Inspect, replace, and align powertrain mounts.	P-2	37, 38

II. AUTOMATIC TRANSMISSION AND TRANSAXLE
C. Off-Vehicle Transmission and Transaxle Repair

Task Number and Description	Priority	Job #s
1. Remove and reinstall transmission/transaxle and torque converter; inspect engine core plugs, rear crankshaft seal, dowel pins, dowel pin holes, and mating surfaces.	P-1	50, 51, 58, 59
2. Inspect, leak test, and flush or replace transmission/transaxle oil cooler, lines, and fittings.	P-1	46
3. Inspect converter flex (drive) plate, converter attaching bolts, converter pilot, converter pump drive surfaces, converter end play, and crankshaft pilot bore.	P-2	50, 51, 52, 53, 55
4. Describe the operational characteristics of a continuously variable transmission (CVT).	P-3	39
5. Describe the operational characteristics of a hybrid vehicle drive train.	P-3	39
6. Disassemble, clean, and inspect transmission/transaxle.	P-2	52, 55
7. Inspect, measure, clean, and replace valve body (includes surfaces, bores, springs, valves, sleeves, retainers, brackets, check valves/balls, screens, spacers, and gaskets).	P-2	52, 53, 55, 56
8. Inspect servo and accumulator bores, pistons, seals, pins, springs, and retainers; determine necessary action.	P-2	53, 56
9. Assemble transmission/transaxle.	P-2	54, 57
10. Inspect, measure, and reseal oil pump assembly and components.	P-2	52, 53, 55, 56
11. Measure transmission/transaxle end play or preload; determine necessary action.	P-1	52, 55
12. Inspect, measure, and replace thrust washers and bearings.	P-2	52, 53 55, 56
13. Inspect oil delivery circuits, including seal rings, ring grooves, and sealing surface areas, feed pipes, orifices, and check valves/balls.	P-2	52, 53, 55, 56
14. Inspect bushings; determine necessary action.	P-2	52, 53, 55, 56
15. Inspect and measure planetary gear assembly components; determine necessary action.	P-2	52, 53, 55, 56
16. Inspect case bores, passages, bushings, vents, and mating surfaces; determine necessary action.	P-2	52, 55
17. Diagnose and inspect transaxle drive, link chains, sprockets, gears, bearings, and bushings; perform necessary action.	P-2	55

II. AUTOMATIC TRANSMISSION AND TRANSAXLE (continued)
C. Off-Vehicle Transmission and Transaxle Repair

Task Number and Description	Priority	Job #s
18. Inspect measure, repair, adjust or replace transaxle final drive components.	P-2	41, 42, 43
19. Inspect clutch drum, piston, check-balls, springs, retainers, seals, and friction and pressure plates, bands and drums; determine necessary action.	P-2	52, 53, 55, 56
20. Measure clutch pack clearance; determine necessary action.	P-1	53, 56
21. Air test operation of clutch and servo assemblies.	P-1	54, 57
22. Inspect roller and sprag clutch, races, rollers, sprags, springs, cages, retainers; determine necessary action.	P-2	52, 55

MANUAL DRIVE TRAIN AND AXLES

III. MANUAL DRIVE TRAIN AND AXLES
A. General: Drive Train Diagnosis

Task Number and Description	Priority	Job #s
1. Identify and interpret drive train concerns; determine necessary action.	P-1	60
2. Research applicable vehicle and service information, fluid type, vehicle service history, service precautions, and technical service bulletins.	P-1	2
3. Check fluid condition; check for leaks; determine necessary action.	P-1	60, 65
4. Drain and refill manual transmission/transaxle and final drive unit.	P-1	77, 78, 79, 80, 86, 89

III. MANUAL DRIVE TRAIN AND AXLES
B. Clutch Diagnosis and Repair

Task Number and Description	Priority	Job #s
1. Diagnose clutch noise, binding, slippage, pulsation, and chatter; determine necessary action.	P-1	71
2. Inspect clutch pedal linkage, cables, automatic adjuster mechanisms, brackets, bushings, pivots, and springs; perform necessary action.	P-1	71, 72
3. Inspect and replace clutch pressure plate assembly, clutch disc, release (throw-out) bearing and linkage, and pilot bearing/bushing (as applicable).	P-1	75
4. Bleed clutch hydraulic system.	P-1	73
5. Check and adjust clutch master cylinder fluid level; check for leaks.	P-1	71, 73
6. Inspect flywheel and ring gear for wear and cracks; determine necessary action.	P-1	75
7. Measure flywheel runout and crankshaft end play; determine necessary action.	P-2	75

III. MANUAL DRIVE TRAIN AND AXLES
C. Transmission/Transaxle Diagnosis and Repair

Task Number and Description	Priority	Job #s
1. Inspect, adjust, and reinstall shift linkages, brackets, bushings, cables, pivots, and levers.	P-2	64, 78, 80
2. Describe the operational characteristics of an electronically-controlled manual transmission/transaxle.	P-3	69
3. Diagnose noise concerns through the application of transmission/transaxle power flow principles.	P-2	60
4. Diagnose hard shifting and jumping out of gear concerns; determine necessary action.	P-2	64, 77
5. Diagnose transaxle final drive assembly noise and vibration concerns; determine necessary action.	P-3	60
6. Disassemble, inspect, clean, and reassemble internal transmission/transaxle components.	P-3	77, 78, 79, 80

III. MANUAL DRIVE TRAIN AND AXLES
D. Drive Shaft and Half Shaft, Universal and Constant-Velocity (CV) Joint Diagnosis and Repair

Task Number and Description	Priority	Job #s
1. Diagnose constant-velocity (CV) joint noise and vibration concerns; determine necessary action.	P-1	67, 68
2. Diagnose universal joint noise and vibration concerns; perform necessary action.	P-2	61, 62, 63, 66
3. Inspect, remove, and replace front-wheel drive (FWD) bearings, hubs, and seals.	P-1	68
4. Inspect, service, and replace shafts, yokes, boots, and universal/CV joints.	P-1	66, 67
5. Check shaft balance and phasing; measure shaft runout; measure and adjust driveline angles.	P-2	61, 62, 63

III. MANUAL DRIVE TRAIN AND AXLES
E. Drive Axle Diagnosis and Repair

E.1 Ring and Pinion Gears and Differential Case Assembly

Task Number and Description	Priority	Job #s
1. Clean and inspect differential housing; check for leaks; inspect housing vent.	P-2	7, 65, 87
2. Check and adjust differential housing fluid level.	P-1	60
3. Drain and refill differential housing.	P-1	86, 89
4. Diagnose noise and vibration concerns; determine necessary action.	P-2	60
5. Inspect and replace companion flange and pinion seal; measure companion flange runout.	P-2	66, 87, 88
6. Inspect ring gear and measure runout; determine necessary action.	P-3	87, 88

III. MANUAL DRIVE TRAIN AND AXLES (continued)
E. Drive Axle Diagnosis and Repair

E.1 Ring and Pinion Gears and Differential Case Assembly

Task Number and Description	Priority	Job #s
7. Remove, inspect, and reinstall drive pinion and ring gear, spacers, sleeves, and bearings.	P-3	87, 88
8. Measure and adjust drive pinion depth.	P-3	88
9. Measure and adjust drive pinion bearing preload.	P-3	88
10. Measure and adjust side bearing preload and ring and pinion gear total backlash and backlash variation on a differential carrier assembly (threaded cup or shim types).	P-3	88
11. Check ring and pinion tooth contact patterns; perform necessary action.	P-3	88
12. Disassemble, inspect, measure, and adjust or replace differential pinion gears (spiders), shaft, side gears, side bearings, thrust washers, and case.	P-3	87
13. Reassemble and reinstall differential case assembly; measure runout; determine necessary action.	P-3	88

E.2 Limited Slip Differential

Task Number and Description	Priority	Job #s
1. Diagnose noise, slippage, and chatter concerns; determine necessary action.	P-3	60
2. Measure rotating torque; determine necessary action.	P-3	89

E.3 Drive Axles

Task Number and Description	Priority	Job #s
1. Inspect and replace drive axle wheel studs.	P-1	108
2. Remove and replace drive axle shafts.	P-1	67, 84, 85
3. Inspect and replace drive axle shaft seals, bearings, and retainers.	P-2	67, 84, 85
4. Measure drive axle flange runout and shaft end play; determine necessary action.	P-2	84
5. Diagnose drive axle shafts, bearings, and seals for noise, vibration, and fluid leakage concerns; determine necessary action.	P-2	68

III. MANUAL DRIVE TRAIN AND AXLES
F. Four-wheel Drive/All-wheel Drive Component Diagnosis and Repair

Task Number and Description	Priority	Job #s
1. Inspect, adjust, and repair shifting controls (mechanical, electrical, and vacuum), bushings, mounts, levers, and brackets.	P-3	83
2. Inspect front-wheel bearings and locking hubs; perform necessary action(s).	P-3	67, 68, 83, 92
3. Check for leaks at drive assembly seals; check vents; check lube level.	P-3	65, 81

III. MANUAL DRIVE TRAIN AND AXLES (continued)
F. Four-wheel Drive/All-wheel Drive Component Diagnosis and Repair

Task Number and Description	Priority	Job #s
4. Identify concerns related to variations in tire circumference and/or final drive ratios.	P-3	81
5. Diagnose noise, vibration, and unusual steering concerns; determine necessary action.	P-3	60, 90
6. Diagnose, test, adjust, and replace electrical/electronic components of four-wheel drive systems.	P-3	83
7. Disassemble, service, and reassemble transfer case and components.	P-3	82

SUSPENSION AND STEERING

IV. SUSPENSION AND STEERING
A. General: Suspension and Steering Systems

Task Number and Description	Priority	Job #s
1. Research applicable vehicle and service information, vehicle service history, service precautions, and technical service bulletins.	P-1	2
2. Identify and interpret suspension and steering system concerns; determine necessary action.	P-1	90

IV. SUSPENSION AND STEERING
B. Steering Systems Diagnosis and Repair

Task Number and Description	Priority	Job #s
1. Disable and enable supplemental restraint system (SRS).	P-1	94, 147
2. Remove and replace steering wheel; center/time supplemental restraint system (SRS) coil (clock spring).	P-1	94, 147
3. Diagnose steering column noises, looseness, and binding concerns (including tilt mechanisms); determine necessary action.	P-2	90, 93
4. Diagnose power steering gear (non-rack and pinion) binding, uneven turning effort, looseness, hard steering, and noise concerns; determine necessary action.	P-2	90, 97
5. Diagnose power steering gear (rack and pinion) binding, uneven turning effort, looseness, hard steering, and noise concerns; determine necessary action.	P-2	90, 97
6. Inspect steering shaft universal-joint(s), flexible coupling(s), collapsible column, lock cylinder mechanism, and steering wheel; perform necessary action.	P-2	93, 94
7. Remove and replace rack and pinion steering gear; inspect mounting bushings and brackets.	P-2	95
8. Inspect rack and pinion steering gear inner tie rod ends (sockets) and bellows boots; replace as needed.	P-2	95

IV. SUSPENSION AND STEERING (continued)
B. Steering Systems Diagnosis and Repair

Task Number and Description	Priority	Job #s
9. Determine proper power steering fluid type; inspect fluid level and condition.	P-1	91, 97
10. Flush, fill, and bleed power steering system.	P-2	97
11. Inspect for power steering fluid leakage; determine necessary action.	P-1	97
12. Remove, inspect, replace, and adjust power steering pump drive belt.	P-1	33, 97
13. Remove and reinstall power steering pump.	P-2	97
14. Remove and reinstall press fit power steering pump pulley; check pulley and belt alignment.	P-2	97
15. Inspect and replace power steering hoses and fittings.	P-2	90, 97
16. Inspect and replace pitman arm, relay (center link/intermediate) rod, idler arm and mountings, and steering linkage damper.	P-2	90, 94, 96
17. Inspect, replace, and adjust tie rod ends (sockets), tie rod sleeves, and clamps.	P-1	90, 95, 96
18. Test and diagnose components of electronically-controlled steering systems using a scan tool; determine necessary action.	P-3	97
19. Identify hybrid vehicle power steering system electrical circuits and safety precautions.	P-2	93
20. Inspect electric power-assisted steering.	P-3	97

IV. SUSPENSION AND STEERING
C. Suspension Systems Diagnosis and Repair

Task Number and Description	Priority	Job #s
1. Diagnose short and long arm suspension system noises, body sway, and uneven ride height concerns; determine necessary action.	P-1	90
2. Diagnose strut suspension system noises, body sway, and uneven ride height concerns; determine necessary action.	P-1	90
3. Inspect, remove, and install upper and lower control arms, bushings, shafts, and rebound bumpers.	P-3	90, 99
4. Inspect, remove, and install strut rods and bushings.	P-3	90, 98
5. Inspect, remove, and install upper and/or lower ball joints (with or without wear indicators).	P-2	90, 99
6. Inspect, remove, and install steering knuckle assemblies.	P-3	90
7. Inspect, remove, and install short and long arm suspension system coil springs and spring insulators.	P-3	90, 100
8. Inspect, remove, and install torsion bars and mounts.	P-3	90
9. Inspect, remove, and install front stabilizer bar (sway bar) bushings, brackets, and links.	P-3	90, 98
10. Inspect, remove, and install strut cartridge or assembly, strut coil spring, insulators (silencers), and upper strut bearing mount.	P-3	90, 101

IV. SUSPENSION AND STEERING (continued)
C. Suspension Systems Diagnosis and Repair

Task Number and Description	Priority	Job #s
11. Inspect, remove, and install track bar, strut rods/radius arms, and related mounts and bushings.	P-3	90, 96, 98, 99
12. Inspect rear suspension system leaf spring(s), bushings, center pins/bolts, and mounts.	P-1	90

IV. SUSPENSION AND STEERING
D. Related Suspension and Steering Service

Task Number and Description	Priority	Job #s
1. Inspect, remove, and replace shock absorbers; inspect mounts and bushings.	P-1	90, 102
2. Remove, inspect, and service or replace front and rear wheel bearings.	P-1	5, 83, 92
3. Describe the function of the power steering pressure switch.	P-3	90

IV. SUSPENSION AND STEERING
E. Wheel Alignment Diagnosis, Adjustment, and Repair

Task Number and Description	Priority	Job #s
1. Diagnose vehicle wander, drift, pull, hard steering, bump steer, memory steer, torque steer, and steering return concerns; determine necessary action.	P-1	103
2. Perform prealignment inspection and measure vehicle ride height; perform necessary action.	P-1	90, 103
3. Prepare vehicle for wheel alignment on alignment machine; perform four-wheel alignment by checking and adjusting front and rear wheel caster, camber and toe as required; center steering wheel.	P-1	103
4. Check toe-out-on-turns (turning radius); determine necessary action.	P-2	103
5. Check SAI (steering axis inclination) and included angle; determine necessary action.	P-2	103
6. Check rear wheel thrust angle; determine necessary action.	P-1	103
7. Check for front wheel setback; determine necessary action.	P-2	103
8. Check front and/or rear cradle (subframe) alignment; determine necessary action.	P-3	103
9. Reset steering angle sensor.	P-2	103

IV. SUSPENSION AND STEERING
F. Wheels and Tires Diagnosis and Repair

Task Number and Description	Priority	Job #s
1. Inspect tire condition; identify tire wear patterns; check for correct tire size and application (load and speed ratings) and adjust air pressure; determine necessary action.	P-1	81, 104
2. Diagnose wheel/tire vibration, shimmy, and noise; determine necessary action.	P-2	103, 104, 106

IV. SUSPENSION AND STEERING (continued)
F. Wheels and Tires Diagnosis and Repair

Task Number and Description	Priority	Job #s
3. Rotate tires according to manufacturer's recommendations.	P-1	106
4. Measure wheel, tire, axle flange, and hub runout; determine necessary action.	P-2	84, 104
5. Diagnose tire pull problems; determine necessary action.	P-2	103
6. Dismount, inspect, and remount tire on wheel; balance wheel and tire assembly (static and dynamic).	P-1	105, 106
7. Dismount, inspect, and remount tire on wheel equipped with tire pressure monitoring system sensor.	P-2	105, 106
8. Inspect tire and wheel assembly for air loss; perform necessary action.	P-1	104
9. Repair tire using internal patch.	P-1	105
10. Identify and test tire pressure monitoring system (indirect and direct) for operation; calibrate system; verify operation of instrument panel lamps.	P-2	105
11. Demonstrate knowledge of steps required to remove and replace sensors in a tire pressure monitoring system.	P-1	105

BRAKES

V. BRAKES
A. General: Brake Systems Diagnosis

Task Number and Description	Priority	Job #s
1. Identify and interpret brake system concerns; determine necessary action.	P-1	107
2. Research applicable vehicle and service information, vehicle service history, service precautions, and technical service bulletins.	P-1	2
3. Describe procedure for performing a road test to check brake system operation; including an anti-lock brake system (ABS).	P-1	107
4. Install wheel and torque lug nuts.	P-1	105, 116, 118

V. BRAKES
B. Hydraulic System Diagnosis and Repair

Task Number and Description	Priority	Job #s
1. Diagnose pressure concerns in the brake system using hydraulic principles (Pascal's Law).	P-1	107, 111
2. Measure brake pedal height, travel, and free play (as applicable); determine necessary action.	P-1	107
3. Check master cylinder for internal/external leaks and proper operation; determine necessary action.	P-1	107
4. Remove, bench bleed, and reinstall master cylinder.	P-1	113

V. BRAKES (continued)
B. Hydraulic System Diagnosis and Repair

Task Number and Description	Priority	Job #s
5. Diagnose poor stopping, pulling or dragging concerns caused by malfunctions in the hydraulic system; determine necessary action.	P-3	111
6. Inspect brake lines, flexible hoses, and fittings for leaks, dents, kinks, rust, cracks, bulging, and wear; check for loose fittings and supports; determine necessary action.	P-1	107, 111
7. Replace brake lines, hoses, fittings, and supports.	P-2	112
8. Fabricate brake lines using proper material and flaring procedures (double flare and ISO types).	P-2	112
9. Select, handle, store, and fill brake fluids to proper level.	P-1	113, 115
10. Inspect, test, and/or replace components of brake warning light system.	P-3	111, 112, 136, 138, 141
11. Identify components of brake warning light system.	P-2	111, 112, 136, 138
12. Bleed and/or flush brake system.	P-1	115
13. Test brake fluid for contamination.	P-1	115

V. BRAKES
C. Drum Brake Diagnosis and Repair

Task Number and Description	Priority	Job #s
1. Diagnose poor stopping, noise, vibration, pulling, grabbing, dragging or pedal pulsation concerns; determine necessary action.	P-1	107
2. Remove, clean, inspect, and measure brake drum diameter; determine necessary action.	P-1	107, 116
3. Refinish brake drum and measure final drum diameter; compare with specifications.	P-1	117
4. Remove, clean, and inspect brake shoes, springs, pins, clips, levers, adjusters/self-adjusters, other related brake hardware, and backing support plates; lubricate and reassemble.	P-1	116
5. Inspect wheel cylinders for leaks and proper operation; remove and replace as needed.	P-2	116
6. Pre-adjust brake shoes and parking brake; install brake drums or drum/hub assemblies and wheel bearings; perform final checks and adjustments.	P-2	116

V. BRAKES
D. Disc Brake Diagnosis and Repair

Task Number and Description	Priority	Job #s
1. Diagnose poor stopping, noise, vibration, pulling, grabbing, dragging, or pulsation concerns; determine necessary action.	P-1	107
2. Remove and clean caliper assembly; inspect for leaks and damage/wear to caliper housing; determine necessary action.	P-1	107, 118

V. BRAKES (continued)
D. Disc Brake Diagnosis and Repair

Task Number and Description	Priority	Job #s
3. Clean and inspect caliper mounting and slides/pins for proper operation, wear, and damage; determine necessary action.	P-1	118
4. Remove, inspect, and replace pads and retaining hardware; determine necessary action.	P-1	118
5. Lubricate and reinstall caliper, pads, and related hardware; seat pads and inspect for leaks.	P-1	118
6. Clean and inspect rotor; measure rotor thickness, thickness variation, and lateral runout; determine necessary action.	P-1	118
7. Remove and reinstall rotor.	P-1	118
8. Refinish rotor on vehicle; measure final rotor thickness and compare with specifications.	P-1	120
9. Refinish rotor off vehicle; measure final rotor thickness and compare with specifications.	P-1	120
10. Retract and re-adjust caliper piston on an integrated parking brake system.	P-3	110
11. Check brake pad wear indicator; determine necessary action.	P-2	118
12. Describe importance of operating vehicle to burnish/break-in replacement brake pads according to manufacturer's recommendations.	P-1	118

V. BRAKES
E. Power-Assist Units Diagnosis and Repair

Task Number and Description	Priority	Job #s
1. Check brake pedal travel with, and without, engine running to verify proper power booster operation.	P-2	114
2. Check vacuum supply (manifold or auxiliary pump) to vacuum-type power booster.	P-1	114
3. Inspect vacuum-type power booster unit for leaks; inspect the check-valve for proper operation; determine necessary action.	P-1	114
4. Inspect and test hydraulically-assisted power brake system for leaks and proper operation; determine necessary action.	P-3	114
5. Measure and adjust master cylinder push rod length.	P-3	114

V. BRAKES
F. Miscellaneous (Wheel Bearings, Parking Brakes, Electrical, Etc.) Diagnosis and Repair

Task Number and Description	Priority	Job #s
1. Diagnose wheel bearing noises, wheel shimmy, and vibration concerns; determine necessary action.	P-3	109
2. Remove, clean, inspect, repack, and install wheel bearings; replace seals; install hub and adjust bearings.	P-1	5, 109

V. BRAKES (continued)
F. Miscellaneous (Wheel Bearings, Parking Brakes, Electrical, Etc.) Diagnosis and Repair

Task Number and Description	Priority	Job #s
3. Check parking brake cables and components for wear, binding, and corrosion; clean, lubricate, adjust, or replace as needed.	P-2	110
4. Check parking brake operation and parking brake indicator light system operation; determine necessary action.	P-1	110
5. Check operation of brake stop light system.	P-1	136
6. Replace wheel bearing and race.	P-2	5, 109
7. Remove and reinstall sealed wheel bearing assembly.	P-2	109
8. Inspect and replace wheel studs.	P-1	108

V. BRAKES
G. Electronic Brake, Traction and Stability Control Systems Diagnosis and Repair

Task Number and Description	Priority	Job #s
1. Identify and inspect electronic brake control system components; determine necessary action.	P-1	121
2. Identify traction control/vehicle stability control system components.	P-3	121
3. Describe the operation of a regenerative braking system.	P-3	111
4. Diagnose poor stopping, wheel lock-up, abnormal pedal feel, unwanted application, and noise concerns associated with the electronic brake control system; determine necessary action.	P-2	124
5. Diagnose electronic brake control system electronic control(s) and components by retrieving diagnostic trouble codes, and/or using recommended test equipment; determine necessary action.	P-2	124
6. Depressurize high-pressure components of an electronic brake control system.	P-3	121
7. Bleed the electronic brake control system hydraulic circuits.	P-1	122
8. Test, diagnose, and service electronic brake control system speed sensors (digital and analog), toothed ring (tone wheel), and circuits using a graphing multimeter (GMM)/digital storage oscilloscope (DSO) (includes output signal, resistance, shorts to voltage/ground, and frequency data).	P-3	123, 124
9. Diagnose electronic brake control system braking concerns caused by vehicle modifications (tire size, curb height, final drive ratio, etc.).	P-3	124

VI. ELECTRICAL/ELECTRONIC SYSTEMS
A. General: Electrical System Diagnosis

Task Number and Description	Priority	Job #s
1. Research applicable vehicle and service information, vehicle service history, service precautions, and technical service bulletins.	P-1	2
2. Demonstrate knowledge of electrical/electronic series, parallel, and series-parallel circuits using principles of electricity (Ohm's Law).	P-1	136
3. Demonstrate proper use of a digital multimeter (DMM) when measuring source voltage, voltage drop (including grounds), current flow and resistance.	P-1	130, 136, 142
4. Demonstrate knowledge of the causes and effects from shorts, grounds, opens, and resistance problems in electrical/electronic circuits.	P-1	130, 138, 142
5. Check operation of electrical circuits with a test light.	P-1	130, 136
6. Check operation of electrical circuits with fused jumper wires.	P-1	125, 138
7. Use wiring diagrams during the diagnosis (troubleshooting) of electrical/electronic circuit problems.	P-1	130, 136, 138
8. Diagnose the cause(s) of excessive key-off battery drain (parasitic draw); determine necessary action.	P-1	125
9. Inspect and test fusible links, circuit breakers, and fuses; determine necessary action.	P-1	125, 130, 138, 139
10. Inspect and test switches, connectors, relays, solenoid solid-state devices, and wires of electrical/electronic circuits; determine necessary action.	P-1	130, 132, 142
11. Replace electrical connectors and terminal ends.	P-1	126
12. Repair wiring harness.	P-1	126
13. Perform solder repair of electrical wiring.	P-1	126
14. Check electrical/electronic circuit waveforms; interpret readings and determine needed repairs.	P-2	130, 134, 164, 170
15. Repair CAN/BUS wiring harness.	P-1	126

VI. ELECTRICAL/ELECTRONIC SYSTEMS
B. Battery Diagnosis and Service

Task Number and Description	Priority	Job #s
1. Perform battery state-of-charge test; determine necessary action.	P-1	127
2. Confirm proper battery capacity for vehicle application; perform battery capacity test; determine necessary action.	P-1	127
3. Maintain or restore electronic memory functions.	P-1	125, 128
4. Inspect and clean battery; fill battery cells; check battery cables, connectors, clamps, and hold-downs.	P-1	127, 128

VI. ELECTRICAL/ELECTRONIC SYSTEMS (continued)
B. Battery Diagnosis and Service

Task Number and Description	Priority	Job #s
5. Perform slow/fast battery charge according to manufacturer's recommendations.	P-1	128
6. Jump-start vehicle using jumper cables and a booster battery or an auxiliary power supply.	P-1	129
7. Identify high-voltage circuits of electric or hybrid electric vehicle and related safety precautions.	P-3	126
8. Identify electronic modules, security systems, radios, and other accessories that require reinitialization or code entry after reconnecting vehicle battery.	P-1	128
9. Identify hybrid vehicle auxiliary (12v) battery service, repair, and test procedures.	P-3	127, 128

VI. ELECTRICAL/ELECTRONIC SYSTEMS
C. Starting System Diagnosis and Repair

Task Number and Description	Priority	Job #s
1. Perform starter current draw tests; determine necessary action.	P-1	130
2. Perform starter circuit voltage drop tests; determine necessary action.	P-1	130
3. Inspect and test starter relays and solenoids; determine necessary action.	P-2	130, 131
4. Remove and install starter in a vehicle.	P-1	133
5. Inspect and test switches, connectors, and wires of starter control circuits; determine necessary action.	P-2	130, 132
6. Differentiate between electrical and engine mechanical problems that cause a slow-crank or a no-crank condition.	P-2	130

VI. ELECTRICAL/ELECTRONIC SYSTEMS
D. Charging System Diagnosis and Repair

Task Number and Description	Priority	Job #s
1. Perform charging system output test; determine necessary action.	P-1	134
2. Diagnose (troubleshoot) charging system for causes of undercharge, no-charge, or overcharge conditions.	P-1	134
3. Inspect, adjust, or replace generator (alternator) drive belts; check pulleys and tensioners for wear; check pulley and belt alignment.	P-1	33, 134
4. Remove, inspect, and re-install generator (alternator).	P-1	134, 135
5. Perform charging circuit voltage drop tests; determine necessary action.	P-1	134

VI. ELECTRICAL/ELECTRONIC SYSTEMS (continued)
E. Lighting Systems Diagnosis and Repair

Task Number and Description	Priority	Job #s
1. Diagnose (troubleshoot) the causes of brighter-than-normal, intermittent, dim, or no light operation; determine necessary action.	P-1	136
2. Inspect interior and exterior lamps and sockets including headlights and auxiliary lights (fog lights/driving lights); replace as needed.	P-1	136
3. Aim headlights.	P-2	137
4. Identify system voltage and safety precautions associated with high-intensity discharge headlights.	P-2	137

VI. ELECTRICAL/ELECTRONIC SYSTEMS
F. Gauges, Warning Devices, and Driver Information Systems Diagnosis and Repair

Task Number and Description	Priority	Job #s
1. Inspect and test gauges and gauge sending units for causes of abnormal gauge readings; determine necessary action.	P-2	139, 140
2. Diagnose (troubleshoot) the causes of incorrect operation of warning devices and other driver information systems; determine necessary action.	P-2	138, 139, 140, 141

VI. ELECTRICAL/ELECTRONIC SYSTEMS
G. Horn and Wiper/Washer Diagnosis and Repair

Task Number and Description	Priority	Job #s
1. Diagnose (troubleshoot) causes of incorrect horn operation; perform necessary action.	P-1	142
2. Diagnose (troubleshoot) causes of incorrect wiper operation; diagnose wiper speed control and park problems; perform necessary action.	P-2	143
3. Diagnose (troubleshoot) windshield washer problems; perform necessary action.	P-2	143

VI. ELECTRICAL/ELECTRONIC SYSTEMS
H. Accessories Diagnosis and Repair

Task Number and Description	Priority	Job #s
1. Diagnose (troubleshoot) incorrect operation of motor-driven accessory circuits; determine necessary action.	P-2	144
2. Diagnose (troubleshoot) incorrect electric lock operation (including remote keyless entry); determine necessary action.	P-2	144, 151
3. Diagnose (troubleshoot) incorrect operation of cruise control systems; determine necessary action.	P-3	145
4. Diagnose (troubleshoot) supplemental restraint system (SRS) problems; determine necessary action.	P-2	148
5. Disable and enable an airbag system for vehicle service; verify indicator lamp operation.	P-1	147
6. Remove and reinstall door panel.	P-1	144

VI. ELECTRICAL/ELECTRONIC SYSTEMS (continued)
H. Accessories Diagnosis and Repair

Task Number and Description	Priority	Job #s
7. Check for module communication errors (including CAN/BUS systems) using a scan tool.	P-2	150
8. Describe the operation of keyless entry/remote-start systems.	P-3	151
9. Verify operation of instrument panel gauges and warning/indicator lights; reset maintenance indicators.	P-1	136, 138, 139, 140, 141
10. Verify windshield wiper and washer operation, replace wiper blades.	P-1	143
11. Diagnose (troubleshoot) radio static and weak, intermittent, or no radio reception; determine necessary action.	P-3	146
12. Diagnose (troubleshoot) body electronic system circuits using a scan tool; determine necessary action.	P-3	149
13. Diagnose the cause(s) of false, intermittent, or no operation of anti-theft systems.	P-3	151
14. Describe the process for software transfers, software updates, or flash reprogramming on electronic modules.	P-3	161

HEATING AND AIR CONDITIONING

VII. HEATING AND AIR CONDITIONING
A. General: A/C System Diagnosis and Repair

Task Number and Description	Priority	Job #s
1. Identify and interpret heating and air conditioning problems; determine necessary action.	P-1	152
2. Research applicable vehicle and service information, vehicle service history, service precautions, and technical service bulletins.	P-1	2
3. Performance test A/C system; identify problems.	P-1	152, 153
4. Identify abnormal operating noises in the A/C system; determine necessary action.	P-2	152
5. Identify refrigerant type; select and connect proper gauge set; record temperature and pressure readings.	P-1	152, 153
6. Leak test A/C system; determine necessary action.	P-1	152
7. Inspect condition of refrigerant oil removed from A/C system; determine necessary action.	P-2	155
8. Determine recommended oil and oil capacity for system application.	P-1	155
9. Using a scan tool, observe and record related HVAC data and trouble codes.	P-3	11

VII. HEATING AND AIR CONDITIONING (continued)
B. Refrigeration System Component Diagnosis and Repair

Task Number and Description	Priority	Job #s
1. Inspect and replace A/C compressor drive belts, pulleys, and tensioners; determine necessary action.	P-1	33, 152, 155
2. Inspect, test, service, or replace A/C compressor clutch components and/or assembly; check compressor clutch air gap; adjust as needed.	P-2	152, 155
3. Remove, inspect, and reinstall A/C compressor and mountings; determine recommended oil quantity.	P-2	155
4. Identify hybrid vehicle A/C system electrical circuits and service/safety precautions.	P-2	152
5. Determine need for an additional A/C system filter; perform necessary action.	P-3	159
6. Remove and inspect A/C system mufflers, hoses, lines, fittings, O-rings, seals, and service valves; perform necessary action.	P-2	152, 156
7. Inspect A/C condenser for airflow restrictions; perform necessary action.	P-1	152
8. Remove, inspect, and reinstall receiver/drier or accumulator/drier; determine recommended oil quantity.	P-2	155, 156
9. Remove, inspect, and install expansion valve or orifice (expansion) tube.	P-1	156
10. Inspect evaporator housing water drain; perform necessary action.	P-1	152
11. Diagnose A/C system conditions that cause the protection devices (pressure, thermal, and PCM) to interrupt system operation; determine necessary action.	P-2	152
12. Determine procedure to remove and reinstall evaporator; determine required oil quantity.	P-2	156
13. Remove, inspect, and reinstall condenser; determine required oil quantity.	P-2	156

VII. HEATING AND AIR CONDITIONING
C. Heating, Ventilation, and Engine Cooling Systems Diagnosis and Repair

Task Number and Description	Priority	Job #s
1. Inspect engine cooling and heater systems hoses; perform necessary action.	P-1	31, 33, 158
2. Inspect and test heater control valve(s); perform necessary action.	P-2	158
3. Diagnose temperature control problems in the heater/ventilation system; determine necessary action.	P-2	153
4. Determine procedure to remove, inspect, and reinstall heater core.	P-2	158

VII. HEATING AND AIR CONDITIONING (continued)
D. Operating Systems and Related Controls Diagnosis and Repair

Task Number and Description	Priority	Job #s
1. Inspect and test A/C-heater blower motors, resistors, switches, relays, wiring, and protection devices; perform necessary action.	P-1	152, 159
2. Diagnose A/C compressor clutch control systems; determine necessary action.	P-2	152
3. Diagnose malfunctions in the vacuum, mechanical, and electrical components and controls of the heating, ventilation, and A/C (HVAC) system; determine necessary action.	P-2	153
4. Inspect and test A/C-heater control panel assembly; determine necessary action.	P-3	153
5. Inspect and test A/C-heater control cables, motors, and linkages; perform necessary action.	P-3	153
6. Inspect A/C-heater ducts, doors, hoses, cabin filters, and outlets; perform necessary action.	P-1	159
7. Identify the source of A/C system odors.	P-2	152
8. Check operation of automatic or semi-automatic heating, ventilation, and air-conditioning (HVAC) control systems; determine necessary action.	P-2	153

VII. HEATING AND AIR CONDITIONING
E. Refrigerant Recovery, Recycling, and Handling

Task Number and Description	Priority	Job #s
1. Perform correct use and maintenance of refrigerant handling equipment according to equipment manufacturer's standards.	P-1	154, 157
2. Identify and recover A/C system refrigerant.	P-1	154
3. Recycle, label, and store refrigerant.	P-1	154, 157
4. Evacuate and charge A/C system; add refrigerant oil as required.	P-1	154

ENGINE PERFORMANCE

VIII. ENGINE PERFORMANCE
A. General: Engine Diagnosis

Task Number and Description	Priority	Job #s
1. Identify and interpret engine performance concerns; determine necessary action.	P-1	160
2. Research applicable vehicle and service information, vehicle service history, service precautions, and technical service bulletins.	P-1	2
3. Diagnose abnormal engine noises or vibration concerns; determine necessary action.	P-3	163
4. Diagnose the cause of excessive oil consumption, coolant consumption, unusual exhaust color, odor, and sound; determine necessary action.	P-2	173

VIII. ENGINE PERFORMANCE (continued)
A. General: Engine Diagnosis

Task Number and Description	Priority	Job #s
5. Perform engine absolute (vacuum/boost) manifold pressure tests; determine necessary action.	P-1	8
6. Perform cylinder power balance test; determine necessary action.	P-2	8
7. Perform cylinder cranking and running compression tests; determine necessary action.	P-1	9
8. Perform cylinder leakage test; determine necessary action.	P-1	9
9. Diagnose engine mechanical, electrical, electronic, fuel, and ignition concerns; determine necessary action.	P-2	8, 9, 10, 130, 134, 160, 164, 167, 170
10. Verify engine operating temperature; determine necessary action.	P-1	31
11. Verify correct camshaft timing.	P-1	28

VIII. ENGINE PERFORMANCE
B. Computerized Controls Diagnosis and Repair

Task Number and Description	Priority	Job #s
1. Retrieve and record diagnostic trouble codes, OBD monitor status, and freeze frame data; clear codes when applicable.	P-1	11, 161
2. Access and use service information to perform step-by-step (trouble-shooting) diagnosis.	P-1	2
3. Perform active tests of actuators using a scan tool; determine necessary action.	P-2	170, 178
4. Describe the importance of running all OBDII monitors for repair verification.	P-1	161
5. Diagnose the causes of emissions or driveability concerns with stored or active diagnostic trouble codes; obtain, graph, and interpret scan tool data.	P-1	11, 160, 161
6. Diagnose emissions or driveability concerns without stored diagnostic trouble codes; determine necessary action.	P-1	160, 161, 165
7. Inspect and test computerized engine control system sensors, powertrain/engine control module (PCM/ECM), actuators, and circuits using a graphing multimeter (GMM)/digital storage oscilloscope (DSO); perform necessary action.	P-2	164, 166, 169, 175, 177
8. Diagnose driveability and emissions problems resulting from malfunctions of interrelated systems (cruise control, security alarms, suspension controls, traction controls, A/C, automatic transmissions, non-OEM installed accessories, or similar systems); determine necessary action.	P-3	160, 163

VIII. ENGINE PERFORMANCE (continued)
C. Ignition System Diagnosis and Repair

Task Number and Description	Priority	Job #s
1. Diagnose (troubleshoot) ignition system-related problems such as no-starting, hard starting, engine misfire, poor driveability, spark knock, power loss, poor mileage, and emissions concerns; determine necessary action.	P-2	160, 161, 164, 165
2. Inspect and test crankshaft and camshaft position sensor(s); perform necessary action.	P-1	28, 29, 165
3. Inspect, test, and/or replace ignition control module, powertrain/engine control module; reprogram as necessary.	P-3	165
4. Remove and replace spark plugs; inspect secondary ignition components for wear and damage.	P-1	164, 165

VIII. ENGINE PERFORMANCE
D. Fuel, Air Induction, and Exhaust Systems Diagnosis and Repair

Task Number and Description	Priority	Job #s
1. Diagnose (troubleshoot) hot or cold no-starting, hard starting, poor driveability, incorrect idle speed, poor idle, flooding, hesitation, surging, engine misfire, power loss, stalling, poor mileage, dieseling, and emissions problems; determine necessary action.	P-2	160, 167, 169
2. Check fuel for contaminants; determine necessary action.	P-2	166
3. Inspect and test fuel pumps and pump control systems for pressure, regulation, and volume; perform necessary action.	P-1	167
4. Replace fuel filter(s).	P-1	167
5. Inspect, service, or replace air filters, filter housings, and intake duct work.	P-1	168
6. Inspect throttle body, air induction system, intake manifold and gaskets for vacuum leaks and/or unmetered air.	P-2	168
7. Inspect and test fuel injectors.	P-2	168, 170
8. Verify idle control operation.	P-1	169
9. Inspect integrity of the exhaust manifold, exhaust pipes, muffler(s), catalytic converter(s), resonator(s), tail pipe(s), and heat shields; perform necessary action.	P-1	173, 174, 177
10. Inspect condition of exhaust system hangers, brackets, clamps, and heat shields; repair or replace as needed.	P-1	173, 174, 177
11. Perform exhaust system back-pressure test; determine necessary action.	P-2	173
12. Check and refill diesel exhaust fluid (DEF).	P-3	173
13. Test the operation of turbocharger/supercharger systems; determine necessary action.	P-3	172

VIII. ENGINE PERFORMANCE (continued)
E. Emissions Control Systems Diagnosis and Repair

Task Number and Description	Priority	Job #s
1. Diagnose oil leaks, emissions, and driveability concerns caused by the positive crankcase ventilation (PCV) system; determine necessary action.	P-3	160, 178
2. Inspect, test, and service positive crankcase ventilation (PCV) filter/breather cap, valve, tubes, orifices, and hoses; perform necessary action.	P-2	178
3. Diagnose emissions and driveability concerns caused by the exhaust gas recirculation (EGR) system; determine necessary action.	P-3	160, 175
4. Diagnose emissions and driveability concerns caused by the secondary air injection and catalytic converter systems; determine necessary action.	P-2	160, 176
5. Diagnose emissions and driveability concerns caused by the evaporative emissions control system; determine necessary action.	P-2	160, 178
6. Inspect and test electrical/electronic sensors, controls, and wiring of exhaust gas recirculation (EGR) systems; perform necessary action.	P-2	175
7. Inspect, test, service, and replace components of the EGR system including tubing, exhaust passages, vacuum/pressure controls, filters, and hoses; perform necessary action.	P-2	175
8. Inspect and test electrical/electronically-operated components and circuits of air injection systems; perform necessary action.	P-3	176
9. Inspect and test catalytic converter efficiency.	P-2	177
10. Inspect and test components and hoses of the evaporative emissions control system; perform necessary action.	P-1	178
11. Interpret diagnostic trouble codes (DTCs) and scan tool data related to the emissions control systems; determine necessary action.	P-3	11, 160, 177, 178

REQUIRED SUPPLEMENTAL TASKS

Shop and Personal Safety

Task Number and Description	
1. Identify general shop safety rules and procedures.	1
2. Utilize safe procedures for handling of tools and equipment.	1
3. Identify and use proper placement of floor jacks and jack stands.	1
4. Identify and use proper procedures for safe lift operation.	1
5. Utilize proper ventilation procedures for working within the lab/shop area.	1
6. Identify marked safety areas.	1
7. Identify the location and the types of fire extinguishers and other fire safety equipment; demonstrate knowledge of the procedures for using fire extinguishers and other fire safety equipment.	1
8. Identify the location and use of eye wash stations.	1
9. Identify the location of the posted evacuation routes.	1
10. Comply with the required use of safety glasses, ear protection, gloves, and shoes during lab/shop activities.	1
11. Identify and wear appropriate clothing for lab/shop activities.	1
12. Secure hair and jewelry for lab/shop activities.	1
13. Demonstrate awareness of the safety aspects of supplemental restraint systems (SRS), electronic brake control systems, and hybrid vehicle high voltage circuits.	1
14. Demonstrate awareness of the safety aspects of high voltage circuits (such as high intensity discharge (HID) lamps, ignition systems, injection systems, etc.).	1, 137
15. Locate and demonstrate knowledge of material safety data sheets (MSDS).	1, 2

Tools and Equipment

Task Number and Description	
1. Identify tools and their usage in automotive applications.	1, 11
2. Identify standard and metric designation.	2, 28, 31, 40, 172
3. Demonstrate safe handling and use of appropriate tools.	1
4. Demonstrate proper cleaning, storage, and maintenance of tools and equipment.	1
5. Demonstrate proper use of precision measuring tools (i.e. micrometer, dial indicator, dial caliper).	16, 18, 21, 27

Preparing Vehicle for Service

Task Number and Description	
1. Identify information needed and the service requested on a repair order.	2
2. Identify purpose and demonstrate proper use of fender covers, mats.	6

Preparing Vehicle for Service (continued)

Task Number and Description	
3. Demonstrate use of the three Cs (concern, cause, and correction).	160
4. Review vehicle service history.	2
5. Complete work order to include customer information, vehicle identifying information, customer concern, related service history, cause, and correction.	2

Preparing Vehicle for Customer

Task Number and Description	
1. Ensure vehicle is prepared to return to customer per school/company policy (floor mats, steering wheel cover, etc.).	6